From: *Natural Earth II*, Tom Patterson

The Brooks/Cole Geography Resource Center

This online tool **provides an array of visual resources** to deepen your understanding of physical geography.

It includes

▶ animations

▶ videos on today's critical topics such as climate change

▶ newsfeeds

▶ Google Earth activities

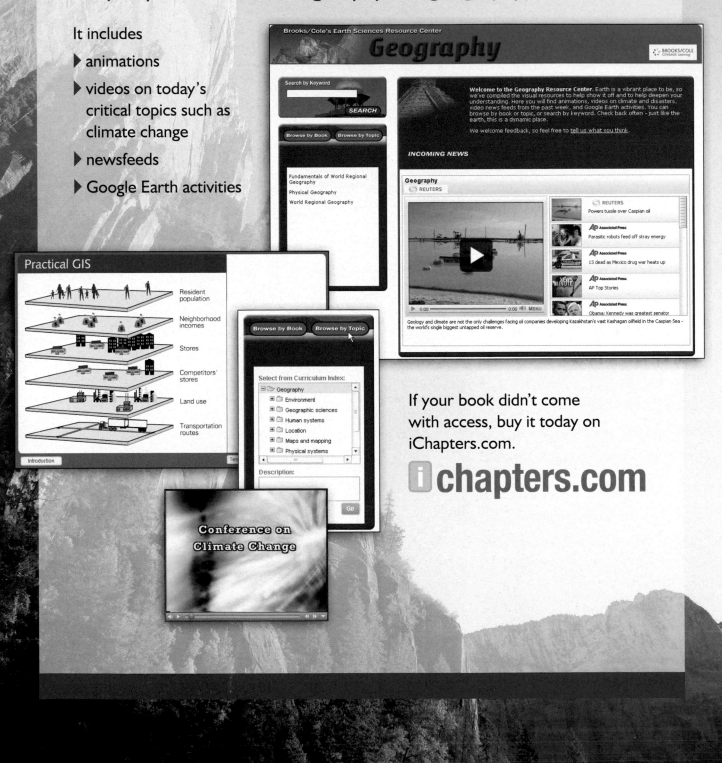

If your book didn't come with access, buy it today on iChapters.com.

ichapters.com

FUNDAMENTALS OF PHYSICAL GEOGRAPHY

James F. Petersen
Texas State University — San Marcos

Dorothy Sack
Ohio University, Athens

Robert E. Gabler
Western Illinois University, Macomb

BROOKS/COLE
CENGAGE Learning™

Australia • Brazil • Japan • Korea • Mexico • Singapore • Spain • United Kingdom • United States

BROOKS/COLE
CENGAGE Learning

Fundamentals of Physical Geography
James F. Petersen, Dorothy Sack,
Robert E. Gabler

Earth Science Editor: Laura Pople

Developmental Editor: Amy K. Collins

Assistant Editor: Samantha Arvin

Editorial Assistants: Jenny Hoang,
   Kristina Chiapella

Media Editor: Alexandria Brady

Marketing Manager: Nicole Mollica

Marketing Assistant: Kevin Carroll

Marketing Communications Manager:
   Belinda Krohmer

Content Project Manager: Hal Humphrey

Art Director: John Walker

Print Buyer: Judy Inouye

Rights Acquisitions Account Manager, Text:
   Timothy Sisler

Rights Acquisitions Account Manager, Image:
   Robyn Young

Production Service: Pre-Press PMG

Text Designer: Diane Beasley and
   Pre-Press PMG

Photo Researcher: Pre-Press PMG

Copy Editor: Christine Hobberlin

Illustrators: Accurate Art, Precision Graphics,
   Rolin Graphics, Pre-Press PMG

Cover Designer: John Walker

Cover Image: Jack Dykinga/Getty Images.
   Geyser, Black Rock Desert, Nevada.

Compositor: Pre-Press PMG

For product information and technology assistance, contact us at
Cengage Learning Customer & Sales Support, 1-800-354-9706

For permission to use material from this text or product,
submit all requests online at **www.cengage.com/permissions**
Further permissions questions can be emailed to
permissionrequest@cengage.com

Library of Congress Control Number: 2009936341

ISBN-13: 978-0-538-73463-9

ISBN-10: 0-538-73463-9

Brooks/Cole
20 Davis Drive
Belmont, CA 94002-3098
USA

Cengage Learning is a leading provider of customized learning solutions with office locations around the globe, including Singapore, the United Kingdom, Australia, Mexico, Brazil, and Japan. Locate your local office at
www.cengage.com/global.

Cengage Learning products are represented in Canada by Nelson Education, Ltd.

To learn more about Brooks/Cole visit **www.cengage.com/brookscole**

Purchase any of our products at your local college store or at our preferred online store **www.ichapters.com.**

Printed in the United States of America
1 2 3 4 5 6 7 13 12 11 10 09

Preface

Our natural, environmental home on planet Earth is a complex system of interacting components that includes climate and the atmosphere, organisms and their communities, water, landforms, and soils. The ways we choose to use and affect the environment today can benefit or endanger our own as well as future generations. More than ever before, people today want and need to know the effects of their actions on the environment at both a local and global scale. Understanding physical geography, that is, what our natural habitat on Earth's surface is like, how it functions, and how it varies over space and time, is critical for making informed decisions regarding the wise use and preservation of our environment and resources. The more we know about the environment on Earth, the more effective we can be in working toward preservation, stewardship, and sustainability. *Fundamentals of Physical Geography* was written to provide students from any academic major with a basic working knowledge of Earth's natural systems and its shared natural habitats, so that they may better understand the consequences of human actions on the environment. This text focuses on presenting the essential content of physical geography to students in a clear, condensed style, which is an excellent format for courses that follow either the semester or quarter system.

Throughout the world, geography is a highly regarded field of inquiry. Recognition of its importance to society is growing along with environmental awareness. Geographical knowledge, skills, and techniques are increasingly valued in the work place. Professional physical geographers use the latest technologies to observe, map, and measure Earth's surface and atmosphere and to model environmental responses and interactions. Physical geographers in the workplace apply satellite imagery, global positioning systems (GPS), computer-assisted mapmaking (cartography), geographic information science (GIS), and other tools for analysis and problem solving. At the college level, physical geography is an ideal science course for any student who would like to make informed decisions that consider environmental limits and possibilities as well as people's wants and needs.

Fundamentals of Physical Geography stresses an appreciation of geography as a discipline worthy of continued study. A focus on relevance is supported by the definition of geography, explanations of useful tools and methodologies, practical applications, as well as the utility of spatial thinking and systems analysis.

Features

Comprehensive View of the Earth System

Fundamentals of Physical Geography introduces all major aspects of the Earth system, identifying physical phenomena and natural processes and stressing their characteristics, relationships, interactions, and distributions. The text covers a wide range of topics, including climate and the atmosphere, water, the solid Earth, and the living components of our planet. With only 17 chapters, *Fundamentals of Physical Geography* provides beginning geography students with a thorough introduction to the essential content of physical geography, which can be easily covered in a single course.

Engaging Graphics

Because the study of geography is greatly enhanced with visual aids, the text includes an array of illustrations and photographs that make the concepts come alive. Selected photographs are accompanied by locator maps to provide spatial context and help students identify the feature's geographic position on Earth. Clear and simple diagrams illuminate challenging concepts, and Environmental System illustrations throughout the text provide a broad view of the features, inputs, and outputs of a complete system, such as storms glaciers, groundwater, or the plate tectonic system.

Clear Explanations

The text uses an easily understandable, narrative style to explain the processes, physical features, and events that occur within, on, or above Earth's surface. The writing style facilitates rapid comprehension and make the study of physical geography meaningful and enjoyable.

Introduction to the Geographer's Tools

Digital technologies have revolutionized our abilities to study the natural environments and processes on Earth's surface. A full chapter is devoted to maps, digital imagery, and other data used by geographers. Illustrations throughout include maps and images, with interpretations provided for the environmental attributes shown in the scenes. There are also introductory discussions of many techniques that geographers use for displaying and analyzing environmental features and processes, including remote sensing, geographic information systems, cartography, and global positioning systems.

Focus on Student Interaction

The text continually encourages students to think, conceptualize, hypothesize, and interact with the subject matter of physical geography. Activities at the end of each chapter can be completed individually, or as a group, and were designed to engage students and promote active learning. Review questions reinforce concepts and prepare students for exams, and practical application assignments require active solutions, such as sketching a diagram, running a calculation, or exploring geographic features using Google Earth. Questions following many figure captions prompt students to think beyond, or to use, the map, graph, diagram, or image, and give further consideration to the topic. Detailed learning objectives at the beginning of the chapters provide a means to measure comprehension of the material.

Three Unique Perspectives This textbook also selectively employs feature boxes that illustrate the major scientific perspectives of physical geography. Through a **spatial perspective,** physical geography focuses on understanding and explaining the locations, distribution, and spatial interactions of natural phenomena. Physical geography also uses a **physical science perspective,** which applies the knowledge and methods of the natural and physical sciences using the scientific method and systems analysis. Through an **environmental perspective,** physical geographers consider impacts, influences, and interactions between human and natural components of the environment, that is, how the environment influences human life and how humans affect the environment.

Map Interpretation Series Because learning map interpretation skills is a priority in physical geography, this text includes activities based on full-color maps printed at their original scale. These maps help students develop valuable map-reading skills while reinforcing the topical material presented. The map interpretation features can be incorporated into lab activities and help to link between lectures, the text, and lab work.

Fundamentals of Physical Geography— Four Major Objectives

To Meet the Academic Needs of the Student

In content and style, *Fundamentals of Physical Geography* was written specifically to meet the needs of students, the end users of this textbook. Students can use the knowledge and understanding obtained through the text and activities to help them make informed decisions involving the environment at the local, regional, and global scale. The book considers the needs of beginning students, with little or no background in the study of physical geography or other Earth sciences. Examples from throughout the world illustrate important concepts and help students bridge the gap between theory and practical application.

To Integrate the Illustrations with the Written Text

The photographs, maps, satellite images, scientific visualizations, block diagrams, graphs, and line drawings clearly illustrate important concepts in physical geography. Text discussions are strongly linked to the illustrations, encouraging students to examine in graphic form and visualize physical processes and phenomena. Some examples of topics that are clearly explained by integrating visuals and text include map and image interpretation (Chapter 2), the seasons (Chapter 3), Earth's energy budget (Chapter 3), wind systems (Chapter 4), storms (Chapter 6), soils (Chapter 9), plate tectonics (Chapter 10), rivers (Chapter 14), glaciers (Chapter 16), and coastal processes (Chapter 17).

To Communicate the Nature of Geography

The nature of physical geography and its three major scientific perspectives (spatial, physical, and environmental) are discussed in Chapter 1. In subsequent chapters, all three perspectives are stressed. For example, location is a dominant topic in Chapter 2 and remains an important theme throughout the text. Spatial distributions are emphasized as the elements of weather and climate are discussed in Chapters 4 through 6. The changing Earth system is a central focus in the text and featured in Chapter 1 and Chapter 8. Characteristics of climate regions and their associated environments are in Chapters 7 and 8. Spatial interactions are demonstrated in discussions of weather systems (Chapter 6), soils (Chapter 9), and volcanic and tectonic activity (Chapter 11). Feature boxes in every chapter present interesting and important examples of each perspective.

To Fulfill the Major Requirements of Introductory Physical Science Courses

Fundamentals of Physical Geography offers a full chapter on the scientific tools and methodologies of physical geography. Earth as a system and the physical processes affecting physical phenomena beneath, at, and above Earth's surface are examined in detail. Scientific method and explanations are stressed. End-of-chapter questions include interpreting graphs of environmental data (or graphing data for study), quantitative analysis, classification, calculating environmental variables, and hands-on map interpretation. Models and systems are frequently cited in discussions of important concepts, and scientific classification is presented in several chapters. Some of these topics include air masses, tornadoes, and hurricanes (Chapter 6), climates (Chapters 7 and 8), biogeography and soils (Chapter 9), water resources (Chapter 13), rivers (Chapter 14), and coasts (Chapter 17).

Physical geography plays a central role in understanding environmental aspects and issues, human–environment interactions, and in approaches to environmental problem solving. The beginning students in this course include the professional geographers of tomorrow. Spreading the message about the importance, relevance, and career potential of geography in today's world is essential to the strength of geography at educational levels from pre-collegiate through university. *Fundamentals of Physical Geography* seeks to reinforce that message.

Ancillaries

Instructors and students alike will greatly benefit from the comprehensive ancillary package that accompanies this text.

For the Student

Geography Resource Center This password-protected site includes interactive maps, animations, and an array of other discipline-related resources to complement your experience with geography. Visit www.iChapters.com to purchase access, if one was not bundled with your text.

For the Instructor

Dynamic Lecture Support

PowerLecture with JoinIn™ A complete all-in-one reference for instructors, the PowerLecture DVD contains PowerPoint® slides with lecture outlines, images from the text, stepped art from the text, zoomable art figures from the text, blank regional maps, videos, Google Earth layers and instructor's manual, and active figures that interactively demonstrate concepts. Besides providing you with fantastic course presentation material, the PowerLecture DVD contains electronic files of the Test Bank and Instructor's Manual, as well as JoinIn, the easiest Audience Response System to use, featuring instant classroom assessment and learning.

Laboratory and GIS Support

GIS Investigations Michelle K. Hall-Wallace, C. Scott Walker, Larry P. Kendall, Christian J. Schaller, and Robert F. Butler.

The perfect accompaniment to any physical geography course, these four groundbreaking guides tap the power of *ArcView*® GIS to explore, manipulate, and analyze large data sets. The guides emphasize the visualization, analysis, and multimedia integration capabilities inherent to GIS and enable students to "learn by doing" with a full complement of GIS capabilities. The guides contain all the software and data sets needed to complete the exercises.

Exploring the Dynamic Earth:
GIS Investigations for the Earth Sciences
ISBN with ArcView CD: 0-534-39138-9
ISBN for use with ArcGIS site license: 0-495-11509-6

Exploring Tropical Cyclones:
GIS Investigations for the Earth Sciences
ISBN with ArcView CD: 0-534-39147-8
ISBN for use with ArcGIS site license: 0-495-11543-6

Exploring Water Resources:
GIS Investigations for the Earth Sciences
ISBN with ArcView CD: 0-534-39156-7
ISBN for use with ArcGIS site license: 0-495-11512-6

Exploring the Ocean Environment:
GIS Investigations for the Earth Sciences
ISBN with ArcView CD: 0-534-42350-7
ISBN for use with ArcGIS site license: 0-495-11506-1

Acknowledgments

Fundamentals of Physical Geography would not have been possible without the encouragement and assistance of editors, friends, and colleagues from throughout the country. Great appreciation is extended to Martha, Emily, and Hannah Petersen; Greg Nadon; and Sarah Gabler; for their patience, support, and understanding.

Special thanks go to the splendid freelancers and staff members of Brooks/Cole Cengage Learning. These include Yolanda Cossio, Publisher; Laura Pople, Sr. Acquisitions Editor; Amy Collins, Development Editor; Samantha Arvin, Assistant Editor; Alexandria Brady, Associate Media Editor; Hal Humphrey, Production Project Manager; illustrators Accurate Art, Precision Graphics, Rolin Graphics, and Pre-Press PMG, Katy Gabel, Pre-Press PMG Project Manager; Jaime Jankowski, Pre-Press PMG Photo Researcher; Jeanne Platt, Editorial Assistant; and Dr. Chris Houser for creating our Google Earth activities.

Photos courtesy of: Rainer Duttmann, University of Kiel; Parv Sethi; Martha Moran, White River National Forest; Mark Muir, Fishlake National Forest; Mark Reid, USGS; Dawn Endico; Gary P. Fleming, Virginia Natural Heritage Program; Tessy Shirakawa, Mesa Verde National Park; Bill Case, Chris Wilkerson, and Michael Vanden Berg, Utah Geological Survey; Center for Cave and Karst Studies, Western Kentucky University; L. Michael Trapasso, Western Kentucky University; Hari Eswaran, USDA Natural Resources Conservation Service; Richard Hackney, Western Kentucky University; David Hansen, University of Minnesota; Susan Jones, Nashville, Tennessee; Bob Jorstad, Eastern Illinois University; Carter Keairns, Texas State University; Parris Lyew-Ayee, Oxford University, UK; L. Elliot Jones, U.S. Geological Survey; Anthony G. Taranto Jr., Palisades Interstate Park–New Jersey Section; Justin Wilkinson, Earth Sciences, NASA Johnson Space Center; Hajo Eicken, Alfred Wegener Institute for Polar and Marine Research; U.S. Fish and Wildlife Service; Loxahatchee National Wildlife Refuge; Philippe Rekacewicz, UNEP/GRID-Arendal *World Atlas of Desertification*. Greg Nadon, Ohio University, L. Michael Trapasso, Western Kentucky University.

Colleagues who reviewed this text and related Physical Geography editions include: Peter Blanken, University of Colorado; J. Michael Daniels, University of Wyoming;

James Doerner, University of Northern Colorado; Greg Gaston, University of North Alabama; Chris Houser, University of West Florida; Debra Morimoto, Merced College; Peter Siska, Austin Peay State University; Richard W. Smith, Harford Community College; Paul Hudson, University of Texas; Alan Paul Price, University of Wisconsin; and Richard Earl and Mark Fonstad, both of Texas State University.

The comments and suggestions of all of the above individuals have been instrumental in developing this text. Countless others, both known and unknown, deserve heartfelt thanks for their interest and support over the years. Despite the painstaking efforts of the reviewers, there will always be questions of content, approach, and opinion associated with the text. The authors wish to make it clear that they accept full responsibility for all that is included in *Fundamentals of Physical Geography*.

James F. Petersen
Dorothy Sack
Robert E. Gabler

Brief Contents

Contents

James F. Petersen   James F. Petersen is Professor of Geography at Texas State University, in San Marcos, Texas. He is a broadly trained physical geographer with strong interests in geomorphology and Earth Science education. He enjoys writing about topics relating to physical geography for the public, particularly environmental interpretation, and has written a landform guidebook for Enchanted Rock State Natural Area in central Texas and a number of field guides. He is a strong supporter of geographic education, having served as President of the NCGE in 2000 after more than 15 years of service to that organization. He has also written or served as a senior consultant for nationally published educational materials at levels from middle school through university, and has done many workshops for geography teachers. Recently, he contributed the opening chapter in an environmental history of San Antonio that explains the physical geographic setting of central Texas.

Dorothy Sack   Dorothy Sack, Professor of Geography at Ohio University in Athens, Ohio, is a physical geographer who specializes in geomorphology. Her research emphasizes arid region landforms, including geomorphic evidence of paleolakes, which contributes to paleoclimate reconstruction. She has published research results in a variety of professional journals, academic volumes, and Utah Geological Survey reports. She also has research interests and publications on the history of geomorphology and the impact of off-road vehicles. Her work has been funded by the National Geographic Society, NSF, Association of American Geographers (AAG), American Chemical Society, and other organizations. She is active in professional organizations, having served as chairperson of the AAG Geomorphology Specialty Group, and several other offices for the AAG, Geological Society of America, and History of Earth Sciences Society. She enjoys teaching and research, and has received the Outstanding Teacher Award from Ohio University's College of Arts and Sciences.

Robert E. Gabler   During his nearly five decades of professional experience, Professor Gabler has taught geography at Hunter College, City of New York, Columbia University, and Western Illinois University, in addition to 5 years in public elementary and secondary schools. At times in his career at Western he served as Chairperson of the Geography and Geology Department, Chairperson of the Geography Department, and University Director of International Programs. He received three University Presidential Citations for Teaching Excellence and University Service, served two terms as Chairperson of the Faculty Senate, edited the *Bulletin of the Illinois Geographical Society,* and authored numerous articles in state and national periodicals. He is a Past President of the Illinois Geographical Society, former Director of Coordinators and Past President of the National Council for Geographic Education, and the recipient of the NCGE George J. Miller Distinguished Service Award.

Physical Geography: Earth Environments and Systems

1

Outline

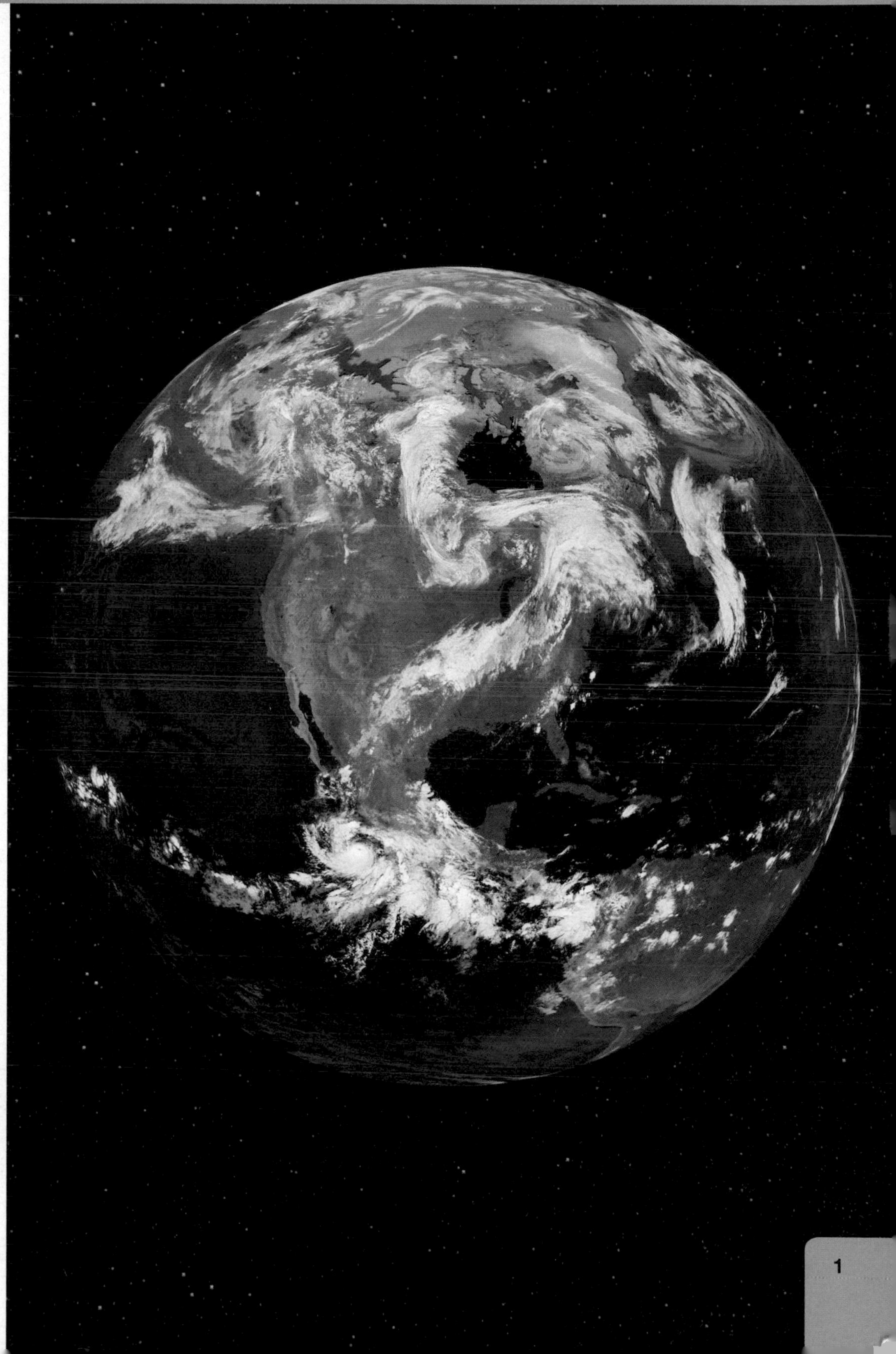

Earth's incredible environmental diversity: An oasis of life in the vastness of space.

NASA

:: Objectives

When you complete this chapter you should be able to:

■ Discuss physical geography as a discipline and profession that considers both the natural world and the human interface with the natural world.

■ Understand how geographic information and techniques are directly applicable in many career fields.

■ Describe the three major perspectives of physical geography: the spatial perspective, the physical science perspective, and the environmental perspective.

■ Conceptualize Earth as a system of interacting parts that respond to both natural and human-induced processes.

■ Explain how interactions between humans and their environments can be both advantageous and risky.

■ Recognize how a knowledge of physical geography invites better understanding of our environment.

Viewed from far enough away to see an entire hemisphere, Earth is both beautiful and intriguing. From this perspective we can begin to appreciate "the big picture," a global view of our planet's physical geography. If we look carefully, we can also recognize geographic patterns, shaped by the processes that make our world dynamic and ever-changing. Characteristics of the oceans, atmosphere, landmasses, and evidence of life, revealed by vegetated regions, are apparent.

From a human perspective, Earth may seem immense and almost limitless. In contrast, viewing the "big picture" reveals Earth's fragile nature—a spherical island of life surrounded by the vast, dark emptiness of space. Except for the external addition of energy from the sun, our planet is a self-contained system that has all the requirements to sustain life. The nature of Earth and its environments provide the life-support systems for all living things. It is important to gain an understanding of the planet that sustains us, to learn about the components and processes that operate to change or regulate the Earth system. Learning the relevant questions to ask is an important step toward finding answers and explanations. Understanding how Earth's features and processes interact to develop the environmental diversity on our planet is the goal of a course in physical geography.

The Study of Geography

Geography refers to the examination, description, and explanation of Earth—its variability from place to place, how places and features change over time, and the processes responsible for these variations and changes. Geography is often called the **spatial science** (the science of locational space) because it includes analyzing and explaining the locations, distributions, patterns, variations, and similarities or differences among phenomena on Earth's surface.

Geographers study processes that influenced Earth's landscapes in the past, how they continue to affect them today, how a landscape may change in the future, and the significance or impact of these changes. Geography is distinctive among the sciences by virtue of its definition and central purpose, and may involve studying any topic related to the scientific analysis of human or natural processes on Earth (■ Fig. 1.1).

■ **FIGURE 1.1**   When conducting research or examining one of society's many problems, geographers are prepared to consider any information or aspect of a topic that relates to their studies.
What advantage might a geographer have when working with other physical scientists seeking a solution to a problem?

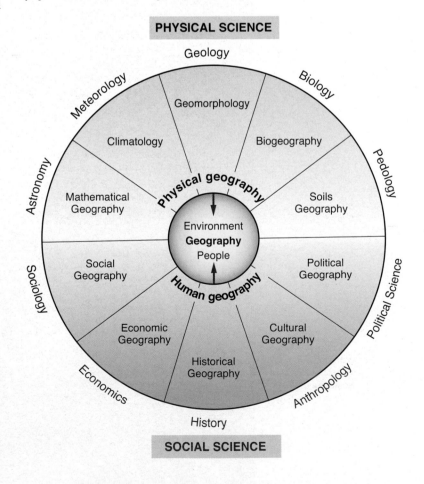

Geographers are also interested in how to divide areas into meaningful **regions,** which are areas identified by distinctive characteristics that distinguish them from surrounding areas. Physical, human, or a combination of factors can define a region. *Regional geography* concentrates on the characteristics of a region (or of multiple regions).

Physical Geography

Physical geography encompasses the processes and features that make up Earth, including human activities where they interface with the environment. Geographers generally take a **holistic approach,** meaning that they often consider both the human and natural phenomena that are relevant to understanding aspects of our planet. Physical geographers are concerned with nearly all aspects of Earth and are trained to view a natural environment in its entirety, as well as how it functions as a unit (■ Fig. 1.2). Yet, most physical geographers focus their expertise on one or two specialties. For example, many *meteorologists* and *climatologists* have studied geography. Meteorologists are interested in the processes that affect daily weather, and they forecast weather conditions. Climatologists are interested in regional climates, the averages and extremes of long-term weather data, understanding climate change and climatic hazards, and the impact of climate on human activities and the environment.

Geomorphology, the study of the nature and development of landforms, is a major subfield of physical geography. Geomorphologists are interested in understanding variations in landforms, and the processes that produce Earth's surface landscapes. *Biogeographers* study plants, animals, and environments, examining the processes that influence, limit, or facilitate their characteristics, distributions, and changes over time. Many *soil scientists* are geographers who map and analyze soil types, determine the suitability of soils for certain uses, and work to conserve soil resources.

Geographers are also widely involved in the study of water bodies and water resources including their processes, movements, impacts, quality, and other characteristics. They may serve as *hydrologists, oceanographers,* or *glaciologists.* Many geographers also function as *water resource managers,* working to ensure that lakes, watersheds, springs, and groundwater sources are adequate in quantity and quality to meet human and environmental needs.

Like other scientists, physical geographers typically apply the **scientific method** as they seek to learn about aspects of Earth. The scientific method involves seeking the answers to questions and determining the validity of new ideas by

■ **FIGURE 1.2** Physical geographers study the elements and processes that affect natural environments. These include rock structures, landforms, soils, vegetation, climate, weather, and human impacts. This is in the White River National Forest, Colorado.
What physical geography characteristics can you observe in this scene?

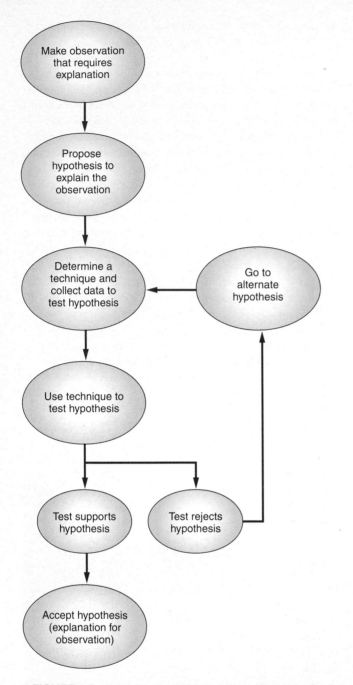

objectively testing all pertinent evidence and facts that affect the issue being studied (■ Fig. 1.3). Using the scientific method, new ideas or proposed answers to questions are only accepted as valid if they are clearly supported by the evidence.

Technology, Tools, and Methods

The technologies that are used for learning about the physical geography of our planet are rapidly changing. The abilities of computer systems to capture, process, model, and display spatial data—functions that can now be performed on a personal computer—were only a dream 30 years ago. Today, the Internet provides access to information and images on virtually any topic. Continuous satellite imaging of Earth has been ongoing for more than 30 years, which has given us a better perspective on environmental changes. Using various energy sources to produce images from space, we are able to see, measure, monitor, and map processes and the effects of certain processes, including many that are invisible to the naked eye. Graphic displays of environmental data and information are becoming more vivid and striking as a result of sophisticated methods of data processing and visual representation. Increased computer power allows the presentation of high-resolution images, three-dimensional scenes, and animated images of Earth's features, changes, and processes (■ Fig. 1.4).

Satellite technology is used to determine the precise location of a positioning receiver on Earth's surface, a capability that has many useful applications for geography and mapping. Today, most mapmaking (*cartography*) and many aspects of map analysis are computer-assisted operations, although the ability to visually interpret a map, a landscape, or an environmental image remains an important geographic skill.

Physical geographers should be able to make observations and gather data in the field, but they must also keep up with new technologies that support and facilitate traditional fieldwork. Technology may provide maps, images, and data, but a person who is knowledgeable about the geographical aspects of the subject being studied is essential to the processes of analysis and

■ **FIGURE 1.3**   The scientific method, widely applicable in physical geography, involves the steps shown here.

1. **Making an observation that requires an explanation.** On a trip to the mountains, you notice that it gets colder as you go up in elevation. Is that just a result of local conditions on the day you were there, or is it a universal relationship?
2. **Restating the observation as a hypothesis.** Here is an example: As we go higher in elevation, the temperature gets cooler. (The answer may seem obvious, but while it is generally true, there are exceptions, depending on environmental conditions that will be discussed in later chapters.)
3. **Determining a technique for testing the hypothesis and collecting necessary data.** The next step is finding a technique for evaluating data (numerical information) and or facts that relate to the hypothesis. In this case, you would gather temperature and elevation data (taken at about the same time for all data points) for the study area.
4. **Applying the technique or strategy to test the validity of the hypothesis.** Here we discover if the hypothesis is supported by adequate evidence, collected under similar conditions to minimize bias. The technique will recommend either acceptance or rejection of the hypothesis. If the hypothesis is rejected, we can test an alternate hypothesis, or we may just discover that our hypothesized relationship is not valid.

problem solving. Many geographers are gainfully employed in positions that apply technology to the problems of understanding our planet and its environments, and their numbers are certain to increase in the future (■ Fig. 1.5).

Image by R. B. Husar, Washington University; the land layer from the SeaWiFS Project; fire maps from the European Space Agency; the sea surface temperature from the Naval Oceanographic Office's Visualization Laboratory; and cloud layer from SSEC, University of Wisconsin

■ **FIGURE 1.4** Complex computer-generated model of Earth, based on data gathered from satellites.
How does this image compare to the Earth image in the chapter opening?

© Ashley Cooper/CORBIS

■ **FIGURE 1.5** A geographer uses computer technology to analyze maps and imagery.
In what ways are computer-generated maps and landscape images helpful in studying physical geography?

Major Perspectives in Physical Geography

Your textbook will demonstrate three major perspectives that physical geography emphasizes: spatial science, physical science, and environmental science. Although the focus on each of these perspectives may vary from chapter to chapter, take note of how each perspective relates to the unique nature of geography as a discipline.

The Spatial Perspective

A central theme in geography is illustrated by its definition as the *spatial science*. Physical geographers have many divergent interests, but they share the common goals of understanding and explaining spatial variations on Earth's surface. The following five examples illustrate spatial factors that geographers typically consider and the problems they address.

Location
Geographic studies often begin with locational information. Features are located using one of two methods: **absolute location,** which is expressed by a coordinate system (or address), or **relative location,** which identifies where a feature exists in relation to something else, usually a fairly well-known location. For example, Pikes Peak, in the Rocky Mountains of Colorado, with an elevation of 4302 meters (14,115 feet), has a location of latitude 38°51' north and longitude 105°03' west. This is an example of an absolute location. However, it could also be stated that Pikes Peak is 36 kilometers (22 miles) west of Colorado Springs (■ Fig. 1.6). This is an example of relative location.

Characteristics of Places
Physical geographers are interested in the environmental features and processes that make a place unique, and also the shared or similar characteristics between places. For example, what physical geographic features make the Rocky Mountains appear as they do? Further, how are the Appalachian Mountains different from the Rockies, and what characteristics are common to these two mountain ranges? Another aspect of the characteristics of places is the analysis of the environmental advantages and challenges that exist in a place.

Spatial Distribution and Pattern
Spatial distribution is a locational characteristic that refers to the extent of the area or areas where a feature exists. For example, where on Earth do we find tropical rainforests? What is the distribution of rainfall in the United States on a particular day? Where on Earth do major earthquakes occur? **Spatial pattern** refers to how features are arranged in space—are they regular or random, clustered together or widely spaced? Population distributions can be either dense or sparse (■ Fig. 1.7). The spatial pattern of earthquakes may be aligned on a map because earthquake faults display similar linear patterns.

GEOGRAPHY'S SPATIAL PERSPECTIVE

:: NATURAL REGIONS

The term *region* has a precise meaning and special significance to geographers. Simply stated, a region is an area that is defined by a certain shared characteristic (or a set of characteristics) existing within its boundaries. The concept of a region is a tool for thinking about and analyzing logical divisions of areas based on their geographic characteristics. Geographers not only study and explain regions, including their locations and characteristics, but also strive to delimit them—to outline their boundaries on a map. An unlimited number of regions can be derived for each of the four major Earth subsystems.

Regions help us understand the arrangement and nature of areas on our planet. Regions can also be divided into subregions. For example, North America is a region, but it can be subdivided into many subregions. Examples of subregions based on natural characteristics include the Atlantic Coastal Plain (similarity of landforms, geology, and locality), the Prairies (ecological type), the Sonoran Desert (climate type, ecological type, and locality), the Pacific Northwest (general locality), and Tornado Alley (region of high potential for these storms).

There are three important points to remember about regions. Each of these points has endless applications and adds considerably to the questions that the process of defining regions based on spatial characteristics seeks to answer.

■ **Natural regions can change in size and shape over time in response to environmental changes.** An example is *desertification*, the expansion of desert regions that has occurred in recent years. Using images from space, we can see and monitor changes in the area covered by deserts, as well as other natural regions.

■ **Boundaries separating different regions tend to be indistinct or transitional, rather than sharp.** For example, on a climate map, lines separating desert from nondesert regions do not imply that extremely arid conditions instantly appear when the line is crossed. If we travel to a desert, it is likely to get progressively more arid as we approach our destination.

■ **Regions are spatial models, devised by humans, for geographic analysis, study, and understanding.** Regions are conceptual models that help us

comprehend and organize spatial relationships and geographic distributions. Learning geography is an invitation to think spatially, and regions provide an essential, extremely useful, conceptual framework in that process.

Understanding regions, through an awareness of how areas can be divided into geographically logical units and why it is useful to do so, is essential in geography. Regions help us to understand, reason about, and make sense of, the spatial aspects of our world.

USDA Forest Service

The Great Basin of the Western United States is a landform region that is clearly defined based on an important physical geographic characteristic. No rivers flow to the ocean from this arid and semiarid region of mountains and topographic basins. The rivers and streams that exist, flow into enclosed basins where the water evaporates away from temporary lakes, or they flow into lakes like the Great Salt Lake, which has no outlet to the sea. Topographic features called *drainage divides* (mountain ridges) form the outer edges of the Great Basin, defining and enclosing this natural region

Pike's Peak   Colorado Springs

© NASA/Goddard Space Flight Center/Earth Observatory

■ **FIGURE 1.6**   A three-dimensional digital model shows the relative location of Pikes Peak to Colorado Springs, Colorado. Because this is a perspective view, the 36 km (22 mi) distance appears to be shorter than its actual ground distance.
What physical geographic characteristics of this place can you extract from the image?

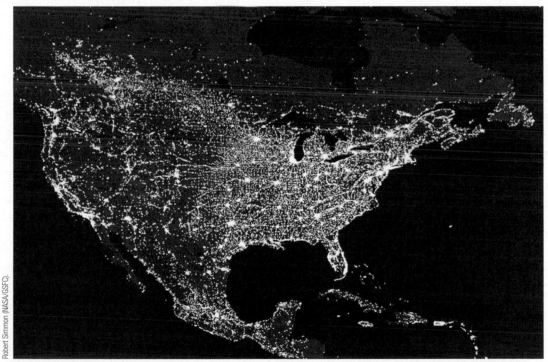

Data courtesy Marc Imhoff (NASA/GSFC) and Christopher Elvidge (NOAA/NGDC). Image by Craig Mayhew (NASA/GSFC) and Robert Simmon (NASA/GSFC).

■ **FIGURE 1.7**   A nighttime satellite image provides good illustrations of distribution and pattern, shown here on most of North America. Spatial distribution is where features are located (or perhaps, absent). Spatial pattern refers to their arrangement. Geographers seek to explain these spatial relationships.
Can you locate and propose possible explanations for two patterns and two distributions in this scene?

Spatial Interaction   Few processes on Earth operate in isolation, because areas on our planet are interconnected. A condition, an occurrence, or a process in one place generally has an impact on other places. Unfortunately, the exact nature of a **spatial interaction**—whether one event actually causes another—is often difficult to establish with certainty.

Examples of observed spatial interaction include the occurrence of abnormally warm ocean waters off South

America's west coast, a condition called *El Niño*, and unusual weather in other parts of the world. Clearing the tropical rainforest may also have a widespread impact on world climates. Geographers work to understand spatial relationships, interactions, and impacts, at local, regional, and global scales.

The Changing Earth

Earth's features and landscapes are continuously changing in a spatial context. Weather maps show where and how weather elements change from day to day, over the seasons, and from year to year. Storms, earthquakes, landslides, and stream processes modify the landscape. Coastlines may change position because of storm waves, *tsunamis*, or changes in sea level. Areas that were once forested have been clear-cut, changing the nature of the environment there. Desert-like conditions seem to be expanding in many arid regions of the world. The cover of sea ice in the polar oceans has expanded and contracted in historic times.

World climates have changed throughout Earth's history, with attendant shifts in the distributions of plant and animal life. Today, changes in Earth's climates and environments are complicated by the impact of human activities. Most of Earth's glaciers are shrinking in response to global warming (■ Fig. 1.8). Earth and its environments are always changing, although at different time scales, so the impact and direction of certain changes can be difficult to determine.

The Physical Science Perspective

Physical geographers observe phenomena, compile data, and seek solutions to problems or the answers to questions that are also of interest to researchers in other physical sciences. However, physical geographers also bring distinctive points of view to scientific study—a holistic perspective and a spatial perspective. By examining the factors, features, and processes that influence an environment, and how these elements work together, we can better understand our planet's dynamic physical geography. We can also appreciate the importance of viewing Earth as a constantly functioning system.

The Earth System

A **system** is any entity that consists of interrelated parts or components. Our planetary environment, the **Earth system,** relies on the interactions among a vast combination of factors that enable it to support life. The individual components of a system, termed **variables,** change through interactions with one another as parts of a functioning unit.

For example, the presence of mountains influences the distribution of rainfall, and variations in rainfall affect the density, type, and variety of vegetation. Plants, moisture, and the underlying rock affect soil that forms in an area. Characteristics of vegetation and soils influence the runoff of water from the land, leading to completion of the circle, because the amount of runoff is a major factor in stream erosion, which eventually can reduce the height of mountains.

Systems can be divided into **subsystems,** which are functioning units of a system that demonstrate strong internal connections. For example, the human body is a system that is composed of many subsystems (for example, the respiratory system, circulatory system, and digestive system). Examining the Earth system as a set of interdependent subsystems facilitates the study of physical geography.

Earth's Four Major Subsystems

There are four major subsystems of the Earth system (■ Fig. 1.9). The **atmosphere** is the gaseous blanket of air that envelops, shields, and insulates Earth. The **lithosphere** makes up the solid Earth—landforms, rocks, soils, and minerals. The **hydrosphere** includes the waters of Earth—oceans, lakes, rivers, and glaciers. The **biosphere** is composed of all living things: people, other animals, and plants.

The characteristics of these subsystems interact to create and nurture the conditions necessary for life on Earth, but the impact and intensity of those interactions are not equal everywhere. This

■ **FIGURE 1.8**   Photographs taken 92 years apart in Montana's Glacier National Park show that Shepard Glacier, like other glaciers in the park, has dramatically receded during that time. This retreat is in response to climate warming and droughts.

What other kinds of environmental change might require long-term observation and recording of evidence?

1913

2005

W. C. Alden (USGS), Courtesy of Glacier National Park Archives

Blase Reardon (USGS), Courtesy of Glacier National Park Archives

Atmosphere

Biosphere

Earth System

Hydrosphere

All, Copyright and photograph by Dr. Parvinder S. Sethi; center inset, NASA

■ **FIGURE 1.9** Earth's four major subsystems
understanding changes in our planet's environme
changes. Earth consists of many interconnected
**How do these systems overlap? For example, h
the hydrosphere, or with the biosphere?**

such as global warming, centers on the
increasing impact that human activities
are exerting on Earth's natural systems.

The Environmental Perspective

Today, we regularly hear about the
environment and we are concerned
about environmental damage caused
by human activities. We also hear news
reports of disasters caused by humans
being exposed to violent natural processes such as earthquakes, floods,
_____ storms. Recent
_____ ers include the
_____ of 2004, Hur-
_____, and Hurricane
_____ adest sense,
_____ e defined as
_____ ade up of all
_____ ral aspects of
_____ growth, our
_____ ng.
_____ are systems
_____ of elements,
_____ that in-
_____ g weather,
_____ lants,
_____ Physical
_____ s well
_____ tanding,

inequality leads to our planet's environmental dive
duces the wide variety of geographic patterns on Ea

Earth Impacts
We are aware that the Earth s
dynamic, responding to continuous changes, and tha
directly observe some of these changes—the seasons,
tides, earthquakes, floods, volcanic eruptions. The interactions
that change our planet function in cycles and processes that
operate at widely varying rates. Many aspects of our planet may
take years, or even more than a lifetime, to accumulate enough
change so that humans can recognize their impact. Long-term
changes in our planet are often difficult to understand or predict with certainty. The evidence must be carefully and scientifically studied to determine what is occurring and what the
potential consequences might be. Changes of this type include
shifts in climates, drought cycles, the spread of deserts, erosion of coastlines, and major changes in river systems. Volcanic
islands have been created in historic times (■ Fig. 1.10), and
a new Hawaiian island is now forming beneath the waters of
the Pacific Ocean. Change may be naturally-caused or human-
induced, or may result from a combination of these factors.
Today, much of the concern about environmental changes,

tlantic,
d this
_____ olcanic eruptions
_____ or processes have reduced the
_____ original size.

**_____ the island formed and cooled, what other environmental
changes should slowly begin to take place?**

Icelandic Ministry for the Environment

GEOGRAPHY'S ENVIRONMENTAL PERSPECTIVE

:: HUMAN–ENVIRONMENT INTERACTIONS

Earth's environmental characteristics support all life on our planet, including human existence. Yet, the effects of human activities on the environment, as well as the impacts of environmental processes on humans, have become topics of increasing concern. Certain environmental processes can be dangerous to human life and property, and certain human activities threaten to cause major, and possibly irrevocable, damage to Earth environments.

Environmental Hazards

The environment becomes a hazard to humans and other life forms when, occasionally and often unpredictably, a natural process operates in an unusually intense or violent fashion. Rain showers may become torrential rains that occur for days or weeks and cause flooding. Some tropical storms gain strength, and reach coastlines as full-blown hurricanes, as Hurricane Katrina did in 2005. Molten rock and gases from deep beneath the Earth move toward the surface and suddenly trigger massive eruptions that can blow apart volcanic mountains. The 2004 tsunami wave that devastated coastal areas along the Indian Ocean provided an example of the potential for the occasional occurrences of natural processes that far exceed our expectable "norm."

When an Earth system operates in a sudden or extraordinary fashion it is a noteworthy environmental event, it is not an environmental hazard unless people or their properties are affected. Many environmental hazards exist because people live where potentially catastrophic environmental events may occur. Nearly every populated area of the world is associated with an environmental hazard or perhaps several hazards. Forested regions are subject to fire; earthquake, landslide, and volcanic activities plague mountain regions; violent storms threaten interior plains; and many coastal regions experience periodic hurricanes or typhoons (the term for hurricanes that strike Asia).

Environmental Degradation

Just as the environment can pose an ever-present danger to humans, through their activities, humans can constitute a serious threat to the environment. Issues such as global warming; acid precipitation; deforestation and the extinction of biological species in tropical areas; damage to the ozone layer of the atmosphere; and desertification have risen to the top of agendas when world leaders meet and international conferences are held. Environmental concerns are recurring subjects

USGS Western Coastal and Marine Geology

Environmental Hazards: Destructive Tsunami. In December of 2004, a powerful undersea earthquake generated a large tsunami, which devastated many coastal areas along the Indian Ocean, particularly in Thailand, Sri Lanka, and Indonesia. Nearly a quarter of a million people were killed, and the homes of about 1.7 million people were destroyed. Here a huge barge was left onshore by the tsunami, which leveled buildings, and stripped the vegetation from the cliffs to a height of 31 meters (102 ft). Some natural–environmental processes, like this one, can be detrimental to humans and their built environment, and others are beneficial. **Can you cite some examples of natural processes that can affect the area where you live?**

of magazine and newspaper articles, books, and television programs.

Much environmental damage has resulted from atmospheric pollution associated with industrialization, particularly in support of the wealthy, developed nations. But as population pressures mount and developing nations struggle to industrialize, human activities are exacting an increasing toll on the soils, forests, air, and waters of the developing world as well. Environmental deterioration is a problem of worldwide concern, and solutions must involve international cooperation in order to be successful. As citizens of the world's wealthiest nation, Americans must seriously consider what steps can be taken to counter major environmental threats related to human activities. What are the causes of these threats? Are the threats real and well documented? What can I personally do to help solve environmental problems? With limited resources on Earth, what will we leave for future generations?

Examining environmental issues from the physical geographer's perspective requires that characteristics of both the environment and the humans involved in those issues be given strong consideration. As it will become apparent in this study of geography, physical environments are changing constantly, and all too frequently human activities result in negative environmental consequences. In addition, throughout Earth, humans live in constant threat from various and spatially distributed environmental hazards such as earthquake, fire, flood, and storm. The natural processes involved are directly related to the physical environment, but causes and solutions are imbedded in human–environmental interactions that include the economic, political, and social characteristics of the cultures involved. The recognition that geography is a holistic discipline—that it includes the study of all phenomena on Earth—requires that physical geographers play a major role in the environmental sciences.

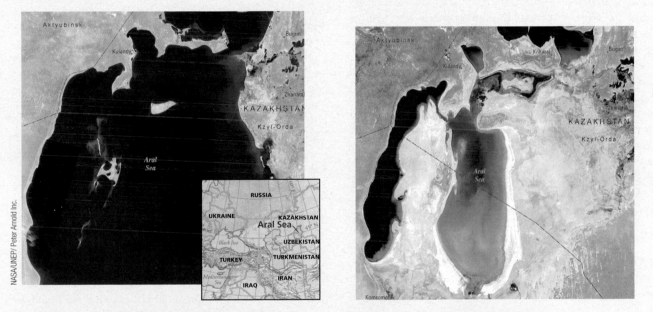

Environmental Degradation: The Shrinking Aral Sea. Located in the Central Asian desert between Kazakhstan and Uzbekistan, the Aral Sea is an inland lake that does not have an outlet stream. The water that flows in is eventually lost by evaporation to the air. Before the 1960s, rivers flowing out of mountain regions supplied enough water to maintain what was the world's fourth largest body of inland water. Since that time, diversion of river water for agriculture has caused the Aral Sea to dramatically shrink. The result has been the disappearance of many species that relied on the lake for survival, along with frequent dust storms, and an economic disaster for the local economy. Without the waters of the lake to moderate temperatures, the winters have become colder and the summers hotter. Today, efforts are underway to restore at least part of the lake and its environments.

What are some examples of how humans have impacted the environment where you live?

because important factors are considered individually and as parts of an environmental system.

The study of relationships between organisms and their environments is a science known as **ecology.** Ecological relationships are complex but naturally balanced "webs of life." The word **ecosystem** (a contraction of *ecological system)* refers to a community of organisms and the relationships of those organisms to each other and to their environment (■ Fig. 1.11). An ecosystem is dynamic in that its various parts are always changing. For instance, plants grow, rain falls, animals eat, and soils develop—all changing the environment of a particular ecosystem. Because each member of the ecosystem belongs to the environment of other parts of that system, a change in one often affects the environment for the others. The ecosystem concept can be applied on almost any scale from local to regional or global, in virtually any geographic location. Your backyard, a farm pond, a grass-covered field, a marsh, a forest, or a portion of a desert can be viewed as an ecosystem. Human activities will always affect the environment in some way, but if we understand the factors and processes involved, we can work to minimize the negative impacts.

A Life-Support System   The most critical and unique attribute of Earth is that it is a **life-support system,** a set of interrelated components that are necessary for the existence of living organisms. On Earth, natural processes produce an adequate supply of oxygen; the sun interacts with the atmosphere, oceans, and land to maintain tolerable temperatures; and photosynthesis or other processes provide food supplies for living things. If a critical part of a life-support system is significantly changed or fails to operate properly, living organisms may no longer be able to survive. Other than the input of energy from the sun, the Earth system provides the necessary environmental constituents and conditions that allow life, as we know it, to exist (■ Fig. 1.12).

Today, we realize that critical parts of our planet's life-support system, **natural resources,** can be abused, wasted, or

NASA

■ **FIGURE 1.12**   The International Space Station functions as a life-support system. Astronauts can venture out on a spacewalk, but they remain dependent on resources like air, food, and water that are shipped in from Earth.
What do the limited resources on space vehicles suggest about our environmental situation on Earth?

exhausted, potentially threatening Earth's ability to support human life. A concern is that humans are rapidly depleting nonrenewable natural resources, like coal and oil, which, once exhausted, will not be replaced. When nonrenewable resources such as these mineral fuels are gone, the alternative resources may be less effective or more expensive.

Besides overconsumption of natural resources, human activities also result in **pollution,** an undesirable or unhealthy contamination in an environment (■ Fig. 1.13). We are aware that critical resources such as air, water, and even land areas can be polluted to the point where they become unusable or even lethal to some life forms. Air pollution has become a serious environmental problem for urban centers throughout the world. What some people do not realize, however, is that pollutants are often transported by winds and waterways hundreds or even thousands of kilometers from their source. Lead from automobile exhaust has been found in the ice of Antarctica, as has the insecticide DDT. Pollution is a worldwide problem that does not stop at political, or even continental, boundaries.

Human–Environment Interactions   Physical geography includes giving special attention to environmental relationships that involve humans and their activities. Interactions between humans and environments are two-way relationships, because the environment influences human behavior and humans affect the environment.

Despite the wealth of resources available on Earth, the capacity of our planet to support the growing numbers of humans may have an

■ **FIGURE 1.11**   Ecosystems are an important aspect of natural environments, which are affected by the interaction of many processes and components.
How do ecosystems illustrate the interactions in the environment?

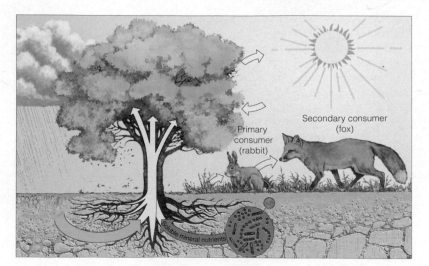

Secondary consumer (fox)

Primary consumer (rabbit)

soluble mineral nutrients

■ **FIGURE 1.13** (a) Denver, Colorado on a clear day, with the Rocky Mountains visible in the background. (b) On a smoggy day from the same location, even the downtown buildings are not visible.

If you were choosing whether to live in a small town, a rural area, or a major city, would pollution affect your decision?

ultimate limit, a population *threshold.* The continually increasing world population has passed 6.8 billion, and more than half the world's people tolerate substandard living conditions and insufficient food. Ultimately, over the long term, the size of the human population cannot exceed the environmental resources necessary to sustain it.

As we consider the importance of sustaining acceptable human living standards for generations to come, it is essential to note that environments do not change their nature to accommodate humans. Humans should alter their behavior to accommodate the limitations and potentials of Earth environments and resources. Geography has much to offer in understanding the factors involved in meeting this responsibility, and in helping us learn more about environmental changes that are associated with human activities. We should all understand the impact of our individual and collective actions on the complex environmental systems of our planet.

Although our current objective is to study physical geography, we should not ignore the information shown in the World Map of Population Density (shown on the inside back cover of this book). Population distributions are highly irregular—from uninhabited to densely settled, typically reflecting the differing capacities of varied environments to support human populations. There are limits to the suitable living space on Earth, and we must use our lands wisely (■ Fig. 1.14).

■ **FIGURE 1.14** (a) As a natural stream channel, Florida's Kissimmee River originally flowed in broad, sweeping bends on its floodplain for 160 km (100 mi) to Lake Okeechobee. (b) In the 1960s and 1970s, the river was artificially straightened to make room for agricultural and urban development, disrupting the previously existing ecosystem at the expense of plants, animals, and human water supplies. Today, because of efforts to restore this habitat, the Kissimmee is reestablishing its flood plain, wetland environments, and its natural channel. (c) One problem facing the restoration project is the invasion of weedy plants that has been occurring since the floodplain was drained, causing a serious fire hazard during the dry season. Controlled burns are necessary to avoid catastrophic wildfires, and to help restore the natural vegetation.

What factors should be considered prior to any attempts to return rivers and wetland habitats to their original condition?

(a)

(b)

(c)

Models and Systems

As physical geographers work to describe and explain the often-complex features of planet Earth and its environments, they support these efforts, as other scientists do, by developing representations of the real world, called *models*. A **model** is a useful simplification of a more complex reality that permits prediction, and each model is designed with a specific purpose in mind. As examples, maps and globes are models—simplified representations that provide us with useful information. Today, many models are computer generated because computers can handle great amounts of data and perform the mathematical calculations that are often necessary to construct and display certain types of information.

There are many kinds of models (■ Fig. 1.15). **Physical models** are solid three-dimensional representations, such as a world globe or a replica of a mountain. **Pictorial/graphic models** include pictures, maps, graphs, diagrams, and drawings. **Mathematical/statistical models** are used to predict possibilities such as river floods or the influence of climate change on daily weather. Words, language, and the definitions of terms or ideas can also serve as models.

Another important type is a **conceptual model**—the mind imagery that we use for understanding our surroundings and experiences. Focus for a moment on the image that the word *mountain* (or *waterfall, cloud, tornado, beach, forest, desert*) generates in your mind. Most likely what you "see" (conceptualize) in your mind is sketchy rather than detailed, but enough information is there to convey a mental idea of a mountain. This image is a conceptual model. For geographers, a particularly important type of conceptual model is the **mental map,** which we use to think about places, travel routes, and the distribution of features in space. How could we even begin to understand our world without conceptual models, and in terms of spatial understanding, without mental maps?

Systems Analysis

Our planet is too complex to permit a single model to explain all of its environmental components and how they affect one another. To begin to comprehend Earth as a whole or to understand most of its environmental components, physical geographers use a powerful strategy called **systems analysis.** Systems analysis suggests that the way to understand how anything works is to use the following strategy: (1) clearly define the system you wish to understand, (2) inventory that system's important parts and processes, (3) examine how each of these parts and processes interact with each other, and how those interactions affect the operation of the system.

Systems analysis often focuses on subsystems. Examples of subsystems examined by physical geographers include the water cycle, climatic systems, storm systems, stream systems, the systematic heating of the atmosphere, and ecosystems. A great advantage of systems analysis is that it can be applied to environments at virtually any spatial scale, from global to microscopic.

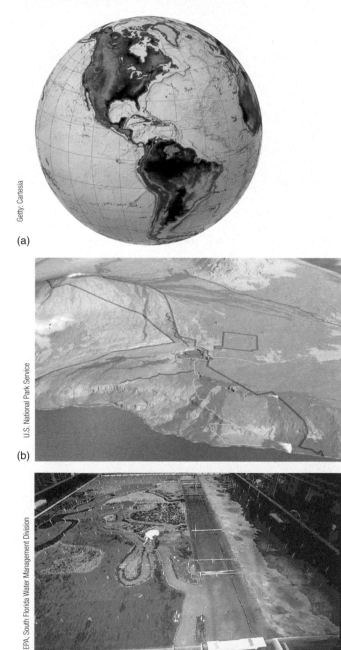

■ **FIGURE 1.15** Models help us understand Earth and its subsystems, by focusing our attention on major features or processes. (a) Globes are physical models that demonstrate many of Earth's characteristics—planetary shape, distributions of landmasses and oceans, and spatial relationships. (b) A digital landscape model shows the environment and terrain of Hawaii Volcanoes National Park. Computer-generated clouds, shadows, and reflections were added to provide "realism" to the scene. (c) This working physical model of the Kissimmee River is used to investigate ways to restore the environment. Proposed modifications can be analyzed on this model before work is done on the actual river (see Figure 1.14).

Getty: Cartesia

(a)

U.S. National Park Service

(b)

EPA, South Florida Water Management Division

(c)

How Systems Work

■ Figure 1.16 shows a systems model in which one can trace the movement of energy or matter into the system (**inputs**), their storage in the system and their movements out of the system (**outputs**), as well as the interactions between components within the system.

A **closed system** is one in which no substantial amount of *matter* crosses its boundaries, although *energy* can go in and out of a closed system (■ Fig. 1.17a). Planet Earth is essentially a closed system. Except for meteorites that reach Earth's surface, the escape of gas molecules from the atmosphere, and a few moon rocks brought back by astronauts, the Earth system is essentially closed to the input or output of matter.

Most Earth subsystems, however, are **open systems** (Fig. 1.17b), because both energy and matter move freely across subsystem boundaries as inputs and outputs. A stream is an excellent illustration of an open subsystem: Matter and energy in the form of soil particles, rock fragments, solar energy, and precipitation enter the stream, and water and sediments leave the stream where it empties into the ocean or some other standing body of water.

When we describe Earth as a system or as a complex set of interrelated systems, we are using conceptual models to help us organize our thinking about what we are observing. Throughout the chapters that follow, we will use the systems concept, as well as many other kinds of models, to help us simplify and illustrate complex features of the physical environment.

Equilibrium in Earth Systems

We often hear about the "balance of nature." What this means is that natural systems have built-in mechanisms that tend to counterbalance, or accommodate, change without affecting the system dramatically. If the inputs entering the system are balanced by outputs, the system is said to have reached a state of **equilibrium.** Most systems are continually shifting slightly one way or another as a reaction to external conditions. This change within a range of tolerance is called **dynamic equilibrium;** that is, the balance is not static but in the long-term changes may be accumulating. A reservoir contained by a dam is a good example of equilibrium in a system (■ Fig. 1.18).

The interactions that cause change or adjustment between parts of a system are called **feedback.** Two kinds of feedback relationships operate in a system. **Negative feedback,** whereby one change tends to offset another, creates a natural counteracting effect that is generally beneficial because it tends to help the system maintain equilibrium.

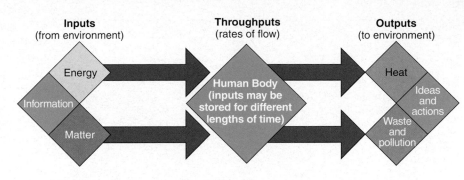

Inputs (from environment)   **Throughputs** (rates of flow)   **Outputs** (to environment)

Energy / Information / Matter → Human Body (inputs may be stored for different lengths of time) → Heat / Ideas and actions / Waste and pollution

■ **FIGURE 1.16**   The human body is an example of a system, with inputs of energy and matter.
What characteristics of the human body as a system are similar to the Earth as a system?

■ **FIGURE 1.17**   (a) Closed systems allow only energy to pass in and out. (b) Open systems involve the inputs and outputs of both energy and matter. Earth is basically a closed system. Solar energy (input) enters the Earth system, and that energy is dissipated (output) to space mainly as heat. External inputs of matter are virtually nil, mainly meteorites, and almost no matter is output from the Earth system. Because Earth is a closed system, humans face limits to their available natural resources. Subsystems on the planet, however, are open systems, with incoming and outgoing matter and energy. Processes are driven by energy.
Think of an example of an open system, and outline some of the matter–energy inputs and outputs involved in such a system.

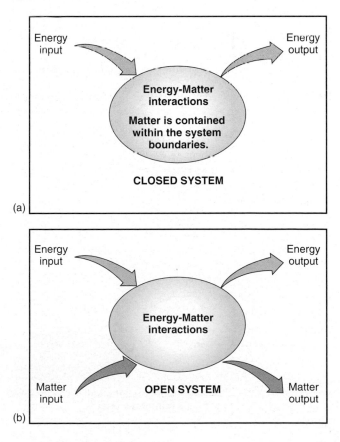

Energy input → Energy-Matter interactions / Matter is contained within the system boundaries. → Energy output

CLOSED SYSTEM

(a)

Energy input / Matter input → Energy-Matter interactions → Energy output / Matter output

OPEN SYSTEM

(b)

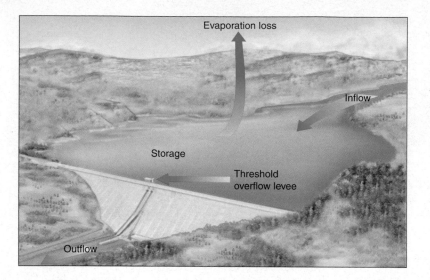

■ **FIGURE 1.18** A reservoir is a good example of dynamic equilibrium in systems. The amount of water coming in may increase or decrease over time, but it must equal the water going out, or the level of the lake will rise or fall. If the input–output balance is not maintained, the lake will get larger or smaller as the system adjusts by holding more or less water in storage. A state of equilibrium (balance) will always exist between inputs, outputs, and storage in the system.

■ **FIGURE 1.19** A feedback loop illustrates how negative feedback tends to maintain system equilibrium. This example shows relationships between the ozone layer, which screens out harmful (cancer related) ultraviolet (UV) energy from the sun, chlorofluorocarbons (CFCs), and potential impacts on life on Earth. CFCs have been used in air conditioning/refrigeration systems and if leaked to the atmosphere they cause ozone depletion. A direct (positive feedback) relationship means that either an increase or a decrease in the first variable will lead to the same effect on the next. Inverse (negative feedback) relationships mean that a change in a variable will cause an opposite change in the next. After one pass through the negative-feedback loop, all subsequent changes in the next cycle will be reversed. A second feedback loop pass (reversing each increase or decrease interaction) illustrates how this works. The last link between skin cancer and human use of CFCs would likely result in people acting to reduce the problem.
What might be the potential (extreme) alternative resulting from a lack of corrective action by humans?

Earth subsystems can also exhibit **positive feedback** sequences for a while—that is, changes that reinforce the direction of an initial change. Animal populations—deer, for example—will adjust naturally to the food supply of their habitats. If the vegetation on which they browse is sparse because of drought, fire, overpopulation, or human impact, deer may starve, reducing the population. However, the smaller deer population may enable the vegetation to recover, and in the next season the deer may increase in numbers. This full process is also an example of a **feedback loop**—a circular set of feedback operations that can be repeated as a cycle (■ Fig. 1.19).

An important factor to consider in systems analysis is the existence of a **threshold,** a condition that, if reached or exceeded (or not met), can cause a fundamental change in a system and the way that it behaves. For example, earthquakes will not occur until the built-up stress reaches a threshold level that overcomes the strength of the rocks to resist breaking. As another example,

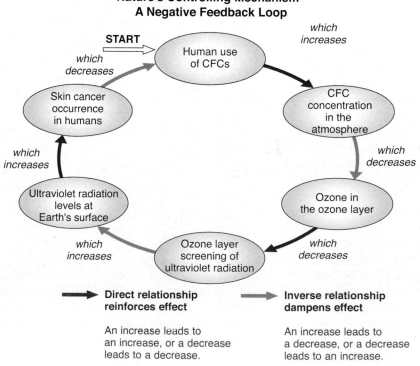

Nature's Controlling Mechanism– A Negative Feedback Loop

fertilizing a plant will help it to grow larger and faster. But if more and more fertilizer is added, will this positive feedback relationship continue forever? Too much fertilizer may poison the plant and cause it to die. With environmental systems, an important question that we often try to answer is how much change a system can tolerate without becoming drastically or irreversibly altered, particularly if the change has negative consequences.

The Earth in Space

We began this book with an image of planet Earth, appearing alone in the vastness of space. It is important to remember, however, that Earth is a component (or subsystem) of our solar system, our galaxy, and our universe. It is dynamic and ever-changing—if we could view an animation or movie instead of a static image, we would see clouds travelling through the atmosphere, as well as Earth's constant motion.

Earth's Movements

Earth undergoes three basic movements: *galactic movement*, *rotation*, and *revolution*. *Galactic movement* is the movement of Earth with the sun and the rest of the solar system in an orbit around the center of the Milky Way Galaxy. This movement has limited effect on the changing environments of Earth and is generally the concern of astronomers. The other two Earth movements, *rotation* on its axis and *revolution* around the sun, are of vital interest to physical geography. The phenomena of day and night, the changing seasons, and variations in the length of daylight hours are consequences of these movements.

Rotation
Earth turning on its *axis*, an imaginary line that extends from the North Pole to the South Pole, is called **rotation.** Earth rotates on its axis at a uniform rate, making one complete turn with respect to the sun in 24 hours.

Earth turns in an eastward direction (■ Fig. 1.20), resulting in the perception of a "rising" sun in the east, which then appears to move westward as it ascends in the sky and then descends toward sunset in the west. Of course, it is actually Earth, not the sun, that is moving, rotating toward the morning sun (that is, turning toward the east).

Rotation accounts for our alternating days and nights. This can be demonstrated by shining a light at a globe while rotating the globe slowly toward the east. You can see that half of the sphere is always illuminated while the other half is not and that new points are continually moving into the illuminated section of the globe (day) while others are moving into the darkened sector (night). This corresponds to Earth's rotation and the sun's energy striking Earth. While one half of Earth receives the light and energy of solar radiation, the other half is in darkness. The dividing line that separates day from night is known as the **circle of illumination,** and it moves from the east toward the west (■ Fig. 1.21).

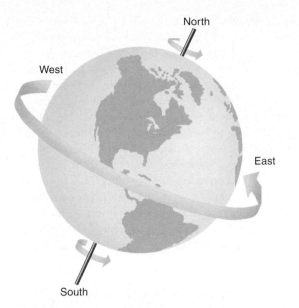

■ **FIGURE 1.20** Earth turns around a tilted axis as it follows its orbit around the sun. Earth's rotation is from west to east, making the stationary sun appear to rise in the east and set in the west.

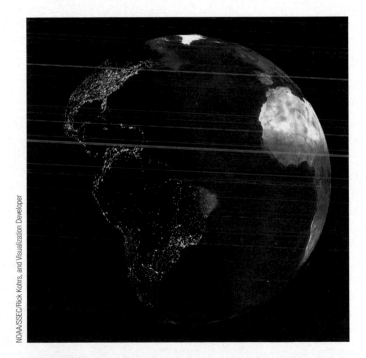

NOAA/SSEC/Rick Kohrs, and Visualization Developer

■ **FIGURE 1.21** The circle of illumination, which separates day from night, is clearly seen on this digital visualization of Earth.

Earth, then, rotates in a direction opposite to the apparent movement of the sun, moon, and stars across the sky. If we look down on a globe from above the North Pole, the direction of rotation is counterclockwise. This eastward direction of rotation not only moves the zone of daylight and nighttime darkness on Earth but also helps define the circulation patterns of the atmosphere and oceans.

The velocity of rotation at the Earth's surface varies with the distance of a given location to the *equator* (the imaginary circle around Earth halfway between the two poles). Every location on Earth undergoes a complete rotation (360°) in 24 hours, or 15° per hour. However, the *linear velocity* depends on the *distance* (not the angle) covered during that 24 hours. The linear velocity at the poles is zero. You can see this by spinning a globe with a postage stamp affixed to the North Pole. The stamp rotates 360° but covers no distance and therefore has no linear velocity. If you place the stamp anywhere between the North and South Poles, however, it will cover a measurable distance during one rotation of the globe. Earth's highest linear velocity is found at the equator, where the distance traveled by a point in 24 hours is greatest. At Kampala, Uganda, near the equator, the velocity is about 460 meters (1500 ft) per second, or approximately 1660 kilometers (1038 mi) per hour (■ Fig. 1.22). In comparison, at St. Petersburg, Russia (60°N latitude), where the distance traveled during one complete rotation of Earth is about half that at the equator, Earth rotates about 830 kilometers (519 mi) per hour.

We are unaware of the speed of rotation because (1) the angular velocity is constant for each place on Earth's surface, (2) the atmosphere rotates with Earth, and (3) there are no nearby objects, either stationary or moving at a different rate with respect to Earth, to which we can compare Earth's movement. Without such references, we cannot perceive the speed of rotation.

Revolution

While Earth rotates on its axis, it also orbits around the sun in a slightly elliptical orbit (■ Fig. 1.23) at an average distance from the sun of about 150 million kilometers (93 million mi). Earth's movement around the sun is called **revolution,** and the time that Earth takes to make one full orbit around the sun determines the length of 1 year. Earth also undergoes 365 ¼ rotations on its axis during the time it takes to complete one revolution of the sun; therefore, a year is said to have 365 ¼ days. Because of the difficulty of dealing with a fraction of a day, it was decided that a year would have 365 days, and every fourth year, called *leap year,* an extra day would be added as February 29.

On about January 3, Earth is closest to the sun, and is said to be at **perihelion** (from Greek: *peri,* close to; *helios,* sun); its distance from the sun then is approximately 147.5 million kilometers (91.5 million mi). At around July 4, Earth is about 152.5 million kilometers (94.5 million mi) from the sun. It is then that Earth has reached its farthest point from the sun and is said to be at **aphelion** (Greek: *ap,* away; *helios,* sun). This annual five million-kilometer distance is relatively insignificant, with little relationship to the seasons and only a minimal effect on the receipt of energy on Earth (a difference of about 3.25 %).

■ **FIGURE 1.22** The speed of rotation of Earth varies with the distance from the equator.

How much faster does a point on the equator move than a point at 60°N latitude?

Plane of the Ecliptic, Inclination, and Parallelism

In its orbit around the sun, Earth moves in a constant plane, known as the **plane of the ecliptic.** Earth's equator is tilted at an angle of 23½° from the plane of the ecliptic, causing Earth's axis to be tilted 23½° from a line perpendicular to the plane (■ Fig. 1.24). In addition to this constant **angle of inclination,** Earth's axis maintains another characteristic called **parallelism.** As Earth revolves around the sun, Earth's axis remains parallel to its former positions. That is, at every position in Earth's orbit, the axis remains pointed toward the same spot in the sky. For the North Pole, that spot is close to the star that we call the North Star, or Polaris.

The characteristics of Earth rotation and revolution can be considered constant in our current discussion, but over the long term, these two movements are subject to change. Earth's axis wobbles over time and will not always remain at an angle of exactly 23½° from a line perpendicular to the plane of the ecliptic. Moreover, Earth's orbit around the sun can change from more circular to more elliptical through periods that can be accurately determined. These and other cyclical changes were calculated and compared by Milutin Milankovitch, a Serbian astronomer during the 1940s, as a possible explanation for the ice ages. Since then, the *Milankovitch Cycles* have often been used when climatologists attempt to explain climatic variations. These variations will be discussed in more detail along with other theories of climate change in Chapter 8.

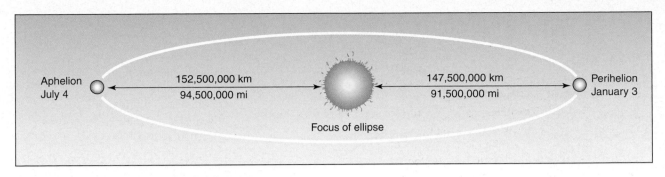

■ FIGURE 1.23 An oblique view of Earth's elliptical orbit around the sun. Earth is closest to the sun at perihelion and farthest away at aphelion. Note that in the Northern Hemisphere summer (July), Earth is farther from the sun than at any other time of the year.
When is Earth closest to the sun?

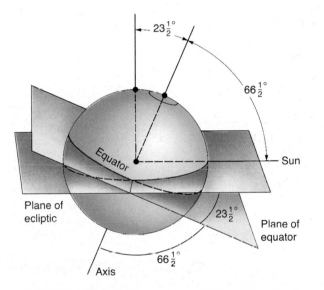

■ FIGURE 1.24 The plane of the ecliptic is defined by the orbit of Earth around the sun. The 23½° inclination of Earth's rotational axis causes the plane of the ecliptic to cut across the plane of the ecliptic.
How many degrees is Earth's axis tilted from the vertical?

Physical Geography and You

The physical environment affects our everyday lives. It is apparent, then, that the study of physical geography and the knowledge of the natural environment that it provides are valuable to all of us. Understanding physical geography helps us to assess environmental conditions, analyze the factors involved, and make informed choices among possible courses of action.

What are the environmental advantages and disadvantages of a particular home site? What sort of environmental impacts might be expected from a proposed development? What potential impacts of natural hazards—flooding, landslides, earthquakes, hurricanes, and tornadoes—should you be aware of where you live? What can you do to minimize potential damage to your household from a natural hazard? What can you do to assure that you and your family are as prepared as possible for the kind of natural hazard that might affect the region where you live, and your home? The study of physical geography will help answer these common questions.

You may be wondering how physical geography might play a role in your future career. By applying their knowledge, skills, and techniques to real-world problems, physical geographers make major contributions to human well-being and to environmental stewardship. A recent publication about geography-related jobs by the United States Department of Labor stated that people in any career field that deals with maps, location, spatial data, or the environment would benefit from an educational background in geography.

Geography is a way of looking at the world and of observing its features. It involves asking questions about the nature of those features as well as appreciating their beauty and complexity. Geography encourages you to seek explanations, gather information, and use geographic skills, tools, and knowledge to solve problems. Just as you see a painting differently after an art course, after this course you should see sunsets, waves, storms, deserts, valleys, rivers, forests, prairies, and mountains with a geographically "educated eye." You will see greater variety in the landscape because you will have been trained to observe Earth differently, with greater awareness and a deeper understanding.

:: Terms for Review

geography	hydrosphere	closed system
spatial science	biosphere	open systems
regions	environment	equilibrium
physical geography	ecology	dynamic equilibrium
holistic approach	ecosystem	feedback
scientific method	life-support system	negative feedback
absolute location	natural resource	positive feedback
relative location	pollution	feedback loop
spatial distribution	model	threshold
spatial pattern	physical model	rotation
spatial interaction	pictorial/graphic model	circle of illumination
system	mathematical/statistical model	revolution
Earth system	conceptual model	perihelion
variable	mental map	aphelion
subsystem	systems analysis	plane of the ecliptic
atmosphere	inputs	angle of inclination
lithosphere	outputs	parallelism

:: Questions for Review

1. Why is geography known as the spatial science? What are some topics that illustrate the role of geography as the spatial science?

2. Why can geography be considered both a physical and a social science? What are some of the subfields of physical geography, and what do geographers study in those areas of specialization?

3. What does a holistic approach mean in terms of thinking about an environmental problem?

4. How do physical geography's three major perspectives make it unique among the sciences?

5. What are the four major divisions of the Earth system, and how do the divisions interact with one another?

6. What is meant by the two-way aspect of human–environment interactions? Why are these interactive relationships falling further out of balance?

7. How do open and closed systems differ? How does feedback affect the dynamic equilibrium of a system?

8. How does negative feedback maintain a tendency toward balance in a system? What is a threshold in a system?

9. Describe briefly how Earth's rotation and revolution affect life on Earth.

10. If the sun is closest to Earth on January 3, why isn't winter in the Northern Hemisphere warmer than winter in the Southern Hemisphere?

:: Practical Applications

1. Give examples from your local area that demonstrate each of the five spatial science topics discussed in the text (location, place characteristics, spatial distributions and patterns, spatial interaction, and the changing Earth).

2. List some potential sources of pollution in your city or town. How could pollution from these sources affect your life? What are some potential solutions to these problems?

3. How can knowledge of physical geography be of value to you now and in the future? What steps should you take if you wish to seek employment as a physical geographer? What advantages might you have when applying for a job?

:: Outline

The San Francisco Bay Area
in a digital "false-color"
satellite image of visible and
near-infrared light. Healthy
vegetation appears red.
This image is similar to those
taken by a digital camera. The
inset is an enlargement of the
airport and shows the pixels
that make up the image.

NASA/GSFC/METI/ERSDAC/JAROS,
and U.S./Japan ASTER Science Team

:: Objectives

When you complete this chapter you should be able to:

- Explain the ways that Earth and its regions, places, and locations can be represented on a variety of visual media—maps, aerial photographs, and other imagery.
- Assess the nature and importance of maps and map-like presentations of the planet, or parts of Earth, citing some examples.
- Find and describe the locations of places using coordinate systems, use topographic maps to find elevations, and understand the three types of map scales.

- Demonstrate a knowledge of techniques that support geographic investigations, including mapping, spatial analysis, satellite and aerial photo interpretation, and data analysis.
- Evaluate the advantages and limitations of different kinds of representations of Earth and its areas.
- Understand how the proper techniques, images, and maps can be used to best advantage in solving geographic problems.
- Recognize the benefits of spatial technologies such as the global positioning system (GPS), geographic information systems (GIS), and remote sensing.

Perhaps as soon as people began to communicate with each other, they began to develop a language of location, using landscape features as directional cues. The earliest known maps were drawn on rock surfaces, clay tablets, metal plates, papyrus, linen or silk, or constructed of sticks. Ancient maps were fundamental to the beginnings of geography. Although many of the basic principles for solving locational problems have been known for centuries, the technologies applied to these tasks are rapidly improving and changing. Through history, maps have become increasingly more common as a result of the appearance of paper, followed by the printing press and the computer.

Today, computer systems allow the generation of complex maps and three-dimensional displays of geographic features that would have been nearly impossible or extremely time-consuming to produce two decades ago. Geographers use these technologies to help them understand spatial relationships and to facilitate locational problem solving. Because maps are so frequently used to convey information, it is important to be able to read and interpret them correctly. An informed knowledge of spatial representation and the ability to communicate locational information is important in our daily lives, and essential in the study of physical geography.

Maps and Location on Earth

Cartography is the science and profession of mapmaking. Geographers who specialize in cartography design maps and globes to ensure that mapped information and data are accurate and effectively presented. Most cartographers would agree that the primary purpose of a map is to communicate spatial information. Maps and globes convey spatial information through graphic symbols, which efficiently relay a vast amount of information. Maps are an essential resource in navigation, political science, community planning, surveying, history, meteorology, geology, and many other career fields. On the television news and in weather reports, maps contribute to our understanding of current events. Think of all the places you encounter maps throughout your daily life. In travel, recreation, education, the media, entertainment, and business, maps are used to communicate important information.

Computer technology has revolutionized cartography. Maps that once had to be hand-drawn (■ Fig. 2.1) are now produced digitally and printed in a short amount of time. Computer-assisted mapping allows easy map revision, which was a time-consuming process when maps were drawn by hand. Information that was once gathered little by little from ground observations and field surveys can now be collected instantly by satellites that flash recorded data back to Earth at the speed of light. Many high-tech locational and mapping technologies are now also in widespread use by the public employing personal computers and satellite-based systems that display locations and directions for use in hiking, traveling, and virtually any means of transportation.

Earth's Shape and Size

Describing global locations and mapping both require a knowledge of the form of our planet and its features. As early as 540 B.C., ancient Greeks theorized that our planet was a sphere. In 200 B.C., Eratosthenes, a philosopher–geographer, estimated Earth's circumference fairly accurately. Earth can generally be considered as a sphere, with an equatorial circumference of 39,840 kilometers (24,900 mi), but the centrifugal force caused by Earth's daily *rotation* bulges the equatorial region outward, and slightly flattens the polar regions, forming a shape that is called an **oblate spheroid.** Yet, at a planetary scale, Earth's deviations from a true sphere are relatively minor. Earth's diameter at the *equator* is 12,758 kilometers (7927 mi), while from pole to pole it is 12,714 kilometers (7900 mi). On a 30.5-centimeter (12-in) globe, this difference of 44 kilometers (27 mi) is about as thick as the wire in a paperclip. This variation from a spherical shape is less than one third of 1%, and is not noticeable in views of Earth from space (■ Fig. 2.2). Nevertheless, people working in very precise navigation, surveying, aeronautics, and cartography must consider Earth's deviations from a perfect sphere.

Landforms also cause departures from true sphericity. Mount Everest in the Himalayas is Earth's highest point at 8850 meters (29,035 ft) above sea level. The lowest point is the Challenger Deep, in the Mariana Trench of the Pacific Ocean southwest of Guam, at 11,033 meters (36,200 ft) below sea level. The difference between these two elevations, 19,883 meters, or just over 12 miles, would also be insignificant on a standard globe,

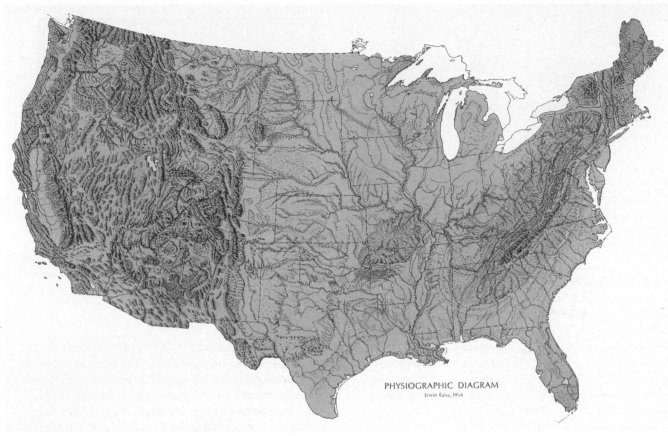

PHYSIOGRAPHIC DIAGRAM
Erwin Raisz, 1954

■ **FIGURE 2.1** When maps had to be hand drawn, artistic talent was required in addition to knowledge of the principles of cartography. Erwin Raisz, a famous and talented cartographer, drew this map of U.S. landforms in 1954 (there were only 48 states at the time).
Are maps like this still valuable for learning about landscapes, or are they obsolete?

■ **FIGURE 2.2** Earth, photographed from space by Apollo 17 astronauts, showing most of Africa and Antarctica. Earth's spherical shape is clearly visible; the bulge of the equatorial regions is too minor to be visible.
What does this suggest about the degree of "sphericity" of Earth?

Globes and Great Circles

Because world globes have essentially the same geometric form as our planet, they represent geographic features and spatial relationships virtually without distortion. A world globe correctly displays the relative shapes, sizes, and comparative areas of Earth features, landforms, water bodies, and distances between locations. Globes also preserve true compass directions. If we want to view the entire world, a globe provides the most accurate representation. Being familiar with the characteristics of a globe helps us understand maps and how they are constructed.

An imaginary circle drawn in any direction on Earth's surface and whose plane passes through the center of Earth is a **great circle** (■ Fig. 2.3a). It is called "great" because this is the largest circle that can be drawn around Earth that connects any two points on the surface. Every great circle divides Earth into equal halves called **hemispheres.** An important example of a great circle is the *circle of illumination,* which divides Earth into light and dark halves—a day hemisphere and a night hemisphere. Any circle on Earth's surface that does not divide the planet into equal halves is called a **small circle** (Fig. 2.3b).

Great circles are useful for navigation, because the trace along any great circle marks the shortest travel route between any two locations on Earth's surface. Connect any two cities, such as Beijing and New York, San Francisco and Tokyo,

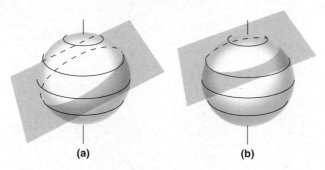

■ **FIGURE 2.3** (a) An imaginary geometric plane, which cuts through Earth and divides it into two equal halves, forms a great circle on Earth's surface. This plane can be oriented in any direction as long as it defines two (equal) hemispheres. (b) The plane shown here slices the globe into unequal parts, so the line of intersection with Earth's surface is a small circle.

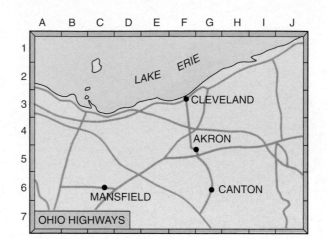

■ **FIGURE 2.4** Using a simple rectangular coordinate system to locate a position. This map employs an alphanumeric location system, similar to that used on many road maps and campus maps. **What are the rectangular coordinates of Mansfield? What is at location F-3?**

New Orleans and Paris, or Kansas City and Moscow, by stretching a large rubber band around a globe so that it touches both cities and divides the globe in half. The rubber band then marks the shortest distance between these two cities. Navigators chart *great circle routes* for aircraft and ships because traveling the shortest distance saves time and fuel. The farther away two points are on Earth, the greater the travel distance savings will be by following the great circle route that connects them.

Latitude and Longitude

Imagine you wish to visit the Football Hall of Fame in Canton, Ohio. Using the Ohio road map, you look up Canton in the map index and find that it is located at "G-6." In box G-6, you locate Canton (■ Fig. 2.4). What you have used is a **coordinate system** of intersecting lines, a system of *grid cells* on the map. A coordinate system must be based on reference points, but defining locations on a spherical planet is difficult because a sphere has no natural beginning and end points. Earth's coordinate system of *latitude and longitude* is based on a set of reference lines that are naturally defined based on its planetary *rotation*, and another set that was arbitrarily defined by international agreement.

Measuring Latitude The **North Pole** and the **South Pole** provide two natural reference points because they mark the opposite positions of Earth's *rotational* axis, around which it turns in 24 hours. The **equator,** halfway between the poles, forms a great circle that separates the Northern and Southern Hemispheres. The equator is 0° latitude, the reference line for measuring **latitude** in degrees north or degrees south. The North Pole (90°N) and the South Pole (90°S) are at the maximum latitudes in each hemisphere.

To locate the latitude of Los Angeles, imagine two lines radiating outward from the center of Earth. One goes straight to Los Angeles and the other goes to the equator at a point directly south of the city. These two lines form a 34° angle that is the latitudinal distance (in degrees) that Los Angeles lies north of the equator, so the latitude of Los Angeles is about 34°N (■ Fig. 2.5a). Because Earth's circumference is approximately 40,000 kilometers (25,000 mi) and there are

360 degrees in a circle, we can divide (40,000 km/360°) to find that 1° of latitude equals about 111 kilometers (69 mi).

One degree of latitude covers a large distance, so degrees are further divided into minutes (') and seconds (") of arc. There are 60 minutes of arc in a degree. Actually, Los Angeles is located at 34°03'N (34 degrees, 3 minutes north latitude). We can get even more precise: 1 minute is equal to 60 seconds of arc. We could locate a different position at latitude 23°34'12"S, which we would read as 23 degrees, 34 minutes, 12 seconds south latitude. A minute of latitude equals 1.85 kilometers (1.15 mi), and a second is about 31 meters (102 ft). The latitude of a location, however, is only half of its global address. Los Angeles is approximately 34° north of the equator, but an infinite number of points exist on the same latitude line.

Measuring Longitude To accurately describe the location of Los Angeles, we must also determine where it is situated along the line of 34°N latitude. To find a location east or west, we use longitude lines, which run from pole to pole, each one forming half of a great circle. The global position of the 0° east–west reference line for longitude is arbitrary, but was established by international agreement in 1884, as the longitude line passing through Greenwich, England (near London). This is the **prime meridian,** or 0° longitude. **Longitude** is the angular distance east or west of the prime meridian.

Longitude is also measured in degrees, minutes, and seconds. Imagine a line drawn from the center of Earth to the point where the north–south running line of longitude that passes through Los Angeles crosses the equator. A second imaginary line will go from the center of Earth to the point where the prime meridian crosses the equator (this location is 0°E or W and 0°N or S). Figure 2.5b shows that these lines drawn from Earth's center define an *angle*, the arc of which is the angular distance that Los Angeles lies west of the prime meridian (118°W longitude). Figure 2.5c shows the latitude and longitude of Los Angeles.

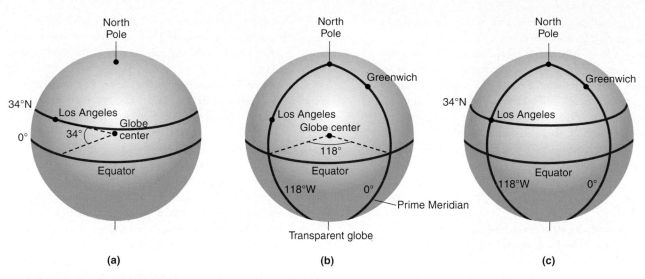

(a) **(b)** **(c)**

■ **FIGURE 2.5** Finding a location by latitude and longitude. (a) The geometric basis for the latitude of Los Angeles, California. Latitude is the angular distance in degrees either north or south of the equator. (b) The geometric basis for the longitude of Los Angeles. Longitude is the angular distance in degrees either east or west of the prime meridian, which passes through Greenwich, England. (c) The location of Los Angeles is 34°N, 118°W.

What is the latitude of the North Pole and does it have a longitude?

Moving both east and west from the prime meridian (0°), longitude increases to a maximum of 180° on the opposite side of the world from Greenwich, in the middle of the Pacific Ocean. Along the Prime Meridian (0° E–W) or the 180° meridian, the E–W designation does not matter, and along the equator (0° N–S), the N–S designation does not matter, and is not needed for indicating location.

Decimal Degrees Instead of using minutes and seconds of arc, **decimal degrees** of longitude and latitude have been used to describe the location of a point. Decimal degrees, expressed to many decimal places, are very precise in pinpointing a location. Also, computer systems (GPS, GIS, Google Earth) handle decimals much more readily than degrees, minutes, and seconds, and some systems require decimal locations. Many computer systems use decimal degrees, without the N, S, E, W letters to indicate cardinal directions. Instead, north latitudes are designated as positive numbers, and south latitudes as negative numbers (with a minus sign). For longitudes, east is positive and west is negative. For example, the Statue of Liberty in New York is located at 40.6894, −74.0447. Latitude is always listed first and the degree symbol (°) is generally not necessary. The practical exercises found at the end of chapters in this book sometimes use decimal coordinates.

The Geographic Grid

Every location on Earth can be located by its latitude north or south of the equator and its longitude east or west of the prime meridian. Our locational reference system is the **geographic grid** (■ Fig. 2.6), the set of imaginary lines that run east and west around the globe to mark latitude, and the lines that run north and south from pole to pole to indicate longitude.

Parallels and Meridians

The east–west lines marking latitude completely circle the globe, are evenly spaced, and are parallel to the equator and each other. Hence, they are known as **parallels.** The equator is the only parallel that is a great circle; all other lines of

■ **FIGURE 2.6** A globe-like representation of Earth, which shows the geographic grid with parallels of latitude and meridians of longitude at 15° intervals.
How do parallels and meridians differ?

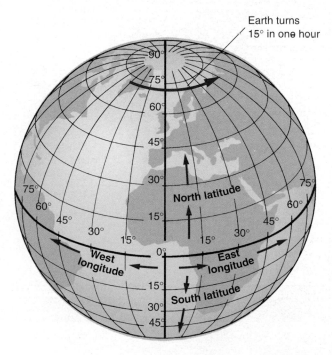

latitude are small circles. One degree of latitude equals about 111 kilometers (69 mi) anywhere on Earth.

Lines of longitude, called **meridians,** run north and south, converge at the poles and measure distances east and west of the prime meridian. Because the meridians converge on the poles, longitude lines get closer together as they run *poleward* from the equator. At the equator, meridians separated by 1° of longitude are about 111 kilometers (69 mi) apart, but at 60°N or 60°S latitude, they are only half that distance apart, about 56 kilometers (35 mi).

Longitude and Time

The world's **time zones** were established based on the relationships between longitude, Earth's *rotation*, and time. Until about 125 years ago, each town or area used what was known as *local time.* **Solar noon** was determined by the precise moment in a day when a vertical stake cast its shortest shadow. This meant that the sun had reached its highest angle in the

sky for that day at that location—noon—and local clocks were set to that time. Because of Earth's *rotation*, noon occurs earlier in a town toward the east, and towns to the west experience noon later.

The international agreement that established the prime meridian at Greenwich (0° longitude) also set standardized time zones. Earth was divided into 24 time zones, one for each hour in a day. Ideally, each time zone spans 15° of longitude, because Earth turns 15° of longitude in an hour (24 × 15° = 360°). The prime meridian is the *central meridian* of its time zone, and every meridian divisible by 15° is the central meridian for a time zone. The time when solar noon occurs at a central meridian was established as noon for all places between 7.5°E and 7.5°W of that meridian. However, as shown in ■ Figure 2.7, time zone boundaries do not follow meridians exactly. In the United States, time zone boundaries commonly follow state lines. It would be very inconvenient to divide a city or town into two time zones—imagine the confusion that would result!

■ **FIGURE 2.7**   World time zones reflect the fact that Earth rotates through 15° of longitude in an hour. Thus, time zones are approximately 15° wide. Political boundaries usually prevent the time zones from following a meridian perfectly.

How many hours of difference are there between the time zone where you live and Greenwich, England? Is it earlier in England or later?

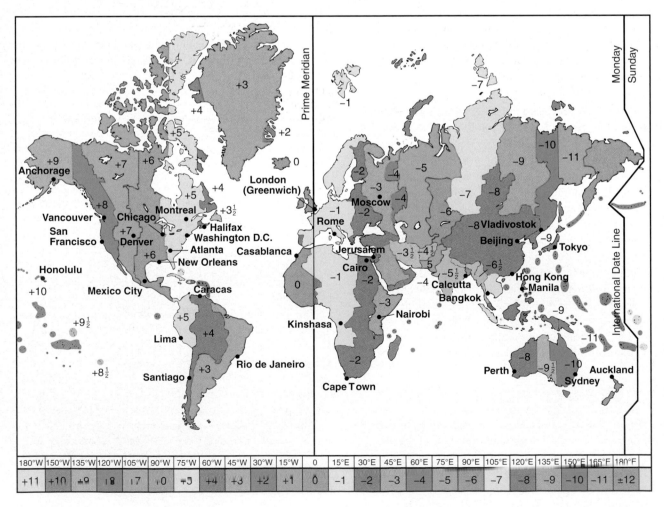

The time of day at the prime meridian, known as *Greenwich Mean Time* (GMT, also called Universal Time, UTC, or Zulu Time), is used as a worldwide reference. Times to the east or west can be easily determined by comparing them to GMT. A place 90°E of the prime meridian would be 6 hours later (90 ÷ 15° per hour), while in the Pacific Time Zone of the United States and Canada, whose central meridian is 120°W, the time would be 8 hours earlier than GMT.

For navigation, longitude can be determined with a *chronometer,* an extremely accurate clock. Two chronometers are used, one set on Greenwich time, and the other on local time. The number of hours between them, earlier or later, determines longitude (1 hour = 15° of longitude). Well before the chronometer was invented, a latitudinal position could easily be determined using a *sextant,* an instrument that measures the angle between the *horizon* and a celestial body such as the noonday sun or the North Star (Polaris).

The International Date Line

The **International Date Line** is a line that generally follows the 180th meridian, except for jogs to separate Alaska and Siberia and to skirt some Pacific islands (■ Fig. 2.8). At the International Date Line, we turn our calendar back a full day if we are traveling east and forward a full day if we are traveling west. Thus, if we are going east from Tokyo to San Francisco and it is 4:30 p.m. Monday just before we cross the International Date Line, it will be 4:30 p.m. Sunday on the other side. If we are traveling west from Alaska to Siberia and it is 10:00 a.m. Wednesday when we reach the International Date Line, it will be 10:00 a.m. Thursday once we cross it. As a way of remembering this relationship, many world maps and globes have Monday and Sunday (M | S) labeled in that order on the opposite sides of the International Date Line. To find the correct day, you just substitute the current day for Monday or Sunday, and use the same relationship.

The need for such a line on Earth to adjust the day was inadvertently discovered by Magellan's crew who, from 1519 to 1521, were the first to circumnavigate the world. Sailing westward from Spain, when they returned from their voyage, it was noticed that one day had apparently been missed in the ship's log. What actually happened was that in going around the world in a westward direction, the crew had experienced one less sunset and one less sunrise than had occurred in Spain during their absence.

The U.S. Public Lands Survey System

The longitude and latitude system locates *points* where those lines intersect. A different system is used in much of the United States to define and locate land *areas*. This is the **U.S. Public Lands Survey System,** or the *Township and Range System,* developed for parceling public lands west of Pennsylvania. The Township and Range System divides land

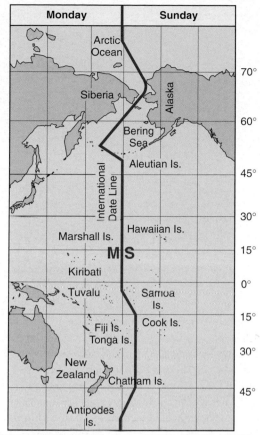

■ **FIGURE 2.8** The International Date Line. West of the line is a day later than east of the line. Maps and globes often have either "Monday | Sunday" or "M | S" shown on opposite sides of the line to indicate the direction of the day change.

Why does the International Date Line deviate from the 180° meridian in some places?

areas into parcels based on north–south lines called *principal meridians* and east–west lines called *base lines*. The meridians are perpendicular to the base lines, but they had to be adjusted (jogged) along their length to accommodate Earth's curvature. If these adjustments were not made, the north–south lines would tend to converge and land parcels defined by this system would be smaller in northern regions of the United States.

The Township and Range System forms a grid of nearly square parcels called *townships* laid out in horizontal *tiers* north and south of the base lines and in vertical *columns* ranging east and west of the principal meridians. A **township** is a square plot 6 miles on a side (36 sq mi, or 93 sq km). Townships are first labeled by their north or south position (■ Fig. 2.9); thus, a township in the third tier south of a base line will be labeled Township 3 South, abbreviated T3S. However, we must also name a township according to its *range*—its location east or west of the principal meridian for the survey area. Thus, if Township 3 South is in the second range east of the principal meridian, its full location can be given as T3S/R2E (Range 2 East).

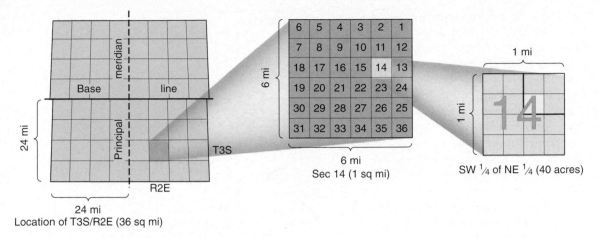

Location of T3S/R2E (36 sq mi)

■ **FIGURE 2.9**   The method of location for *areas* of land according to the U.S. Public Lands Survey System.

How would you describe the extreme southeastern 40 acres of section 20 in the middle diagram?

The Public Lands Survey System divides townships into 36 **sections** of 1 square mile, or 640 acres (2.6 sq km, or 259 ha). Sections are designated by numbers from 1 to 36 beginning in the northeastern most section with section 1, snaking back and forth across the township, and ending in the southeast corner with section 36. Sections are divided into four *quarter sections*, named by their location within the section—northeast, northwest, southeast, and southwest, each with 160 acres (65 ha). Quarter sections are further subdivided into four *quarter–quarter sections,* sometimes known as *forties,* each with an area of 40 acres (16.25 ha). These quarter–quarter sections are also named after their position in

the quarter: the northeast, northwest, southeast, and southwest forties. Thus, we can describe the location of the 40-acre tract that is shaded in Figure 2.9 as being in the SW ¼ of the NE ¼ of Sec. 14, T3S/R2E. The order is consistent from smaller division to larger, and township location is always listed before range (T3S/R2E).

The Township and Range System exerts an enormous influence on landscapes in the Midwest and West, and gives these regions a checkerboard appearance from the air or from space (■ Fig. 2.10). Road maps in states that use this survey system strongly reflect its grid, because many roads follow the regular and angular boundaries between square parcels of land.

■ **FIGURE 2.10**   Rectangular field patterns result from the U.S. Public Lands Survey System that is used in the Midwest and Western United States. Note the slight jog in the field pattern to the right of the farm buildings near the lower edge of the photo.

How do you know this photo was not taken in the Midwestern United States?

The Global Positioning System

The **global positioning system (GPS)** is a technology for determining locations on Earth. This high-tech system was created for military applications but today, it is being adapted to many public uses, from surveying to navigation. The global positioning system uses radio signals, transmitted by a network of satellites orbiting 17,700 kilometers (11,000 mi) above Earth (■ Fig. 2.11). GPS is based on the principle of *triangulation,* which means that if we can find the distance to our position, measured from three or more different locations (in this case, satellites) we can determine our location. The distances are calculated by measuring the time it takes for a signal, broadcast at the speed of light from a satellite, to arrive at the receiver. A GPS receiver displays a location in latitude, longitude, and elevation, or on a map display. Small GPS receivers are useful to travelers, hikers, and backpackers who need to keep track of their location (■ Fig. 2.12).

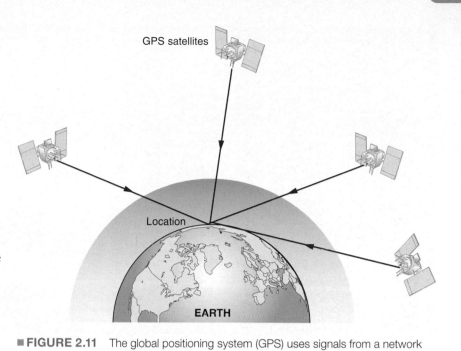

■ FIGURE 2.11　The global positioning system (GPS) uses signals from a network of satellites to determine a position on Earth. A GPS receiver on the ground calculates the distances from several satellites (a minimum of three) to find its location by longitude, latitude, and elevation. With the distance from three satellites, a position can be located within meters, but with more satellite signals and sophisticated GPS equipment, the position can be located very precisely.

■ FIGURE 2.12　A GPS receiver provides a readout of its latitudinal and longitudinal position based on signals from a satellite network. Small handheld units provide an accuracy that is acceptable for many uses and, like this one, can also display locations on a map. **What other uses can you think of for a small GPS unit like this that displays its longitude and latitude as it moves from place to place?**

In mountain areas, hikers can also use a GPS to understand the relationship between changes in elevations and environments. Map-based GPS systems are becoming popular and are widely used in vehicles, boats, and aircraft. GPS applications also support map making from data gathered in the field. With sophisticated GPS equipment and techniques, it is possible to find locational coordinates within small fractions of a meter (■ Fig. 2.13).

Maps and Map Projections

Maps are extremely versatile—they can be reproduced easily, can depict the entire Earth or show a small area in great detail, are easy to handle and transport, and can be displayed on a computer monitor. Yet, it is impossible for one map to fit all uses. The many different varieties of maps all have qualities that can be either advantageous or problematic, depending on the application. Knowing some basic concepts concerning maps and cartography will greatly enhance a person's ability to effectively use a map, and to select the right map for a particular task.

Advantages of Maps

If a picture is worth a thousand words, then a map is worth a million. Because they are graphic representations and use symbolic language, maps show spatial relationships and portray geographic information with great efficiency.

USGS/Mike Poland

■ **FIGURE 2.13** A scientist monitoring volcanoes in Washington state uses a professional GPS system to record a precise location by longitude, latitude, and elevation. This is the view from Mt. St. Helens, with another volcano, Mt. Adams, in the distance.

Maps supply an enormous amount of information visually that would take many pages to describe in words (probably less successfully). Imagine trying to tell someone about all of the information that a map of your city, county, state, or campus provides: sizes, areas, distances, directions, street patterns, railroads, bus routes, hospitals, schools, libraries, museums, highway routes, business districts, residential areas, population centers, and so forth. Maps can display the best route from one place to another, the shapes of Earth features, and they can be used to measure distances and areas. Cartographers can produce maps to illustrate almost any relationship in the environment. The potential applications of maps are practically infinite, even "out of this world," because our space programs have produced detailed maps of the moon and other extraterrestrial features (■ Fig. 2.14). For many reasons, whether it is presented on paper, on a computer screen, or as a mental concept in the mind, the map is the geographer's most important tool.

■ **FIGURE 2.14** Lunar geography. A detailed map of the moon shows a major crater that is 120 km in diameter (75 mi). Even the side of the moon that never faces Earth has been mapped in considerable detail.
How were we able to map the moon in such detail?

NASA

Limitations of Maps

On a globe, we can directly compare the size, shape, and area of Earth features, and we can measure distance, direction, shortest routes, and true directions. Yet, because of the distortion inherent in maps, we can never compare or measure all of these properties on a single map. It is impossible to present a spherical planet on a flat (two-dimensional) surface and accurately maintain all of its geometric properties. This process has been likened to trying to flatten out an eggshell.

On maps that show large regions or the world, Earth's curvature causes apparent and pronounced distortion, but when a map depicts only a small area, the distortion should be negligible. If we use a state park map while hiking, the distortion will be too small to affect us. To be skilled map users, we must know which properties a certain map depicts accurately, which features it distorts, and for what purpose a map is best suited. By being aware of these map characteristics, we can make accurate comparisons and measurements on maps and better understand the information that a map conveys.

Examples of Map Projections

Transferring a spherical grid onto a flat surface produces a **map projection.** Although maps are not actually made this way, certain projections can be demonstrated by putting a light inside a transparent globe so that the grid lines are projected onto a flat surface (plane), a cone, a cylinder, or other geometric forms that are flat or can be cut and flattened out (■ Fig. 2.15). Today, map projections are developed mathematically, using computers to fit the geographic grid to a surface. Map projections will always distort the shape, area, direction or distance of map features, or some combination thereof, so it is important for mapmakers to choose the best projection for the task.

Projecting the grid lines onto a plane, or flat surface, produces a map called a *planar projection* (Fig. 2.15a). These maps are most often used to show the polar regions, with the pole centrally located on the circular map, which displays one hemisphere.

Maps of middle-latitude regions, such as the contiguous United States, are typically based on *conic projections* because they portray these latitudes with minimal distortion. In a simple conic projection, a cone is fitted over the globe with its pointed top centered over a pole (Fig. 2.15b). Parallels of latitude on a conic projection are arcs that become smaller toward the pole, and meridians appear as straight lines radiating toward the pole.

A well-known example of a cylindrical projection (Fig. 2.15c) is the **Mercator projection**, commonly used in schools and textbooks, although less so in recent years. The Mercator world map is a mathematically adjusted *cylindrical projection* on which meridians appear as parallel lines instead of converging at the poles. Obviously, there is enormous

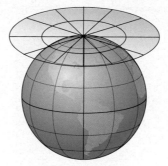

(a) Planar projection

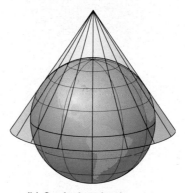

(b) Conical projection

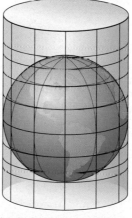

(c) Cylindrical projection

■ **FIGURE 2.15**   The theory behind the development of (a) planar, (b) conic, and (c) cylindrical projections. Although projections are not actually produced this way, they can be demonstrated by projecting light from a transparent globe. **Why do we use different map projections?**

east–west distortion of areas in the high latitudes because the distances between meridians are stretched to the same width that they are at the equator (■ Fig. 2.16). The spacing of parallels on a Mercator projection is also not equal, as they are on Earth. This projection does not display all areas accurately, and size distortion increases toward the poles.

Gerhardus Mercator devised this map in 1569 to provide a property that no other world projection has. A straight line

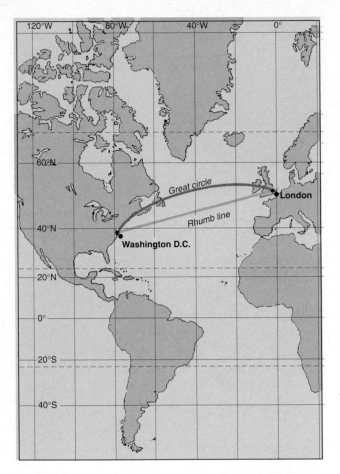

■ FIGURE 2.16 The Mercator projection was designed for navigation, but has often been misused as a general-purpose world map. Its most useful property is that lines of constant compass heading, called *rhumb* lines, are straight lines. **Compare the sizes of Greenland and South America on this map to their proportional sizes on a globe. Is the distortion great or small?**

drawn anywhere on a Mercator projection is a line of true compass direction, called a *rhumb line,* which was very important in navigation (see again Fig. 2.16). On Mercator's map, navigators could draw a straight line between their location and the place where they wanted to go, and then follow a constant compass direction to get to their destination.

Properties of Map Projections

The geographic grid has four important geometric properties: (1) Parallels of latitude are always parallel, (2) parallels are evenly spaced, (3) meridians of longitude converge at the poles, and (4) meridians and parallels always cross at right angles. Because no map projection can maintain all four of these properties at once, cartographers must decide which properties to preserve at the expense of others. Closely examining a map's grid system to determine how these four properties are affected will help you discover areas of greatest and least distortion.

Area Cartographers are able to create a world map that maintains correct area relationships; that is, areas on the map have the same size proportions to each other as they have in reality. Thus, if we cover any two parts of the map with, for example, a quarter, no matter where the quarter is placed it will cover equivalent areas on Earth. Maps drawn with this property, called **equal-area maps,** should be used if size comparisons are being made between two or more areas. The property of equal area is also essential when examining spatial distributions. As long as the map displays equal area and a symbol represents the same quantity throughout the map, we can get a good idea of the distribution of any feature—for example—people, churches, cornfields, hog farms, or volcanoes. However, equal-area maps distort the shapes of mapped features (■ Fig. 2.17) because it is impossible to show both equal areas and correct shapes on the same map.

Shape Flat maps cannot depict large regions of Earth without distorting either their shape or their comparative sizes in terms of area. However, using the proper map projection will depict the true shapes of continents, regions, mountain ranges, lakes, islands, and bays. Maps that maintain the correct shapes of areas are **conformal maps.** To preserve the shapes of Earth features on a conformal map, meridians and parallels always cross at right angles just as they do on the globe.

The Mercator projection presents correct shapes, so it is a conformal map, but areas away from the equator are exaggerated in size. The Mercator projection's distortions led generations of students to believe incorrectly that Greenland is about equal in size to South America (compare Fig. 2.16 to Fig. 2.17), but South America is actually about eight times larger.

Distance No flat map can maintain a constant distance scale over Earth's entire surface. The scale on a map that depicts a large area cannot be applied equally everywhere

■ FIGURE 2.17 An equal-area world projection map. This map preserves area relationships but distorts the shape of landmasses. **What world map would you prefer, one that preserves area or one that preserves shape, and why?**

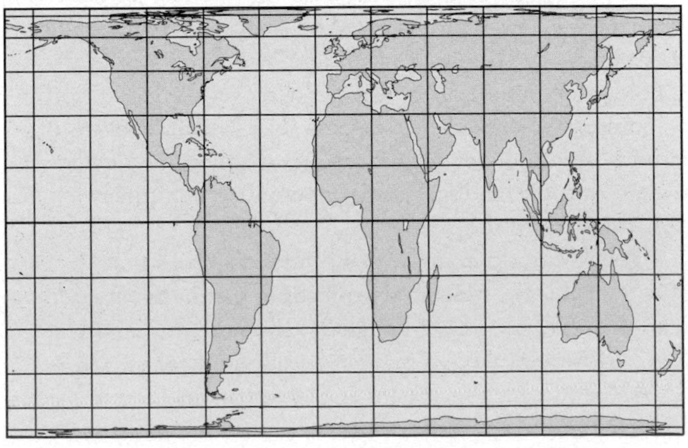

on that map. On maps of small areas, however, distance distortions will be minor, and the accuracy will usually be sufficient for most purposes. Maps can be made with the property of **equidistance** in specific instances. That is, on a world map, the equator may have equidistance (a constant scale) along its length, and all meridians may have equidistance, but not the parallels. On another map, all straight lines drawn from the center may have equidistance, but the scale will not be constant unless lines are drawn from the center.

Direction

Because longitude and latitude directions run in straight lines, not all flat maps can show true directions as straight lines. Thus, lines of latitude or longitude that curve on maps are not drawn as true compass directions. An example of a map that shows true direction is the **azimuthal map** (■ Fig. 2.18), one kind of planar projection. These are drawn with a central focus, and all straight lines that pass through that center are true compass directions.

Compromise Projections

In developing a world map, one cartographic strategy is to compromise by creating a map that shows both area and shape fairly well but is not really correct for either property. These world maps are **compromise projections** that are neither conformal nor equal area, but an effort is made to balance distortion to produce an "accurate looking" global map (■ Fig. 2.19a). An *interrupted*

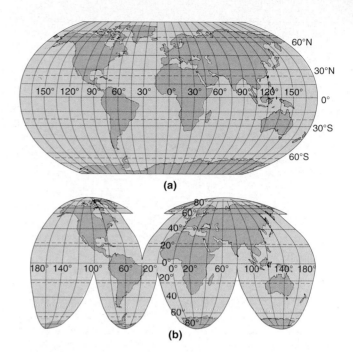

■ **FIGURE 2.19** The Robinson projection (a) is a compromise projection because it departs from equal area to better depict the shape of the continents, but seeks to show both area and shape reasonably well, although not truly accurately. Distortion in projections can be also reduced by interruption (b)—that is, by having a central meridian for each segment of the map.
What is a disadvantage of (b) in terms of usage?

projection can also be used to reduce the distortion of land-masses (Fig. 2.19b) by moving much of the distortion to the oceanic regions.

Map Basics

Maps not only contain spatial information and data, but they also display essential information about the map itself. This information and certain graphic features (often in the margins) are intended to facilitate using the map. Among these items are the map title, date, legend, scale, and direction. A map should have a *title* that tells what area is depicted and what subject the map concerns. For example, "Yellowstone National Park: Trails and Camp Sites." Most maps should also indicate when they were published and the date to which its information applies, so users know if the map information is current or outdated, or whether the map is intended to show historical data.

Legend

A map should have a **legend**—a key to symbols used on the map (see Appendix B). For example, if one dot represents 1000 people or the symbol of a pine tree represents a park, the legend should explain this information. If color shading is used on the map to represent elevations, different climatic regions, or other factors, then a key to the color-coding should be provided.

■ **FIGURE 2.18** Azimuthal map centered on the North Pole. Although a polar view is the conventional orientation of such a map, it could be centered anywhere on Earth. Azimuthal maps show true directions between all points from the center, but can only show one hemisphere.

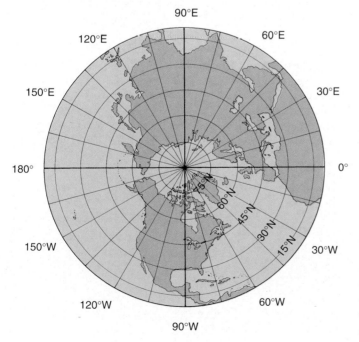

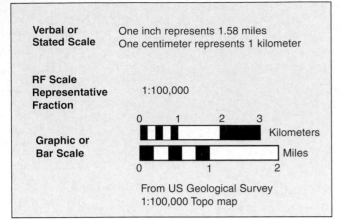

Verbal or Stated Scale	One inch represents 1.58 miles One centimeter represents 1 kilometer
RF Scale Representative Fraction	1:100,000
Graphic or Bar Scale	*[graphic bar scales]* From US Geological Survey 1:100,000 Topo map

■ **FIGURE 2.20** Map scales. A *verbal scale* states the relationship between a map measurement and the corresponding distance that it represents on Earth. Verbal scales generally mix units (centimeters/ kilometer or inches/mile). A *representative fraction (RF)* scale is a ratio between a distance on a map (1 unit) and its actual length on the ground (here, 100,000 units). An RF scale requires that measurements be in the same units both on the map and on the ground. A *graphic scale* is a device used for measuring distances on the map in terms of distances on the ground.

Scale

Obviously, maps depict features smaller than they actually are. If the map is used for measuring sizes or distances, or if the size of the area represented might be unclear to a map user, it is essential to indicate the map scale (■ Fig. 2.20). A map **scale** is an expression of the relationship between a distance on the ground and the same distance as it appears on the map. Knowing the map scale is essential for measuring distances and for determining areas. Map scales can be conveyed in three basic ways.

A **verbal scale** is a statement on the map that indicates, for example, "1 centimeter to 100 kilometers" (1 cm represents 100 km) or, "1 inch to 1 mile" (1 inch on the map represents 1 mile on the ground). Stating a verbal scale tends to be how most of us would refer to a map scale in conversation. If written on a map, however, a verbal scale will no longer be correct if the original map is reduced or enlarged. When stating a verbal scale it is acceptable to use different map units (centimeters, inches) to represent another measure of true length it represents (kilometers, miles).

A **representative fraction (RF) scale** is a ratio between a unit of distance on the map to the distance that unit represents in reality (expressed in the same units). Because a ratio is also a fraction, units of measure, being the same in the numerator and denominator, cancel each other out. An RF scale is therefore free of units of measurement and can be used with any unit of linear measurement—meters, centimeters, feet, or inches—as long as the same unit is used on both sides of the ratio. As an example, a map may have an RF scale of 1:63,360, which can also be expressed 1/63,360. This RF scale can mean that 1 inch on the map represents 63,360 inches on the ground. It also means that 1 cm on the map represents 63,360 cm on the ground. Knowing that 1 inch on the map represents 63,360 inches on the ground may be difficult to conceptualize unless we realize that 63,360 inches is equal to 1 mile. Thus,

the representative fraction 1:63,360 means the map has the same scale as a map with a verbal scale of 1 inch to 1 mile.

A **graphic scale,** or **bar scale,** is used for making distance measurements on a map. Graphic scales are graduated lines (or bars) marked with map distances that are proportional to distances on the Earth. To use a graphic scale, take a straight edge of a piece of paper, and mark the distance between any two points on the map. Then use the graphic scale to find the equivalent distance on Earth's surface. Graphic scales have two major advantages:

1. It is easy to determine distances on the map, because the graphic scale can be used like a ruler to make measurements.
2. Graphic scales are applicable even if the map is reduced or enlarged, because the scale (on the map) will also change proportionally in size. The map and the graphic scale, however, must be enlarged or reduced together (the same amount) for the graphic scale to be accurate.

Direction

The orientation and geometry of the geographic grid give us an indication of direction because parallels of latitude are east–west lines and meridians of longitude run directly north–south. Many maps have an arrow pointing to north as displayed on the map. A north arrow may indicate either *true north* or *magnetic north*—or two north arrows may be given, one for true (geographic) north and one for magnetic north.

Earth has a magnetic field that makes the planet act like a giant bar magnet, with magnetic north and south poles, each with opposite charges. Although the magnetic poles shift position slightly over time, they are located in the Arctic and Antarctic regions and do not coincide with the geographic poles. A compass needle points toward the magnetic north pole, not the geographic north pole. If we know the **magnetic declination,** the angular difference between magnetic north and true geographic north, we can compensate for this difference (■ Fig. 2.21). Thus, if our compass points north and the

■ **FIGURE 2.21** Map symbol showing true north, symbolized by a star representing Polaris (the North Star), and magnetic north, symbolized by an arrow. The example indicates 20°E magnetic declination.

In what circumstances would we need to know the magnetic declination of our location?

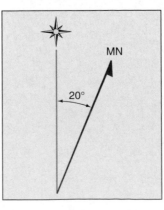

magnetic declination for our location is 20°E, we can adjust our course knowing that our compass is pointing 20°E of true north. To do this, we should turn 20°W from the direction indicated by our compass in order to face true north. Magnetic declination varies from place to place and also changes over time. For this reason, magnetic declination maps are revised periodically, so using a recent map is very important. Despite the existence of electronic locational systems, magnetic compasses remain important for direction finding in isolated areas, because they require no batteries and have no electronic parts that could fail.

Thematic Maps

Maps designed to focus attention on the spatial extent and distribution of one feature (or a few related ones) are called **thematic maps.** Examples include maps of climate, vegetation, soils, earthquake epicenters, or tornadoes. There are two major types of spatial data, *discrete* and *continuous* (■ Fig. 2.22).

Discrete data means that either the feature is located at a particular place or it is not there—for example, hot springs, tropical rainforests, rivers, tornado paths, or earthquake faults. Discrete data are represented on maps by point, area, or line symbols to show their locations and distributions (Fig. 2.22a–c). The locations, distributions, and patterns of discrete features are of great interest in understanding spatial relationships. *Regions* are areas that exhibit a common characteristic or set of characteristics within their boundaries, and are typically represented by different colors or shading. Physical geographic regions include areas of similar soil, climate, vegetation, landforms, or other characteristics (see the world and regional maps throughout this book).

Continuous data means that a measurable numerical value exists everywhere on Earth (or within the area of interest displayed) for a certain characteristic; for example, every location has a measurable elevation (or temperature, or air pressure, or population density). The distribution of continuous data is often shown using **isolines**—lines on a map that

■ **FIGURE 2.22** Discrete and continuous spatial data (variables). *Discrete variables* represent features that are present at certain locations but do not exist everywhere. Discrete variables can be (a) *points* as shown by locations of large earthquakes in Hawaii (or places where lightning has struck or locations of water-pollution sources), (b) *lines* as in the path taken by Hurricane Rita (or river channels, tornado paths, or earthquake fault lines), (c) *areas* like the land burned by a wildfire (or clear-cuts in a forest, or the area where an earthquake was felt). *A continuous variable* means that every location has a certain measurable characteristic; for example, everywhere on Earth has an elevation, even if it is zero (at sea level) or below (a negative value). The map (d) shows the continuous distribution of temperature variation in part of eastern North America. Changes in a continuous variable over an area can be represented by isolines, shading, colors, or with a 3-D appearance.

Can you name other environmental examples of discrete and continuous variables?

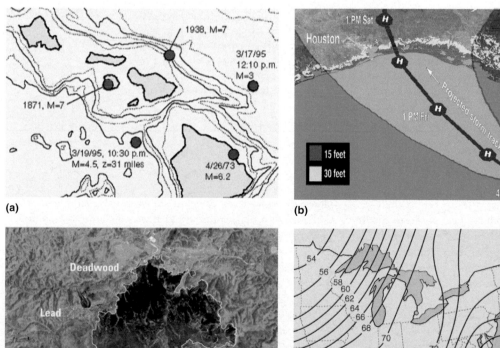

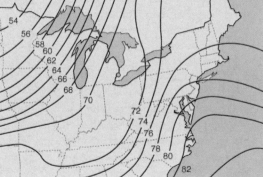

(a)

(b)

(c)

(d)

GEOGRAPHY'S SPATIAL PERSPECTIVE

:: USING VERTICAL EXAGGERATION TO PORTRAY TOPOGRAPHY

Most maps present a landscape as if viewed from directly overhead, looking straight down.

This perspective is referred to as a *map view* or *plan view* (like architectural house plans). Illustrating terrain

on a flat map and showing differences in elevation requires some sort of symbol to display elevations. Topographic

This digital elevation model (DEM) of Anatahan Island (145°40′ E, 16°22″ N) and the surrounding Pacific Ocean floor is presented in 3-D and colorized according to elevation and seafloor depth relative to sea level. This image of Anatahan has three times vertical exaggeration. The vertical scale has been stretched three times compared to the horizontal scale.
The vertical scale bar represents a distance of 3800 meters, so taking the vertical exaggeration into account, what horizontal distance would the same scale bar length represent in meters?

connect points with the same numerical value (Fig. 2.22d). Isoline types include *isotherms,* which connect points of equal temperature; *isobars,* which connect points of equal barometric pressure; *isobaths* (also called bathymetric contours), which connect points with equal water depth; and *isohyets,* which connect points receiving equal amounts of precipitation.

Topographic Maps

Topographic maps portray the land surface and elevational information, along with certain important natural and cultural landscape features. **Topographic contour lines** are map isolines, which connect points that are at the same elevation

above mean sea level (or below sea level such as in Death Valley, California). For example, if we walk around a hill following the 1200-foot contour line on a map, we would always be maintaining a constant elevation of 1200 feet and walking on a level line. Contour lines are excellent for showing elevation changes and the land surface on a map. The spacing and shapes of the contours give a map reader a mental image of the terrain (■ Fig. 2.23).

Figure 2.24 illustrates how contour lines portray the land surface. The bottom portion of the diagram is a simple contour map of an asymmetrical hill. Note that the elevation difference between adjacent contour lines on this map is 20 feet. The constant difference in elevation between adjacent contour

maps use contour lines, which can also be enhanced by relief shading (see the Map Interpretation feature in Chapter 11, Volcanic Landforms, for an example).

For many purposes, either a side view (profile) view, or an oblique (perspective) view of the terrain helps us visualize the landscape. Block diagrams, 3-D models of Earth's surface, show the topography from a perspective view and information about the subsurface can be included. They provide a perspective with which most of us are familiar, similar to looking out an airplane window or from a high vantage point. But such diagrams are not always intended for

making accurate measurements, and many block diagrams represent hypothetical or stylized, rather than actual, landscapes.

A topographic profile (see again ■ Fig. 2.24) illustrates the shape of a land surface as if viewed directly from the side. Profiles are graphs of elevation changes over distance along a transect line. Elevation and distance information collected from a topographic map or from other elevation data can be used to draw a topographic profile to show the terrain. If the geology of the subsurface is represented as well, such profiles are called *geologic cross sections*.

Block diagrams, profiles, and cross sections are most often drawn

with *vertical exaggeration*, which enhances changes in elevation. This makes mountains appear taller, valleys deeper, slopes steeper, and the terrain appears more rugged. Vertical exaggeration is used to make subtle changes in the terrain more noticeable. Most profiles and block diagrams should indicate how much the vertical presentation has been stretched, so that there is no misunderstanding. Three times vertical exaggeration means that the features presented appear to be three times higher than they really are, but the horizontal scale is correct.

USGS/ digital elevation model by Steve Schilling; geo-referenced by Frank Trusdell

Compare the vertically exaggerated elevation model to this natural scale (not vertically exaggerated) version. This is how the island and the seafloor actually look in terms of slope steepness and relief.

lines is called the **contour interval.** If we hiked from point A to point B what kind of terrain would we cover? We start from at sea level at point A and immediately begin to climb. We cross the 20-foot contour line, then the 40-foot, the 60-foot, and, near the top of our hill, the 80-foot contour level. After walking over a relatively broad summit that is above 80 feet but not as high as 100 feet (or we would cross another contour line), we once again cross the 80-foot contour line, which means we must be starting down. During our descent, we cross each lower level in turn until we arrive back at sea level (point B).

In the top portion of Figure 2.24, a **profile** (side view) helps us to visualize the topography we covered in our walk.

We can see why the trip up the mountain was more difficult than the trip down. Closely spaced contour lines near point A represent a steeper slope than the more widely spaced contour lines near point B. Actually, we have discovered something that is true of all isoline maps: The closer together the lines are on the map, the steeper the **gradient** (the greater the rate of change per unit of horizontal distance).

Topographic maps use symbols to show many other features in addition to elevations (see Appendix B)—for instance, water bodies such as streams, lakes, rivers, and oceans or cultural features such as towns, cities, bridges, and railroads. The USGS produces topographic maps of the United States at several different scales (see Appendix B). Many of these maps use

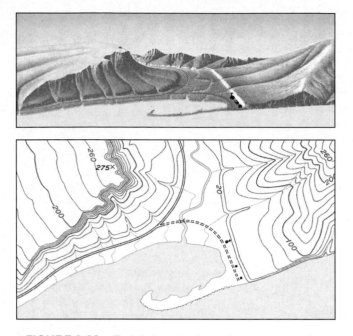

■ **FIGURE 2.23** (Top) A view of a river valley and surrounding hills, shown on a shaded-relief diagram. Note that a river flows into a bay partly enclosed by a sand spit. The hill on the right has a rounded surface form, but the one on the left forms a cliff along the edge of an inclined but flat upland. (Bottom) The same features represented on a contour map.
If you only had a topographic map, could you visualize the terrain shown in the shaded-relief diagram?

■ **FIGURE 2.24** A topographic profile and contour map. Topographic contours connect points of equal elevation relative to mean sea level. The upper part of the figure shows the topographic profile (side view) of an island. Horizontal lines mark 20-foot intervals of elevation above sea level. The lower part of the figure shows how these contour lines look in map view.
What is the relationship between the spacing of contour lines and steepness of slope?

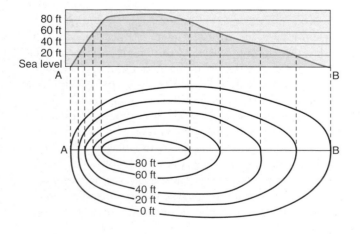

English units for their contour intervals, but more recent maps use metric units. Contour maps that show undersea topography are called *bathymetric charts*, which in the United States are produced by the National Ocean Service.

Modern Mapmaking

Today, the vast majority of mapmakers employ computer technologies. For most mapping projects, computer systems are faster, more efficient, and less expensive than the hand-drawn cartographic techniques they have replaced. Spatial data representing elevations, depths, temperatures, or populations can be stored in a digital database, accessed, and displayed on a map. The database for a map may include information on coastlines, political boundaries, city locations, river systems, map projections, and coordinate systems. In digital form, maps can be easily revised because they do not have to be manually redrawn with each revision or major change. Computer generated map revision is essential for updating rapidly changing phenomena such as weather systems, air pollution, ocean currents, volcanic eruptions, and forest fires. Digital maps can be instantly disseminated and shared via the Internet. However, it is still important to understand basic cartographic principles to make a good map. A computer mapping system will draw only what an operator instructs it to draw.

Geographic Information Systems

A **geographic information system (GIS)** is a versatile innovation that stores geographic databases, supports spatial data analysis, and facilitates the production of digital maps. A GIS is a computer-based technology that assists the user in the entry, analysis, manipulation, and display of geographic information derived from combining any number of digital *map layers*, each composed of a specific *thematic map* (■ Fig. 2.25). A GIS can be used to make the scale and map projection of these map layers identical, thus allowing the information from several or all layers to be combined into new and more meaningful composite maps. GIS is especially useful to geographers as they work to address problems that require large amounts of spatial data from a variety of sources.

What a GIS Does
Imagine that you are in a giant map library with hundreds of paper maps, all of the same area, but each map shows a different aspect of the same location: one map shows roads, another highways, another trails, another rivers (or soils, or vegetation, or slopes, or rainfall, and so on almost to infinity). The maps were originally produced at different sizes, scales, and projections (including some maps that do not preserve shape or area). These cartographic factors make it very difficult to *visually* overlay and compare the spatial information among these different maps. You also have digital terrain models and satellite images that you would

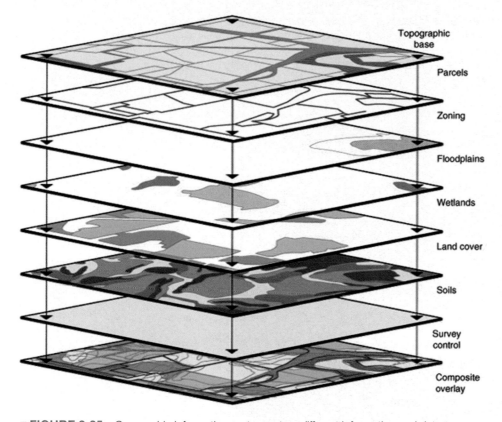

Topographic base
Parcels
Zoning
Floodplains
Wetlands
Land cover
Soils
Survey control
Composite overlay

■ **FIGURE 2.25**   Geographic information systems store different information and data as individual map layers. GIS technology is widely used in geographic and environmental studies in which several different variables need to be assessed and compared spatially to solve a problem. **Can you think of other applications for geographic information systems?**

you want to display new data in your map, such as the locations of earthquake-related calls each station receives, that can be done easily by creating a new layer using software tools.

Many geographers are employed in careers that apply GIS technology. The capacity of a GIS to integrate and analyze a wide variety of geographic information, from census data to landform characteristics, makes it useful to both human and physical geographers. With nearly unlimited applications in geography and other disciplines, GIS will continue to be an important tool for spatial analysis.

Computer-generated three-dimensional (3-D) views of elevation data, called **digital elevation models (DEMs),** are particularly useful for displaying topography (■ Fig. 2.26a). Digital elevation data, when input as a layer in a GIS, can be used to make many types of terrain displays and maps, including shaded relief maps and contour maps, which can also be assigned a certain color between contours. Created with a GIS or

like to compare to the maps. Further, because few aspects of the environment involve only one factor or exist in spatial isolation, you want to be able to combine a selection of these geographic aspects on a single composite map. You have a spatial–geographic problem, and to solve that problem, you need a way to make several representations of a part of Earth directly comparable. What you need is a GIS and the knowledge of how to use this system.

Each map is digitally scanned and stored as a layer of spatial information that represents an individual thematic map layer as a separate digital file (see again Fig. 2.25). A GIS can display any layer or any combination of layers, geometrically registered (fitted) to any map projection and at any scale that you specify. The maps, images, and data sets can now be directly compared at the same size, based on the same map projection and map scale. A GIS can digitally *overlay* any set of thematic map layers that are needed. If you want to see the locations of *homes* on a *river floodplain,* a GIS can be used to provide an instant map by retrieving, combining, and displaying the *home* and *floodplain* map layers simultaneously. If you want to see *earthquake faults* and *artificial landfill areas* in relation to locations of *fire stations* and *police stations,* that composite map will require four layers, but this is no problem for a GIS to display. And if

other digital technology, **visualization models** (*visualizations*) are computer-generated image models designed to illustrate and explain complex processes and features. Many visualizations are presented as three-dimensional images and/or as animations, based on environmental data and satellite images or air photos. An example is shown in Figure 2.26c where a DEM and a satellite image were combined in a GIS to produce a 3-D landscape model of the Rocky Mountain front at Salt Lake City, based on actual image and elevation data. This process is called *draping* (like draping cloth over some object). These models help us understand and conceptualize many environmental processes and features. The detail represented by elevation change can be enhanced by stretching the height of features, using a technique called **vertical exaggeration** (illustrated in a box in this chapter).

Actually, any geographic factor represented by continuous data can be displayed either as a two-dimensional contour map or as a 3-D surface to illustrate and enhance the visibility of the spatial variation that it conveys (■ Fig. 2.27). Today, the products and techniques of cartography are very different from their beginning forms, but the goal of making a representation of Earth remains the same—to effectively communicate geographic and spatial knowledge in a visual format.

■ **FIGURE 2.26** (a) A digital elevation model (DEM) of Salt Lake City, Utah, and the Rocky Mountain front displays elevations in a grid with a three-dimensional appearance. (b) A digital satellite image of the same area. (c) These two data "layers" can be combined in a GIS to produce a digital 3-D model of the landscape at Salt Lake City. Digital landscape models are useful for studying many aspects of the environment. The examples here are enlarged to show the pixel resolution.

USGS

(a)

NASA

(b)

NASA/JPL/NIMA

(c)

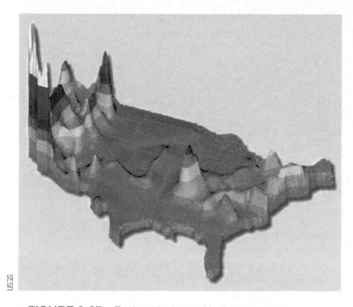

USGS

■ FIGURE 2.27 Earthquake hazard in the contiguous United States: a continuous variable displayed as a surface in 3-D perspective, making it easy to understand how potential earthquake danger varies spatially across this part of the United States.

Other than where you might expect to find a high level of earthquake hazard, are there locations with a level of this characteristic you find surprising?

Remote Sensing of the Environment

Remote sensing is the collection of information and data about distant objects or environments, typically from aerial and space images, which have many map-like qualities. Using remote sensing systems, we can detect objects and scenes that are not visible to humans and display them as images.

Digital Imaging and Photography

In technical terms, a *photograph* is taken with a camera on *film*. Digital cameras or image scanners produce a **digital image**—an image that is converted into numerical data. A digital image is similar to a mosaic, made up of grid cells with varying colors or tones that form a picture. Most images returned from space are digital, because digital data can be easily broadcast back to Earth. Digital imagery also offers the advantages of computer-assisted data processing, image enhancement, and image sharing, and can provide a thematic layer in a GIS. Digital images consist of **pixels,** short for "picture element," the smallest area resolved in a digital picture (as seen in the enlarged inset of the San Francisco International Airport in the chapter opening image). A key factor in digital images is **spatial resolution,** expressed as how much area each pixel represents. For example, the satellite image of the San Francisco Bay Area (chapter opener) has a resolution where each pixel side represents 15 meters on Earth. If the pixel

size is small enough on the finished image, the mosaic effect will be either barely noticeable or invisible to the human eye. Satellites that image an entire hemisphere at once, or large continental areas, use coarse resolutions to produce a more generalized scene.

Even before the invention of the airplane, aerial photographs provided us with "birds-eye" views of our environment via kites and balloons. Both air photos and digital images may be *oblique* (■ Fig. 2.28a), which means that they are taken at an angle other than perpendicular to Earth's surface, or *vertical* (Fig. 2.28b), looking straight down. Image interpreters use aerial photographs and digital imagery to analyze relationships among objects on Earth's surface. A *stereoscope* allows overlapped pairs of images (typically aerial photos) taken from different positions to allow viewing of features in three dimensions.

■ FIGURE 2.28 (a) Oblique photos provide a "natural view," like looking out of an airplane window. This oblique aerial photograph shows farmland, countryside, and forest. (b) Vertical photos provide a maplike view (as in this view of Tampa Bay, Florida).

What are the benefits of an oblique view, compared to a vertical view?

USDA

(a)

USGS

(b)

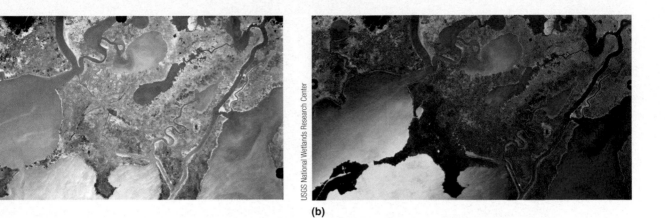

USGS National Wetlands Research Center

(a)   **(b)**

■ **FIGURE 2.29**   Aerial photographs to compare. (a) A natural color aerial photograph. (b) The same scene in false color near-infrared. Red tones indicate vegetation; dark blue—clear, deep water; and light blue—shallow or muddy water. This is a wetlands area on the coast of Louisiana.
If you were asked to make a map of vegetation or water features, which image would you prefer to use and why?

Near-infrared (NIR) energy, light energy at wavelengths that are too long for our eyes to see, cuts though atmospheric haze better than visual light does. An incorrect but widely held notion of NIR techniques is that they image heat, or temperature variations. Near-infrared energy is actually light reflected off of surfaces, and not radiated heat energy. Normal color photographs taken from very high altitudes or from space tend to have low contrast and can appear hazy (■ Fig. 2.29a). Near-infrared photographs and digital images tend to provide very clear images when taken from high altitude or space. Color NIR photographs and digital images are sometimes referred to as "false color" pictures, because on NIR, healthy grasses, trees, and most plants will show up as bright red, rather than green (Fig. 2.29b and see again the chapter opening satellite image). Near-infrared photographs and images have many applications for environmental study, particularly for water resources, vegetation, and crops.

Specialized Remote Sensing

There are many different remote sensing systems, each designed for specific imaging applications. Remote sensing may use ultraviolet light, visible light, NIR light, thermal infrared energy (heat), lasers, and microwaves (radar) to produce images.

Thermal Infrared   **Thermal infrared (TIR)** images show patterns of heat and temperature instead of light and can be taken either day or night. Thermal sensors convert heat patterns into a digital image. TIR images record temperature differences—hot objects show up in light tones, and cool objects will be dark. Generally, a computer is used to emphasize heat differences by colorizing the image. Thermal imaging is used to find volcanic hot spots and geothermal sites, and to locate forest fires through dense smoke.

Weather satellites use thermal infrared imaging. We have all seen these TIR images on television when the meteorologist says, "Let's see what the weather satellite shows." Clouds are depicted in black on the original thermal image because they are colder than the surface of Earth below. Because we don't like to see black clouds, the image tones are reversed, like a photo negative, so that the clouds appear white. These images may also be colorized to show cloud heights, because clouds are progressively colder at higher altitudes (■ Fig. 2.30).

Radar   **Radar** (from **ra**dio **d**etection **a**nd **r**anging) transmits radio waves and produces an image of the energy signals that are reflected back. Radar systems can operate day or night. **Weather radar** uses a ground-based system to monitor and track thunderstorms, hurricanes, or tornadoes. Radar penetrates most clouds (day or night) but reflects off of raindrops and other precipitation, producing map-like images of precipitation patterns (■ Fig. 2.31). There are also several kinds of radar systems that produce a map-like image of the surface (topography, rock, water, ice, sand dunes, etc.).

Sonar   **Sonar** (from **so**und **na**vigation and **r**anging) uses the reflection of emitted sound waves to probe the ocean depths. Much of our understanding of seafloor topography, and mapping of the seafloor, has been a result of sonar applications.

Multispectral Remote Sensing   This type of sensing uses and compares more than one kind of image of the same place, for example, radar and TIR images, or NIR and normal color photos. Common on satellites, *multispectral scanners* sense many kinds of energy simultaneously and relay them to receiving stations as separate digital images. Each part of the energy spectrum yields different information about aspects of the environment. The separate images, like thematic

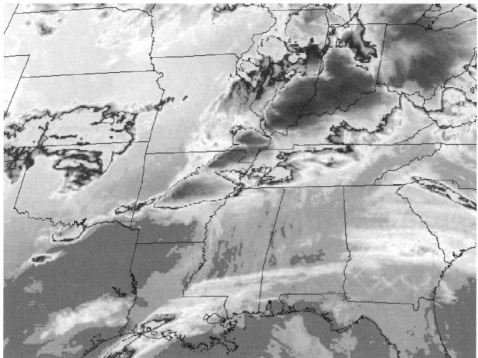

NOAA/GOES Satellite Image

■ **FIGURE 2.30**   Thermal infrared weather images show patterns of heat and cold. This image shows part of the Southeastern United States beamed back from a U.S. weather satellite. Original thermal images are black and white, but here the stormy areas have been colorized. Reds, oranges, and yellows show where the storm is most intense, and blues less intense.
Why are the storm patterns on thermal weather images important to us?

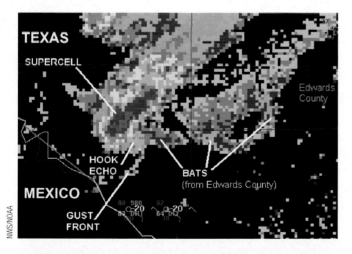

NWS/NOAA

■ **FIGURE 2.31**   Weather radar shows a severe thunderstorm with a hook shaped pattern that is associated with tornadoes. Colors show rainfall intensity: green—light rainfall; yellow—moderate; and orange-red—heavy. The radar has also picked up reflections from huge groups of flying bats, among the millions that live in this region.
How are weather radar images helpful during hazardous weather conditions?

map layers in a GIS, can be combined, depending on which ones are needed for analysis.

Digital technologies for mapping, modeling, and imaging our planet's features continue to provide us with data and information that contribute to our understanding of the Earth system and to help us address environmental concerns. Through continuous monitoring, global, regional, and even local changes can be detected and mapped. Digital mapping, GPS, GIS, and remote sensing have revolutionized the field of geography, but the fundamental principles concerning maps and cartography remain basically unchanged. Whether they are on paper, displayed on a computer monitor, hand drawn in the field, or stored as a mental image, maps and various kinds of representations of Earth continue to be essential tools for geographers and other scientists. Many geographers are gainfully employed in positions that apply spatial technologies to understanding our planet and its environments, and their numbers are certain to increase in the future.

:: Terms for Review

cartography	township	topographic map
oblate spheroid	section	topographic contour line
great circle	global positioning system (GPS)	contour interval
hemisphere	map projection	profile
small circle	Mercator projection	gradient
coordinate system	equal-area map	geographic information system (GIS)
North Pole	conformal map	digital elevation model (DEM)
South Pole	equidistance	visualization model
equator	azimuthal map	vertical exaggeration
latitude	compromise projection	remote sensing
prime meridian	legend	digital image
longitude	scale	pixel
decimal degree	verbal scale	spatial resolution
geographic grid	representative fraction (RF) scale	near-infrared (NIR)
parallel	graphic (bar) scale	radar
meridian	magnetic declination	thermal infrared (TIR)
time zone	thematic map	weather radar
solar noon	discrete data	sonar
International Date Line	continuous data	multispectral remote sensing
U.S. Public Lands Survey System	isoline	

:: Questions for Review

1. Why is a great circle useful for navigation?
2. Why were the time zones established, and why are they generally centered on even 15° meridians of longitude?
3. If you fly across the Pacific Ocean from the United States to Japan, how will the International Date Line affect the day change?
4. In terms of the kinds of locations they describe, how is longitude and latitude different from the U.S. Public Lands Survey System?
5. Why is it impossible for maps to provide a completely accurate representation of Earth's surface? What is the difference between a conformal map and an equal-area map?
6. What is the difference between an RF and a verbal map scale?
7. What does the concept of thematic map layers mean in a geographic information system?
8. What specific advantages do computers offer to the map-making process?
9. What is the difference between a photograph and a digital image?
10. What does a weather radar image show in order to help us understand weather patterns?

:: Practical Applications

1. Select a place within the United States that you would most like to visit for a vacation. You have with you a highway map, a USGS topographic map, and a satellite image of the area. What kinds of information could you get from one of these sources that are not displayed on the other two? What spatial information do they share (visible on all three)?
2. If you were an applied geographer and wanted to use a geographic information system to build an information database about the environment of a park (pick a state or national park near you), what are the five most important layers of mapped information that you would want to have? What combinations of two or more layers would be particularly important to your purpose?
3. If it is 2:00 a.m. Tuesday in New York (EST), what time and day is it in California (PST)? What time is it in London (GMT)? What is the date and time in Sydney, Australia (151° East)?
4. If 10 centimeters (3.94 in.) on a map equal 1 kilometer (3281 ft) on the ground, what is the RF scale of the map? You can round the answer to the nearest thousand. This is the formula to use for scale conversions of this kind: MD/ED = 1/RFD (Map distance/Earth distance = 1/representative fraction denominator).

5. Using the Search window in Google Earth, fly to the heart of the following cities and identify the latitude and longitude. Measure the latitude and longitude using decimal degrees with two decimal places (e.g., 41.89 N as opposed to 41°88'54" N). Make sure that you correctly note whether the latitude is North (N) or South (S) of the equator and whether the longitude is East (E) or West (W) of the prime meridian.

   a. London, England
   b. Paris, France
   c. New York City
   d. San Francisco, California
   e. Buenos Aires, Argentina
   f. Cape Town, South Africa
   g. Moscow, Russia
   h. Beijing, China
   i. Sydney, Australia
   j. Your hometown

   Enter the following coordinates into Google Earth to identify the locations. Go to the Google Earth preferences and select decimal degrees. Review: Latitude is always listed first, and if there are no N, S, E, W, designations, positive numbers mean N or E and negative numbers mean S or W.

   a. 41.89N, 12.492E
   b. 33.857S, 151.215E
   c. 29.975N, 31.135E
   d. 90.0, 0
   e. −90.0, −90.0
   f. 27.175, 78.042
   g. 27.99 N, 86.92E
   h. 40.822N, 14.425E
   i. 48.858N, 2.295E

MAP INTERPRETATION
TOPOGRAPHIC MAPS

The Map

A topographic map is a widely used tool for graphically depicting variations in elevation within an area. A contour line connects points of equal elevation above some reference datum, usually mean sea level. A vast storehouse of information about the relief and the terrain can be interpreted from these maps by understanding the spacing and configuration of contours. For example, elevations of mountains and valleys, steepness of slopes, and the direction of stream flow can be determined by studying a topographic map. In addition to contour lines, many standard symbols are used on topographic maps to represent mapped features, data, and information (a guide to these symbols is in Appendix B).

The elevation difference represented by adjacent contour lines depends on the map scale and the relief in the mapped area, and is called the *contour interval*. Contour intervals on topographic maps are typically in elevation measurements divisible by ten. In mountainous areas, wider intervals are needed to keep the contours from crowding and visually merging together. A flatter locality may require a smaller contour interval to display subtle relief features. It is good practice to note both the map scale and the contour interval when first examining a topographic map.

Keep in mind several important rules when interpreting contours:

- Closely spaced contours indicate a steep slope, and widely spaced contours indicate a gentle slope.
- Evenly spaced contours indicate a uniform slope.
- Closed contour lines represent a hill or a depression.
- Contour lines never cross but may converge along a vertical cliff.
- A contour line will bend upstream when it crosses a valley.

Interpreting the Map

1. What is the contour interval on this map?
2. The map scale is 1:24,000. One inch on the map represents how many feet on the Earth's surface?
3. What is the highest elevation on the map? Where is it located?
4. What is the lowest elevation on the map? Where is it located?
5. Note the mountain ridge between Boat and Emerald Canyons (C-4). Is it steeper on its east side or its west side? What led you to your conclusion?
6. In what direction does the stream in Boat Canyon flow? What led you to your conclusion?
7. The aerial photograph below depicts a portion of the topographic map on the opposite page. What area of the air photo does the map depict? How well do the contours represent the physical features seen on the air photo?
8. Identify some cultural features on the map. Describe the symbols used to depict these features. The map shown is older than the aerial photograph. Can you identify some cultural features on the aerial photograph not depicted on the contour map?

You can also compare this topographic map to the Google Earth presentation of the area. Find the map area by zooming in on these latitude and longitude coordinates: 33.565556N, 117.803889W.

Aerial photograph of the coast at Laguna Beach, California.

Opposite:
Laguna Beach, California
Scale 1:24,000
Contour interval = 20 feet
U.S. Geological Survey

Solar Energy and Atmospheric Heating

3

:: Outline

The Earth–Sun System

Characteristics of the Atmosphere

Heating the Atmosphere

Air Temperature

Our sun, the ultimate energy source for Earth–atmosphere systems.

Courtesy of SOHO/[instrument] consortium. SOHO is a project of international cooperation between ESA and NASA.

48

:: Objectives

When you complete this chapter you should be able to:

- Describe how the receipt and distribution of solar energy on Earth are affected by Earth–sun relationships.
- Recall the major gases found in the atmosphere.
- Conceptualize how Earth's receipt of solar energy is utilized and affected by the major atmospheric gases.

- Discuss how solar energy that reaches Earth's surface is transferred to the atmosphere.
- Explain why water plays such an important role in heat energy transfer.
- List the characteristics of vertical atmospheric temperature layers.
- Describe the controls on horizontal distribution of Earth's surface temperatures.

In this chapter, we study the atmospheric systems that produce the environmental conditions known as *weather* and *climate*. These are familiar terms, however, in physical geography, we must carefully distinguish between them. **Weather** refers to the conditions of atmospheric elements at a given time and for a specific area. That area could be as large as the Chicago metropolitan area or as small and specific as a weather observation station. The study of the weather and changing atmospheric conditions is the science of **meteorology.**

Examining weather observations that occurred at a place over 30 years or more provides us with a good idea of its climate. **Climate** often describes an area's average weather over the seasons, but it also considers departures from the normal or average that are likely to occur and why. Climates also consider extreme weather events that may affect a place or region. Average temperatures and precipitation throughout the seasons can describe the climate of the southeastern United States, but we would also include the likelihood of events such as hurricanes or snowstorms and when they could occur. **Climatology** is the study of the varieties of climates, both past and present, and the processes that produce the different climates on Earth. Climatology also concerns classifying climate types, their seasonal weather characteristics, their distribution, and the extent of climate regions.

Five basic atmospheric *elements* are the "ingredients" of weather and climate: (1) solar energy (*insolation*), (2) temperature, (3) pressure, (4) wind, and (5) precipitation. Weather forecasts generally include the present temperature, the probable temperature range, a description of the cloud cover, the chance of precipitation, air pressure, and the wind speed and direction. All of these elements are important for understanding and categorizing weather and climate.

Solar energy drives the atmospheric systems, so the *insolation* a place receives is the most important factor, because the other four elements depend in part on the intensity and duration of solar energy. We will first focus our attention in this chapter on relationships between Earth and the sun, and on temperature, which is the initial product of insolation. The other three elements will be examined in subsequent chapters.

The Earth–Sun System

The sun's energy comes from fusion (thermonuclear reaction) that takes place under extremely high pressure and at temperatures exceeding 15,000,000°C (27,000,000°F). Two hydrogen atoms fuse together to form a helium atom in a process similar to the explosion of a hydrogen bomb (■ Fig. 3.1). Energy leaves

■ **FIGURE 3.1**   (a) The fireball explosion of a hydrogen bomb is created by thermonuclear fusion.
(b) This same reaction powers the sun.
What elements drive a fusion reaction?

(a)

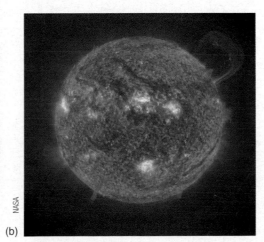
(b)

GEOGRAPHY'S ENVIRONMENTAL PERSPECTIVE
:: PASSIVE SOLAR ENERGY

When we think of using solar energy to heat buildings or to generate electricity, we often look toward complex technology. Long before photovoltaic cells were invented, however, people sought to moderate temperatures in their homes through strategies that are referred to as passive uses of solar energy. Today these strategies are still important to energy conservation. For buildings, it is essential to know which cardinal direction each side of a structure faces, and how the daily and seasonally changing sun angles illuminate that structure.

The concept is rather simple. First, sunlight should flood your home with solar energy in the wintertime, adding more heat during the cold season. Then, limit the amount of insolation entering the home during the summer, to keep the interior cooler during the hottest months, while still allowing daylight to illuminate the interior. Environmentally-conscious home designers do this by adjusting the number and sizes of windows in a home, considering the direction that each window faces, and designing an appropriate roof overhang on sides that face the sun. The idea is to allow fairly direct sunlight through the windows when the sun is lower in the sky in winter, and to shade those same windows when the sun is higher in the sky in summer. In the northern hemisphere midlatitudes, south- and west-facing windows are the ones that need the most shading in summer.

Knowing both the latitude of the dwelling or building and how the sun angle will change over the year at that locale is a first step. Figure 1 shows the maximum and minimum sun angles experienced by a location at 40° N latitude.

In the Southwest, early Native Americans, like many early cultures, were aware of the benefits of passive solar design for their living accommodations. The Cliff Palace in Mesa Verde, Colorado, is a wonderful example of an 800-year old cliff-dwelling. Here the cave roof and overhang perform the same service as the environmentally-conscious home design. Direct sunlight enters the structures during winter and the cliff overhang provides shade for the dwelling during summer.

Using this knowledge, we can do some simple things to save money on heating and air conditioning costs. Considering passive solar principles and using window curtains, shades, or blinds can do a lot to conserve energy and save on energy bills.

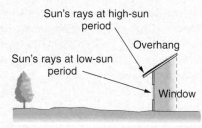

Modern house designs take seasonal changes in sun angles into account. The diagram shows maximum and minimum noon sun angles at 40° N latitude. In the summer when the sun is at a high angle, the window is in the shade, but direct sunlight comes in during the window when the noon sun is at a lower angle.

National Park Service Geologic Resources/D. Luchsinger

The Cliff Palace at Mesa Verde National Park in Colorado shows that early Native Americans understood the use of passive solar energy in locating their cliff dwellings under natural overhangs.

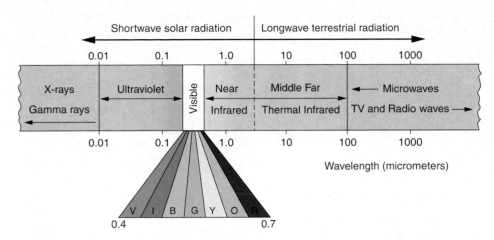

■ **FIGURE 3.2** Radiation from the sun travels toward Earth in a wide spectrum of wavelengths, which are measured in micrometers (μm). One μm is one millionth of a meter. Human vision sees wavelengths between 0.4–0.7 micrometers (visible light). Solar radiation is in shortwave radiation (light), whereas terrestrial (Earth) radiation is in long wavelengths (heat, or thermal infrared).

Are radio signals considered longwave or shortwave radiation?

the sun in the form of **electromagnetic energy,** which travels through empty space at the speed of light in a spectrum of varying wavelengths (■ Fig. 3.2). It takes about 8.3 minutes for the sun's energy to reach Earth. Approximately 9% of solar energy is made up of *gamma rays*, *X-rays*, and *ultraviolet radiation*, all of which are shorter in wavelength than visible light. These invisible wavelengths can affect tissues in the human body. Absorbing too many X-rays can be dangerous, and exposure to excessive ultraviolet light gives us sunburn and is a primary cause of skin cancer. About 41% of the solar spectrum comes in the form of light that is visible to humans, and each color is distinguished by a band of wavelengths. About 49% of the sun's radiant energy is in wavelengths that are longer than those of visible light; of these, the shorter wavelengths are known as *near-infrared light*. Energy emitted in much longer waves, also called *thermal infrared*, is felt as heat. The last 1% of solar radiation falls into the band regions of microwave, television, and radio wavelengths.

Collectively, gamma rays, X-rays, ultraviolet rays, visible light, and near-infrared light are often referred to as **shortwave radiation.** Starting from thermal infrared, longer wavelengths of energy are considered **longwave radiation.** Through our advances in technology, we have learned to harness some electromagnetic wavelength bands for our own uses. In communications, we employ radio waves, microwaves, and television signals; in health care, we use X-rays. In remote sensing and national defense, visible light is necessary for photography and visible satellite imagery. Radar uses microwaves to detect weather patterns, aircraft, and many other objects. Thermal infrared sensors help us to detect differences in temperature caused by wildfires (even through smoke), atmospheric or oceanic temperatures, volcanic activity, and other environmental variations related to temperature contrasts.

The sun radiates energy into space at an almost steady rate. At its outer edge, Earth's atmosphere intercepts slightly less than 2 calories per square centimeter per minute of solar

energy. A **calorie** is the energy required to raise the temperature of 1 gram of water 1°C. This can also be expressed in units of power—in this case, around 1370 watts per square meter. The rate of a planet's receipt of solar energy is known as the **solar constant** and has been measured outside Earth's atmosphere by satellites. The atmosphere affects the amount of solar radiation received on Earth's surface because clouds absorb part of that energy, some is reflected back to space, and some is diffused to and away from Earth. If Earth did not have an atmosphere, the solar energy received at a particular location for a particular time would be a constant value determined by the latitude.

Sun Angle, Duration, and Insolation

Recognizing the consistency of the solar constant leads us directly into a discussion of why the intensity of the sun's rays varies from place to place and during the seasons. Seasonal variations in temperature must be due primarily to differences in the amount and intensity of solar radiation received by various places on Earth, known as **insolation** (for **in**coming **sol**ar radi**ation**). Insolation is not equal everywhere on Earth for many reasons, but two important influences result from our planet *rotating* on its axis and *revolving* around the sun in the manner that it does: the duration of daylight hours and the angle of the solar rays. The length of daylight controls the duration of solar radiation, and the angle of the sun's rays affects the intensity of solar radiation received. Together, the intensity and duration of radiation are the major factors affecting the insolation received at any location on Earth's surface.

Therefore, a location will receive more insolation if (1) the sun shines more directly, (2) the sun shines longer, or (3) both. The intensity of solar radiation received at any one time varies from place to place because Earth presents a spherical surface to insolation. Therefore, only one line of latitude on the Earth's rotating surface can receive radiation at right angles, while the rest receive insolation at less than a 90° angle. As seen in ■ Figure 3.3, the solar energy that strikes a location nearly vertically is more intense and is spread over less area than an equal amount that strikes the surface at an oblique angle. In addition, atmospheric gases interact with solar energy and diminish the insolation that arrives at the surface. Oblique rays pass through a greater distance of atmosphere than vertical rays, so more insolation will be lost in the process. ■ Figure 3.4 shows the intensity of total solar energy at various latitudes when the most direct radiation (from 90° angle rays) strikes on the equator.

No insolation is received at night, and the duration of solar energy is related to the daylight hours received at various

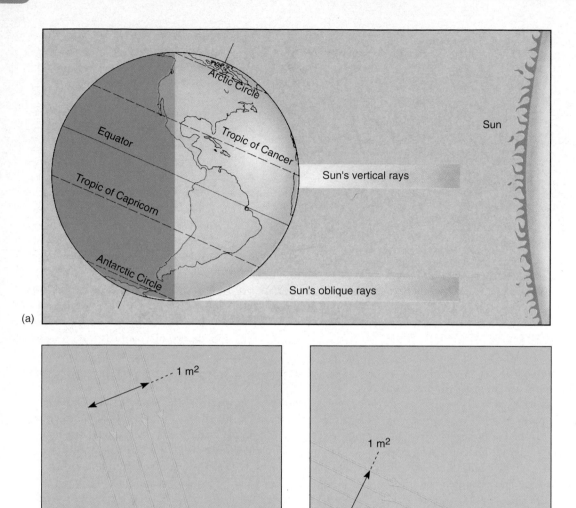

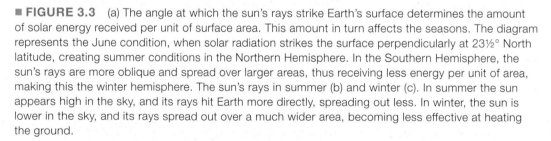

■ **FIGURE 3.3** (a) The angle at which the sun's rays strike Earth's surface determines the amount of solar energy received per unit of surface area. This amount in turn affects the seasons. The diagram represents the June condition, when solar radiation strikes the surface perpendicularly at 23½° North latitude, creating summer conditions in the Northern Hemisphere. In the Southern Hemisphere, the sun's rays are more oblique and spread over larger areas, thus receiving less energy per unit of area, making this the winter hemisphere. The sun's rays in summer (b) and winter (c). In summer the sun appears high in the sky, and its rays hit Earth more directly, spreading out less. In winter, the sun is lower in the sky, and its rays spread out over a much wider area, becoming less effective at heating the ground.

places on Earth (Table 3.1). Obviously, the longer the period of daylight, the greater the amount of solar radiation that will be received at that location. Periods of daylight hours vary in length through the seasons, as well as from place to place.

The Seasons

As we begin our discussion of seasons it is recommended that you review the section on *Plane of the Ecliptic, Inclination, and Parallelism* in Chapter 1. As we will soon see, seasons are caused by the 23½° tilt of Earth's equator to the plane of the ecliptic (see again Fig. 1.24) and the parallelism of the axis that is maintained as Earth orbits the sun. About June 21, Earth is in an orbital position where the north polar axis is inclined toward the sun at an angle of 23½°. On this date, at noon, at 23½°N latitude the sun's rays will be directly overhead, and strike the surface at 90°. In the Northern Hemisphere, this day during Earth's orbit is called the summer **solstice.** In ■ Figure 3.5 position A, note that the Northern and Southern Hemispheres receive unequal amounts of light

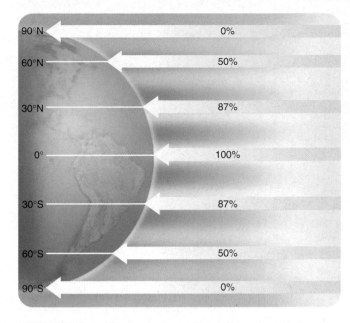

■ FIGURE 3.4 The percentage of incoming solar radiation (insolation) striking various latitudes when the direct rays strike the equator.

How much less solar energy is received at 60° latitude than that received at the equator?

from the sun, because as Earth rotates under these conditions, a larger portion of the Northern Hemisphere remains in daylight. Conversely, a larger portion of the Southern Hemisphere remains in darkness. Thus, Repulse Bay, Canada, north of the Arctic Circle, experiences a full 24 hours of daylight at the June solstice. On the same day, someone in New York City will experience a longer period of daylight than darkness. However, Buenos Aires, Argentina, will have a longer period

of darkness than daylight on that day. This day is called the winter solstice in the Southern Hemisphere. Thus, June 21 is the longest day of the year in the Northern Hemisphere (with the highest yearly sun angles), and in the Southern Hemisphere it is the shortest day, with the lowest sun angles of the year.

Imagine Earth moving from its June solstice position toward its position a quarter of a year later, in September. As Earth moves toward that new position, imagine the changes that will be taking place in our three cities. In Repulse Bay, there will be an increasing amount of darkness through July, August, and September. In New York, sunset will be arriving earlier. In Buenos Aires, the situation will be reversed; as Earth moves toward its position in September, the periods of daylight in the Southern Hemisphere will begin to get longer, the nights shorter.

On or about September 22, Earth will reach a position known as an **equinox** (Latin: *aequus*, equal; *nox*, night). On this date (the autumnal equinox for the Northern Hemisphere), day and night will be of equal length at all locations on Earth. Thus, on the equinox, conditions are identical for both hemispheres. As you can see in ■ Figure 3.6, position B, Earth's axis points neither toward nor away from the sun (imagine the axis is pointed at the reader); the circle of illumination passes through both poles, and it cuts Earth in half along its axis.

Imagine again the revolution and rotation of Earth while moving from around September 22 toward a new position another quarter of a year later in December. We can see that in Repulse Bay the nights will be getting longer until, on the winter solstice, which occurs on or about December 21, this northern town will experience 24 hours of darkness (Fig. 3.5, position C). The only natural light at all in Repulse Bay will be a faint glow at noon refracted from the sun below the horizon.

TABLE 3.1
Duration of Daylight for Certain Latitudes

Length of Day (Northern Hemisphere) (read down)			
LATITUDE (IN DEGREES)	**MAR. 20/SEPT. 22**	**JUNE 21**	**DEC. 21**
0.0	12 hr	12 hr	12 hr
10.0	12 hr	12 hr 35 min	11 hr 25 min
20.0	12 hr	13 hr 12 min	10 hr 48 min
23.5	12 hr	13 hr 35 min	10 hr 41 min
30.0	12 hr	13 hr 56 min	10 hr 4 min
40.0	12 hr	14 hr 52 min	9 hr 8 min
50.0	12 hr	16 hr 18 min	7 hr 42 min
60.0	12 hr	18 hr 27 min	5 hr 33 min
66.5	12 hr	24 hr	0 hr
70.0	12 hr	24 hr	0 hr
80.0	12 hr	24 hr	0 hr
90.0	12 hr	24 hr	0 hr
LATITUDE	**MAR. 20/SEPT. 22**	**DEC. 21**	**JUNE 21**
Length of Day (Southern Hemisphere) (read up)			

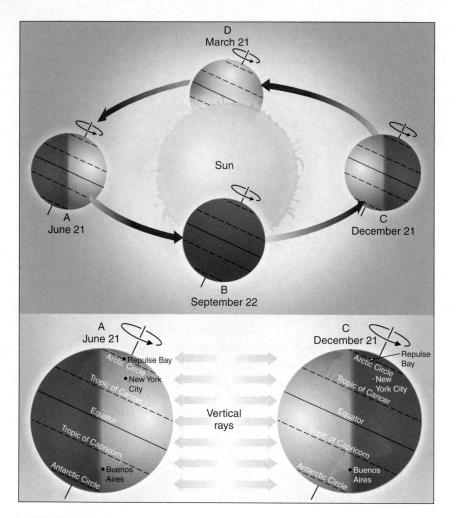

■ **FIGURE 3.5** The geometric relationships between Earth and the sun during the solstices in June and December. Note the differing day lengths at the summer and winter solstices in the Northern and Southern Hemispheres.

on the Antarctic Circle in the Southern Hemisphere will experience a winter solstice similar to that which Repulse Bay had around December 21 (Fig. 3.5, position A). There will be no daylight in 24 hours, except what appears at noon as a glow of twilight in the sky.

Latitude Lines Delimiting Solar Energy

Looking at the diagrams of Earth in its various positions as it revolves around the sun, we can see that the angle of inclination is important. On June 21, the plane of the ecliptic is directly on 23½°N latitude. The sun's rays can reach 23½° beyond the North Pole, bathing it in sunlight. The Arctic Circle, an imaginary line drawn around Earth 23½° from the North Pole (or 66½° north of the equator) marks this limit. We can see from the diagram that all points on or north of the **Arctic Circle** will experience no darkness on the June solstice and that all points south of the Arctic Circle will have some darkness on that day. The **Antarctic Circle** in the Southern Hemisphere (23½° north of the South Pole, or 66½° south of the equator) marks a similar limit.

Furthermore, it can be seen from the diagrams that the sun's **vertical (direct) rays** (rays that strike Earth's surface at right angles) also shift position relative to the poles and the equator as Earth revolves around the sun. At the time of the June solstice, the sun's rays are vertical, or directly overhead, at noon at 23½° north of the equator. This imaginary line around Earth marks the northernmost position at which the solar rays will ever be directly overhead during a full revolution of our planet around the sun. The imaginary line marking this limit is the **Tropic of Cancer** (23½°N latitude). Six months later, at the time of the December solstice, the solar rays are vertical, and the noon sun is directly overhead 23½° *south* of the equator. The imaginary line marking this limit is the **Tropic of Capricorn** (23½°S latitude). During the March and September equinoxes, the vertical solar rays will strike directly at the equator; the noon sun is directly overhead at all points on that line (0° latitude).

Note also that on any day of the year the sun's rays will strike Earth at a 90° angle at only one latitudinal position, either on or between the two lines of the tropics. On that day, all other positions will receive the sun's rays at an angle of less than 90° (or may receive no sunlight at all). The latitude at which the noon sun is directly overhead is known as the sun's

In New York, too, the days will get shorter, and the sun will set earlier. Again, we can see that in Buenos Aires the situation is reversed. Around December 21, that city will experience its summer solstice; conditions will be much as they were in New York City in June.

Moving from late December through another quarter of a year to late March, Repulse Bay will have longer periods of daylight, as will New York, while in Buenos Aires the nights will be getting longer. Then, on or about March 21, Earth will again be in an equinox position (the vernal equinox in the Northern Hemisphere) similar to the one in September (Fig. 3.6, position D). Again, days and nights will be equal all over Earth (12 hours each).

Finally, moving through another quarter of the year toward the June solstice where we began, Repulse Bay and New York City are both experiencing longer periods of daylight than darkness. The sun is setting earlier in Buenos Aires until, on or about June 21, Repulse Bay and New York City will have their longest day of the year and Buenos Aires its shortest. Further, we can see that around June 21, a point

Three distinct patterns occur in the latitudinal distribution of the seasonal receipt of solar energy in each hemisphere. These patterns serve as the basis for recognizing six latitudinal zones, or bands, of insolation and temperature that circle Earth (■ Fig. 3.8).

In the Northern Hemisphere, we take the Tropic of Cancer and the Arctic Circle as the dividing lines for three of these distinctive zones. The area between the equator and the Tropic of Cancer can be called the *north tropical zone*. Here, insolation is always high but is greatest at the time of the year that the sun is directly overhead at noon. This occurs twice a year, and these dates vary according to latitude (see again Fig. 3.7). The *north middle-latitude zone* is the wide band between the Tropic of Cancer and the Arctic Circle. In this belt, insolation is greatest on the June solstice when the sun reaches its highest noon angle and the period of daylight is longest. Insolation is least at the December solstice when the sun is lowest in the sky and the period of daylight the shortest. The *north polar zone*, or *Arctic zone*, extends from the Arctic Circle to the pole. In this region, insolation is greatest at the June solstice, but ceases during the period that the sun's rays are blocked entirely by the tilt of Earth's axis. This period lasts for six months at the North Pole but is as short as a single day directly on the Arctic Circle.

Similarly, there is a *south tropical zone*, a *south middle-latitude zone*, and a *south polar zone*, or *Antarctic zone*, all separated by the Tropic of Capricorn and the Antarctic Circle in the Southern Hemisphere. These areas get their greatest amounts of insolation at opposite times of the year from the northern zones.

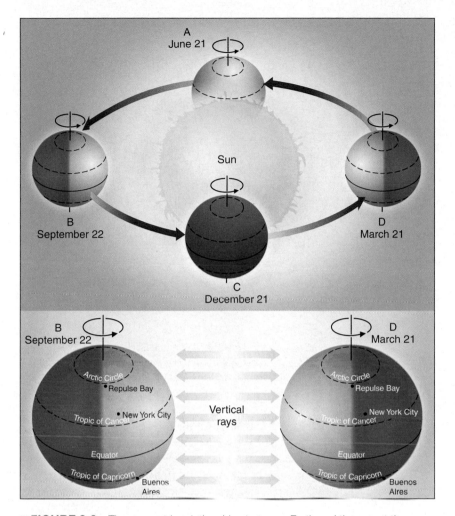

■ **FIGURE 3.6** The geometric relationships between Earth and the sun at the March and September equinoxes. Daylight and darkness periods are 12 hours everywhere because the circle of illumination crosses the equator at right angles and cuts through both poles.

If Earth were not inclined on its axis, would there still be latitudinal temperature variations? Would there be seasons?

declination. ■ Figure 3.7 is an example of an *analemma*, a graph shaped like a figure 8, which is often drawn on globes to show the sun's declination throughout the year.

Variations of Insolation with Latitude

Neglecting for the moment the atmosphere's influence on variations in insolation during a 24-hour period, the amount of energy received begins after dawn and increases as Earth rotates toward the time of solar noon. A place will receive its greatest insolation at solar noon when the sun has reached its zenith, or highest point in the sky, for that day. The insolation then decreases as the sun's angle lowers toward the next period of darkness. Obviously, at any location, no insolation is received during the hours of darkness.

Characteristics of the Atmosphere

The atmosphere has a profound effect on the amount of solar energy received to heat Earth and power its environmental systems. A knowledge of the structure and composition of the atmosphere will help us understand the relationships involved as insolation interacts with the air and Earth's surface.

Composition of the Atmosphere

The atmosphere extends to approximately 480 kilometers (300 mi) above Earth's surface. Its density decreases rapidly with altitude; in fact, 97% of the air is concentrated in the first

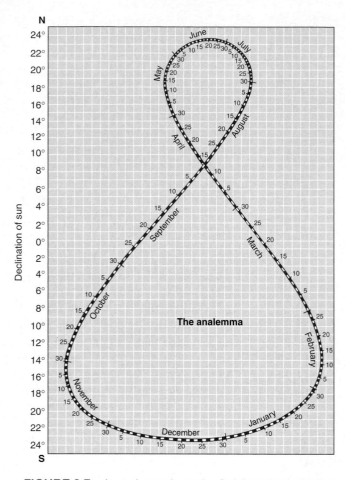

■ **FIGURE 3.7** An analemma is used to find the solar declination (latitudinal position) of the vertical noon sun for each day of the year. **What is the declination of the sun on October 30th?**

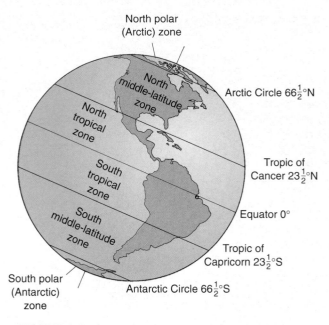

■ **FIGURE 3.8** The equator, the Tropics of Cancer and Capricorn, and the Arctic and Antarctic Circles define six latitudinal zones that have distinctive insolation characteristics. **Which zone(s) would have the least annual variation in insolation? Why?**

Abundant Gases

Of the gases in the atmosphere, nitrogen (N_2) makes up the largest proportion of air. Nitrogen is of major importance in supporting plant growth. In addition, some of the other atmospheric gases are vital to the development and maintenance of life. One of the most important of these gases is oxygen (O_2), which humans and all other animals use to breathe and oxidize (burn) food that they eat. *Oxidation*, the chemical combination of oxygen with other substances to create new products, occurs in situations outside animal life as well. Rapid oxidation takes place, for instance, when we burn fossil fuels or wood and thus release large amounts of heat energy. The decay of certain rocks or organic debris and the development of rust are examples of slow oxidation. All of these processes depend on the presence of oxygen in the atmosphere. The third most abundant gas in our atmosphere is Argon (Ar). It is not a chemically active gas and therefore neither helps nor hinders life on Earth.

25 kilometers (16 mi) or so. The atmosphere is composed of numerous gases (Table 3.2). Most of these gases remain in the same proportions regardless of the atmospheric density. A bit more than 78% of the atmosphere's volume is made up of nitrogen, and nearly 21% consists of oxygen. Argon comprises most of the remaining 1%. The percentage of carbon dioxide in the atmosphere has risen through time, but is a little less than 0.04% by volume. There are traces of other gases as well: ozone, hydrogen, neon, xenon, helium, methane, nitrous oxide, and others.

TABLE 3.2
Composition of the Atmosphere Near Earth's Surface

Permanent Gases			Variable Gases	
Gas	**Symbol**	**Percent (by Volume) Dry Air**	**Gas (and Particles)**	**Symbol**
Nitrogen	N_2	78.08	Water vapor	H_2O
Oxygen	O_2	20.95	Carbon dioxide	CO_2
Argon	Ar	0.93	Methane	CH_2
Neon	Ne	0.0018	Nitrous oxide	N_2O
Helium	He	0.0005	Ozone	O_3
Hydrogen	H_2	0.0006	Particles (dust, soot, etc.)	
Xenon	X_e	0.000009	Chlorofluorocarbons	

Water, Particulates, and Aerosols

Water as water vapor is the most variable gas in the atmosphere. It ranges from 0.02% by volume in a cold, dry climate to more than 4% in the humid tropics. Ways of expressing the amount of water vapor in the atmosphere will be discussed later under the topic of humidity, but it is important to note that the variations in this percentage over time and place are an important consideration in the examination and comparison of climates.

Water vapor also absorbs heat in the lower atmosphere, retarding rapid heat loss from Earth. Thus, water vapor plays a large role in the insulating action of the atmosphere. In addition to gaseous water vapor, liquid water exists in the atmosphere as rain and as fine droplets in clouds, mist, and fog. Solid water exists in the atmosphere as ice crystals, snow, sleet, and hail.

Particulates are solids suspended in the atmosphere. *Aerosols* are solids or liquids suspended in the atmosphere—they include particulates, but they also include tiny liquid droplets and/or ice crystals composed of chemicals other than water. For example, sulfur dioxide crystals (SO_2) are atmospheric aerosols. Particulates can be considered as aerosols, but all aerosols are not necessarily particulate matter. Particulates and aerosols can be pollutants from transportation and industry, but the majority are substances that exist naturally in our atmosphere (■ Fig. 3.9). Particles such as dust, smoke, pollen and spores, volcanic emissions, bacteria, and salts from ocean spray can all play an important role in absorption of energy and in the formation of raindrops.

Carbon Dioxide

Excluding water vapor, carbon dioxide is the fourth most abundant atmospheric gas. The involvement of carbon dioxide in the system known as the *carbon cycle* has been studied for generations. Plants, through a process known as **photosynthesis,** use sunlight (mainly ultraviolet radiation) as the driving force to combine carbon dioxide and water to produce carbohydrates (sugars and starches), in which energy, derived originally from the sun, is stored and used by vegetation (■ Fig. 3.10). Oxygen is given off as a by-product. Animals then use the oxygen to oxidize the carbohydrates, releasing the stored energy. A by-product of this process in animals is the release of carbon dioxide, which completes the cycle when it is in turn used by plants in photosynthesis.

In recent years, earth scientists have directly linked carbon dioxide with long-term changes in Earth's atmospheric temperatures and climates. Scientists are concerned with the relationship between carbon dioxide and Earth's temperatures

NASA International Space Station

■ **FIGURE 3.9** Volcanic eruptions, like this one at Mount Etna in Italy, add a variety of gases, particulates, aerosols, and water vapor into our atmosphere. This image was taken by astronauts from the orbiting International Space Station.
What other ways are particles added to the atmosphere?

because of changes in the **greenhouse effect,** which is also the primary reason for the moderate temperatures on Earth. A greenhouse (glass structure that houses plants) will behave in a somewhat similar manner to a closed vehicle parked in the sun (■ Fig. 3.11). Insolation (shortwave radiation) goes through the transparent glass roof and walls of the greenhouse and helps the plants inside to thrive, even in a cold outdoor environment. After being absorbed by the materials in the greenhouse, the shortwave energy is re-radiated as longwave heat energy, which cannot escape rapidly, thus warming the greenhouse interior.

Like the glass of a greenhouse, carbon dioxide and water vapor (and other greenhouse gases) in the atmosphere are largely transparent to incoming solar radiation, but can impede the escape of longwave radiation by absorbing it and then radiating it back to Earth. For example, carbon dioxide emits about half of its absorbed heat energy back to Earth's surface. Of course, although the results are similar, the processes involving the glass of a car or greenhouse and the

■ **FIGURE 3.10** The equation of photosynthesis shows how solar energy (mainly UV radiation) is used by plants to manufacture sugars and starches from atmospheric carbon dioxide and water, liberating oxygen in the process. The stored food energy is then eaten by animals, which also breathe the oxygen released by photosynthesis.

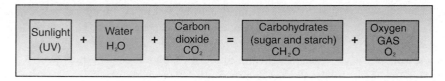

| Sunlight (UV) | + | Water H_2O | + | Carbon dioxide CO_2 | = | Carbohydrates (sugar and starch) CH_2O | + | Oxygen GAS O_2 |

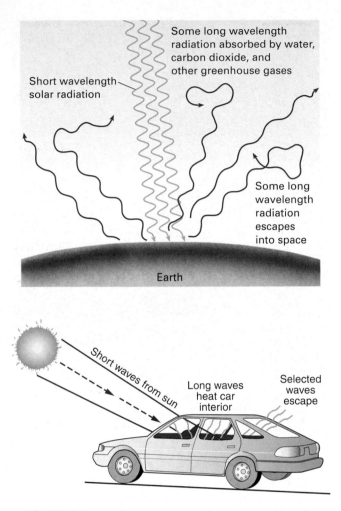

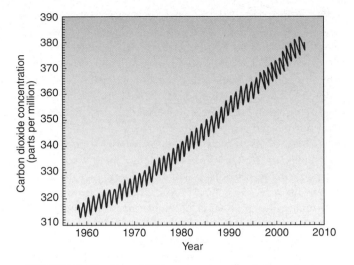

■ **FIGURE 3.12**   Since 1958, measurements of atmospheric carbon dioxide recorded at Mauna Loa, Hawaii, have shown an upward trend.
Why do you suppose the line zigzags each year?

■ **FIGURE 3.11**   (a) Greenhouse gases in our atmosphere allow short-wavelength solar radiation (sunlight) to penetrate Earth's atmosphere relatively unhampered, while some of the long-wavelength radiation (heat) is kept from escaping into outer space. (b) A similar sort of heat buildup occurs in a closed car. The light energy penetrates the car windows and heats the interior, but the glass prevents some of the longwave heat radiation from escaping.
How might you prevent your car interior from becoming so hot on a summer day?

atmosphere are significantly different. The heat of a closed car, or a greenhouse, increases because the air is trapped and cannot circulate to the outside air. Our atmosphere is free to circulate, but is selective as to which wavelengths of energy it will transmit. The greenhouse effect in Earth's atmosphere is not a bad thing, for, without any greenhouse gases in the atmosphere, Earth's surface would be too cold to sustain human life. The greenhouse process helps maintain the warmth of the planet and is a factor in Earth's *heat energy budget* (discussed later in this chapter).

However, a serious environmental issue arises when increasing concentrations of greenhouse gases cause measurable increases in worldwide temperatures. Since the Industrial Revolution, humans have been adding more and more carbon dioxide to the atmosphere through the burning of fossil (car-

bon) fuels. At the same time, Earth has undergone massive *deforestation* (the removal of forests) for urban, agricultural, commercial, and industrial development. Vegetation uses large amounts of carbon dioxide in photosynthesis, so removing the vegetation causes more carbon dioxide to remain in the atmosphere. ■ Figure 3.12 shows how these two human activities have combined to increase carbon dioxide in the atmosphere through time. Carbon dioxide absorbs the longwave heat energy radiated from Earth's surface, restricting its escape to space, so rising amounts of carbon dioxide in the atmosphere increase the greenhouse effect and help produce a global rise in temperatures. This issue will be discussed in more detail in Chapter 8.

Ozone   Another vital gas in Earth's atmosphere is **ozone.** The ozone molecule (O_3) is related to the oxygen molecule (O_2), except it is made up of three oxygen atoms whereas oxygen gas consists of only two. Ozone is formed in the upper atmosphere when an oxygen molecule is split into two oxygen atoms (O) by the sun's ultraviolet radiation.

In the lower atmosphere, ozone is formed by electrical discharges (like high-tension power lines and lightning) as well as by incoming shortwave solar radiation. It is a toxic pollutant and a major component of urban smog, which can cause sore and watery eyes, soreness in the throat and sinuses, and difficulty in breathing. Near the surface, ozone is a menace and can hurt life forms. However, in the upper atmosphere, ozone is essential to living organisms because it absorbs large amounts of the sun's UV radiation that would otherwise reach Earth's surface.

Without the ozone layer of the upper atmosphere, excessive UV radiation reaching Earth would severely burn human skin, increase the incidence of skin cancer and optical cataracts, destroy certain microscopic forms of marine life, and damage plants. Ultraviolet radiation is also responsible for suntans and painful sunburns, depending on an individual's skin tolerance and exposure.

For many years, there has been evidence that human activities, especially the addition of chlorofluorocarbons (CFCs) and nitrogen oxides (NO_X) to the atmosphere, may damage Earth's fragile ozone layer. When the CFC's and nitrogen oxides reach the upper atmosphere, they produce chemical reactions that attack the ozone layer and reduce the amount of ozone serving as a natural UV filter.

Scientists have been increasingly concerned about a so-called "hole" in the ozone layer centered over Antarctica. From Earth's surface to the outer atmosphere, there are traces of ozone present, so the ozone hole is really an area where the ozone level is considerably less than it should be (■ Fig. 3.13).

Evidence of damage to the ozone layer is well documented. Atmospheric ozone levels are measured in Dobson Units (du), established by G. M. B. Dobson along with his Dobson Spectrometer in the late 1920s. A range of 300 to 400 du indicates a sufficient amount of ozone to prevent damage to Earth's life forms. To understand these units better, consider that 100 du equals 1 millimeter of thickness that ozone would have at sea level. Ozone measured inside the "hole" has dropped as low as 95 du in recent years, and the area of the ozone-deficient atmosphere (a more accurate description than a "hole") has exceeded the size of North America.

Vertical Layers of the Atmosphere

There are several systems used to divide the atmosphere into vertical layers. One system is based on the protective function that the layers provide. An example of a layer in this system is the *ozonosphere*, another name for the ozone layer. A second system, often used by chemists and physicists, divides the atmosphere into layers based on chemical composition. A third system, most often used by meteorologists and climatologists, identifies four layers divided according to differences in temperature and rates of temperature change (■ Fig. 3.14).

In the system based on temperature characteristics, the lowest layer is the **troposphere** (from Greek: *tropo*, turn—the turning or mixing zone). The troposphere extends about 8–16 kilometers (5–10 mi) above the surface. Its thickness, which tends to vary seasonally, is least at the poles and greatest at the equator. Virtually all of Earth's weather and climate takes place within the troposphere.

The troposphere has two distinct characteristics that differentiate it from other atmospheric layers; first, water vapor is rarely found above the troposphere. The other characteristic is that temperature normally decreases with increased altitude in the troposphere. The average rate at which temperatures in the troposphere decrease with altitude is called the **environmental lapse rate** (or the **normal lapse rate**); it amounts to 6.5° C per 1000 meters (3.6° F/1000 ft).

The altitude at which the temperature ceases to drop with increased altitude is called the *tropopause*. It is the boundary that separates the troposphere from the **stratosphere**—the next layer of the atmosphere. The temperature of the lower stratosphere remains fairly constant (about −57° C, or −70° F) to

■ **FIGURE 3.14** Vertical temperature changes in Earth's atmosphere are the basis for its subdivision into the troposphere, stratosphere, mesosphere, and thermosphere.

■ **FIGURE 3.13** For decades, satellite sensors have produced images of the ozone hole (shaded in purple) over Antarctica. This shows the spatial extent of the ozone hole in September of 2007. **What are the potential effects of ozone depletion on the world's human population?**

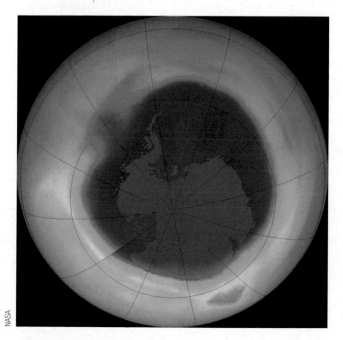

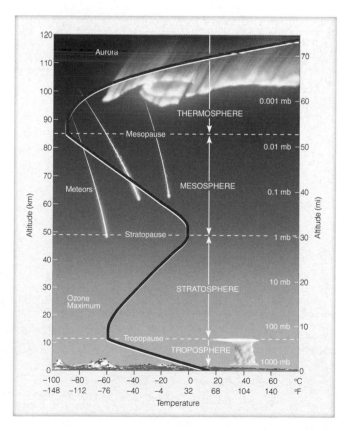

an altitude of about 32 kilometers (20 mi). It is in the stratosphere that we find the ozone layer. As the ozone absorbs UV radiation, the absorbed energy results in releases of heat, and thus temperatures increase in the upper stratosphere. Temperatures at the *stratopause* (another boundary), which is about 50 kilometers (30 mi) above Earth, are about the same as temperatures found on Earth's surface, although little of that heat can be transferred because the air is so thin.

Above the stratopause is the *mesosphere*, where temperatures tend to drop with increased altitude; the mesopause (the last boundary) separates the mesosphere from the *thermosphere*, where temperatures increase until they approach 1100° C (2000° F) at noon. Again, the air is so thin at this altitude that there is practically a vacuum and little heat can be transferred.

Atmospheric Effects on Solar Radiation

As solar energy passes through Earth's atmosphere, more than half of its intensity is lost through various processes. In addition, the amount of insolation received at a particular location also depends on the latitude, time of day, season, and atmospheric thickness (all of which are related to the angle of the sun's rays). The transparency of the atmosphere (or the amount of cloud cover, moisture, carbon dioxide, and solid particles in the air) also plays a vital role.

When the sun's energy passes through the atmosphere, several things happen to it (the following figures represent approximate averages for the entire Earth; at any one location or time, they may differ): (1) 26% of the energy is reflected directly back to space by clouds and the ground; (2) 8% is *scattered* by minute atmospheric particles and returned to space as diffuse radiation; (3) 19% is *absorbed* by the ozone layer and water vapor in the clouds of the atmosphere; (4) 20% reaches Earth's surface as diffuse radiation after being scattered; and (5) 27% reaches Earth's surface as direct radiation (■ Fig. 3.15). In other words, on a worldwide average, 47% of the incoming solar radiation eventually reaches the surface, 19% is retained in the atmosphere, and 34% is returned to space. Because Earth's energy budget is in equilibrium, the 47% received at the surface is ultimately returned to the atmosphere by processes that we will now examine.

■ **FIGURE 3.15** **Environmental Systems: Earth's Radiation Budget** From one year to the next, Earth's overall average temperature varies little. This indicates that a long-term global balance, or equilibrium, exists between solar energy received and then re-radiated away by the Earth system. Note that only 47% of the incoming solar energy reaches and is absorbed by Earth's surface. Eventually, the energy gained by the atmosphere is lost to space. However, the radiation budget is a dynamic one. As a result, there is growing concern that one of the elements, human activity, will cause the atmosphere to absorb more Earth-emitted energy, thus raising global temperatures.

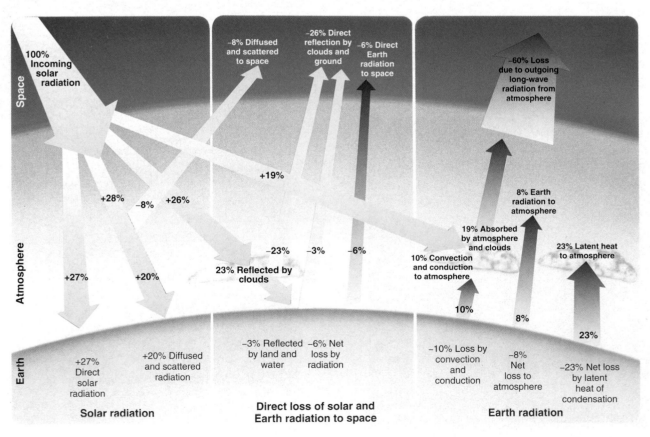

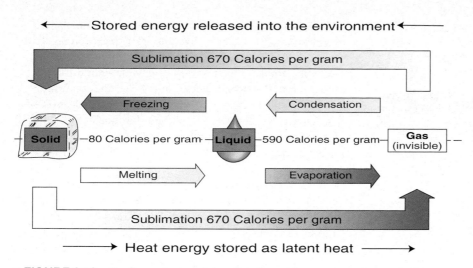

← Stored energy released into the environment ←

Sublimation 670 Calories per gram

Freezing ←   Condensation ←

Solid — 80 Calories per gram — **Liquid** — 590 Calories per gram — **Gas (invisible)** —

Melting ⇒   Evaporation ⇒

Sublimation 670 Calories per gram

→ Heat energy stored as latent heat →

■ **FIGURE 3.16** The three physical states of water and the energy exchanges between them. As you read the diagram, consider where heat energy comes from and where it goes. For example, when water freezes into ice, heat energy must come out of the water as it freezes, and thus enters the environment. To evaporate water into vapor, the heat must go into the water for it to evaporate, and therefore must come from the surrounding environment. **Why do you suppose that some of the energy in these exchanges is referred to as "latent heat"?**

Water and Heat Energy

As it penetrates our atmosphere, some of the incoming solar radiation is involved in several energy exchanges. One exchange involves how water is altered from one state to another. Water is the only substance that can exist in all three states of matter—as a solid, a liquid, and a gas—within the normal temperature range on Earth. In the atmosphere, water exists as a clear, odorless gas called *water vapor*. It is also a *liquid* in the atmosphere (as clouds, fog, and rain), in the oceans, and in other water bodies on and beneath the surface. Liquid water is also contained within vegetation and animals. Finally, water exists as a *solid* in snow and ice in the atmosphere, as well as on and under the surface of the colder parts of Earth.

Not only does water exist in all three states of matter, but it also can change from one state to another, as illustrated in ■ Figure 3.16. In doing so, it becomes involved in Earth's heat energy system. The molecules of a gas move faster than do those of a liquid. During the process of *condensation*, when water vapor changes to liquid water, its molecules slow down and some of their energy is released into the environment, about 590 calories per gram (cal/g). The molecules of a solid move even more slowly than those of a liquid, so during the *freezing* process, when water changes to ice, additional energy is released into the environment, this time 80 calories per gram. When the process is reversed, heat must be added to the ice. Thus, *melting* ice requires the addition of 80 calories per gram to the ice, from the surrounding environment. Further, *evaporation* requires the addition of 590 calories per gram be added to the liquid water, from the environment. This added energy is stored in the water as *latent* (or hidden) *heat*. The **latent heat of fusion** refers to the 80 calories per

gram released into the environment when water freezes into ice; to melt ice into water, this heat energy comes from the environment. The **latent heat of evaporation** (590 cal/g) is added to the water, from the environment, to form water vapor, and the **latent heat of condensation** (also 590 cal/g) is removed from the water vapor, and released into the environment as it condenses into liquid water. The last is **latent heat of sublimation** at 670 calories per gram (the addition of 590 cal/g + 80 cal/g). *Sublimation* is the process where ice turns to vapor or vapor turns to ice without going through a liquid phase. Snowflakes and frost are formed by sublimation.

Some of these energy exchanges can be easily demonstrated. For example, if you hold an ice cube in your hand, your hand feels cold because the heat removed from your hand (the surrounding environment in this case) is needed to melt the ice. We are cooled by evaporating perspiration because heat is absorbed by the evaporating perspiration, thereby lowering skin temperature.

Heating the Atmosphere

The 19% of direct solar radiation that is retained by the atmosphere is "locked up" in the clouds and the ozone layer and thus is not available to heat the troposphere. Other sources must be found to explain the creation of atmospheric warmth. The explanation lies in the 47% of incoming solar energy reaching Earth's surface (on both land and water) and in the transfer of heat energy from Earth back to the atmosphere. This is accomplished through radiation, conduction, convection, advection, and the latent heat of condensation (■ Fig. 3.17).

Processes of Heat Energy Transfer

Radiation The process by which electromagnetic energy is transferred from the sun to Earth is called **radiation.** We should be aware that all objects with a temperature above absolute zero emit electromagnetic radiation. The characteristics of that radiation depend on the temperature of the radiating body. The warmer the object, the more energy it will emit, and the shorter the wavelengths at peak emission. Because the sun's absolute temperature is 20 times that of Earth, the sun emits much more energy, and at shorter wavelengths, than Earth. The sun's energy output per square meter is approximately 160,000 times that of Earth! Further, the majority of solar energy is emitted shortwave

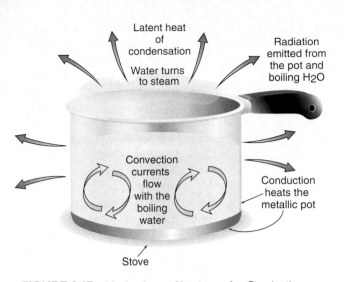

■ **FIGURE 3.17** Mechanisms of heat transfer. Conduction occurs when heat travels from the heat source to the pot and then also to the water. Convection occurs as the hotter water flows upward, and the cooler water sinks, forming a convective current in the boiling water. Radiation, emitted as heat energy, flows outward into the surrounding air from the boiling water, the pot and the heat source. Lastly, heat of condensation is released as water vapor turns back into a liquid as steam.
How might we add advection to this small system?

energy, whereas Earth's energy is radiated as longwaves. Thus, *light energy* from the sun is absorbed by Earth and heats its surface, which, being cooler than the sun, gives off energy in the form of heat energy (thermal infrared). It is this *longwave thermal radiation* from Earth's surface that heats the lower atmospheric layers and accounts for the heat of the day.

Conduction
The means by which heat is transferred from one part of a body to another or between two touching objects is called **conduction.** Heat flows from the warmer to the cooler object in an attempt to equalize temperature. Conduction is what makes anything that is hotter than your skin feel warm or hot to the touch.

Atmospheric conduction occurs at the interface (zone of contact) between the atmosphere and Earth's surface. However, heat transfer by conduction is minor in terms of warming the atmosphere because it affects only the air closest to the surface, because air is a poor conductor of heat. Air is the opposite of a good conductor; it is a good insulator, and a layer of air is sometimes put between two panes of glass to help insulate a window. Air is also used as an insulation layer in sleeping bags and cold-weather parkas. In fact, if air was a good conductor of heat, our kitchens would become unbearable when we turned on the stove or oven.

Convection
As parcels of air near the surface are heated, they expand in volume, become less dense than the surrounding air, and therefore rise. This vertical transfer of heat through the atmosphere is called **convection,** the same process by which boiling water circulates in a pot on a stove.

The water near the bottom is heated first, becoming lighter and less dense as it is heated. As this water rises, colder, denser surface water flows down to replace it. As this descending water is warmed, it too flows upward while additional colder water moves downward. These convective currents set into motion by the heating of a fluid (liquid or gas) make up a *convectional system.* Such systems account for much of the vertical transfer of heat in the atmosphere and the oceans, and are a major cause of clouds and precipitation.

Advection
Advection is the term applied to horizontal heat transfer. There are two major advection agents within the Earth–atmosphere system: winds and ocean currents. Both agents help transfer energy horizontally between the equatorial and polar regions, thus maintaining the energy balance in the Earth–atmosphere system (■ Fig. 3.18).

Latent Heat of Condensation
When water evaporates, a significant amount of energy is stored in the water vapor as latent heat (see again Fig. 3.16). This water vapor is then transported by advection or convection to new locations where condensation takes place and the stored energy is released. This is a major process of energy transfer within the Earth system. The *latent heat of evaporation* helps cool the atmosphere while the *latent heat of condensation* helps warm the atmosphere and is also a source of energy for storms.

The Heat Energy Budget
The Heat Energy Budget at Earth's Surface
Now that we are familiar with the various means of heat transfer, we should examine what happens to the 47% of solar

■ **FIGURE 3.18** Latitudinal variation in the energy budget. Low latitudes receive more insolation than they lose by re-radiation and have an energy surplus. High latitudes receive less energy than they lose, and therefore have an energy deficit.
How do you think the surplus energy in the low latitudes is transferred to higher latitudes?

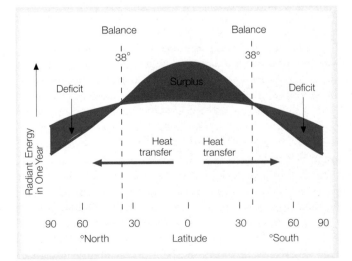

energy that reaches Earth's surface (see again Fig. 3.15). Approximately 14% of this energy is emitted by Earth as thermal infrared (heat) radiation. This 14% includes a net loss of 6% (of the total) directly to outer space, and the other 8% is absorbed by the atmosphere. In addition, there is a transfer back to the lower atmosphere (by conduction and convection) of 10 of the 47% that reached Earth. The remaining 23% returns to the atmosphere through releases of latent heat of condensation. Thus, the 47% of the insolation that reached Earth's surface is returned to other segments of the system and dissipated to space, with no long-term gain or loss. Therefore, at Earth's surface, the heat energy budget is in balance.

Examination of the heat energy budget of Earth's surface helps us understand the energy system that heats the atmosphere. The input to the system is the incoming shortwave solar radiation (light) that reaches Earth's surface; this is balanced by Earth's output of longwave (heat) radiation back to the atmosphere and to space.

Of course, it should be noted that the percentages mentioned earlier are estimates, and simplified in that they refer to *net* losses that occur over a long period of time. In the shorter term, heat may be passed from Earth to the atmosphere and then back to Earth in a chain of cycles before it is finally released into space. The absorption and reflection of incoming solar radiation, and the emission of outgoing terrestrial radiation, can all be affected by the kind of ground cover at the surface.

The Heat Energy Budget in the Atmosphere
About 60% of the solar energy intercepted by the Earth system is temporarily retained by the atmosphere. This includes 19% of solar radiation absorbed by the clouds and the ozone layer, 8% emitted by *longwave radiation* from the Earth's surface, 10% transferred from the surface by *conduction* and *convection*, and 23% released by the *latent heat of condensation*. Some of this energy is recycled back to the surface for short periods of time, but eventually all of it is lost into outer space as more solar energy is received. Hence, just as was the case at Earth's surface, the heat energy budget in the atmosphere is in balance over long periods of time—a dynamically stable system. However, many scientists believe that an imbalance in the heat energy budget, with possible negative effects, could develop due to the *greenhouse effect*.

Variations in the Heat Energy Budget
The figures we have seen for the heat energy budget are averages for the whole Earth over many years. For any *particular* location, the heat energy budget is most likely not balanced. Some locations have a surplus of incoming solar energy over outgoing energy loss, and others have a deficit. The main causes of these variations are differences in latitude and seasonal fluctuations.

As we have noted previously, the amount of insolation received is directly related to latitude (see again Fig. 3.18). In the tropical zones, where insolation is high throughout the year, more solar energy is received at Earth's surface and in the atmosphere than can be emitted back into space. In the Arctic and Antarctic zones, however, there is so little insolation during the winter, when Earth is still emitting longwave radiation, that there is a large deficit for the year. Locations in the middle-latitude zones have lower deficits or surpluses, but only at about latitude 38° is the budget balanced. If it were not for the heat transfers within the atmosphere and the oceans, the tropical zones would get hotter and the polar zones would get colder through time.

At any location, the heat energy budget varies throughout the year according to the seasons, with a tendency toward a surplus in the summer or high-sun season and a tendency toward a deficit six months later. Seasonal differences may be small near the equator, but they are great in the middle-latitude and polar zones.

Air Temperature
Temperature and Heat

Although heat and temperature are highly related, they are not the same. **Heat** is a form of energy—the total kinetic energy of all the atoms that make up a substance. All substances are made up of molecules that are constantly in motion (vibrating and colliding), so they possess kinetic energy—the energy of motion. This energy is manifested as heat. **Temperature** is the average kinetic energy of individual molecules in a substance. When something is heated, its atoms vibrate faster, and its temperature increases. The amount of heat energy depends on the mass of the substance being considered, whereas the temperature refers to the energy of individual molecules. Thus, a burning match has a high temperature but minimal heat energy; the oceans have moderate temperatures but high heat energy content.

Temperature Scales

Two different scales are generally used for measuring temperature. The one that Americans are most familiar with is the **Fahrenheit scale,** devised in 1714 by Daniel Fahrenheit, a German scientist. On this scale, the temperature at which water boils at sea level is 212° F, and the temperature at which water freezes is 32° F. This scale is used in the English system of measurements.

The **Celsius scale** (also called the **centigrade scale**) was devised in 1742 by Anders Celsius, a Swedish astronomer. It is part of the metric system. The temperature at which water freezes at sea level on this scale was set at 0° C, and the temperature at which water boils was designated as 100° C.

The United States is one of the few countries that still make widespread use of the Fahrenheit scale and even in the United States, a majority of the scientific community uses the Celsius scale. For these reasons, this book presents comparable Celsius and Fahrenheit figures given side by side for temperatures. Similarly, whenever important figures for distance, area, weight, or speed are given, we use the metric system followed by the English system. Appendix A provides comparisons and conversions between the two systems.

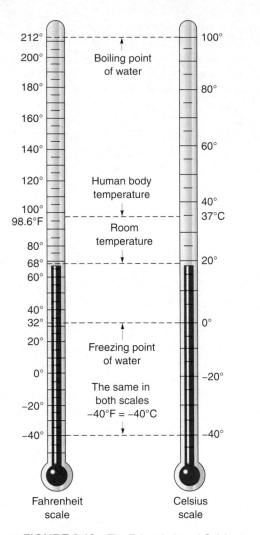

■ FIGURE 3.19 The Fahrenheit and Celsius temperature scales. The scales are aligned to permit direct conversion of readings from one to the other.
When it is 70°F, what is the temperature in Celsius degrees?

■ Figure 3.19 can help you compare the Fahrenheit and Celsius systems as you encounter temperature figures outside this book. In addition, the following formulas can be used for conversion from Fahrenheit to Celsius or vice versa:

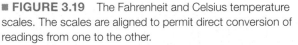

$$°C = (°F − 32) ÷ 1.8$$

$$°F = (°C × 1.8) + 32$$

Short-Term Variations in Temperature

Local changes in atmospheric temperature can have several causes. These are related to the processes of receiving and dissipating energy from the sun and to various properties of Earth's surface and the atmosphere.

The Daily Effects of Insolation As we noted earlier, the amount of insolation at any particular location varies both throughout the year (annually) and throughout the day

(diurnally). Annual fluctuations are associated with the sun's changing declination and hence with the seasons. Diurnal changes are related to Earth's daily rotation. Each day, insolation receipt begins at sunrise, reaches its maximum at noon (local solar time), and returns to zero after sunset.

Although insolation is greatest at noon, you probably know that temperatures usually do not reach their maximum until 2–4 p.m. (■ Fig. 3.20). This is because the insolation received by Earth from sunrise until the afternoon hours exceeds the energy being lost through Earth radiation. Sometime around 3–4 p.m., when outgoing Earth radiation begins to exceed insolation, temperatures start to fall. The daily lag of Earth radiation and temperature behind insolation is accounted for by the time it takes for Earth's surface to be heated to its maximum and for this energy to be radiated to the atmosphere.

Insolation receipt ends after sunset, but much of the energy that has been stored in Earth's surface layer during the day is lost during the night. The lowest temperatures occur around dawn, when the maximum amount of energy has been emitted and before replenishment from the sun can occur. Thus, if we disregard other factors for the moment, we can see that there is a predictable hourly change in temperature called the **daily march of temperature.** There is a gentle decline from mid-afternoon until dawn and a rapid increase in the 8 hours or so from dawn until the next maximum is reached.

■ FIGURE 3.20 Diurnal changes in air temperature are controlled by insolation and outgoing Earth radiation. Where incoming energy exceeds outgoing energy (orange), the air temperature rises. Where outgoing energy exceeds incoming energy (blue), air temperature drops.
Why does temperature rise even after solar energy declines?

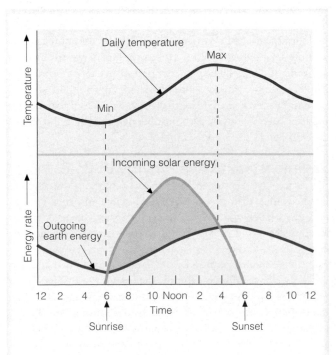

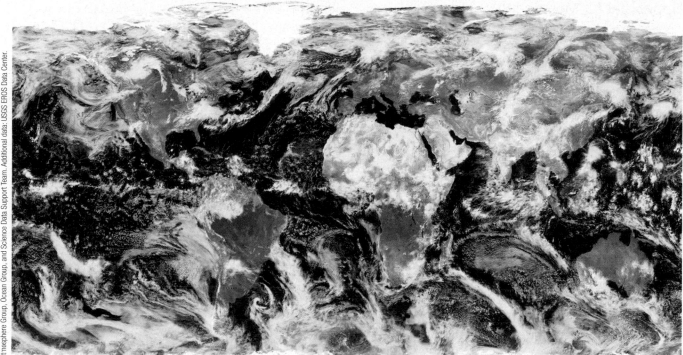

■ **FIGURE 3.21**   This composite of several satellite images shows a variety of cloud cover and storm systems across Earth.

In general, which are the cloudiest latitude zones and which are the zones with the clearest skies?

Cloud Cover   The extent and density of cloud cover is another factor that affects the temperature of Earth's surface and the atmosphere. Weather satellites have shown that, at any time, clouds cover about 50% of Earth (■ Fig. 3.21). A heavy cloud cover reduces the amount of insolation a place receives, causing lower daytime temperatures on a cloudy day. In contrast, we also have the greenhouse effect, in which clouds, composed mainly of water droplets, absorb heat energy radiating from Earth. Clouds keep nighttime temperatures near Earth's surface warmer than they would otherwise be. The general effect of cloud cover is to moderate temperatures by lowering the maximum and raising the minimum temperatures. In other words, cloud cover makes for cooler days and warmer nights.

Differential Heating of Land and Water   For reasons we will later explain in detail, bodies of water heat and cool more slowly than the land. The air above Earth's surface is heated or cooled in part by the surface beneath it. Therefore, temperatures over bodies of water or on land subjected to ocean winds (**maritime** locations) tend to be more moderate than those of land-locked places at the same latitude. Thus, the greater the **continentality** of a location (the distance removed from a large body of water), the less its temperature pattern will be modified.

Reflection   The capacity of a surface to reflect the sun's energy is called its **albedo;** a surface with a high albedo has a high percentage of reflection. The more solar energy that

is reflected back into space by Earth's surface, the less that is absorbed for heating the atmosphere or the surface. Temperatures will be higher at a given location if its surface has a low albedo rather than a high albedo.

Snow and ice are good reflectors with an albedo of 90–95%. This means that only 5–10% of the incoming solar radiation is absorbed by snow and ice, as most of the solar energy is reflected away. This is one reason why glaciers on high mountains do not melt away in the summer or why there may still be snow on the ground on a sunny day in the spring. A forest has an albedo of only 10–15% (or 85–90% absorption), which is good for the trees because they need solar energy for photosynthesis. The albedo of cloud cover varies, from 40 to 80%, according to the thickness of the clouds. The high albedo of many clouds is why much solar radiation is reflected directly back into space by the atmosphere.

The albedo of water varies greatly, depending on the water depth and the angle of the sun's rays. If the angle of the sun's rays is high, smooth water will reflect little. In fact, if the sun is vertical over a calm ocean, the albedo will be only about 2%. However, a low sun angle, such as just before sunset, causes an albedo of more than 90% from the same ocean surface. Likewise, a snow surface in winter, when solar angles are low, can reflect up to 95% of the energy striking it; because of this, skiers must constantly be aware of the danger of severe sunburns and possible snow blindness from reflected solar radiation.

Horizontal Air Movement We have already seen that advection is the major mode of horizontal transfer of heat and energy over Earth's surface. Any movement of air due to the wind, whether on a large or small scale, can have a significant effect on the temperatures of a location. Thus, wind blowing from an ocean to land will generally bring cooler temperatures in summer and warmer temperatures in winter. Large parcels of air moving from polar regions into the middle latitudes can cause sharp drops in temperature, whereas air moving poleward will usually bring warmer temperatures.

Vertical Distribution of Temperature

Environmental Lapse Rates We have learned that Earth's atmosphere is primarily heated from the ground up as a result of longwave terrestrial radiation, conduction, and convection. Thus, temperatures in the troposphere are usually highest at ground level and decrease with increasing altitude. As noted earlier in the chapter, this decrease in the free air of approximately 6.5° C per 1000 meters (3.6° F/1000 ft) is known as the environmental lapse rate.

The lapse rate at a particular place can vary for a variety of reasons. Lower lapse rates can exist if denser and colder air flows into a valley from a higher elevation or if advectional winds bring air in from a cooler region at the same altitude. In each case, the air near the surface is cooled. In contrast, if the surface is heated strongly on a hot summer afternoon, the air near Earth will be disproportionately warm, and the lapse rate will increase. Fluctuations in lapse rates due to abnormal temperature conditions at various altitudes can play an important role in the weather a place may have on a given day.

Temperature Inversions Under certain circumstances, the normal observed decrease of temperature with increased altitude might be reversed; temperature may actually *increase* for several hundred meters. This is called a **temperature inversion.**

Some inversions take place 1000–2000 meters (3280–6560 ft) above the surface of Earth where a layer of warmer air interrupts the normal decrease in temperature with altitude (■ Fig. 3.22). Such inversions tend to stabilize the air, causing less turbulence and discouraging both precipitation and the development of storms. Upper air inversions may occur when air settles slowly from the upper atmosphere. Air is compressed as it sinks, rising in temperature, and becoming more stable and less buoyant. Inversions caused by descending air are common at about 30–35° north and south latitudes.

An upper air inversion common to the coastal area of California results when cool marine air blowing in from the Pacific Ocean moves under stable, warmer, and lighter air aloft created by subsidence and compression. An inversion layer tends to maintain itself; that is, the cold underlying air is heavier and cannot rise through the warmer air above. Not only does the cold air resist rising or moving, but pollutants, such as smoke, dust particles, and automobile exhaust, created at Earth's surface, also fail to disperse. They will accumulate in the lower atmosphere, below the inversion layer. This situation is particularly acute in the Los Angeles area, which is a basin surrounded by higher mountainous areas (■ Fig. 3.23). Cooler air blows into the basin from the ocean and then cannot escape horizontally because of the landform barriers, or vertically, because of the inversion.

Some of the most noticeable temperature inversions are those that occur near the surface when Earth cools the lowest layer of air through conduction and radiation (■ Fig. 3.24).

■ **FIGURE 3.22** (Left) Temperature inversion caused by subsidence of air. (Right) Lapse rate associated with the column of air (A) in the left-hand drawing.
Why is the pattern (to the right) called a temperature inversion?

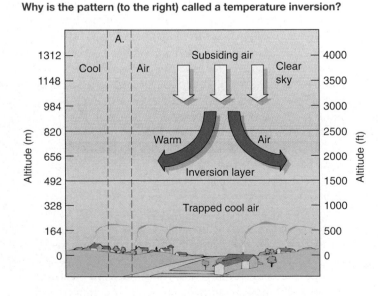

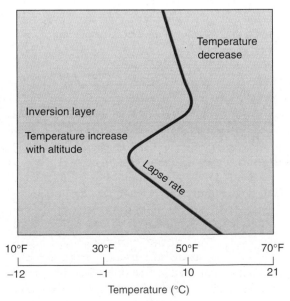

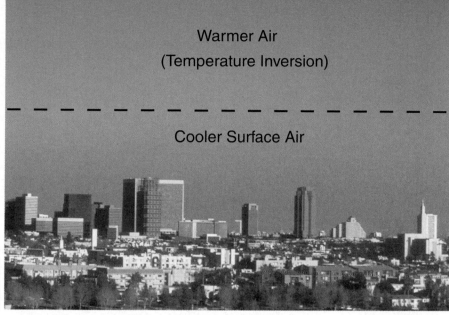

■ **FIGURE 3.23** Conditions producing smog-trapping inversion in the Los Angeles area.
Why is the air clear above the inversion?

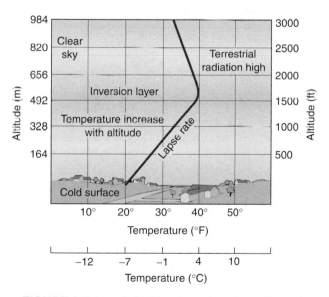

■ **FIGURE 3.24** Temperature inversion caused by the rapid cooling of the air above the cold surface of Earth at night.
What is the significance of an inversion?

In this situation, the coldest air is nearest the surface and the temperature rises with altitude. Inversions near the surface most often occur on clear, cold nights in the middle latitudes. Snow cover or the recent advection of cool, dry air into the area can enhance inversions. Such conditions produce rapid cooling of Earth's surface at night through radiation of heat energy gained during the daytime. Then the layers of the atmosphere that are closest to Earth are cooled by conduction, leaving warmer air aloft. Calm air conditions near the surface

help produce and partially result from these temperature inversions.

Surface Inversions: Fog and Frost Fog and frost will be discussed in detail in Chapter 5, but for now you should understand that they often occur as the result of a surface inversion. Especially where the land surface is hilly, cold, dense surface air will tend to flow downslope and accumulate in the valleys. The colder air on the valley floors and other low-lying areas sometimes produces fog or, if it is cold enough, a frost. Farmers use a variety of methods to prevent such frosts from destroying their crops. For example, fruit trees in California are often planted on the warmer hillsides instead of in the valleys. Farmers may also put blankets of straw, cloth, or some other insulation over their plants. This prevents the escape of Earth's heat radiation to space, keeping the plants warmer. Large fans and helicopters are sometimes used in an effort to mix the air layers and break up the inversion (■ Fig. 3.25). Huge orchard heaters that warm the air can also be used to disturb the temperature layers.

Controls of Earth's Surface Temperatures

Variations in temperatures over Earth's surface are caused by several *controls*. The major controls are (1) latitude, (2) land and water distribution, (3) ocean currents, (4) altitude, (5) landform barriers, and (6) human activity.

Latitude Latitude is the most important control of temperature variation involved in weather and climate. Recall that there are distinct patterns in the latitudinal distribution of the seasonal and annual receipt of solar energy. These variations in receipt of solar energy have a direct effect on temperatures. In general, annual insolation tends to decrease from lower latitudes to higher latitudes (see again Fig. 3.4). Table 3.3 shows the average annual temperatures for several locations in the Northern Hemisphere. We can see that, responding to insolation (with one exception), a poleward decrease in temperature exists for these locations. The exception is near the equator. Because of the heavy cloud cover in equatorial regions, annual temperatures there tend to be lower than at places slightly to the north or south, where skies are clearer.

Land and Water Distribution The world's oceans and seas are storehouses of water for the Earth system, but they also store tremendous amounts of heat energy.

© R. Sager

■ **FIGURE 3.25**   Propellers mix the air, breaking up inversions to protect these apple orchards in Washington from frost.

TABLE 3.3
Average Annual Temperature

Location	Latitude	(°C)	(°F)
Libreville, Gabon	0°23'N	26.5	80
Ciudad Bolivar, Venezuela	8°19'N	27.5	82
Bombay, India	8°58'N	26.5	80
Amoy, China	24°26'N	22.0	72
Raleigh, North Carolina	35°50'N	18.0	66
Bordeaux, France	44°50'N	12.5	55
Goose Bay, Labrador, Canada	53°19'N	−1.0	31
Markova, Russia	64°45'N	−9.0	15
Point Barrow, Alaska	71°18'N	−12.0	10
Mould Bay, NWT, Canada	76°17'N	−17.5	0

The widespread distribution of the oceans makes them an important atmospheric control in modifying the atmospheric elements. Different substances heat and cool at different rates. Land heats and cools faster than water. There are three main reasons for this phenomenon. First, the *specific heat* of water is greater than that of land. Specific heat refers to the amount of heat necessary to raise the temperature of 1 gram of any substance 1° C. Water, with a specific heat of 1 calorie/gram degree C, must absorb more heat energy than land with specific heat values of about 0.2 calories/gram degree C, to be raised the same number of degrees in temperature.

Second, water is *transparent* and solar energy can penetrate through the surface into the layers below, whereas with *opaque* materials like soil and rock, solar energy is concentrated on the surface. Thus, a given unit of heat energy will spread through a greater volume of water than land. Third, because liquid water circulates and mixes, it can transfer heat to deeper layers within its mass. The result is that as summer changes to winter, the land cools more rapidly than bodies of water, and as winter becomes summer, the land heats more rapidly. Because the air gets much of its heat from the surface, the differential heating of land and water surfaces produces inequalities in the air temperature above these two surfaces.

Not only do water and land heat and cool at different rates, but so do various land surface materials. Soil, forest, grass, and rock surfaces all heat and cool differentially and thus vary the temperatures of the overlying air.

Ocean Currents   Surface ocean currents are large movements of water driven by the winds, and affected by many other processes. They may flow from a place of warm temperatures to one of cooler temperatures and vice versa. These movements result from the attempt of Earth systems to reach a balance; in this instance, a balance of temperature and density.

Earth rotation affects the movements of the winds, which in turn affect the movement of the ocean currents. In general, ocean currents move in a circular, clockwise direction in the Northern Hemisphere and in a counterclockwise direction in the Southern Hemisphere (■ Fig. 3.26). Because the ocean temperature greatly affects the temperature of the air above it, an ocean current that moves warm equatorial water toward the poles (a warm current), or cold polar water toward the equator (a cold current) can significantly modify the air temperatures of those locations. If the currents pass close to land and are accompanied by ocean breezes, they can have a significant impact on the coastal climate.

The Gulf Stream, with its extension, the North Atlantic Drift, is an example of an ocean current that moves warm water poleward. This warm water keeps the coasts of Great Britain, Iceland, and Norway ice free in wintertime and moderates the climates of nearby land areas (■ Fig. 3.27). We can see the effects of the Gulf Stream if we compare the winter conditions of the British Isles with those of Labrador in northeastern Canada. Though both are at the same latitude, the average temperature in Glasgow, Scotland, in January is 4° C (39° F), while during the same month it is −21.5° C (−7° F) in Nain, Labrador.

The California Current off the United States West Coast helps moderate the climate of that coastal region as it brings cold water south. As the current swings southwest away from

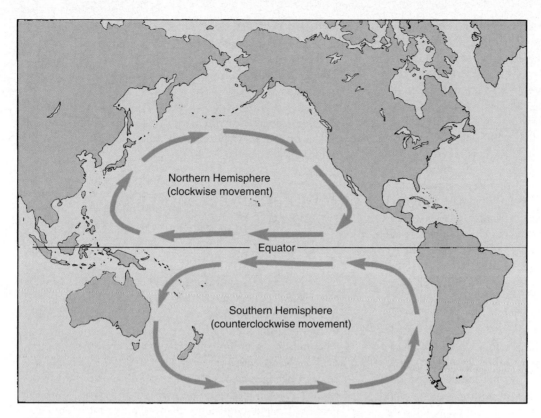

■ **FIGURE 3.26**   A simplified map of currents in the Pacific Ocean shows their basic rotary pattern. Major currents move clockwise in the Northern Hemisphere and counterclockwise in the Southern Hemisphere. A similar pattern exists in the Atlantic.

What direction might a hurricane forming off western Africa take as it approached the United States?

■ **FIGURE 3.27**   The Gulf Stream (the North Atlantic Drift farther north and eastward) is a warm current that moderates the climate of northern Europe.

Use this figure and the information gained in Figure 3.26 to discuss the route sailing ships would follow from the United States to England and back.

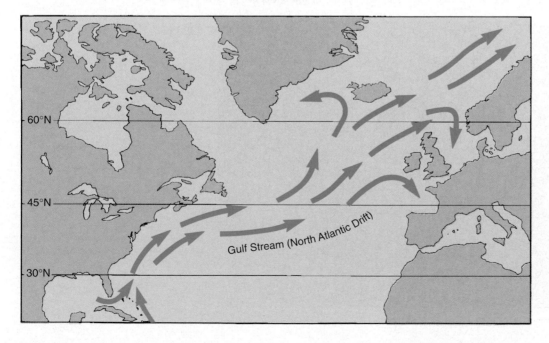

the coast, cold bottom water is drawn to the surface causing further chilling of the air above. San Francisco's cool summers (July Average: 14° C, or 58° F) show the effect of this current.

Altitude

As we have seen, temperatures within the troposphere decrease with increasing altitude. In Southern California, you can find snow for skiing if you go to an altitude of 2400–3000 meters (8000–10,000 ft) in winter. Mount Kenya, 5199 meters (17,058 ft) high and located at the equator, is still cold enough to have glaciers. Anyone who has hiked upward 500, 1000, or 1500 meters in midsummer has experienced a decline in temperature with increasing altitude. Even if it is hot in the valley below, you may need a sweater once you climb a few thousand meters (■ Fig. 3.28). The city of Quito, Ecuador, only 1° south of the equator, has an average temperature of only 13° C (55° F) because it is located at an altitude of about 2900 meters (9500 ft). Altitude as a factor will be discussed again when dealing with highland climates in Chapter 8.

Landform Barriers

Landform barriers, especially large mountain ranges, can block air movement from one place to another and thus affect the temperatures of an area. For example, the Himalayas keep cold, wintertime Asiatic air out of India, giving much of the Indian subcontinent a year-round tropical climate. Mountain orientation can create some significant differences as well. In North America, for example, southern slopes face the sun and tend to be warmer than the shady north-facing slopes. Snowcaps on the south-facing slopes may have less snow and may exist at a higher elevation. North-facing slopes usually have more snow, and it extends to lower elevations.

Human Activities

Deforestation, draining swamps, or creating large reservoirs are human activities that can significantly affect local climatic patterns and, possibly, global temperature patterns as well. The building and expansion of cities around the world have created pockets of warm temperatures that are known as *urban heat islands*. In each of these examples, human activities have changed the surface landscape and surface cover, which affect the surface *albedo* and available moisture for *latent heat exchanges*.

Temperature Distribution at Earth's Surface

To display the distribution of surface temperatures on a map, we use **isotherms.** Isotherms (from Greek: *isos*, equal; *therm*, heat) are lines on a map that connect points of equal temperature. When constructing isothermal maps showing temperature distribution, we need to account for elevation by adjusting temperature readings to what they would be at sea level. This adjustment means adding 6.5° C for every 1000 meters of elevation (the environmental lapse rate). The rate of temperature change on an isothermal map is called the *temperature gradient*. Closely spaced isotherms indicate a steep temperature gradient (a rapid temperature change over a shorter distance), and widely spaced lines indicate a weak one (a slight temperature change over a longer distance).

■ Figure 3.29a and 3.29b show the horizontal distribution of temperatures for the world during January and July, when the seasonal extremes of high and low temperatures are most obvious in the Northern and Southern Hemispheres.

■ **FIGURE 3.28**   Snow-capped mountains show the visual evidence that temperatures decrease with altitude. This mountain is in Grand Teton National Park in Wyoming, named after the dramatic range of jagged peaks, such as this one.
At what rate per 1000 meters do temperatures decrease with height in the troposphere?

National Park Service (NPS Photo)

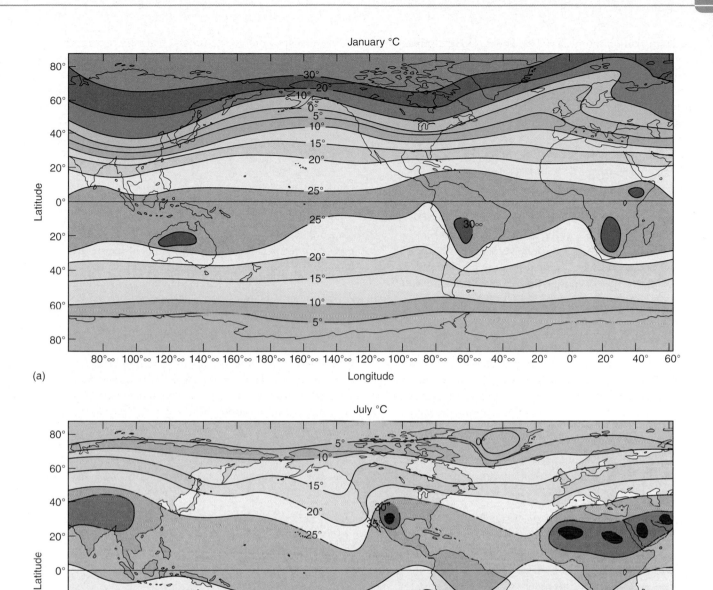

■ **FIGURE 3.29** (a) Average sea-level temperatures in January (°C). (b) Average sea-level temperatures in July (°C).

Observe the temperature gradients between the equator and northern Canada in January and July. Which is greater and why?

The easiest feature to recognize on both maps is the general orientation of the isotherms; they run nearly east–west around Earth, as do the parallels of latitude.

A more detailed study of Figures 3.29a, b and a comparison of the two maps reveal several important features. The highest temperatures in January are in the Southern Hemisphere; in July, they are in the Northern Hemisphere. Comparing the latitudes of Portugal and southern Australia can demonstrate this point. Note on the July map that Portugal in the Northern Hemisphere is nearly on the 20° C isotherm, whereas in southern Australia in the Southern Hemisphere the average July temperature is around 10° C, even though the two locations are approximately the same distance from the equator. The temperature differences between the two hemispheres are again a product of insolation, this time changing as the sun shifts north and south across the equator between its positions at the two solstices.

Note that the greatest deviation from the east–west trend of temperatures occurs where the isotherms leave large landmasses to cross the oceans. As the isotherms leave the land, they usually bend rather sharply toward the pole in the hemisphere experiencing winter and toward the equator in the summer hemisphere. This pattern in the isotherms is a direct reaction to the differential heating and cooling of land and water. The continents are hotter than the oceans in the summer and colder in the winter. Other interesting features on the January and July maps can be mentioned briefly. The isotherms poleward of 40° latitude are much more regular in their east–west orientation in the Southern than in the Northern Hemisphere. This is because in the Southern Hemisphere (often called the "water hemisphere") there is little land south of 40°S latitude to produce land and water contrasts. Note also that the temperature gradients are much steeper in winter than in summer in both hemispheres. The reason for this can be understood when you recall that the tropical zones have high temperatures throughout the year, whereas the polar zones have large seasonal differences. Hence, the difference in temperature between tropical and polar zones is much greater in winter than in summer.

As a final point, observe the especially sharp swing of the isotherms off the coasts of eastern North America, southwestern South America, and southwestern Africa in January and off Southern California in July. In these locations, the normal bending of the isotherms due to land-water temperature differences is increased by the presence of warm or cool ocean currents.

Annual March of Temperatures

Isothermal maps are commonly plotted for January and July because there is a lag of about 30–40 days from the solstices, when the amount of insolation is at a minimum or maximum (depending on the hemisphere), to the time of minimum or maximum temperatures. This **annual lag of temperature** behind insolation is similar to the daily lag of temperature. It is a result of the changing relationship between incoming solar radiation and outgoing Earth radiation.

Temperatures continue to rise for a month or more after the summer solstice because insolation continues to exceed Earth's radiation loss. Temperatures continue to fall after the winter solstice until the increase in insolation finally matches Earth's radiation. In short, the lag exists because it takes time for Earth to heat or cool and for those temperature changes to be transferred to the atmosphere.

The annual changes of temperature for a location can be plotted in a graph. The mean temperature for each month in a place such as Peoria, Illinois, or Sydney, Australia, is recorded and a line drawn to connect the 12 temperatures (■ Fig. 3.30).

■ **FIGURE 3.30** The annual march of temperature at Peoria, Illinois and Sydney, Australia.
Why do these two locations have opposite temperature curves?

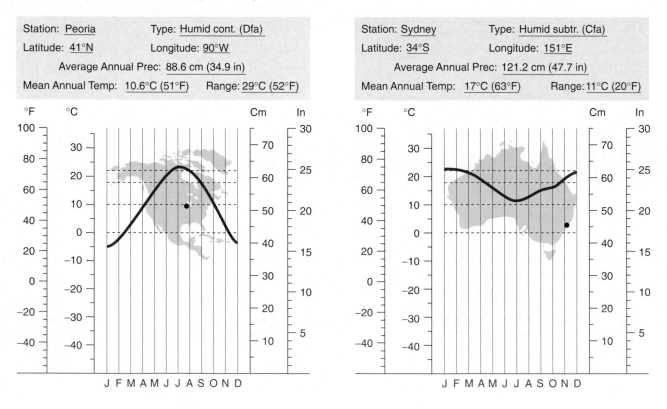

The mean monthly temperature is the average of the daily mean temperatures recorded at a weather station during a month. The daily mean temperature is the average of a 24-hour day's high and low temperatures. The curve that connects the 12 monthly temperatures depicts the **annual march of temperature** and shows changes in solar radiation as reflected by temperature changes over the year that result from seasonal variations in solar radiation.

This chapter has been designed to show the variations of Earth's energy systems and the nature of Earth's dynamic energy balances. These variations result from complex interrelationships between the characteristics of Earth and its atmosphere, and the energy gained and lost by the planet's environments. The variations are both horizontal across the surface and vertical in our atmosphere. Further, they vary both daily and seasonally.

Variations of Earth's energy systems impose both diurnal and annual rhythms on our agricultural activities, recreational pursuits, clothing styles, architecture, and energy bills. Human activities are constantly influenced by temperature changes, which reflect the input-output patterns of Earth's energy systems.

:: Terms for Review

weather
meteorology
climate
climatology
electromagnetic energy
shortwave radiation
longwave radiation
calorie
solar constant
insolation
solstice
equinox
Arctic Circle
Antarctic Circle
vertical (direct) rays

Tropic of Cancer
Tropic of Capricorn
declination
photosynthesis
greenhouse effect
ozone
troposphere
environmental lapse rate (normal lapse rate)
stratosphere
latent heat of fusion
latent heat of evaporation
latent heat of condensation
latent heat of sublimation
radiation

conduction
convection
advection
heat
temperature
Fahrenheit scale
Celsius (centigrade) scale
daily march of temperature
maritime
continentality
albedo
temperature inversion
isotherm
annual lag of temperature
annual march of temperature

:: Questions for Review

1. List the five basic elements of weather and climate. Which is most important and why is this so?
2. The electromagnetic spectrum consists of various types of energy by their wavelengths. Where is the division between longwave and shortwave energy?
3. Identify the two major factors that cause regular variation in insolation throughout the year. How do they combine to cause the seasons?
4. What processes in the atmosphere prevent insolation from reaching Earth's surface? What percentages of insolation do reach Earth's surface?
5. What processes transfer heat from Earth's surface to the atmosphere? Why is water so important in energy exchange?
6. What is meant by Earth's heat energy budget and how does it stay in balance?
7. What are some causes of short-term temperature variation within a local area?
8. Give several reasons why temperature inversions occur.
9. List the important controls of temperature variation and distribution over Earth's surface. Which control is most important and why is this so?
10. What factors cause the greatest deviations from an east-west trend in the isotherms on Figures 3.29a and 3.29b? What factors cause the greatest differences between January and July maps?

:: Practical Applications

1. Use the analemma presented in Figure 3.7 to determine the latitude where the noon sun will be directly overhead on February 12, July 30, November 2, and December 30.

2. Imagine you are at the equator on March 21. The noon sun would be directly overhead. However, for every degree of latitude that you travel to the north or south, the noon solar angle would decrease by the same amount. For example, if you travel to 40°N latitude, the solar angle would be 50°.

    a. Explain this relationship.
    b. Develop a formula or set of instructions to generalize this relationship.
    c. What would be the noon solar angle at 40°N on June 21? On December 21?

3. With respect to incoming solar radiation, how are albedo and absorption related? Develop a mathematical relationship between these two processes. If the albedo of a grassy lawn is 23% and that of a blacktop driveway is 4%, what is the difference in the absorption between these two surfaces?

Atmospheric Pressure, Winds, and Circulation

:: Outline

The swirling circulation patterns in Earth's atmosphere are created by changes in pressure and winds.

NASA/GSFC

:: Objectives

When you complete this chapter you should be able to:

- Explain why atmospheric pressure declines with altitude, and generally varies with latitude because of air temperature differences and vertical air movement.
- Associate precipitation, clouds, and windy conditions with the rising air of low pressure systems, and clear, calm conditions with the subsiding air of high pressure systems.
- Explain why latitudinal changes in sun angle and seasonal changes in daylight hours cause temperature variations that influence atmospheric pressure.

- Provide examples of how differences in atmospheric pressure affect wind velocity and direction.
- Understand why the Coriolis effect apparently causes winds and ocean currents to bend to the right of the direction of motion in the Northern Hemisphere and to the left in the Southern Hemisphere.
- Outline the major latitudinal pressure systems and wind belts and their influence on the circulation of global winds and ocean currents.
- Discuss examples of how the jet stream winds and interchanges between the atmosphere and oceans influence weather systems.

An individual gas molecule in the atmosphere weighs almost nothing. So it may be surprising to learn that as huge numbers of air molecules collide with anything, they exert an average pressure of 1034 grams per square centimeter (14.7 lb/sq in) at sea level. The reason why people are not crushed by this atmospheric pressure is that the air and water inside us—in our blood, tissues, and cells—exerts an equal outward pressure that balances the atmospheric pressure.

Atmospheric pressure variations exert a major influence on our weather and climate. Differences in *atmospheric pressure* circulate the air, and create our *winds,* which cycle water from the oceans to the landmasses, and are an important factor in driving the world's ocean currents. Wind movement disperses seeds and pollen in the biosphere, and transports dust, soil particles, and sand.

In 1643, Evangelista Torricelli, a student of Galileo, performed an experiment that was the basis for inventing the mercury barometer, an instrument that measures atmospheric (also called *barometric*) pressure. Torricelli filled a long glass tube, closed on one end, with mercury and inverted it in an open pan of mercury. The mercury inside the tube fell until it was at a height of about 76 centimeters (29.92 in) above the mercury in the pan, leaving a vacuum bubble at the closed, upper end of the tube. The pressure exerted by the atmosphere on the mercury in the open pan was equal to the pressure from the mercury trying to drain from the tube. As the atmospheric pressure increased, it pushed the mercury to a higher level in the tube, and as the air pressure decreased, the mercury level in the column dropped proportionately.

Variations in Atmospheric Pressure

In the strictest sense, a mercury barometer does not actually measure the pressure exerted by the atmosphere, but instead measures the response to atmospheric pressure. That is, when the atmosphere exerts a specific pressure, the mercury will

respond by rising to a specific height (■ Fig. 4.1). Meteorologists typically work with actual pressure units, most often the millibar (mb). **Standard sea-level pressure** is equal to 1013.2 millibars and will support a level of 76 centimeters (29.92 in) of mercury in a barometer. These values are

■ **FIGURE 4.1** A simple mercury barometer. Standard sea-level pressure of 1013.2 millibars will cause the mercury to rise 76 centimeters (29.92 in) in the tube.

When air pressure increases, what happens to the mercury in the tube?

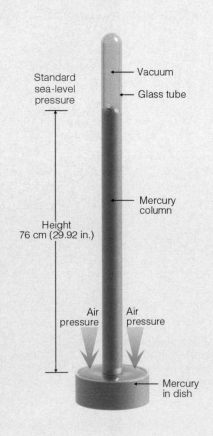

important, because they provide the cutoff measurements that are used to define areas (often called *cells*) of low pressure and areas of high pressure.

Air Pressure and Altitude

Air pressure decreases with increasing elevation on Earth and with altitudes above it, because the higher we go, the air molecules become more widely spaced and diffused. The increased space between gas molecules in the air results in lower air density and lower air pressure (■ Fig. 4.2). In fact, at the top of Mount Everest (elevation 8850 m, or 29,035 ft), the air pressure is only about one-third of the pressure at sea level.

People are usually not sensitive to small, gradual changes in air pressure. However, when we climb to high elevations or fly to altitudes significantly above sea level, we become aware of the effects of air pressure on our bodies. While flying at 10,000–12,000 meters (33,000–35,000 ft) commercial airliners are pressurized to maintain air pressure at safe levels for the passengers and crew. At cruising altitude, airliner cabins are typically pressurized to an equivalent pressure that would be experienced at an elevation of about 2100–2400 meters (7000–8000 ft). Still, pressurization may vary, so our ears may pop as they adjust to rapid pressure changes when ascending or descending. Hiking or skiing in locations that are a few thousand meters in elevation will affect us if we are used to sea-level pressure. Reduced air pressure means that there are fewer air molecules in a given volume of air and less oxygen will be contained in each breath. Thus, we get out of breath much more easily at high elevations until our bodies

adjust to the reduced air pressure and corresponding drop in oxygen level.

Altitude, however, is not the only factor that causes air pressure to change. On Earth's surface, variations in pressure are also related to the intensity of heating from insolation, local humidity, and global or regional air circulation. Changes in air pressure at a given location often indicate a change in the weather is occurring or coming.

Cells of High and Low Pressure

An area where the pressure is lower than standard sea-level pressure is often simply called a **low,** but a **cyclone** is also a general term for a low pressure area. A high pressure cell is called a **high,** or an **anticyclone.** Low and high pressure areas are often referred to as *cells*, and are represented by the capital letters **L** and **H** on weather maps that we often see on the Internet, on TV, and in the newspaper.

A low, or cyclone, is an area where air is rising. As air moves upward away from the surface, it relieves pressure from that surface. In this case, barometer readings will fall. The situation in a high, or anticyclone, is just the opposite. Under conditions of high pressure, air descends toward the surface and barometer readings will rise, indicating an increased atmospheric pressure on the surface. Lows and highs are illustrated in ■ Figure 4.3.

Winds blow toward the center of a cyclone, that is, they *converge on a low pressure cell.* The center of a low pressure system serves as the focus for **convergent wind circulation.** In contrast, the winds blow outward, and away from the center of an anticyclone, *diverging from the high pressure cell.* In a high pressure system, the center of the cell serves as the source for **divergent wind circulation.** Given existing pressure differences, we would expect that converging and diverging winds would move in straight paths as shown in Figure 4.3. Yet, wind motions are much more complex because they are influenced by other factors, which will be explained in the section on wind.

Horizontal Pressure Variations

There are two main causes of horizontal variations in air pressure. One is *thermal* (determined by air temperature) and the other is *dynamic* (related to atmospheric air motion).

Thermally-induced changes in air pressure are relatively easy to understand. Both air movement and air density are related to temperature differences that result from the unequal distribution of insolation, differential heating of land and water, and the varying albedos that exist on Earth's surface. A scientific law states that the pressure and density of a gas will vary inversely with its temperature. Thus, as daytime heating warms the air in contact with Earth's surface, the air expands in volume and decreases in density. As the density of the heated air decreases, it will rise and consequently lower the atmospheric pressure at the surface. Thermally-induced rising of warm air contributes to the typically low pressures that dominate the equatorial regions.

■ **FIGURE 4.2**   Both air pressure and air density decrease rapidly with increasing altitude.

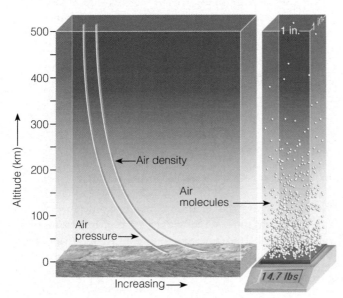

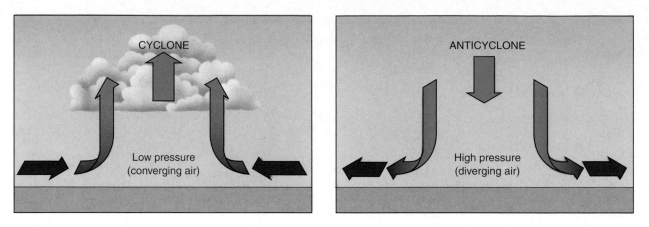

■ **FIGURE 4.3** Winds converge and ascend in cyclones (low pressure centers) and descend and diverge from anticyclones (high pressure centers).

How is temperature related to the density of air?

If air becomes cold, there will be an increase in density and a decrease in volume, which causes the air to sink, increasing the atmospheric pressure. For these reasons, polar regions regularly experience high pressure. The low pressure in the equatorial zone and the high pressures at the polar regions are considered thermally-induced, because the air temperature plays a dominant role in creating these pressure conditions.

Given this knowledge, we might expect that a gradual increase in air pressure would accompany the general decline in temperature from the equator to the poles. However, sea level barometer readings indicate that pressure does not increase in a regular pattern poleward from the equator. Instead, there are regions of high pressure in the subtropics, and low pressure in the subpolar regions, that develop through dynamic air movements.

The dynamic processes that affect air pressure are related to broad patterns of atmospheric circulation. For example, at the equator, when rising air encounters the tropopause it splits into two air currents flowing in opposite poleward directions (north and south). When these high level air currents reach the subtropical regions, they encounter similar air currents flowing equatorward from the middle latitudes. As these opposing upper-air currents merge, the air aloft will "pile up" and descend, producing high pressure in the subtropical regions.

At the surface in the subpolar regions, air flowing out of the polar high pressure zones encounters air flowing out of the subtropical regions. The collision of these merging winds causes the air to rise, creating a dynamically-induced zone of low pressure that is common in the subpolar regions. Both the subtropical high and subpolar low pressure regions are dominantly the results of dynamic air movement.

Mapping Pressure Distribution

Maps are the best medium for geographers and meteorologists when they analyze the existing and changing spatial patterns that influence our weather. But air pressure is also strongly influenced by elevation, in addition to the spatial variations in air pressure. When atmospheric pressure is mapped, or presented in weather broadcasts, the air pressure reported for every surface location is adjusted to reflect what it would be if that place was at sea level. Adjusting air pressure to its sea level equivalent is important because the variations due to altitude are far greater than those caused by changes in the weather or in the air temperature. Without this adjustment, elevation differences would mask the regional differences, which are more important to understanding the weather. For example, if meteorologists did not adjust barometer readings to the sea level equivalent, Denver, Colorado (the "Mile High City") would always be reporting low pressure conditions, yet the barometric pressure there varies up and down, just like it does at every other place.

Isobars (from Greek: *isos*, equal; *baros*, weight) are lines drawn on maps that connect points with equal values of air pressure. When isobars are closely spaced, they portray a significant difference in pressure over a short distance, hence a strong **pressure gradient.** Isobars that are widely spaced indicate a weak pressure gradient. On weather maps that show variations in atmospheric pressure, centers of high or low pressure are outlined by roughly concentric isobars, which form a closed system of isobars around those cells. In map view, cells of high and low pressure vary in shape from roughly circular to elongated.

Wind

Pressure Gradients and Wind

Wind is the movement of air in response to differences in atmospheric pressure. Winds are the atmosphere's means of attempting to balance uneven pressure distributions and they vary in velocity, duration, and direction. The velocity and strength of a wind depends on the intensity of the pressure gradient that produces the wind. As we noted previously, the

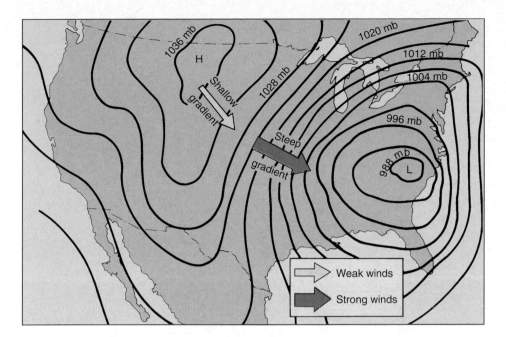

■ **FIGURE 4.4** The relationship of wind to the pressure gradient: The steeper the pressure gradient, the stronger will be the resulting wind.

Where else on this figure (other than the area indicated) would there be strong winds?

pressure gradient is the rate of change in atmospheric pressure between two points. When the pressure gradient is steep, with a large pressure change over a short distance, the winds will be fast and strong (■ Fig. 4.4). Winds tend to flow down a pressure gradient from high pressure to low pressure, just as water flows downslope from a high point to a low one. A useful little rhyme, "Winds always blow, from high to low," will remind you of the direction of winds. Yet, the wind does not generally flow in a straight line directly from high to low.

The wind also plays a major role in correcting the imbalances in radiational heating and cooling that occur on Earth. On the average, locations equatorward of 38° latitude receive more energy from the sun than they re-radiate directly back to space, whereas locations poleward of 38° radiate away more than they gain from solar energy (see again Fig. 3.18) Earth's planetary wind system transports energy poleward to help maintain a global energy balance. The wind system also strongly influences the ocean currents, which also transport great quantities of heat energy from areas that receive a surplus to regions where a deficit exists. Thus, without winds and ocean currents, the equatorial regions would continually get hotter and the polar regions continually colder through time.

In addition to the advectional (horizontal) transport of heat energy, winds also carry water vapor from the air above water bodies, where it has evaporated, to land surfaces where it condenses and precipitates. Without these winds, land areas would be arid and barren. In addition, winds influence evaporation rates. Furthermore, as we become more aware of our energy needs, harnessing the power of the wind is becoming more important, along with other natural energy sources such as solar and water power.

Wind Terminology

Winds are named after the direction or location that they come from. Thus, a wind that comes out of the northeast is called a *northeast wind*. A wind from the south, even though blowing toward the north, is called a *south* (or *southerly*) *wind*. It is helpful to use the phrase "out of" when describing a wind direction, to help people understand the correct direction, and avoid confusion about the origin of the wind.

The side of any object that faces the direction from which a wind is coming is called the **windward** side. Thus, a windward slope is the side of a mountain that the wind blows against (■ Fig. 4.5). **Leeward** means the sheltered, downwind side that faces in the direction the wind is blowing toward.

■ **FIGURE 4.5** Windward means facing into the wind and leeward means facing away from the wind.

How might vegetation differ on the windward and leeward sides of an island?

Thus, when the winds are coming out of the west, the leeward slope of a mountain would be the eastern slope. Although winds can blow from any direction, in some places, or during certain seasons at a particular place, winds may tend to regularly blow more from one direction than any other. These winds are referred to as **prevailing winds.**

The Coriolis Effect and Wind

Two factors that are related to Earth rotation greatly influence winds. First, our fixed-grid system of latitude and longitude is constantly rotating. Thus, our frame of reference for tracking the path of any free-moving object—whether it is an aircraft, a missile, an ocean current, or the wind—is constantly changing its position. Second, the Earth's rotational speed increases as we move equatorward and decreases as we move toward the poles. For example, someone in St. Petersburg, Russia (60° north latitude), where the distance around a parallel of latitude is about half of that at the equator, moves at about 830 kilometers per hour (519 mph) as Earth rotates, while someone in Kampala, Uganda, near the equator, moves at about 1660 kilometers per hour (1038 mph).

Because of Earth rotation, anything moving horizontally appears to be deflected to the right of its direction of travel in the Northern Hemisphere and to the left in the Southern Hemisphere. This apparent deflection is the **Coriolis effect.** The amount of deflection, or apparent curvature of the travel path, is a function of the object's speed and its latitudinal position. As the latitude increases, so does the impact of the Coriolis effect (■ Fig. 4.6). The Coriolis effect decreases at lower latitudes, and it has no impact along the equator. Also, as the distance of travel increases so does the apparent deflection of the travel path resulting from the Coriolis effect.

The flow of winds and ocean currents both experience this apparent Coriolis deflection. Winds in the Northern Hemisphere moving from high to low pressure are apparently deflected to the right of their expected path (and to the left in the Southern Hemisphere). In addition, when considering winds at Earth's surface, we must take into account another factor. **Friction** also interacts with the pressure gradient and the Coriolis effect.

At altitudes of about 1000 m or more above Earth's surface, frictional drag is of little consequence to the winds. At this level, with virtually no frictional drag, the wind will initially flow down the pressure gradient but then turn 90° in response to the Coriolis effect. When the Coriolis effect is counter-balanced by the pressure gradient force, the resulting wind, termed a **geostrophic wind,** will flow parallel to the isobars (■ Fig. 4.7).

At or near Earth's surface, where the wind encounters obstacles like trees, buildings, topography, and slower moving layers of air, frictional drag becomes an additional factor because it reduces the wind speed. A lower wind speed reduces the Coriolis effect, but the pressure gradient is not affected. With the pressure gradient and Coriolis effect no longer in balance, the wind does not flow between the isobars like its upper-level counterpart. Instead, a surface wind flows obliquely (at about a 30 degree angle) across the isobars and into an area of low pressure.

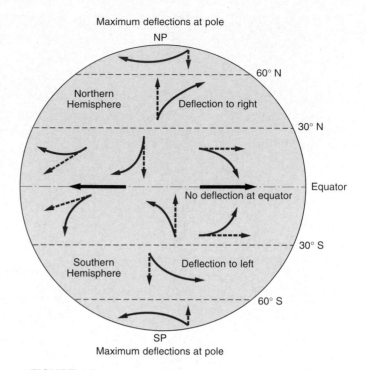

■ **FIGURE 4.6** Schematic illustration of the apparent deflection (Coriolis effect) caused by Earth's rotation when an object (or the wind) moves north, south, east, or west in both hemispheres. **If no Coriolis effect exists at the equator, where would the maximum Coriolis effect be located?**

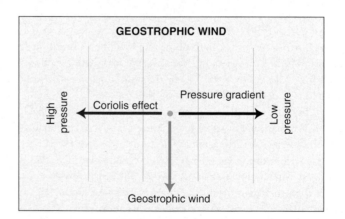

■ **FIGURE 4.7** This Northern Hemisphere example illustrates that in a geostrophic wind, the Coriolis effect causes it to veer to the right until the pressure gradient and Coriolis effect reach an equilibrium and the wind flows parallel to the isobars.

Cyclones, Anticyclones, and Wind Direction

Imagine a high pressure cell (anticyclone) in the Northern Hemisphere in which the air is moving outward from the center in all directions in response to the pressure gradient. As it moves, the air will be deflected to the right, no matter which direction it was originally going. Therefore, the winds moving out of an anticyclone in the Northern Hemisphere will move

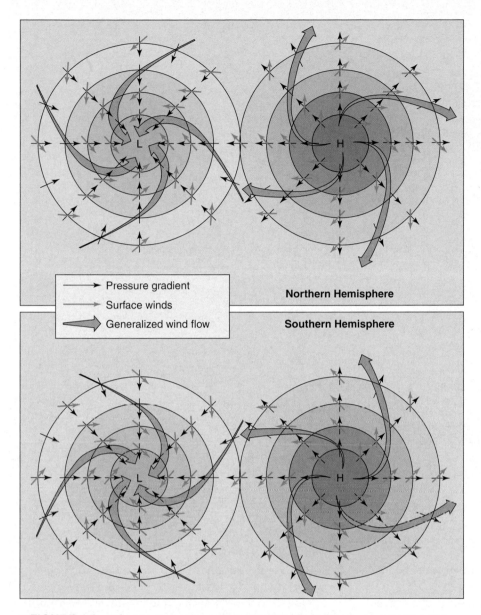

■ FIGURE 4.8   Movement of surface winds associated with low pressure centers and high pressure centers in the Northern and Southern Hemispheres.
What do you think might happen to the diverging air of an anticyclone if there is a cyclone nearby?

away from the center of high pressure in a clockwise spiral (■ Fig. 4.8).

In response to the pressure gradient, air from all directions tends to flow toward the center of a low pressure area (cyclone). Despite the fact that winds are apparently deflected to the right in the Northern Hemisphere, strong pressure gradients in a low pressure cell cause the winds to flow into the center of the low in a counterclockwise spiral. These spirals are reversed in direction in the Southern Hemisphere, where winds and currents are apparently deflected to the left. Thus, in the Southern Hemisphere, winds moving away from an anticyclone do so in a counterclockwise spiral, and winds moving into a cyclone move in a clockwise spiral.

Global Pressure and Wind Systems

A Model of Global Pressure

Using what we have learned about pressure on Earth's surface, we can understand a simplified model of the world's pressure belts (■ Fig. 4.9). Later, we see how actual conditions depart from this model and examine why these differences occur.

Centered approximately over the equator is a belt of low pressure, or a **trough.** This is the region on Earth with the greatest annual heating, so we can conclude that this low pressure area, the **equatorial low (equatorial trough),** is

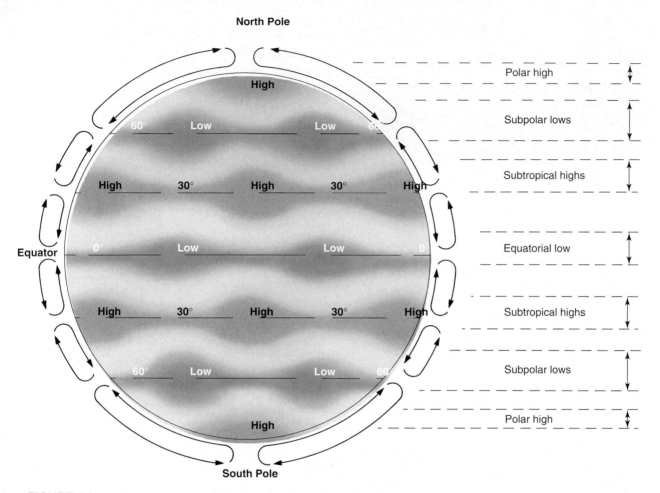

■ **FIGURE 4.9** Idealized world pressure belts. Note the arrows on the perimeter of the globe that illustrate the cross-sectional flow associated with the surface pressure belts.

Why do some of these pressure belts occur in pairs?

determined primarily by thermal factors, which cause the air to rise.

North and south of the equatorial low, centered on about 30°N and 30°S, cells of relatively high pressure dominate. These are the **subtropical highs,** which result from dynamic air motion related to the sinking of convectional cells initiated at the equatorial low.

Poleward of the subtropical highs in both hemispheres large belts of low pressure extend along the upper-middle latitudes, called the **subpolar lows.** Dynamic factors dominate the formation of the subpolar lows, as opposing winds collide and cause air to rise.

The Arctic and Antarctic regions are dominated by high pressure systems called the **polar highs.** Extremely cold temperatures and the consequent sinking of dense cold air create the higher pressures found in the polar regions.

This system of pressure belts is a generalized model, but it is still very useful for understanding global wind circulations, and dominant pressure patterns. Yet, both temperatures and atmospheric pressures change from month to month, day to day, or hour to hour at any location. This model of pressure belts does not reflect these smaller changes, but it does give a

general idea of the latitudinal patterns of surface atmospheric pressure that influence Earth's weather and climate regions.

Processes in the atmosphere tend to form latitudinal belts of high and low pressure, but the simplified model does not consider the alternating influences of ocean basins and continents that these belts traverse. In latitudes where continental landmasses are separated by ocean basins, pressure belts tend to break up to form cellular pressure systems. These cells of high and low pressure develop because the belts are affected by the differential heating of land and water. Landmasses also affect air movement and the development of pressure systems due to surface friction and air flowing up or down mountains.

Seasonal Variations in Pressure Distribution

In general, pressure belts shift northward in July and southward in January, following the migration of the sun's direct rays between the Tropics of Cancer and Capricorn. Thus, thermally-induced seasonal variations affect the pressure patterns, as seen in ■ Figure 4.10. Seasonal differences tend to be minimal at low latitudes, where little temperature variation

Average sea-level pressure (January)

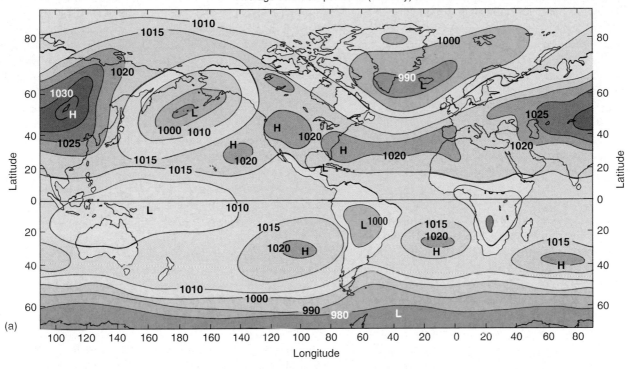

(a)

Average sea-level pressure (July)

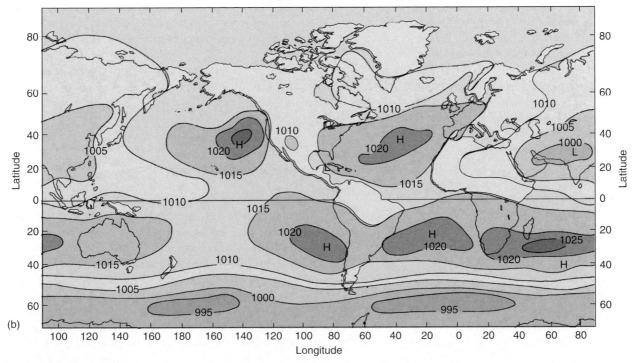

(b)

■ **FIGURE 4.10**   (a) Average sea-level pressure (in millibars) in January. (b) Average sea-level pressure (in millibars) in July.
What is the difference between the January and July average sea-level pressures at your location? Why do they vary?

occurs. At high latitudes, seasonal temperature differences are larger because of the greater annual variation in daylight length and angle of the sun's rays. Landmasses also alter the pattern of seasonal pressure variations for a particular latitude, especially in the Northern Hemisphere, where land accounts for 40% of the surface area, as opposed to less than 20% in the Southern Hemisphere.

January During the northern hemisphere winter, middle and high latitude continents become much colder than the surrounding oceans. Figure 4.10a shows that in the Northern Hemisphere this variation leads to the development of high pressure cells over the land areas. In contrast, the subpolar lows develop over the oceans because they are comparatively warmer. Over eastern Asia, there is a strongly developed anticyclone during the winter months, known as the **Siberian High.** Its equivalent in North America, the **Canadian High,** is not as strong because North America is smaller than the Eurasian continent.

Two low pressure centers also develop: the **Icelandic Low** in the North Atlantic**,** and the **Aleutian Low** in the North Pacific. These low pressure cells result from the clash of winds flowing out of the polar high to the north and from the subtropical highs to the south. The Aleutian and Icelandic lows are associated with cloudy, unstable weather and are a major source of winter storms, whereas Canadian and Siberian Highs are associated with clear, blue-sky days; calm, starry nights; and cold, stable weather. Therefore, during the winter months, cloudy and sometimes dangerously stormy weather tends to be associated with the two oceanic lows and clear, but cold, weather with the continental highs.

We can also see that in January the polar high in the Northern Hemisphere is well developed, primarily because of thermal cooling during the coldest time of the year. The subtropical highs of the Northern Hemisphere have moved slightly south of their average annual position, as the sun's rays migrate toward the Tropic of Capricorn. The equatorial low also shifts south of its average annual position, which is at the equator.

In the Southern Hemisphere during January (where it is summer), the subtropical high pressure belt breaks into three cells centered over the oceans because the warmer continents produce lower pressures compared to those over the oceans. Because there is virtually no land between 45°S and 70°S latitude, the subpolar low circles Earth over the Southern Ocean as an unbroken belt. There is little seasonal movement in this belt of low pressure other than in January, when it is located a few degrees equatorward of its July (winter) position.

July The high pressure over the North Pole weakens during the summer, primarily because of the lengthy (24-hour daylight) heating in that region (Fig. 4.10b). The Aleutian and Icelandic Lows also weaken and shift poleward from their winter position. North America and Eurasia, which developed high pressure cells during the cold winter months, develop extensive low pressure cells slightly to the south during the summer. The subtropical highs in the Northern Hemisphere are strong in the summer,

and they migrate poleward from their winter position. The North Pacific subtropical high is termed the **Pacific High (**or the **Hawaiian High),** a pressure system that greatly affects the climates of the west coast of North America. North Americans call the corresponding high pressure cell in the North Atlantic the **Bermuda High,** but it is the **Azores High** to Europeans and West Africans. The equatorial low moves north in July, following the seasonal migration of the sun's rays, and the subtropical highs of the Southern Hemisphere are equatorward of their January (summer) locations.

We have seen that there are essentially seven belts of pressure (two polar highs, two subpolar lows, two subtropical highs, and one equatorial low), which form into cells of pressure in latitudes where oceans are separated by large landmasses. These belts and cells vary in size, intensity, and locational shifting with the seasons following the migration of the sun's vertical rays. These global-scale pressure systems form a fairly regular latitudinal distribution but also migrate by latitude with the seasons, so they are sometimes referred to as *semipermanent pressure systems*.

Winds, the major means of transport for energy and moisture through the atmosphere, can also be examined on a global scale. However, for now, we will disregard the influences of land and water differences, variations in elevation, and seasonal changes. These factors will be addressed later. This simplification allows us to construct a basic model of the atmosphere's global circulation. This model can also help us explain specific climate features such as the rain and snow of the Sierra Nevada and Cascade Range and the arid regions directly to the east of those mountains. A basic wind model will provide insights for understanding the movement of ocean currents, which are driven, in part, by the global wind systems.

A Model of Atmospheric Circulation

Because winds are caused by pressure differences, a system of global winds can be based on the model of atmospheric pressures (see again Fig. 4.9). Convergence and divergence are very important to understanding global wind patterns. Knowing that surface winds flow out of highs and toward low pressure areas, we can use the global model of pressure belts to develop a global wind system model (■ Fig. 4.11). This model, although simplified, takes into account differential heating, Earth rotation, and atmospheric dynamics. The Coriolis effect is also considered, as winds will not blow along straight north–south paths. It is important to remember that winds are named after the direction from which they come.

The full idealized global model includes six wind belts in addition to the seven pressure zones. Two wind belts, one in each hemisphere, are located where winds blow from the polar highs toward the subpolar lows. As these winds are strongly deflected by the Coriolis effect to the right in the Northern Hemisphere and to the left in the Southern, they become the **polar easterlies.**

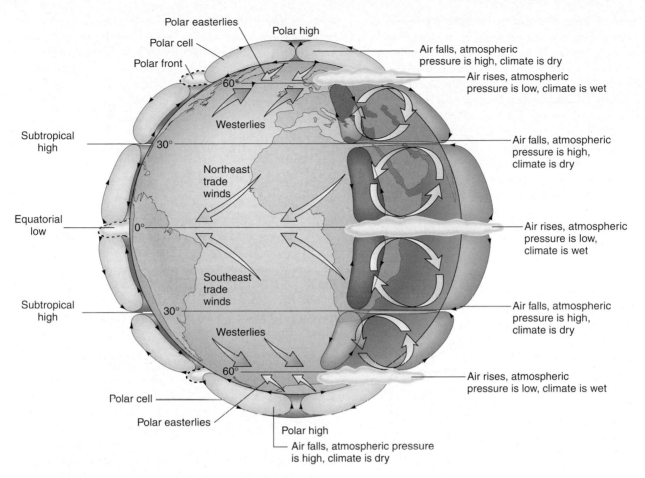

■ FIGURE 4.11 The general circulation of Earth's atmosphere.

Divergent circulation out of the subtropical highs, flowing both equatorward and poleward, provides the source for the remaining four wind belts. In each hemisphere, winds flow out of the poleward sides of the subtropical highs into the subpolar lows. Because of the great apparent bending of winds flowing from the higher latitude sides of the subtropical highs, the general wind movement is from the west. These winds of the upper-middle latitudes are the **westerlies.** The winds blowing from the subtropical highs toward the equator are the **trade winds.** Because the bending from the Coriolis effect in lower latitudes is minimal, they are the **northeast trades** in the Northern Hemisphere and the **southeast trades** south of the equator.

This model provides a basic concept of the global atmospheric circulation, although many other factors, both local and regional, also influence the winds. The pressure systems, and consequently the winds, move in response to heating differences that change with the seasonal position of the sun. Also continent and ocean differences between the Northern and Southern hemisphere affect high and low pressure zones, and thus the winds.

Conditions within Latitudinal Zones

Trade Winds
The trade winds blow out of the subtropical highs toward the equatorial low in both the Northern and Southern Hemispheres between latitudes 5° and 25°. Because

of the Coriolis effect, the northern trades move away from the subtropical high in a clockwise direction out of the northeast. In the Southern Hemisphere, the trades diverge out of the subtropical high toward the equatorial trough from the southeast, as their movement is counterclockwise. Because the trades from both hemispheres tend to blow out of the east, they are also known as the *tropical easterlies.*

The trade winds tend to be constant, steady, and consistent in their direction. The area of the trades varies somewhat during the seasons, moving a few degrees of latitude north and south with the sun. Near their source in the subtropical highs, the weather of the trades is clear and dry, but after crossing large expanses of ocean, the trades have a high potential for stormy weather. Early Spanish sailing ships depended on the northeast trade winds to drive their galleons from Europe to destinations in Central and South America in search of gold, spices, and new lands. Going eastward toward home, navigators usually tried to plot a course using the westerlies to the north.

The Intertropical Convergence Zone
The *equatorial low,* where the trade winds converge, coincides with the world's latitudinal belt of heaviest precipitation and most persistent cloud cover. This area is called the **intertropical convergence zone (ITCZ or ITC)** because it is where the

trade winds from the tropics of both hemispheres converge on the equatorial region. Here the air, which is very humid and heated by the sun, tends to expand and rise, maintaining the low pressure of the area and high potential for rainfall.

Close to the equator, roughly between 5°N and 5°S, the strongest belt of convergence is located, with rising air, heavy rainfall, and calm winds that have no prevailing direction. This zone is known as the *doldrums.* Sailing ships often remained becalmed in the doldrums for days. It is interesting to note that the word doldrums in English means a listless condition or a depressed state of mind. The sailors were in the doldrums in more ways than one.

Subtropical Highs The areas of subtropical high pressure, generally located between latitudes 25° and 35°N and S, are the source of winds blowing poleward as the westerlies and equatorward as the trade winds. The high pressure results from air sinking and settling from higher altitudes. These subtropical belts of variable or calm winds have been called the "horse latitudes." This name comes from the occasional need by sailors to eat their horses or throw them overboard in order to conserve drinking water and lighten the weight when their sailing ships were becalmed in these latitudes. The centers of the subtropical highs are areas, like the doldrums, where there are no strong prevailing winds. Weather conditions are typically clear, sunny, dry, and rainless, especially over the eastern sides of the ocean basins (along west coasts) where the high pressure cells are strongest.

Westerlies The winds that flow poleward out of the subtropical highs in the Northern Hemisphere are strongly deflected to the right and thus blow from the southwest. Those in the Southern Hemisphere are strongly deflected to the left and blow out of the northwest. Thus, these winds are correctly labeled the westerlies. They tend to be less consistent in direction than the trade winds, but they are usually stronger winds and may be associated with stormy weather. The westerlies occur between about 35° and 65°N and S latitudes. In the Southern Hemisphere, very little land exists in these latitudes to affect the development of the westerlies so they blow with great consistency and strength. Much of western Europe, Canada, and most of the United States (except Florida, Hawaii, and northern Alaska) are influenced by weather brought by the westerlies.

Polar Winds Accurate observations of pressure and wind are sparse in the polar regions; therefore, we must rely on satellite imagery and data for much of our information. Pressures tend to be consistently high throughout the year at the poles. The polar highs feed prevailing winds that circle the polar regions and blow easterly toward the subpolar low pressure systems.

Despite our limited knowledge of the wind systems of the polar regions, we do know that the winds can be highly variable, blowing at times with great speed and intensity. When the cold air flowing out of the polar regions meets the warmer air of the westerlies, they do so like two warring armies: One does not absorb the other. Instead, the denser, heavier cold air pushes the

warm air upward, forcing it to rise. The line along which these two great wind systems battle is appropriately known as the *polar front,* basically the zone of the *subpolar low.* The weather that results from the meeting of cold polar air and warmer air from the subtropics can be very stormy. In fact, most of the storms that move slowly through the middle latitudes in the path of the prevailing westerlies have developed at the polar front.

Latitudinal Migration with the Seasons

Just as insolation, temperature, and pressure systems migrate north and south, Earth's wind systems also migrate with the seasons. During the summer months in the Northern Hemisphere, maximum insolation is received north of the equator. This condition causes the pressure belts to move north as well, and the wind belts of both hemispheres shift accordingly. Six months later, when maximum heating is taking place south of the equator, the wind systems have migrated south in response to the migration of the pressure systems. Thus, the seasonal migration of winds and pressure cells is an example of how actual atmospheric circulation differs from our idealized model.

The regions that are most strongly influenced by these seasonal migrations are the boundary zones between two wind or pressure systems. During the winter, these regions are subject to the influence of one system. As summer approaches, the system that dominated in the winter will migrate poleward and the equatorward system will move in to influence the region. Two of these boundary zones in each hemisphere experience these distinctive summer-to-winter fluctuations in climate. The first lies between latitudes 5° and 15°, where the wet equatorial low of the high-sun season (summer) alternates with the dry subtropical high of the low-sun season (winter). The second occurs between 30° and 40°, where the subtropical high dominates in summer but is replaced by the wetter westerlies and the polar front in winter.

California is an example of a region located within a zone of transition between two wind and pressure systems (■ Fig. 4.12). During the winter, this region is under the influence of the westerlies blowing out of the Pacific High. These winds, full of moisture from the ocean, bring winter rains and storms to "sunny" California. As summer approaches, however, the polar front and the westerlies move north. As California comes under the influence of the calm and steady subtropical high pressure system, it experiences the climate for which it is famous: day after day of warm and dry, clear, blue, skies. This alternation of moist winters and dry summers is typical of the western sides of all landmasses between 30° and 40° latitude.

Longitudinal Variation in Pressure and Wind

We have seen that there are important latitudinal differences, and shifts, in pressure and winds. Significant longitudinal variations also occur, especially in the region of the subtropical highs.

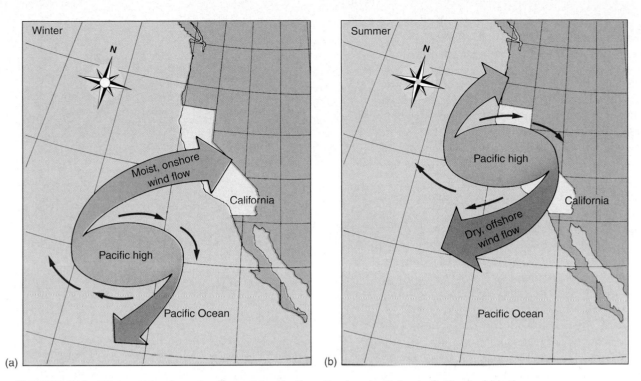

■ FIGURE 4.12   Winter and summer positions of the Pacific anticyclone in relation to California. (a) In the winter, the anticyclone lies to the south and feeds the westerlies that bring cyclonic storms and rain from the North Pacific to California. (b) In the summer, the anticyclone brings high pressure, with warm, sunny, and dry conditions.

In what ways would the seasonal migration of the Pacific anticyclone affect agriculture in California?

As was previously noted, the subtropical high pressure cells, which are generally centered over the oceans, are much stronger on their eastern sides than on their western sides. Thus, in the subtropics, over the eastern sides of ocean basins (west coasts of continents) subsidence and divergence are especially noticeable. The upper level temperature inversions, associated with the subtropical highs, develop air that is clear and calm. Air flowing equatorward from this eastern side of the high produces steady trade winds with clear, dry weather.

Over the western sides of the ocean basins (east coasts of the continents), conditions are markedly different. In its passage over the ocean, the diverging air is warmed and moistened; thus, turbulent and stormy weather conditions are likely to develop. As indicated in ■ Figure 4.13, wind movement in the western portions of the anticyclones, influenced by the Coriolis effect, bends poleward and toward landmasses. The trade winds in these areas are especially weak or nonexistent much of the year.

Figures 4.10 and 4.11 illustrate that there are great land–sea contrasts in temperature and pressure throughout the year in the higher latitudes, especially in the Northern Hemisphere. In the cold continental winters, the land is associated with pressures that are higher than those over the oceans, and thus there are strong, cold winds from the land to the sea. In the summer, the situation changes, with

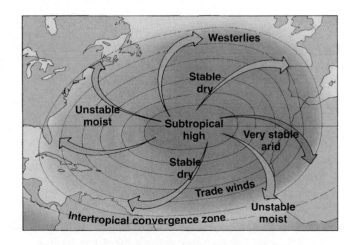

■ FIGURE 4.13   Circulation pattern in a Northern Hemisphere subtropical anticyclone. Subsidence of air is strongest in the eastern part of the anticyclone, producing calm air and arid conditions over adjacent land areas. The southern margin of the anticyclone feeds the northeast trade winds.

What wind system is fed by the northern margin?

relatively low pressure existing over the continents because of higher temperatures. Wind directions are thus greatly affected, and the pattern is reversed so that winds flow from the sea toward the land.

Upper Air Winds and Jet Streams

Thus far, we have examined the atmosphere's surface wind patterns, but upper level winds are also important—in particular, the winds at altitudes above 5000 meters (16,500 ft), and at higher levels in the upper troposphere. The formation, movement, and decay of cyclones and anticyclones in the middle latitudes depend to a great extent on flows of air in the atmosphere that are at high altitudes.

The upper air wind circulation is less complex compared to the circulation of surface winds. In the upper troposphere, an average westerly flow, the upper air westerlies, is maintained poleward of about 15°–20° latitude in both hemispheres. Because of the reduced frictional drag, the upper air westerlies blow much faster than their surface counterparts. Upper air winds became apparent during World War II when high-altitude bombers flying eastward covered distances faster than when they flew westward. Pilots had encountered the upper air westerlies, or perhaps even the **jet streams**—very strong air currents embedded within the upper air westerlies. The jet streams are high altitude examples of geostrophic winds, flowing parallel between isobars in response to a balance between the Coriolis effect and the pressure gradient.

The best known jet stream is the *polar front jet stream*, which flows in the tropopause, above the polar front, the area of the subpolar low. Ranging from 40–160 kilometers (25–100 mi) in width and up to 2 or 3 kilometers (1–2 mi) in depth, the polar front jet stream is a faster, internal current of air within the upper air westerlies. While the polar front jet stream flows over the middle latitudes, another westerly *subtropical jet stream* flows above the subtropical highs in the lower-middle latitudes (■ Fig. 4.14a). Figure 4.14b shows the position of these jet streams as they relate to the vertical and surface circulation of the atmosphere. Both jet streams are best developed in winter when temperatures exhibit their steepest gradient, and in the summer they weaken in intensity. In winter the subtropical jet stream frequently disappears completely and the polar front jet stream tends to migrate northward.

In general, the upper air westerlies and the associated polar front jet stream flow in a fairly smooth pattern (■ Fig. 4.15a). At times, however, the upper air westerlies develop a wave form, termed long waves, or **Rossby waves,** named after the Swedish meteorologist who discovered their existence (Fig. 4.15b). Rossby waves result in cold polar air pushing into the lower latitudes and forming troughs of low pressure, while warm tropical air moves into higher latitudes, forming ridges of high pressure. Eventually, the upper air Rossby waves become so elongated that "tongues" of air are cut off, forming warm and cold cells in the upper air (Fig. 4.15c and d). This process helps maintain a

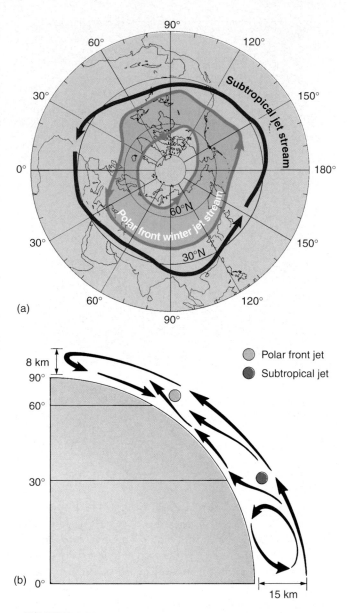

(a)

(b)

■ **FIGURE 4.14** (a) Approximate location of the subtropical jet stream and area of activity of the polar front jet stream (shaded) in the Northern Hemisphere winter. (b) A vertical schematic of atmospheric circulation and locations of the jet streams. **Which jet stream is most likely to affect your home state?**

net poleward flow of energy from equatorial and tropical areas. The cells eventually dissipate, and the normal pattern returns (Fig. 4.15a).

In addition to their influence on weather, jet streams are important for other reasons. They can carry pollutants, such as radioactive or volcanic dust, over great distances and at relatively rapid rates. The polar jet stream carried ash from the 1980 Mount St. Helens eruption in Washington state hundreds of kilometers eastward across the United States

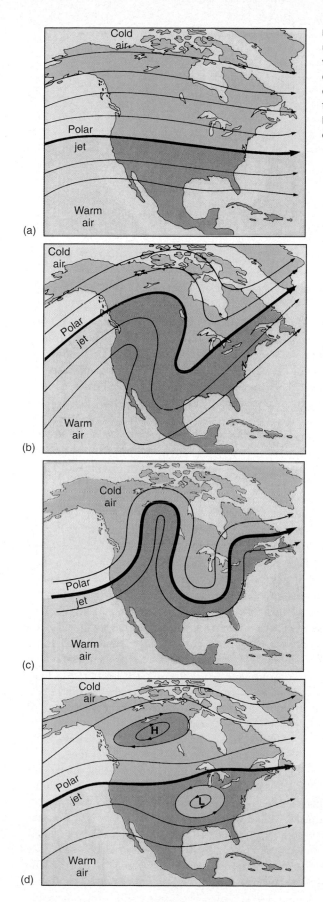

(a)

(b)

(c)

(d)

■ **FIGURE 4.15** Development and dissipation of Rossby waves in the upper air westerlies. (a) A fairly smooth flow prevails. (b) Rossby waves form, with a ridge of warm air extending into Canada and a trough of cold air extending down to Texas. (c) The trough and ridge may begin to turn back on themselves. (d) The trough and ridge are cut off and will dissipate. The flow will then return to a pattern similar to (a).

How are Rossby waves closely associated with the changeable weather of the central and eastern United States?

and Southern Canada. Nuclear fallout from the Chernobyl incident in the former Soviet Union was monitored for days as it crossed the Pacific Ocean and the United States, in the jet stream. Pilots flying eastward—for example, from North America to Europe—take advantage of the jet stream. Flying times with the wind are significantly shorter than when flying against this strong and fast wind.

Regional and Local Wind Systems

The global wind system illustrates general circulation patterns that reflect latitudinal imbalances in temperature. On a regional or local scale, additional wind systems develop in response to similar temperature conditions. Monsoon winds are an example that are sub-continental in size and develop in response to seasonal variations in temperature and pressure. Many regions experience differences in wind directions and conditions over the seasons. At the smallest scale are local winds, which develop in response to the diurnal (daily) variations in heating and other local effects upon pressure and winds.

Monsoon Winds

The term *monsoon* comes from the Arabic word *mausim,* meaning season. Arabic sailors have used this word for many centuries to describe seasonal changes in wind direction across the Arabian Sea between Arabia and India. As a meteorological term, **monsoon** refers to the directional reversal of winds from one season to the next. Usually, a monsoon occurs when humid winds from the ocean blow toward the land in the summer, but shift to dry, cooler winds in winter that blow seaward off the land. A monsoon typically involves a full 180° seasonal change in the wind direction.

The monsoon is most characteristic of southern Asia although it also occurs on other continents. In the winter, as Asia's giant landmass becomes much colder than the surrounding oceans, the continent develops a strong high pressure cell from which there is a strong outflow of air (■ Fig. 4.16). These cold dry winds blow southward from the continent toward the tropical low that exists over the warmer oceans. The winter monsoon is a dry season as air is coming from a dry continental area.

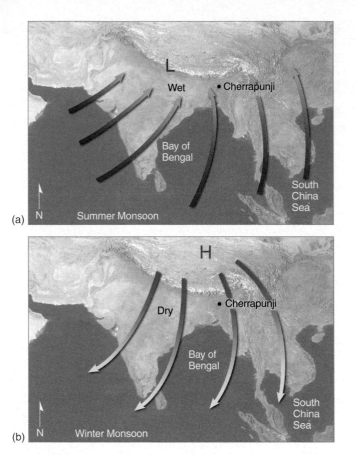

■ **FIGURE 4.16** Seasonal reversals of wind direction create the Asiatic monsoon system. (a) The "wet monsoon," with its onshore flow of tropical humid air in summer, is characterized by heavy precipitation. (b) The offshore flow of dry continental air in winter creates the "dry monsoon" and drought conditions in southern Asia.
How do the seasonal changes of wind direction in Asia differ from those of the southern United States?

In summer, the Asian continent becomes very warm and develops a large low pressure center that attracts warm, moist air from the oceans. Convective uplift or landform barriers make this hot humid air rise and cool to bring heavy precipitation. Extremely heavy rainfall and flooding can occur during the summer monsoon in the foothills of the Himalayas, in other parts of India, and in areas of Southeast Asia. Northern Australia is also a true monsoon region that experiences a full wind reversal from summer to winter. The southern United States and West Africa have "monsoonal tendencies," because of seasonal wind shifts, but do not experience monsoons in the true meaning of the term.

Local Winds

Despite affecting a much smaller area, local winds are also important. These local winds are often a response to local terrain, or land–water differences in heating and cooling, much like larger wind systems.

One type of local wind is known by several names in different parts of the world—for example, **chinook** in the

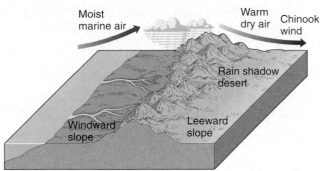

■ **FIGURE 4.17** Chinook (or Foehn) winds result when air descends a mountain barrier, and its relative humidity lowers as the air compresses and warms. This produces the relatively warm, dry conditions with which these winds are associated.
The term Chinook means "snow eater." Can you offer an explanation for how this name came about?

Rocky Mountain region and **foehn** (pronounced "fern") in the Alps. Chinook-type winds occur when air must pass over a mountain range. After crossing the mountains, as these winds flow down the leeward slope, the air is compressed and heated at a greater rate than it had cooled when it ascended the windward slope (■ Fig. 4.17). Thus, the air flows into the valley below as warm, dry winds. The rapid temperature rise brought about by such winds has been known to damage crops, increase forest-fire hazard, and trigger avalanches.

An especially hot and dry wind is the **Santa Ana** of Southern California. It forms when high pressure develops over the desert regions of Southern California and Nevada. The clockwise circulation out of the high drives warm dry air from the desert over the mountains of Eastern California, and the air becomes warmer and more arid as it moves down the western slopes. The hot, dry Santa Ana winds are notorious for fanning forest and brush fires, which plague the southwestern United States, especially in California.

Also known as **katabatic winds, drainage winds** are local to mountainous regions and occur under calm, clear conditions. Cold, dense air from a high plateau or mountainous area flows down valleys and pours onto the land below. Drainage winds can be extremely cold and strong, especially when they result from cold air emanating from the glaciers that cover Greenland and Antarctica.

The **land breeze–sea breeze** is a diurnal (daily) cycle of local winds that occurs in response to the differential heating of land and water (■ Fig. 4.18). During the day, the land—and the air above it—heats quickly to a higher temperature than an adjacent body of water (ocean, sea, or large lake), and the air over the land expands and rises. This process creates a local low pressure on the land, and the rising air is replaced by denser, cooler air from over the water. Thus, a cool, moist sea breeze blows in over the land during the day. The sea breeze is one reason why seashores are so popular in summer, because cooling winds alleviate the heat. These winds can cause a 5°C–9°C (9°F–16°F)

GEOGRAPHY'S SPATIAL PERSPECTIVE
:: THE SANTA ANA WINDS AND FIRE

Wildfires require three factors to occur: *oxygen, fuel,* and an *ignition source*. The conditions for all three factors vary geographically, so their spatial distributions are not equal everywhere. In locations where all three factors exist, the danger from wildfires is high. Oxygen in the atmosphere is constant, but winds, which supply more oxygen as a fire consumes it, vary with location, weather, and terrain. High winds spread fires rapidly and make them difficult to extinguish. Fuel for wildfires is usually supplied by dry vegetation (leaves, branches, and dry annual grasses). Certain environments have more of this fuel than others. Dense vegetation tends to support the spread of fires. Vegetation can also become dried out by transpiration losses during a drought or an annual dry season. Also, once a fire becomes large, extreme heat in areas where it is spreading causes vegetation along the edges of the fire to lose its moisture through evaporation. Ignition sources are the means by which a fire is started.

Lightning and human causes, such as campfires and trash fires, provide the main ignition sources for wildfires.

Southern California offers a regional example of how conditions combine with the local physical geography to create an environment that is conducive to wildfire hazard. This is also a region where many people live in forested or scrub-covered locales or along the urban–wildland fringe—areas that are very susceptible to fire. High pressure, warm weather, and low relative humidity dominate Southern California's Mediterranean climate for much of the year. This region experiences high fire potential because of the warm dry air and the vegetation that has dried out during the arid summer season.

The most dangerous circumstances for wildfires in Southern California occur when high winds are sweeping the region. When a strong cell of high pressure forms east of Southern California, the clockwise (anticyclonic) circulation directs winds from the north and east toward the coast. These

warm, dry winds (called Santa Ana winds) blow down from nearby high-desert regions, becoming adiabatically warmer and drier as they descend into the coastal lowlands. The Santa Ana wind speeds can be 50–90 kilometers per hour (30–50 mph) with stronger local wind gusts reaching 160 kilometers per hour (100 mph). Just like blowing on a campfire to get it started, the Santa Ana winds produce fire weather that can cause a wildfire to spread extremely rapidly. Most people take great care during these times to avoid or strictly control any activities that could cause a fire to start, but occasionally accidents, acts of arson, or lightning strikes ignite a wildfire.

Ironically, although the Santa Ana winds create dangerous fire conditions, they also provide some benefits because the winds tend to blow air pollutants offshore and out of the urban region. In addition, because they are strong winds flowing opposite to the direction of ocean waves, surfers can enjoy higher than normal waves when Santa Ana winds are blowing.

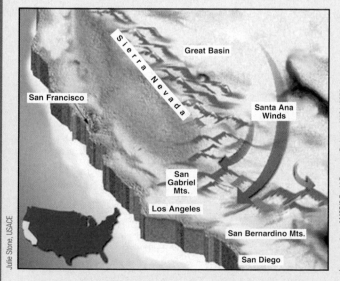

The geographic setting and wind direction for Santa Ana winds.

This satellite image shows strong Santa Ana winds from the northeast fanning wildfires in Southern California, and blowing the smoke offshore for many kilometers.

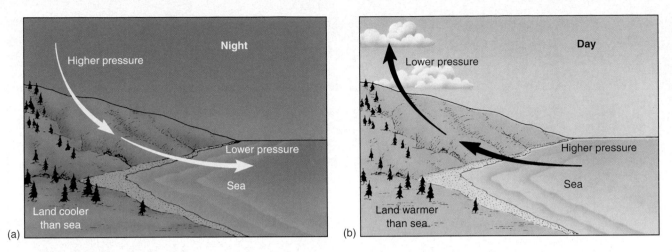

■ **FIGURE 4.18** Land and sea breezes. This day-to-night reversal of winds is a consequence of the different rates of heating and cooling of land and water areas. (a) The land becomes colder than the sea during the night. (b) During the daytime the land becomes warmer than the sea. The air flows from the cooler to the warmer area.

What is the impact on daytime coastal temperatures of the land and sea breeze?

reduction in temperature on the coast, as well as a lesser influence on land perhaps as far from the sea as 15–50 kilometers (9–30 mi).

During hot summer days, a sea breeze cools cities like Los Angeles, Chicago, and Milwaukee. At night, the land and the air above it cool more quickly and to a lower temperature than the water body and the air above it. Consequently, the pressure builds higher over the land and air flows out toward the lower pressure over the water, creating a land breeze. For thousands of years, sailboats have left their coasts at dawn, when there is still a land breeze, and have returned with the sea breeze of the late afternoon.

In mountainous areas under the calming influence of a high pressure system, there is a daily **mountain**

breeze–valley breeze cycle (■ Fig. 4.19). During the day, the sun heats the high mountain slopes faster than the valleys, which are shaded by the mountains. The warm air at higher elevations expands and rises, drawing air from the valley up the mountain slopes. This warm daytime breeze is the valley breeze, named for its place of origin. Clouds, which often cling to mountain peaks, are the visible evidence of condensation occurring in the warm air as it rises from the valleys. Both the valley and the mountains are cooled at night, but the mountains lose more terrestrial radiation to space because of thin air at higher elevation and get much colder than the valleys. This cold, dense air from the high mountains flows downslope into the valleys as a cool, nighttime, mountain breeze.

■ **FIGURE 4.19** Mountain and valley breezes. This daily reversal of winds results from heating of mountain slopes during the day (a), and their cooling at night (b). Warm air is drawn up slopes during the day, and cold air drains down the slopes at night.

How might a green, shady valley floor and a bare, rocky mountain slope contribute to these changes?

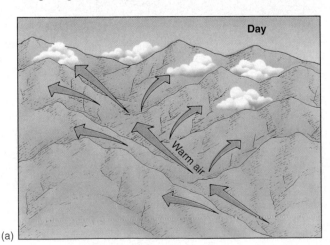

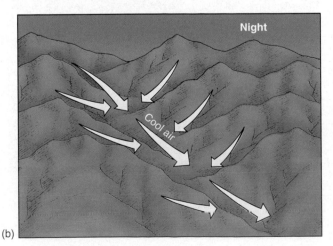

Ocean–Atmosphere Interactions

Most of Earth's surface acts as a dynamic interface between two fluids, air in the atmosphere and water in the oceans. Gases in the atmosphere and waters in the oceans behave in many ways that are similar, but a major difference is their densities. Water molecules are much more closely packed together, so water's density is 800 times higher than the density of air. Yet, air motion in the atmosphere can cause or strongly affect movements in the oceans, and the oceans also affect the atmosphere in many ways. Some of these interactions only affect local areas, some are regional, and others can have a global impact.

Among the best known relationships are the winds that create waves and help to drive the major ocean currents. Because of the high density of water compared to air, faster movements in the atmosphere are reflected as much slower movements in the oceans. Ocean–atmosphere interactions exist at many scales with respect to both time and geographical area, and it will take many years of study before they are fully understood.

Ocean Currents

Like the global wind systems, ocean currents play a significant role in helping to equalize the imbalances of heat energy between the tropical and polar regions. Surface **ocean currents** are fairly steady flows of seawater that move in a prevailing direction, somewhat like rivers in the ocean. The surface water temperatures of ocean currents flowing along the coasts have a great influence on the climate of coastal locations.

Earth's wind system is a primary factor in the flow of surface ocean currents. Other factors include the Coriolis effect, and the size, shape, and depth of the basin of a sea or ocean. Ocean currents are also influenced and driven by variations in density that result from temperature and salinity differences, the tides, and wave action.

Most of the major ocean surface currents move in broad circulatory patterns, called **gyres,** which flow around the subtropical highs. Because of the Coriolis effect, and the direction of flow around a cell of high pressure, the gyres follow a clockwise direction in the Northern Hemisphere and a counterclockwise direction in the Southern Hemisphere (■ Fig. 4.20). Most surface currents do not cross the equator, where the impact of the Coriolis effect is minimal.

The trade winds drive currents near the equator in a westward flow of ocean water called the *Equatorial Current.* At the western margins of ocean basins along the eastern coastlines, its warm tropical waters are deflected poleward. As warm water flows into higher latitudes, they move through regions of cooler waters so they are identified as **warm currents,** and cooler water flowing equatorward form **cold currents** (■ Fig. 4.21). It is important to know that the terms warm and cold, applied to ocean currents, only means warmer or colder than the adjacent water that a current flows through.

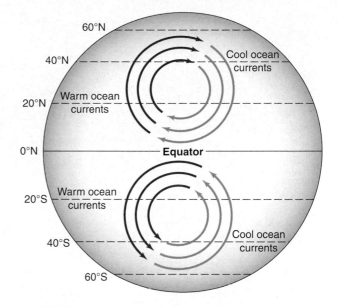

■ **FIGURE 4.20** The major ocean currents flow in broad gyres in opposite directions in the Northern and Southern Hemispheres. **What influences the direction of these gyres?**

In the Northern Hemisphere, warm currents, such as the Gulf Stream and the Kuroshio (Japan) Current, are deflected strongly to the right (or east) because of the increasing apparent impact of the Coriolis effect that occurs with higher latitudes. At about 40°N, the winds of the westerlies begin to drive these warm waters eastward across the ocean, forming the North Atlantic Drift and the North Pacific Drift. Eventually, these currents encounter landmasses at the eastern margin of the ocean, and are deflected toward the equator. After flowing across the ocean basin in higher latitudes, the currents lose warmth, becoming cooler than the adjacent waters as they flow equatorward into the subtropical latitudes. They now have become cold currents. Nearing the equator, these currents complete the circulation pattern when they rejoin the westward-moving Equatorial current.

On the east side of the North Atlantic, the North Atlantic Drift flows north of the British Isles and around Scandinavia, keeping those areas warmer than their latitudes would suggest. Some Norwegian ports, located north of the Arctic Circle, remain ice free because of this relatively warm water. Cold polar water in the Labrador Current and Oyashio (Kurile) Current flows southward into the ocean basins of the Atlantic and Pacific and along the western margins of the continents.

Oceanic circulation in the Southern Hemisphere is comparable to that in the Northern except that the gyres flow in a counterclockwise direction. Also, because in the Southern Hemisphere there is little land poleward of 40°S, the West Wind Drift (or Antarctic Circumpolar Drift) circles around Antarctica as a cold current across the Southern Ocean almost without interruption. It is cooled by the influence of its high latitudinal location and cold air from the Antarctic ice sheet.

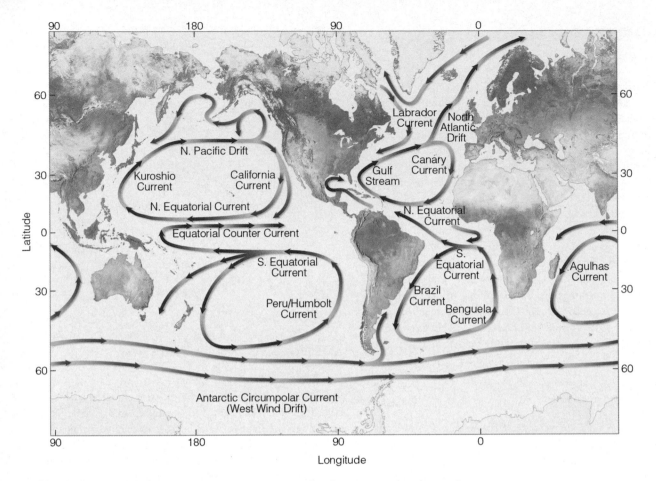

■ **FIGURE 4.21**   Map of the major world ocean currents, showing warm and cool currents.
How does this map of ocean currents help explain the mild winters in London, England?

In general, warm currents flow poleward along east coasts of continents as they carry tropical waters into the cooler waters of higher latitudes (for example, the Gulf Stream or the Brazil Current). Cold currents flow equatorward along west coasts of continents (for example, the California Current and the Humboldt Current). Warm currents tend to bring humidity and warmth to the east coasts of continents along which they flow, and cool currents tend to have a drying and cooling effect on the west coasts. Subtropical highs on the west coast of continents are in contact with cold ocean currents, which cool the air and stabilize and strengthen the eastern side of a subtropical high. On the east coasts of continents, contacts with warm ocean currents cause the western sides of subtropical highs to be less stable and weaker.

The general circulation patterns of oceanic surface currents are consistent throughout the year. However, the position of the currents may respond to seasonal changes in atmospheric heating and circulation. **Upwelling,** a process where deep cold water rises to the surface, reinforces the cold currents along west coasts in subtropical latitudes, adding to the strength and effect of the California, Humboldt (Peru), Canary, and Benguela Currents.

El Niño

As you can see in Figure 4.21, the cold Humboldt Current flows equatorward along the coasts of Ecuador, Chile, and Peru. When the current approaches the equator, upwelling brings nutrient-rich cold water along the coast.

In some years, usually during the months of November and December, a weak warm flow of tropical waters from the east called a *countercurrent* replaces the normally cold coastal waters. Fishing is a major industry along this coastline, and regional fishermen have known of the phenomenon for hundreds of years. Without the upwelling of nutrients from below to feed the fish, fishing comes to a standstill. The residents of the region have named this occurrence **El Niño,** which is a Spanish reference to the Christ child ("the boy") because this phenomenon occurs about Christmastime.

The warm-water countercurrent usually lasts for 2 months or less, but occasionally can last for several months. In these situations, water temperatures are raised not just along the coast, but also for thousands of kilometers offshore (■ Fig. 4.22). Over the past decade, the term

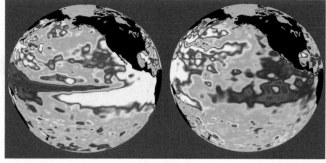

NASA/GSFC

■ **FIGURE 4.22**   These thermal infrared satellite images show El Niño (left) and La Niña (right) episodes in the Tropical Pacific. The red and white shades display the warmer sea surface temperatures, while the blues and purples mark areas of cooler temperatures. **From what continent does an El Niño originate?**

El Niño has come to describe these exceptionally strong episodes. During the past 50 years, approximately 18 years experienced these El Niño conditions (with the ocean water warmer than normal for six consecutive months). Not only do the El Niños affect the temperature of the equatorial Pacific, but also the strongest of them have an impact on global weather patterns.

El Niño and the Southern Oscillation
Understanding the processes that produce an El Niño requires that we examine conditions across the Pacific, not just in the waters off of South America. In the 1920s, Sir Gilbert Walker, a British scientist, discovered a connection between surface pressure readings at weather stations on the eastern and western sides of the Pacific basin. He noted that a rise in pressure in the eastern Pacific is usually accompanied by a fall in pressure in the western Pacific and vice versa. He called this seesaw pattern the **Southern Oscillation.** The link between El Niño and the Southern Oscillation is so great that together they are often referred to as ENSO (El Niño/Southern Oscillation).

During a typical year, the eastern Pacific (along the west coast of South America) has a higher pressure than the western Pacific. This east-to-west pressure gradient enhances the Pacific trade winds, producing a surface current that moves from east to west at the equator. The western Pacific develops a warm layer of water, while in the eastern Pacific the cold Humboldt Current is enhanced by upwelling (■ Fig. 4.23a).

When the Southern Oscillation swings in the opposite direction, the normal conditions described above change dramatically, with pressure increasing in the western Pacific and decreasing in the eastern Pacific. This pressure change causes the trade winds to weaken or, in some cases, reverse. This reversal causes warm water in the western Pacific to flow eastward, increasing sea-surface temperatures in the central and eastern Pacific. This eastward shift signals the beginning of El Niño (■ Fig. 4.23b).

In contrast, at times the trade winds will intensify into more powerful winds that reinforce strong upwelling, and

sea-surface temperatures will be colder than normal. This condition is known as **La Niña** (in Spanish, "the girl," but scientifically simply the opposite of El Niño). La Niña episodes typically bring about the opposite effects of an El Niño episode.

El Niño and Global Weather
Cold ocean waters impede cloud formation, except for coastal fogs. Thus, under normal conditions, clouds tend to develop over the warm waters of the western Pacific but not over the cold waters of the eastern Pacific. During an El Niño, when warm water migrates eastward, widespread clouds develop over the equatorial region of the Pacific (see again Fig. 4.23b). These clouds can build to heights of 18,000 meters (59,000 ft) and can disrupt the high-altitude winds, trigger other wind changes, and affect the global weather.

Scientists have tried to document past El Niño events by piecing together historic evidence, such as sea-surface temperature records, observations of pressure and rainfall, fisheries' records, and the writings of people living along the west coast of South America dating back to the 15th century. Additional evidence from the region comes from the growth patterns of coral and trees.

Based on historical evidence, we know that El Niños have occurred as far back as records go. One disturbing fact, however, is that they appear to be occurring more often. Over the past few decades, El Niños have been occurring, on average, every 2.2 years. Even more alarming is the fact that they appear to be getting stronger. The record-setting El Niño of 1982–1983, was surpassed by another El Niño event in 1997–1998, which brought heavy and damaging rainfall to the southern United States, from California to Florida. Snowstorms in the northeastern United States were more frequent and stronger than in most years. In recent years, scientists have become better able to monitor and forecast El Niño and La Niña events, which can help us prepare for weather events that are associated with these conditions.

North Atlantic Oscillation

Our improved observation skills have led to the discovery of the **North Atlantic Oscillation (NAO)**—a relationship between the Azores (subtropical) High and the Icelandic (subpolar) Low. The east-to-west, see-saw motion of the Icelandic Low and the Azores High control the strength of the westerly winds and the direction of storm tracks across the North Atlantic. There are two recognizable NAO phases.

A positive NAO phase is identified by higher than average pressure in the Azores High and lower than average pressure in the Icelandic Low. The increased pressure difference between the two systems results in stronger and more frequent winter storms that follow a more northerly track (■ Fig. 4.24a). Then, warm and wet winters occur in Europe, but in Canada and Greenland the winters are cold and dry. The Eastern United States may experience a mild, wet winter.

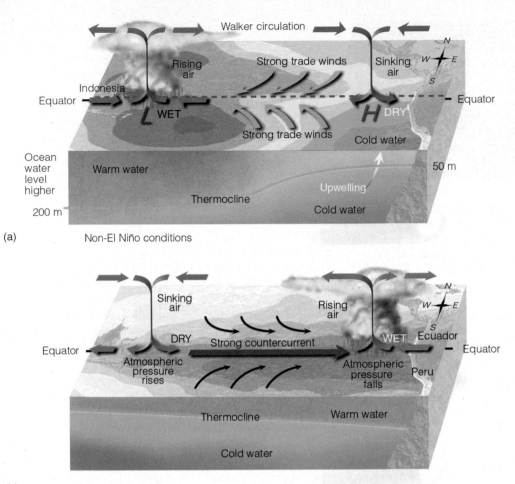

(a) Non-El Niño conditions

(b) El Niño conditions

■ **FIGURE 4.23**   (a) In a non-El Niño (normal) year, strong trade winds and increased upwelling of cold sea water occur along the west coast of South America, bringing rains to Southeast Asia. (b) During El Niño, the easterly trade winds weaken, allowing the central Pacific to warm and the rainy area to migrate eastward.

Near what country or countries does El Nino begin?

■ **FIGURE 4.24**   Positions of the pressure systems and winds involved with the (a) positive and (b) negative phases of the North Atlantic Oscillation (NAO).

Which two pressure systems are used to establish the NAO phases?

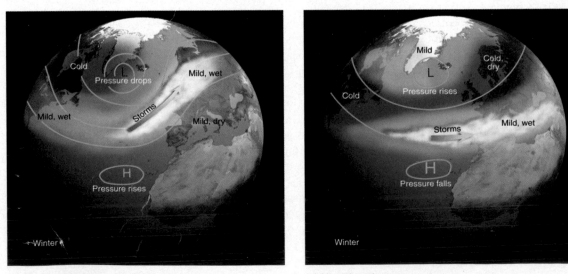

(a) Positive phase

(b) Negative phase

The negative NAO phase occurs with a weak Azores High and a weak Icelandic Low. The smaller pressure gradient between these cells moderates the westerlies, resulting in fewer and weaker winter storms (Fig. 4.24b). Northern Europe will experience cold air and moist air moves into the Mediterranean. The East Coast of the United States will experience more cold air and snowy winters. The NAO varies from year to year, but also has a tendency to stay in one phase for several years in a row.

Scientists continue working to better understand atmosphere–ocean interactions and their impacts on global weather patterns, and as technology improves, our forecasting ability will increase. Studying the close association between the atmosphere and hydrosphere has given us a better understanding of the complex relationships that exist between storm systems and global climate patterns.

:: Terms for Review

standard sea-level pressure
cyclone (low)
anticyclone (high)
convergent wind circulation
divergent wind circulation
isobar
pressure gradient
wind
windward
leeward
prevailing wind
Coriolis effect
friction
geostrophic wind
trough
equatorial low (equatorial trough)
subtropical high

subpolar low
polar high
Siberian High
Canadian High
Icelandic Low
Aleutian Low
Pacific High (Hawaiian High)
Bermuda High (Azores High)
polar easterlies
westerlies
trade winds
northeast trades
southeast trades
intertropical convergence zone (ITCZ)
jet stream
Rossby wave
monsoon

chinook
foehn
Santa Ana
drainage wind (katabatic wind)
land breeze–sea breeze
mountain breeze–valley breeze
ocean current
gyre
warm current
cold current
upwelling
El Niño
Southern Oscillation
La Niña
North Atlantic Oscillation (NAO)

:: Questions for Review

1. What is atmospheric pressure at sea level? How do you suppose Earth's gravity is related to atmospheric pressure?
2. Horizontal variations in air pressure are caused by thermal or dynamic factors. How do these two factors differ?
3. What kind of pressure (high or low) would you expect to find in the center of an anticyclone? Describe and diagram the wind patterns of anticyclones and cyclones in the Northern and Southern Hemispheres.
4. How do landmasses affect the development of belts of atmospheric pressure over Earth's surface?
5. Why do Earth's wind systems and pressure belts migrate with the seasons?
6. How are the land breeze–sea breeze and monsoon circulations similar? How are they different?
7. What effect on valley farms could a strong drainage wind have?
8. What is the relationship between ocean currents and global surface wind systems? How does the gyre in the Northern Hemisphere differ from the one in the Southern Hemisphere?
9. Where are the major warm and cold ocean currents located with respect to Earth's continents? Which currents have the greatest effect on North America?
10. What is an El Niño? What are some impacts that it has on global weather?

:: Practical Applications

1. Look at the January (Fig. 4.10a) and July (Fig. 4.10b) maps of average sea-level pressure. Answer the following questions:
   a. Why is the subtropical high pressure belt more continuous (linear, not cellular) in the Southern Hemisphere than in the Northern Hemisphere in July?
   b. During July, what area of the United States exhibits the lowest average pressure? Why?

2. The amount of power that can be generated by wind is determined by the equation

$$p = \tfrac{1}{2} D \times S^3$$

   where P is the power in watts, D is the density, and S is the wind speed in meters per second (m/sec). Because $D = 1.293 \text{ kg/m}^3$, we can rewrite the equation as

$$P = 0.65 \times S^3$$

   How much power (in watts) is generated by the following wind speeds: 2 meters per second, 6 meters per second, 10 meters per second, 12 meters per second?

3. Atmospheric pressure decreases at the rate of 0.036 millibar per foot as one ascends through the lower portion of the atmosphere.
   a. The Willis Tower (formerly the Sears Tower), in Chicago, Illinois, is one of the world's tallest buildings at 1450 feet. If the street-level pressure is 1020.4 millibars, what is the pressure at the top of the tower?
   b. If the difference in atmospheric pressure between the top and ground floor of an office building is 13.5 millibars, how tall is the building?

Humidity, Condensation, and Precipitation

5

:: Outline

This large thunderstorm demonstrates that the important process of transferring moisture from the atmosphere to the land can be dramatic and occasionally hazardous.

NOAA/NWS/Greg Lundeen

:: Objectives

When you complete this chapter you should be able to:

- Explain why available freshwater remains a limited and precious resource even though Earth's surface is dominated by water.
- Outline the processes in the hydrologic cycle, including how water circulates among, and interacts with, the lithosphere, atmosphere, hydrosphere, and biosphere.
- Understand that relative humidity is a percentage of moisture saturation in the air, and why it is dependent on air temperature and moisture content.
- Explain why if no change in moisture content occurs, as air warms the relative humidity will fall, whereas cooling air causes the relative humidity to rise.

- Determine what processes cause the air to reach the dew point temperature and attain a relative humidity of 100%, a condition that can lead condensation and precipitation.
- Apply adiabatic lapse rates to determine temperature changes in air that rises, expands, and cools as well as to air that sinks, compresses, and warms.
- Describe the atmospheric conditions of temperature, humidity, pressure, and winds that influence precipitation potential and the kinds of precipitation that may result.
- Provide examples of the great geographic variations in precipitation, evaporation, and water availability that exist on Earth.

Water is vital to life on Earth. Although some living things can survive without air, no organism can survive without water. Water is necessary for photosynthesis, soil formation, and the absorption of nutrients by animals and plants.

Water can dissolve so many substances that it has been called the *universal solvent,* thus it is almost never found in a pure state. Even rain contains impurities picked up in the atmosphere, impurities that also facilitate the development of clouds and precipitation. Because rain contains some dissolved carbon dioxide from the air, most rainwater is a very weak form of carbonic acid. The normal acidity of rainwater, however, should not be confused with the environmentally damaging *acid rain,* which is at least ten times more acidic.

In addition to dissolving and transporting minerals, water transports solid particles in suspension and by other means. It carries minerals and nutrients in streams, through the soil, through openings in subsurface rocks, and through plants and animals. Water deposits solid matter on streambeds and floodplains, in river deltas, and on the ocean floor.

The surface tension of water and the behavior of water molecules in drawing together cause **capillary action**—the ability of water to pull itself upward through small openings despite the pull of gravity. Capillary action transports dissolved material in an upward direction through rock and soil. Capillary action also moves water into the stems and leaves of plants—even to the uppermost needles of the great California redwoods and to the tops of tall rainforest trees. Another important property of water is that it expands when it freezes. Ice is therefore less dense than water and consequently will float on water, as do ice floes and icebergs (■ Fig. 5.1).

Finally, water is a substance that is slow to heat and slow to cool, compared to most materials. Bodies of water are reservoirs that store heat, which can moderate the cold of winter; yet the same water body will have a cooling effect in summer. This effect on temperature can be experienced in the vicinity of lakes as well as on seacoasts.

Earth's water—the *hydrosphere* (from Latin: *hydros,* water)—is found in all three states: as a liquid in rivers, lakes, oceans, and rain; as a solid in snow and ice; and as water vapor (a gas) in our atmosphere. About 73% of Earth's surface is covered by water, with the largest proportion in the world's oceans; most of Earth's freshwater is in the polar glaciers (■ Fig. 5.2). In all, the total water content of the Earth system, whether liquid, solid, or vapor, is about 1.36 billion cubic kilometers (326 million cu mi), and the vast majority is salt water in the oceans (■ Fig. 5.3).

■ **FIGURE 5.1**   Huge icebergs off the coast of Antarctica float like ice in a drinking glass.
If the floating ice melts completely before you can drink from the glass, will the liquid level rise, fall, or remain the same as before? Why?

National Science Foundation Jeffrey Kietzmann

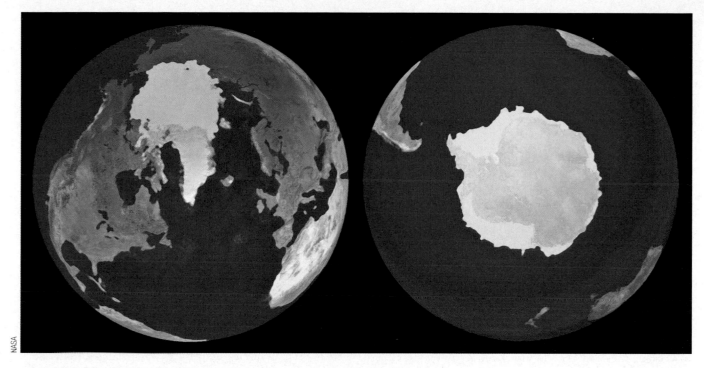

■ **FIGURE 5.2**   Earth's surface is mainly covered by the oceans. Most of Earth's fresh water is held in the glacial ice of the polar regions, as seen in these images centered on the north and south poles.
Can you distinguish between the Greenland and Antarctic ice sheets and the seasonal (pack ice) that has formed on the oceans' surface?

■ **FIGURE 5.3**   Earth's water sources. The vast majority of water in the hydrosphere is seawater in the world's oceans. The fresh water supply stored in polar ice sheets is relatively unavailable for use.
How might global warming or cooling alter this figure?

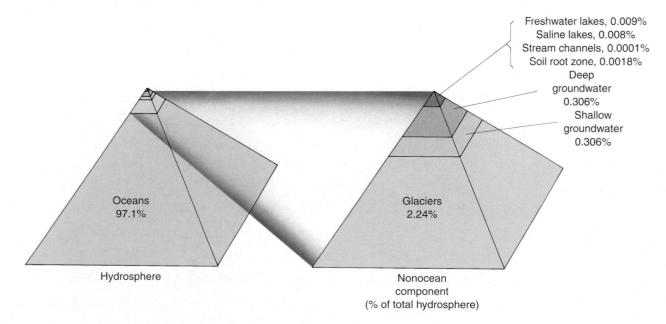

Freshwater lakes, 0.009%
Saline lakes, 0.008%
Stream channels, 0.0001%
Soil root zone, 0.0018%
Deep groundwater 0.306%
Shallow groundwater 0.306%

Oceans 97.1%

Glaciers 2.24%

Hydrosphere

Nonocean component (% of total hydrosphere)

The Hydrologic Cycle

Although water cycles into and out of the atmosphere, lithosphere, and biosphere, the total amount of water in the hydrosphere remains constant. The **hydrologic cycle** is the circulation of water from one of Earth's systems to another. The atmosphere contains water vapor gained by evaporation from water bodies and land surfaces. In the hydrologic cycle, water is continually transferred from one state to another— as a liquid, vapor, or solid. When water vapor condenses into liquid water and falls as precipitation, several things may happen to it. First, it may go directly into a body of water—a lake, river, pond, or ocean. Alternatively, it may fall on the land surface, where it runs off to form rivers, streams, ponds, or lakes. Or, it may be absorbed into the ground and be contained in soil, or flow in open spaces that exist in loose rock fragments and voids in solid rock. Ultimately, most of the water in the ground or on the surface reaches the oceans. Some of the water that fell and accumulated as snow will become part of the massive ice cover held in storage on Greenland and Antarctica, or in mountain glaciers, while some snow will thaw in spring and feed streams. Other water, used by plants and animals, temporarily becomes a part of living things. In short, there are six storage areas for water in the hydrologic cycle: the atmosphere, the oceans, bodies of fresh water, plants and animals, snow and glacial ice, and open spaces beneath Earth's surface.

Evaporation returns liquid water to the atmosphere as a gas. Water evaporates from all bodies of water, from plants and animals, and from soils; and it can even evaporate from falling precipitation. Once evaporation returns liquid water to the atmosphere as gaseous water vapor, the cycle can be repeated through condensation and precipitation.

The hydrologic cycle is basically a continuous cycle of evaporation, condensation, and precipitation, with transportation of water over the land, in water bodies, and in the ground (■ Fig. 5.4). All of these processes are constantly operating. The hydrologic cycle for the Earth system *as a whole* can be considered a closed system, because energy flows into and out of the system, but there is no gain or loss of water (matter). Although Earth's hydrologic cycle is a closed system, its subsystems (like the hydrologic cycle of a region or water body) operate as open systems with energy and matter flowing both in and out (see again the reservoir in Fig. 1.18). Earth's overall hydrosphere and its subsystems are not static; they are dynamic as water changes from one state to another and is transported from the atmosphere to Earth's surface and back.

■ **FIGURE 5.4   Environmental Systems: The Hydrologic Cycle** The hydrologic cycle concerns the circulation of water from one part of the Earth system to another. Largely through condensation, precipitation, and evaporation, water is cycled continually between the atmosphere, the soil, subsurface storage, lakes and streams, plants and animals, glacial ice, and the oceans.
Can you explain whether Earth's overall hydrologic cycle is a closed system or an open system?

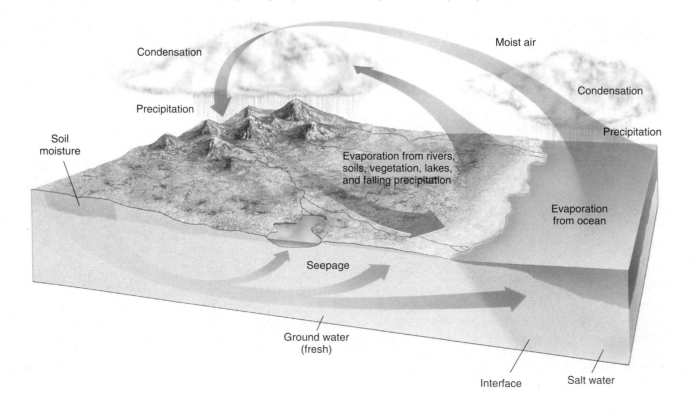

Water in the Atmosphere

The Water Budget

We are most familiar with water in its liquid form as it pours from a tap, falls as rain, or when we enjoy the recreational opportunities and scenic beauty of water bodies. In the atmosphere, water can exist as ice (snow, hail), or as tiny liquid droplets that form clouds and fog, but it also exists as a tasteless, odorless, and transparent gas known as water vapor. The troposphere contains 99% of the water vapor in the atmosphere. Water vapor makes up a small but highly variable percentage of the atmosphere by volume. Through the exchange of water among its states by evaporation, condensation, precipitation, melting, and freezing, water plays a significant role in regulating and modifying temperatures locally and globally. In addition, as we noted in Chapter 3, water vapor in the atmosphere both reflects and absorbs significant portions of incoming solar energy and outgoing terrestrial radiation, as well as reradiating some of that absorbed heat energy from the atmosphere back to Earth. Also, the insulating qualities of water vapor reduce the loss of heat from Earth's surface, which helps to maintain the moderate ranges of temperature found on this planet.

Because Earth's hydrosphere is a *closed system,* no water is received from outside the Earth system nor lost from it. Thus, an increase in water within one hydrologic subsystem must be accounted for by a loss in another. Put another way, we say that the Earth system operates on a *water budget,* in which the total quantity of water remains the same and in which deficits must balance gains throughout the entire system. The amount of water associated with any one component of the hydrosphere changes constantly over time and place, particularly where water is stored. For example, about 24,000 years ago, during the last major ice age, glaciers expanded, sea level dropped, and evaporation and precipitation were greatly reduced. Less water was stored in the oceans, but this was balanced by increased storage on the land as glacial ice.

We know that the atmosphere gives up a great deal of water, most obviously by condensation into clouds, fog, dew, and through several forms of precipitation (rain, snow, hail, sleet). If the quantity of water in the atmosphere globally remains at the same level through time, the atmosphere must be absorbing water from other parts of the system in an amount equal to that which it is giving up. During a single minute, the atmosphere gives up more than one billion tons of water through precipitation or condensation, while another billion tons are evaporated and absorbed as water vapor into the atmosphere.

Solar energy is used in evaporation and is then stored in water vapor, to be released during condensation. Although the energy transfers involved in evaporation and condensation account for a small portion of the total heat energy budget, the actual energy is significant. Imagine the amount of energy released every minute when a billion tons of water condense out of the atmosphere. This vast storehouse of energy, the latent heat of condensation, is a major source of power for Earth's storms: hurricanes, tornadoes, and thunderstorms.

A very important determinant of the amount of water vapor that can be held by the air is temperature. The warmer air is, the greater the quantity of water vapor it can hold. Therefore, we can make a generalization that air in the polar regions can hold far less water vapor (approximately 0.2% by volume) than the hot air of the tropical and equatorial regions of Earth, where the air can contain as much as 5% by volume.

Saturation and Dew Point Temperature

When air of a given temperature holds all the water vapor that it can, it is said to be in a state of **saturation** and to have reached its **capacity.** If a constant air temperature is maintained, but enough water vapor is added, the air will be saturated and unable to hold any more water vapor. For example, when you take a shower, the air in the room becomes increasingly humid until a point is reached at which the air cannot contain more water. At that point, excess water vapor begins condensing onto the colder mirrors and walls.

■ Figure 5.5 illustrates the connection between higher moisture capacity and rising temperatures. If a parcel of air at 30°C is saturated, then it will be at its capacity and will contain

■ **FIGURE 5.5**   This graph shows the maximum amount of water vapor that can be contained in a cubic meter of air over a wide range of temperatures.
Compare the change in capacity if the air temperature is raised from 0°C to 10°C, to a change from 20°C to 30°C. What does this indicate about the relationship between temperature and capacity?

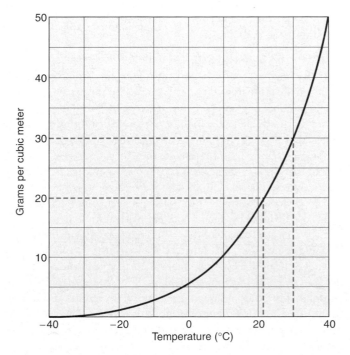

30 grams of water vapor in each cubic meter of air (30 g/m³). Now suppose the temperature of the air increases to 40°C *without* increasing the water vapor content. The parcel is no longer saturated because air at 40°C can hold more than 30 grams per cubic meter of water vapor (actually, 50 g/m³). Conversely, if we decrease the temperature of saturated air from 30°C (which contains 30 g/m³ of water vapor) to 20°C (which has a water vapor capacity of only 17 g/m³), 13 grams of the water vapor will condense out of the air because of the reduced capacity.

If an unsaturated parcel of air is cooled, it may eventually reach a temperature at which the air will become saturated. This critical temperature is known as the **dew point**—the temperature at which condensation is imminent. For example, if a parcel of air at 30°C contains 20 g/m³ of water vapor, it is not saturated because it can hold 30 g/m³. However, if that parcel of air cooled to 21°C, it would become saturated because the capacity of air at 21°C is 20 g/m³. Thus, that parcel of air at 30°C has a dew point temperature of 21°C. It is the cooling of air to below its *dew point temperature* that brings about the condensation that must precede any precipitation.

Because the capacity of air to hold water vapor increases with rising temperatures, air in the equatorial regions has a higher dew point temperature than air in the polar regions. Thus, because the atmosphere can hold more water in the equatorial regions, there is greater potential for large quantities of precipitation than in the polar regions. Likewise, in the middle latitudes, summer months, because of their higher temperatures, have more potential for heavy precipitation than do winter months.

Humidity

The amount of water vapor in the air at any one time and place is called **humidity.** There are three common ways to express humidity. Each method provides information that contributes to our discussion of weather and climate.

Absolute and Specific Humidity
The measure of the mass of water vapor that exists within a given volume of air is called **absolute humidity.** It is expressed in the metric system as the number of grams per cubic meter (g/m³) or in the English system as grains per cubic foot (gr/ft³). **Specific humidity** is the mass of water vapor (given in grams) per mass of air (given in kilograms). Both measures indicate the actual amount of water vapor that the air contains at a certain place and time. Because most water vapor gets into the air by the evaporation of water from Earth's surface and air is cooler at higher altitudes, absolute and specific humidity tend to decrease with altitude.

We have also learned that air is compressed as it sinks and expands as it rises. Thus, a parcel of air changes its volume as it moves vertically, but there may be no change in the amount of water vapor in that quantity of air. We can see, then, that absolute humidity, although it measures the amount of water vapor, can vary as a result of vertical movements that change the volume of an air parcel. In contrast, specific humidity changes *only* as the quantity of the water vapor changes.

For this reason, specific humidity is the preferred measurement among geographers and meteorologists.

Relative Humidity
The best-known means of describing the water vapor in the atmosphere—which we commonly encounter in newspaper, television, and radio weather reports—is **relative humidity.** It is the ratio between the amount of water vapor in air of a given temperature and the maximum amount of vapor that the air could hold at that temperature. Relative humidity is a percentage that indicates how close the air is to saturation. It is important to know that both the air temperature and the amount of moisture in the air influence the relative humidity. If the air temperature goes up or down, or the amount of moisture in the air goes up or down, the relative humidity will also change.

If the temperature and absolute humidity of an air parcel are known, its relative humidity can be determined by using Figure 5.5. For instance, if a parcel of air has a temperature of 30°C and an absolute humidity of 20 grams per cubic meter, we can look at the graph and determine that if air at that temperature were saturated, its capacity would be 30 grams per cubic meter. To determine relative humidity, all we do is divide 20 grams (actual content) by 30 grams (content at capacity) and multiply by 100 (to get an answer in percentage):

$$(20 \text{ grams} \div 30 \text{ grams}) \times 100 = 67\%$$

The relative humidity in this case is 67%. In other words, the air is holding only two thirds of the water vapor it could contain at 30°C; it is only at 67% of its capacity.

Two important factors affect the geographic variation of relative humidity. One of these is moisture availability. For example, because there is more water available for evaporation from a water body, the air there typically contains more moisture than air of similar temperature over land. Conversely, the air overlying an inland region, like the central Sahara Desert, may be very dry because it is far from the oceans and little water is available to be evaporated. A second factor in the geographic variation of relative humidity is temperature. In regions of higher temperature, relative humidity for air containing the same amount of water vapor will be lower than it would be in a cooler region.

Relative humidity varies if the amount of water vapor increases due to the evaporation of moisture into the air *or* if the temperature increases or decreases. Thus, although the quantity of water vapor may not change through a day, the relative humidity will vary with the daily temperature cycle. As air temperature increases from the overnight low at around sunrise to its maximum in mid-afternoon, the relative humidity decreases because the warmer air is capable of holding greater quantities of water vapor. Then, as the air becomes cooler, decreasing toward its minimum temperature around sunrise, the relative humidity progressively increases (■ Fig. 5.6). Even if no water vapor is gained or lost by the air at a particular location, relative humidity will increase at night because of cooling and decrease during the daytime because of warming.

Relative humidity affects our comfort through its relationship to the rate of evaporation. When people perspire, evaporation causes cooling because the heat used to evaporate

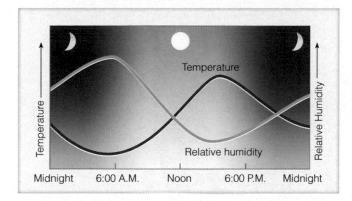

■ **FIGURE 5.6** This graph illustrates the relationship between air temperature and relative humidity as the variables change throughout a typical 24-hour period. Even with no change in the absolute moisture content of the air, as the day warms the relative humidity drops and at night, when it is cooler, the relative humidity rises.

How is the relationship between air temperature and relative humidity applied when using a hair dryer?

perspiration becomes stored in water vapor as latent heat and is subtracted from the skin. This is why on a hot August day when the temperature approaches 35°C (95°F), you will be more uncomfortable in Atlanta, Georgia, where the relative humidity is 90%, than in Tucson, Arizona, where it may be only 15% at the same temperature. Your perspiration will evaporate at a faster rate at the lower relative humidity of 15%, and you will benefit from evaporative cooling. When the relative humidity is 90%, the air is nearly saturated, so less evaporation can take place and less heat is drawn from your skin.

Sources of Atmospheric Moisture

Water evaporates into the atmosphere from many different sources, most importantly from bodies of water, especially the oceans. Water also evaporates from ground surfaces and soils,

from droplets of moisture on vegetation, from vehicles, pavement, roofs, other surfaces, and from falling precipitation.

Vegetation provides a source of water vapor through another process, known as **transpiration,** where plants lose moisture to the air. In some parts of the world—notably tropical rainforests of heavy, lush vegetation—transpiration accounts for a significant amount of atmospheric humidity. **Evapotranspiration,** the combined term for evaporation and transpiration, accounts for virtually all water vapor in the atmosphere.

Rates of Evaporation

Evaporation rates are affected by several factors. The first factors include the amount and temperature of accessible water. Table 5.1 shows that evapotranspiration rates tend to be greater over the oceans than over the continents. The only place this generalization is not true is in equatorial regions between 0° and 10°N and S, where the vegetation is so lush on the land that transpiration provides a large amount of moisture to the air.

Second is the degree to which the air is saturated with water vapor. The lower the relative humidity, the greater the rate of evaporation will be, given equal water availability. Compare the length of time it takes your swimsuit to dry on a hot, humid day with how long it takes on a day when the air is dry.

Third is the wind, which also affects evaporation. If there is no wind, the air that overlies a water surface may approach saturation, and once saturation is reached, evaporation will cease. However, if it is windy, the wind will blow saturated or nearly saturated air away from the evaporating surface, replacing it with air of lower humidity. This allows evaporation to continue as long as the wind keeps blowing humid air away and bringing in drier air. Anyone who has gone swimming on a windy day has experienced the chilling effects of rapid evaporation.

Air temperature also strongly influences evaporation rates; as air temperature increases, so does the water temperature at the evaporation source. An increase in temperature assures that more energy is available to the water molecules for their change from a liquid to a gaseous state. Consequently,

TABLE 5.1
Distribution of Actual Mean Evapotranspiration

			Latitude			
Zone	**60°–50°**	**50°–40°**	**40°–30°**	**30°–20°**	**20°–10°**	**10°–0°**
Northern Hemisphere						
Continents	36.6 cm (14.2 in.)	33.0 (13.0)	38.0 (15.0)	50.0 (19.7)	79.0 (31.1)	115.0 (45.3)
Oceans	40.0 (15.7)	70.0 (27.6)	96.0 (37.8)	115.0 (45.3)	120.0 (47.2)	100.0 (39.4)
Mean	38.0 (15.0)	51.0 (20.1)	71.0 (28.0)	91.0 (35.8)	109.0 (42.9)	103.0 (40.6)
Southern Hemisphere						
Continents	20.0 cm (7.9 in.)	NA	51.0 (20.1)	41.0 (16.1)	90.0 (35.4)	122.0 (48.0)
Oceans	23.0 (9.1)	58.0 (22.8)	89.0 (35.0)	112.0 (44.1)	119.0 (47.2)	114.0 (44.9)
Mean	22.5 (8.8)	NA	NA	99.0 (39.0)	113.0 (44.5)	116.0 (45.7)

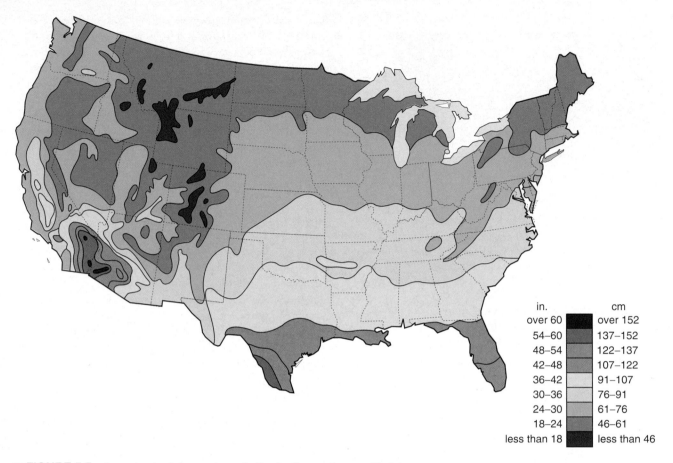

in.		cm
over 60	■	over 152
54–60	■	137–152
48–54	■	122–137
42–48	■	107–122
36–42	■	91–107
30–36	■	76–91
24–30	■	61–76
18–24	■	46–61
less than 18	■	less than 46

■ **FIGURE 5.7** Annual potential evapotranspiration for the contiguous 48 states.
Why is potential evapotranspiration so high in the southwestern desert?

more water can evaporate. Also, as the temperature of the air increases, so does its capacity to contain moisture.

Potential Evapotranspiration

So far, we have discussed *actual* evapotranspiration (evaporation and transpiration). However, geographers and meteorologists are also concerned with **potential evapotranspiration** (■ Fig. 5.7), which refers to the amount of evapotranspiration that *could* occur *if* an unlimited moisture supply was available. Formulas are used to estimate the potential evapotranspiration at a location because evapotranspiration is difficult to measure directly. These formulas commonly consider temperature, latitude, vegetation, and soil character (permeability, water-retention ability) as factors that could affect the potential evapotranspiration.

In places where precipitation exceeds potential evapotranspiration, there is a surplus of water for storage in the ground and in water bodies, allowing water to flow in streams and rivers away from those areas. Water can also be exported to drier places by artificial means, if systems of canals or pipelines are feasible. When potential evapotranspiration exceeds precipitation, as it does during the dry summer months in California, and in the arid West, then no water is available for

storage; in fact, the water stored during previous rainy months evaporates quickly into the warm, dry air (■ Fig. 5.8). Soil becomes dry and the vegetation dries out and turns brown. For this reason, fires are a potential hazard during the late summer months in California.

Condensation, Fog, and Clouds

Condensation, the process by which a gas is changed to a liquid, occurs when air saturated with water vapor is cooled. Once the air temperature cools until it has a relative humidity of 100% (the air has reached the dew point temperature), condensation will occur with additional cooling. It follows, then, that condensation depends on (1) the relative humidity and (2) the degree of cooling. In the arid air of Death Valley, California, a great amount of cooling must take place to reach the dew point temperature. In contrast, on a humid summer afternoon in Biloxi, Mississippi, a minimal cooling will bring on saturation and condensation.

This cooling process is how droplets of water form on the side of a cold drink glass on a warm afternoon.

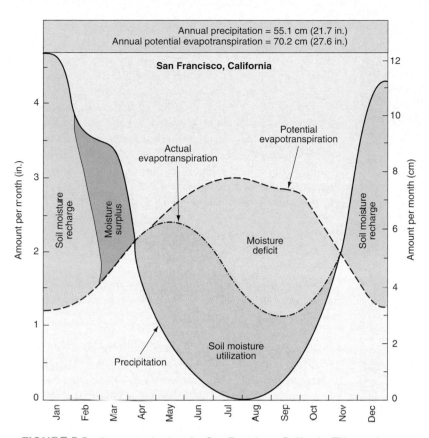

Annual precipitation = 55.1 cm (21.7 in.)
Annual potential evapotranspiration = 70.2 cm (27.6 in.)

San Francisco, California

■ **FIGURE 5.8** The water budget for San Francisco, California. This graph illustrates the water budget system, which "keeps score" of the balance between water input by precipitation, and water loss to evaporation and transpiration, permitting month-by-month estimates of both runoff and soil moisture. **When would irrigation be necessary at this site?**

The temperature of the air is lowered by contact with the cold glass. If air touching the glass is cooled sufficiently, its relative humidity will reach 100%, and further cooling beyond the dew point will result in condensation, forming water droplets on the glass.

Condensation Nuclei

For condensation to occur in the atmosphere, another factor is important: the presence of **condensation nuclei.** These are minute particles in the atmosphere that provide a surface upon which condensation can take place. Sea-salt particles in the air are common condensation nuclei that come from the evaporation of saltwater spray and ocean water. Other common nuclei include dust, smoke, pollen, and volcanic material. In many instances, these nuclei are chemical particles that are the by-products of industrialization. The condensation that takes place on such chemical nuclei is often corrosive and dangerous to human health; when it is, we know it as *smog* (a term that combines *smoke* and *fog*).

Fog, clouds, and dew are all results of condensation of water vapor. The cooling that produces this condensation can occur as a result of radiational cooling, through advection, through convection, or through a combination of these processes.

Fog

Fog and clouds appear when water vapor condenses on nuclei and a large number of these droplets form a mass. Not being transparent to light in the way that water vapor is, these masses of condensed water droplets appear as fog or clouds, in shades of white or gray.

On a worldwide basis, fog is a minor form of condensation, but in certain regions it has important climatic effects. The "drip factor" helps sustain vegetation and animals along desert coastlines where fog occurs. Fog also plays havoc with transportation systems. Navigation on the seas is more difficult in fog, and air travel can be impeded when fog causes airports to shut down until visibility improves. Highway travel is also greatly hampered by heavy fogs, which can lead to huge, chain-reaction vehicle pileups.

Radiation Fog Radiational cooling can produce **radiation fog,** also called *temperature-inversion fog,* or *ground fog.* This kind of fog typically occurs on initially cold, clear, calm nights, and lasts until morning. Clear, calm conditions allow for massive amounts of terrestrial radiation to be lost from the ground. With no incoming radiation at night, the ground becomes cold as it gives up much of the heat that it received during the day. In turn, the air directly above the surface is cooled by conduction through contact with the cold ground. Because the cold surface can cool only the lower few meters of the atmosphere, a temperature inversion is created as air near the surface becomes colder than the warmer air above. If this cold layer of air at the surface is cooled to a temperature below its dew point, then condensation will occur, often in the form of a low-lying fog.

The chances of a temperature-inversion fog occurring are increased in valleys and depressions, where cold air drains down from higher areas into the lowlands. During a cold night, this air can be cooled below its dew point, resulting in a fog that forms like a pond in the valley bottom (■ Fig. 5.9a). It is common in mountainous areas to see early morning radiation fog in the valleys while snow-capped mountaintops shine against a clear blue sky. Radiation fog has a diurnal cycle, forming during the night and becoming densest around sunrise when temperatures are lowest. It then "burns off" during the day when solar energy slowly penetrates the fog and warms the ground surface. The ground in turn warms the air directly above it, increasing its temperature and its capacity to hold water vapor, causing the fog to evaporate.

(a)

© Richard Hamilton Smith/ CORBIS

(b)

© Craig Aurness/ CORBIS

(c)

M. Trapasso

■ **FIGURE 5.9**   Types of fog. (a) Radiation fog typically forms during cold, clear nights under calm high pressure conditions, and lasts until morning. (b) Advection fog is caused by warm moist air passing over colder water or a colder coastal surface. (c) Upslope fog is caused by moist air adiabatically cooling as it rises up a mountain slope.

What unique problems might coastal residents face as a result of fog?

Advection Fog

Another common type of fog is **advection fog,** which occurs when warm, moist air moves over a colder land or water surface. When the warm air is cooled below its dew point through heat loss by conduction to the colder surface below, condensation produces fog. Advection fog is usually less localized than radiation fog. It is also less likely to have a diurnal cycle, though if not too thick, it can be burned off during the day to return again in the evening. More common, however, is the persistent advection fog that spreads itself over a large area for days at a time.

During the summer months, advection fog may form as warm air moves over cool water in large lakes or the oceans. These fogs are common during the summer along the West Coast of the United States. During the summer months, the Pacific subtropical high moves north, and winds flow toward the coast where they pass over the cold California Current. When condensation occurs, fogs form and flow inland over the shore, pushed from behind by the eastward movement of air and pulled by the low pressure of the warmer land (Fig. 5.9b). Advection fogs also occur in New England, especially along the coasts of Maine and the Canadian Maritime Provinces, when warm, moist air from above the Gulf Stream flows north over the colder waters of the Labrador Current.

Upslope Fog

Another type of fog clings to windward sides of mountain slopes and is known as **upslope fog.** Its appearance is sometimes the source of geographic place names—for example, the Great Smoky Mountains where this type of fog is common. During early morning hours in middle-latitude locations, moist air may ascend a slope and cool to the dew point, forming a fog blanket (Fig. 5.9c). In wet tropical areas, mountain slopes may be covered in a misty fog at any time of day. Because of the very humid air in those regions, reaching the dew point may result from only a minor drop in temperature.

Dew and Frost

Dew is made up of tiny water droplets formed by the condensation of water vapor on cool surfaces, like plants, buildings, and metal objects. Dew collects on surfaces that are good radiators of heat (such as cars or blades of grass) that give up large amounts of heat during the night. If air cools below its initial dew point when it comes in contact with these cold surfaces, water droplets form in beads on the surface. If that temperature is below 0°C (32°F), **frost** forms. It is important to note that frost is not frozen dew, but results from a *sublimation* process—water vapor changing directly from the gaseous to the frozen state (see again Figure 3.16).

Clouds

Clouds are the source of all precipitation. **Precipitation** consists of atmospheric water, either liquid or solid, that falls to Earth. Of course, not all clouds produce precipitation, but precipitation will not occur without the formation of a cloud first. Clouds are important factors in the heat energy budget. They absorb some of the incoming solar energy, reflect some of that energy back to space, and scatter or diffuse other wavelengths of energy to and away from Earth. In addition, clouds absorb some of Earth's radiation and reradiate that heat energy back to the surface. Clouds are an ever-changing, and often beautiful, aspect of our environment.

Cloud Forms

Clouds are composed of billions of tiny water droplets and/or ice crystals so small (some measured in 1000ths of a millimeter) that they remain suspended in the atmosphere. Clouds can appear white, shades of gray, or even black. The thicker a cloud is, the more sunlight it is able to absorb and block from our view, and the darker it will appear.

Cloud names generally (but not always) consist of two parts. The first part refers to the cloud's height: low-level clouds, below 2000 meters (6500 ft), are called **strato;** middle-level clouds, from 2000 to 6000 meters (6500–19,700 ft), are named **alto;** and high-level clouds, above 6000 meters (19,700 ft), are termed **cirro.**

The second part of the name concerns the form, or shape, of the clouds. The three basic shapes are termed *cirrus, stratus,* and *cumulus.* Classification systems categorize cloud formations into many subtypes, but most clouds are variations of the three basic shapes. ■ Figure 5.10 illustrates the appearance and the general heights of common clouds; ■ Figure 5.11 provides images of the major cloud types.

Cirrus clouds (from Latin: *cirrus,* a lock or wisp of hair) form at very high altitudes, normally 6000–10,000 meters (19,800–36,300 ft), and are made up of ice crystals (Fig 5.11a). They are thin, stringy, white clouds that trail like feathers across the sky. When associated with fair weather, cirrus clouds are scattered white patches in a clear blue sky.

Cumulus clouds (from Latin: *cumulus,* heap or pile) develop vertically rather than forming the more horizontal structures of the cirrus and stratus types (Fig. 5.11b). Cumulus are puffy and rounded, usually with a flat base, which can be anywhere from 500 to 12,000 meters (1650–39,600 ft) above sea level. From this base, they develop into great rounded structures, often with tops like cauliflower. Cumulus clouds provide visible evidence of an unstable atmosphere; and their base is the point where condensation has begun in a column of air as it rises upward.

Stratus clouds (from Latin: *stratus,* layer) appear at lower altitudes from the surface up to almost 6000 meters (19,800 ft). The basic characteristic of stratus clouds is their horizontal appearance, in layers of fairly uniform thickness (Fig. 5.11c). The horizontal configuration indicates that they form in stable atmospheric conditions, which inhibit vertical development.

Often stratus clouds cover the entire sky with gray cloud layers, producing dull, gray, overcast skies. A stratus cloud

■ **FIGURE 5.10** Clouds are named based on their height and their form.
Observe this figure and Figure 5.11; what cloud type is present in your area today?

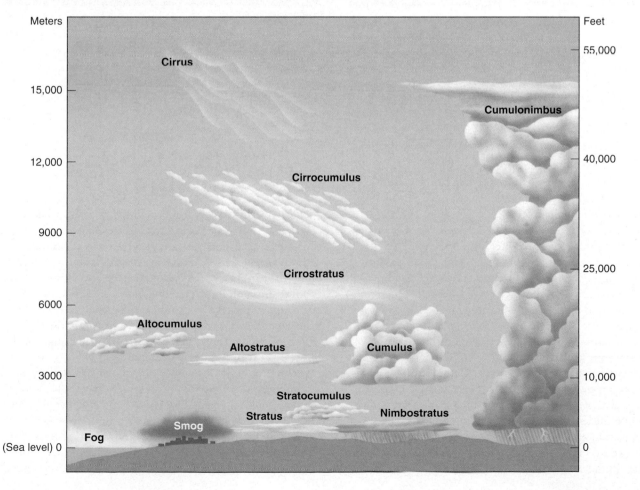

(a) Cirrus

(b) Cumulus

(c) Stratus

(d) Cumulonimbus

■ **FIGURE 5.11**   Four basic cloud types. (a) Cirrus clouds are feathery looking clouds consisting of ice crystals at high altitude. (b) Cumulus clouds are puffy-looking clouds, often with flat bottoms that indicate where rising air has cooled to the dew point temperature. (c) Stratus clouds are low altitude layered clouds that, if dense, are often associated with gray skies and overcast weather. (d) Cumulonimbus clouds are towering rain clouds that are undergoing strong and high vertical uplift that can bring heavy hail and gusty winds.

formation may lie over an area for days, and any precipitation will be light but steady and persistent.

Refer again to Figure 5.10 to become familiar with the basic cloud types. Some cloud shapes exist in all three levels—for example, *stratocumulus* (*strato* = low level + *cumulus* = a rounded shape), *altocumulus,* and *cirrocumulus* all share the rounded appearance of cumulus clouds. *Altostratus* (*alto* = middle level + *stratus* = layered shape) and *cirrostratus* have two-part names, but low-level layered clouds are simply called *stratus.* Lastly, thin, stringy *cirrus* clouds are found only at high levels, so the term *cirro* (meaning high-level) is not necessary.

Other terms used in describing clouds include **nimbo,** or **nimbus,** meaning precipitation (rain) is falling. A *nimbostratus* cloud may bring a long-lasting drizzle. **Cumulonimbus** is the thunderstorm cloud, with a flat top, called an anvil head, as well as a relatively flat base, and it becomes darker as it grows higher and thicker, blocking the incoming sunlight (Fig. 5.11d). Cumulonimbus clouds are the source of many atmospheric concerns including strong winds, torrential rain, flash flooding, thunder, lightning, hail, and possibly tornadoes.

Adiabatic Heating and Cooling

Clouds typically develop from cooling that results when a parcel of air rises. The rising parcel of air will expand as it encounters decreasing atmospheric pressure with altitude. This expansion allows the air molecules to spread out, which causes the parcel's air temperature to decrease. This is known as **adiabatic cooling;** the temperature decreases at the lapse rate of approximately 10°C per 1000 meters (5.6°F/1000 ft). Air descending through the atmosphere is compressed by the increasing pressure and undergoes **adiabatic heating,** which increases air temperature by the same rate.

However, the rising and cooling air parcels will eventually reach the dew point temperature and water vapor will condense and form cloud droplets. After condensation occurs, the adiabatic cooling of a rising parcel will decrease at a lower rate because latent heat of condensation is being released into the air. To differentiate between these two adiabatic cooling rates, we refer to the pre-condensation rate (10°C/1000 m) as the **dry adiabatic lapse rate** and the lower, post-condensation rate as the **wet adiabatic lapse rate.** The latter rate averages 5°C per 1000 meters (3.2°F/1000 ft).

A rising air parcel will cool at either the dry or the wet adiabatic rate. Which rate is operating depends on whether condensation is occurring (wet adiabatic rate) or is not occurring (dry adiabatic rate). As air descends the temperature will continually warm by compression, increasing its capacity to hold water vapor and preventing condensation. Thus, the temperature of descending air that is being compressed always increases at the dry adiabatic rate. Adiabatic temperature changes result from changes in volume and do not involve the addition or subtraction of heat from external sources.

It is extremely important to differentiate between the *environmental lapse rate* (where a temperature measuring *device* is moving up or down) and the *adiabatic lapse rates* (where the *air* is moving up or down). In Chapter 3 we learned that, in general, atmospheric temperatures decrease with increasing altitude. This is the environmental lapse rate (also called the normal lapse rate), which is variable, but averages 5.5°C per 1000 meters (3.6°F/1000 ft) and is measured by meteorological instruments sent aloft. The environmental lapse rate reflects the vertical temperature structure of the atmosphere. Adiabatic lapse rates indicate temperature changes that result when air moves up or down, whether or not condensation is occurring (■ Fig. 5.12).

Instability and Stability

An air parcel will be buoyant and rise as long as it is warmer than the surrounding air. When it reaches a layer of the atmosphere that is the same temperature as itself, it will stop rising. Air parcels that rise because they are warmer than the surrounding atmosphere are said to be *unstable* (also *instable*). In contrast, air that is colder than the surrounding atmosphere tends to sink to lower levels. Sinking air is said to be *stable.*

Determining the stability or instability of an air parcel involves answering a fairly simple question. If an air parcel were lifted to a specific altitude (cooling at an *adiabatic* lapse rate), would it be warmer, colder, or the same temperature as the surrounding air (as determined by the *environmental* lapse rate) at that same altitude?

If the air parcel is warmer than the air at the specified altitude, then the parcel will be unstable and will continue to rise, because warmer air is less dense and therefore buoyant. This condition is called **instability** (■ Fig. 5.13). An air parcel that is colder than the surrounding air would sink back toward Earth as a result of its greater density, causing conditions of **stability.**

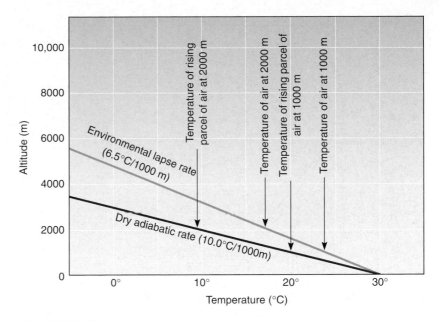

■ **FIGURE 5.12**   Comparison of the dry adiabatic lapse rate and the environmental, or normal, lapse rate. The environmental lapse rate is the average vertical change in temperature. Air displaced upward will cool (at the dry adiabatic rate) because of expansion.

In this example, using the environmental lapse rate, what is the air temperature at 2000 meters?

Whether an air parcel will be stable or unstable is related to the cooling and heating of air at Earth's surface. With cooling of air through radiation and conduction on a cool, clear night, air near the surface will be relatively close in temperature to the air aloft, thus enhancing stability. With rapid surface heating on a hot summer day, the air near the surface becomes much warmer than that above, and will become unstable. Pressure zones can also be related to atmospheric stability. In areas of high pressure, stability is maintained by slowly subsiding air from aloft. In low pressure regions instability is promoted by the tendency for air to rise.

Precipitation Processes

Condensed water droplets float around within clouds and do not fall to Earth because they are so tiny (0.02 mm, or less than 1000th of an inch) that gravity does not overcome the buoyant effects of air and the currents and updrafts that exist in clouds. ■ Figure 5.14 shows the relative sizes of a condensation nucleus, a cloud droplet, and a raindrop. It takes about a million cloud droplets to form one raindrop.

Precipitation occurs when droplets of water or ice crystals become too large and heavy to be held aloft, and fall as rain, snow, sleet, or hail. The type of precipitation depends largely on how it formed and what the temperature is in the cloud and below as it falls to Earth. The most widely accepted theories for how precipitation develops include the **collision–coalescence**

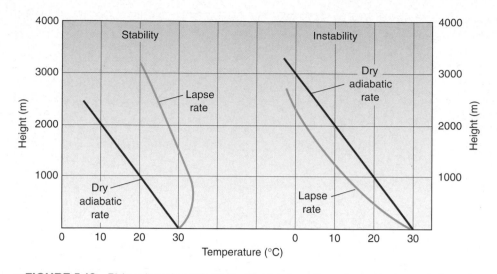

■ **FIGURE 5.13**   Rising air cools adiabatically. Whether it continues to rise or does not depends on whether adiabatic cooling is less rapid or more rapid than the normal lapse rate. Stability occurs where the adiabatic cooling rate exceeds the normal lapse rate (left graph). Thus, lifted air will become colder than the surrounding air and tend to sink. Unstable air occurs where the adiabatic cooling rate is less than the normal lapse rate (right graph). Thus, lifted air will be warmer than its surroundings, buoyant, and continue to rise even after the original lifting force is removed. **In these examples, what would the air temperature be at 2000 meters if the air at the surface rose to this level?**

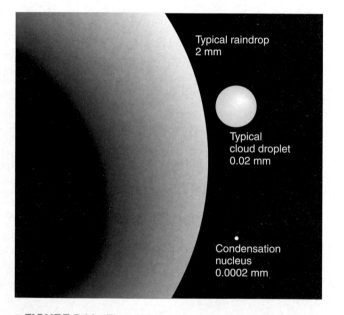

■ **FIGURE 5.14**   The relative sizes of raindrops, cloud droplets, and condensation nuclei.

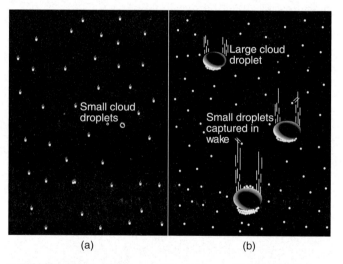

■ **FIGURE 5.15**   Collision and coalescence. (a) Tiny cloud droplets falling at about the same speed are unlikely to collide and coalesce, and if they do collide, they tend to bounce off of each other because of the surface tension of water. (b) Large droplets falling rapidly can capture some of the smaller droplets.

process for warm clouds and the **Bergeron** (or **ice crystal**) **process** for cold clouds.

Precipitation in the tropics and in warm clouds is likely to form by collision–coalescence, a process that is well described by its name. Water is quite cohesive (able to stick to itself), so as water droplets collide while circulating in a cloud, they tend to coalesce (or grow together) until they are large and heavy enough to fall. In falling, larger droplets overtake smaller, more buoyant droplets and capture them to form larger raindrops. This process occurs in the warmer clouds where the moisture exists as liquid water (■ Fig. 5.15).

At higher latitudes, many storm clouds possess three distinctive layers. The lowermost is a warm layer where the temperatures are above the freezing point of 0°C (32°F) and water droplets are liquid. Above this, the second layer is composed of some ice crystals but mainly **supercooled water** (liquid water cooler than 0°C). In the uppermost layer of these tall clouds, if temperatures are lower than or equal to –40°C (–40°F), ice crystals will dominate (■ Fig. 5.16). It is in relation to these layered clouds that Scandinavian meteorologist Tor Bergeron presented his explanation.

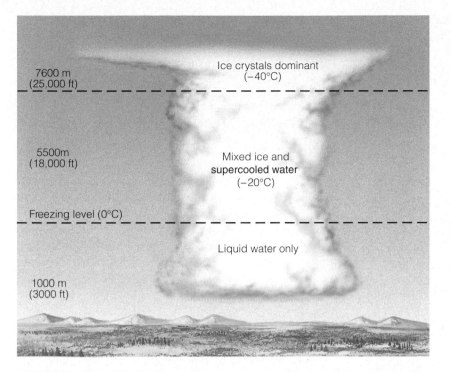

7600 m
(25,000 ft)

Ice crystals dominant
(−40°C)

5500m
(18,000 ft)

Mixed ice and
supercooled water
(−20°C)

Freezing level (0°C)

Liquid water only

1000 m
(3000 ft)

■ **FIGURE 5.16** The distribution of water, supercooled water, and ice crystals in a cumulonimbus storm cloud.

What is the difference between water and supercooled water?

The Bergeron (or ice crystal) process begins at great heights in the ice crystal and supercooled water layers of the clouds. Supercooled water has a tendency to freeze on any available surface. It is for this reason that aircraft flying through middle- to high-latitude thunderstorms run the risk of severe icing and potential disaster. The ice crystals become *freezing nuclei* upon which the supercooled water can freeze to form growing ice crystals. This process can also create snow. If the frozen precipitation falls through lower cloud layers that have above-freezing temperatures, the ice crystals melt and fall as liquid rain. Finally, as raindrops fall through the warmer section of the cloud, the collision–coalescence process may cause the raindrops to grow larger.

Forms of Precipitation

Rain, consisting of droplets of liquid water, is by far the most common form of precipitation. Raindrops vary in size but are generally about 2–5 millimeters (approximately 0.1–0.25 in.) in diameter (see again Figure 5.14). As we all know, rain can come in many ways: as a brief afternoon shower, as a steady rainfall, or in the deluge of a heavy rainstorm. When the temperature of an air mass is only slightly below its initial dew point, the raindrops may be very small (about 0.5 mm or less in diameter). The result is a fine mist called **drizzle.**

Snow is the second most common form of precipitation. When water vapor freezes directly into a solid without first passing through a stage as liquid water, it forms minute ice crystals around the freezing nuclei. These grow into hexagonal ice crystals that make up the intricate six-sided form of snowflakes. Snow will reach the ground only if the cloud and the air below

maintain subfreezing temperatures (below 0°C or 32°F).

Sleet is rain that freezes as it falls through a thick layer of subfreezing air near the surface. The cold air causes raindrops to freeze into small solid particles of clear or milky ice.

Hail is a less common form of precipitation than rain, snow, or sleet. It occurs most often during the spring and summer months as a result of thunderstorm activity. Hail forms as lumps of ice, called *hailstones,* which can vary in size from 5 millimeters (0.2 in.) in diameter to larger than a baseball (■ Fig. 5.17). The United States record is a hailstone of 17.8 cm (7 in.) in diameter with a circumference of 47.6 cm (18.75 in.) that fell in Aurora, Nebraska, in 2003. Dropping from the sky, hailstones can be highly destructive to livestock, crops, and other vegetation, as well as to vehicles and buildings.

Hail forms when ice crystals are lifted by strong updrafts in cumulonimbus (thunderstorm) clouds. As these ice crystals circulate within the cloud, they collide with supercooled water droplets that freeze onto the ice, which grows in accumulating layers. The resulting hailstones, made up of concentric ice layers, have a frosty, opaque appearance after breaking out of the strong updrafts in the cloud and falling to Earth. The larger the hailstone, the more times it has accumulated additional frozen layers.

On occasion, raindrops can have a temperature below freezing yet maintain liquid form. These supercooled droplets will instantly freeze if they fall onto a surface that is also at a subfreezing temperature. The resulting ice that may coat just certain objects or the entire landscape is known as **freezing rain** (or **glaze**). People usually call this kind of precipitation an *ice storm* (■ Fig. 5.18). Because of the weight of ice, glazing can break tree branches that bring down telephone and power lines. Surface ice accumulations cause extremely slippery conditions

■ **FIGURE 5.17** Hailstones can be the size of golf balls, or even larger.

What gives them their spherical appearance?

■ **FIGURE 5.18** An ice storm covers the landscape with a dangerous glaze of ice.

Why are power failures a common occurrence with ice storms?

that make driving dangerous and some roads impassable. Yet, ice storms can also produce a beautiful natural landscape. Sunlight glitters on the ice, reflecting and making a diamond-like surface that covers vegetation, buildings, and vehicles.

Factors Necessary for Precipitation

Three factors are necessary for precipitation to develop. The first is the presence of *moist air,* which provides a moisture source (for precipitation) and energy (as latent heat of condensation). Second are the *condensation nuclei* around which the water vapor can condense. Third is an **uplift mechanism** that forces the air to rise enough to cool (by the dry adiabatic rate) to the dew point temperature. Precipitation results from one of four major uplift mechanisms that force parcels of air to rise and condense: *convectional, frontal, cyclonic (convergence),* and *orographic lifting.* (■ Fig. 5.19).

Convectional Precipitation Convection occurs as air, heated near the surface, expands, becomes lighter, and rises. **Convectional precipitation** is most common in the hot and humid tropical and equatorial areas, and during the summer in many middle latitude locations. Once condensation begins in a convectional column of air, further lifting will be encouraged by additional energy from the release of latent heat of condensation. Convectional uplift can cause the heavy precipitation, thunder, lightning, and tornadoes of spring and summer afternoon thunderstorms. The strong convectional updrafts that occur in towering cumulonimbus clouds frequently produce hail along with the thundershowers.

■ **FIGURE 5.19** Four uplift mechanisms for air. The principal cause of precipitation is upward movement of moist air resulting from convectional, frontal, cyclonic, or orographic lifting.

What kind of air movement is common to the depictions in all four diagrams?

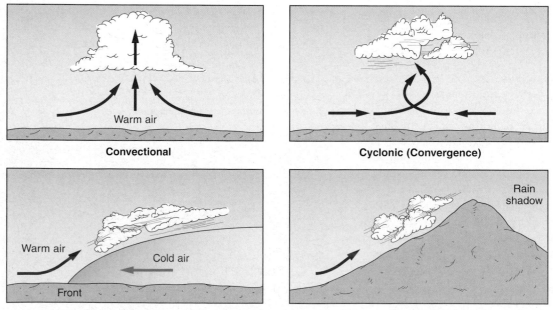

Convectional

Cyclonic (Convergence)

Frontal

Orographic

Frontal Precipitation

The zone of contact between relatively warm and relatively cold bodies of air is known as a **front**. As was mentioned in the previous chapter, the concept of a weather *front* comes from the term for the line of contact between opposing armies. When two large bodies of air that differ in temperature, humidity, and density collide, the warmer, less dense air mass is lifted above the colder body of air. Collision causes uplift and uplift results in cooling—producing condensation and precipitation. **Frontal precipitation** develops as warm moisture-laden air collides with cold air and rises above the front. Depending on many factors, especially the temperature and moisture content of the clashing *air masses*, frontal precipitation can produce many kinds of weather, from cloudy overcast conditions to rain, snow, or ice storms.

To better understand fronts, we will later examine what causes unlike bodies of air to collide, as well as the weather conditions that are associated with different kinds of fronts. These topics will be discussed in Chapter 6.

Cyclonic Precipitation

Cyclonic uplift (convergence) was introduced in Chapter 4 (see Fig. 4.3) and involves air interacting with a cell of low pressure (*cyclone*). Air will flow from all directions into a low pressure system that is roughly circular; the direction of inflow is counterclockwise in the Northern Hemisphere. When air *converges* on a cyclone, it is pulled into the rising air of the low pressure cell. Therefore, cloudy conditions and precipitation are common around the center of a cyclone. Hurricanes and the rain associated with these storms (**cyclonic precipitation**) are fed by air flowing into the convergent uplift around a circular low pressure system, and the energy resulting from the release of latent heat of condensation.

Orographic Precipitation

When land barriers—such as a mountain range, a hilly region, or the escarpment (steep edge) of a plateau—lie in the path of prevailing winds, air is forced to rise above these barriers. Masses of air are cooled by expansion as they ascend over a topographic barrier, and condensation takes place. The resultant precipitation is termed **orographic precipitation** (from Greek: *oros*, mountains).

As orographic precipitation falls on the windward side of a mountain, the air parcel and the clouds lose some of their moisture content (the *absolute* or *specific humidity* declines). However, continued cooling as the air rises can maintain the relative humidity at 100%; precipitation continues as long as the air contains adequate moisture and keeps rising. As the air descends the leeward slope, its temperature warms (at the dry adiabatic rate), and condensation ceases. The leeward side is in what is called the **rain shadow** (■ Fig. 5.20a). Just as being in

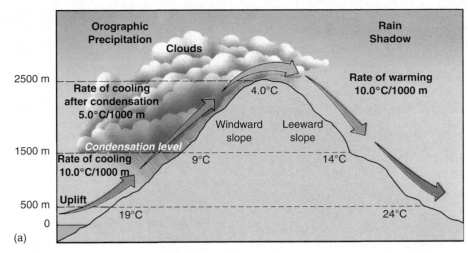

(a)

R. Gabler

(b)

R. Gabler

(c)

■ **FIGURE 5.20** Orographic precipitation and the rain shadow effect. (a) Orographic uplift over the windward (western) slope of the Sierras produces condensation, cloud formation, and precipitation, resulting in (b) dense stands of forest. (c) Semiarid or rain-shadow conditions occur on the leeward (eastern) slope of the Sierras.
Can you identify a mountain range in Eurasia on its leeward side?

a shadow means that you are not receiving any direct sunlight, being in the rain shadow means that an area does not receive much rain (or other precipitation). If you live near a mountain range, you can see the effects of orographic precipitation and the rain shadow in the vegetation patterns (Figs. 5.20b and c). The windward side of mountains (for example, the Sierra

:: THE LIFTING CONDENSATION LEVEL (LCL)

When you look at clouds, you can see that certain cloud types have relatively flat bases. Cloud tops may be quite irregular, but cloud bases are often flat and lie at the same altitude. This level is the altitude to which the air must be lifted (and cooled at the dry adiabatic rate) before saturation is reached. Any additional lifting and clouds will form and build upward.

The height at which clouds form from rising air is the lifting condensation level (LCL) and can be estimated by the equation:

$$\text{LCL (in meters)} = 125 \text{ meters} \times (\text{Celsius temperature} - \text{Celsius dew point temperature})$$

For example, if the surface temperature is 7.2°C (45°F) and the dew point temperature is 4.4°C (40°F), then the LCL is estimated at 350 meters (1148 ft) above the surface.

Note: Different layers of clouds may exist at the same time. Low, middle, and high clouds may all appear on the same day. These clouds may have formed in other regions and be only passing overhead. The formula presented here is best used with the lowest level of cloud cover that appears overhead.

J. Peterson

The flat bottom of these clouds shows the lifting condensation level (LCL), the altitude where the dew point temperature is reached, 100% relative humidity exists, and condensation has occurred.

Nevada in California) will be heavily forested. The opposite slopes in the rain shadow will be drier, usually with a sparse cover of vegetation.

Distribution of Precipitation

Distribution Over Time To understand the significance of a location's precipitation, we consider *average annual precipitation* to get an impression of the mois-

ture that a region gets during a year. We can also look at the annual or monthly number of *rain days*—days when 1.0 millimeter (0.01 in.) or more of rain was received during a 24-hour period. Less than this amount is known as a *trace* of rain.

We should also consider the *average monthly precipitation*. By examining the precipitation data for all twelve months, we can get an idea of seasonal variations in precipitation for a location (■ Fig. 5.21). For instance, in describing the climate of California, average annual precipitation would not give the full story because an annual average would not show the distinct wet and dry seasons that

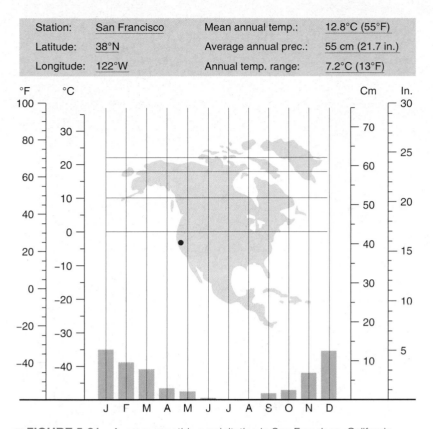

Station:	San Francisco	Mean annual temp.:	12.8°C (55°F)
Latitude:	38°N	Average annual prec.:	55 cm (21.7 in.)
Longitude:	122°W	Annual temp. range:	7.2°C (13°F)

■ **FIGURE 5.21** Average monthly precipitation in San Francisco, California, is represented by colored bars. A graph of monthly precipitation gives a good impression of the seasonal weather in a place. If we had only the annual precipitation total, we would not know that nearly all the precipitation occurs in only half of the year.

How would this kind of rainfall pattern affect agriculture?

characterize this region; only monthly averages would give us that information.

Latitudinal Distribution

Great geographic variability exists in the global distribution of precipitation. ■ Figure 5.22 shows the average annual precipitation of Earth's land areas. Latitude has a strong impact on precipitation distribution because the occurrence, absence, and variations of many weather and climate factors are related to latitude. For example, because warmer air can hold more water vapor and colder air can hold less, there is a general decrease of precipitation from the equator to the poles.

The equatorial zone is generally an area of high precipitation—typically more than 200 cm (79 in.) annually. High temperatures and instability in the equatorial region lead to a general pattern of rising air, which generates precipitation. This tendency is reinforced by the convergence of the trade winds as they flow toward the equator from opposite hemispheres. In fact, the *Intertropical Convergence Zone* is one of the two great zones where air masses converge. The other is along the *Polar Front* within the westerlies.

In general, the air of the trade wind zones is stable compared with the instability in the equatorial region. Under the influence of these steady winds, there are few atmospheric disturbances that would lead to convergent or convectional lifting. However, because the trade winds have an easterly flow, when they move onshore along east coasts or islands with high elevations, they carry moisture from the oceans. Thus, within the trade wind belt, east coasts tend to be wetter than west coasts.

In fact, where the air of the equatorial and trade wind regions—with its high temperatures and vast amounts of moisture—moves onshore from the ocean and meets a landform barrier, record rainfalls can be measured. The windward slope of Mount Waialeale on Kauai, Hawaii, at approximately 22°N latitude, holds the world's record for greatest average annual rainfall—1198 cm (471 in.).

Moving poleward from the trade wind belts, we enter the subtropical high pressure zones where the air is subsiding. As it sinks, it is warmed adiabatically, increasing its moisture-holding capacity and reducing the precipitation in this area. In fact, if we look at ■ Figure 5.23, we can see a drop in precipitation amounts that corresponds to the latitudes of the subtropical high pressure cells. These zones of subtropical high pressure are where we find most of the great deserts of the world: in northern and southern Africa, Arabia, North America, and Australia. The exceptions to this subtropical aridity occur along the east sides of continents, where the subtropical high pressure cells are weak and wind direction is onshore. This exception is especially true of the monsoon regions.

In the zones of the westerlies, from about 35° to 65°N and S latitude, precipitation occurs largely from the collision of cold, dry polar air masses with warm, humid subtropical air masses along the polar front. Thus, there is much frontal precipitation in this zone.

In the middle latitudes, the continental interiors are drier than the coasts because they are farther away from the oceans. However, where air in the prevailing westerlies is forced to rise, as it does when it crosses the Cascades Range and Sierra Nevada of the Pacific Northwest and California, especially during the winter months, there is heavy orographic precipitation. Thus, in the middle latitudes, continental west coasts tend to be wet, and precipitation decreases eastward toward the continental interiors. Along eastern coasts within the westerlies, precipitation usually increases once again because of proximity to humid air from the oceans. Here, convection and the convergence associated with hurricanes bring precipitation, mainly in the summer months.

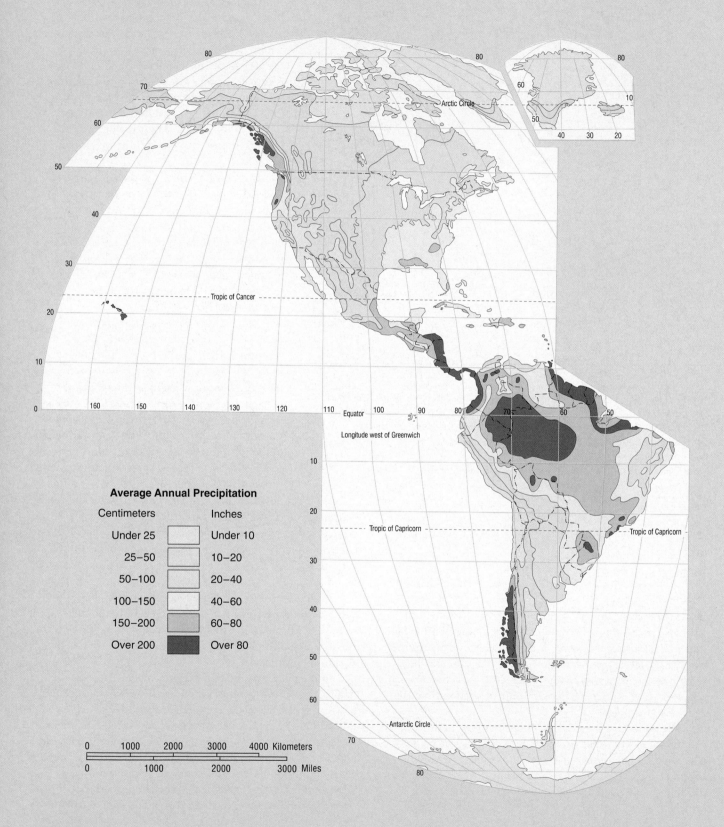

Average Annual Precipitation

Centimeters		Inches
Under 25		Under 10
25–50		10–20
50–100		20–40
100–150		40–60
150–200		60–80
Over 200		Over 80

■ **FIGURE 5.22** World map of average annual precipitation.
In general, where on Earth's surface does the heaviest rainfall occur? Why?

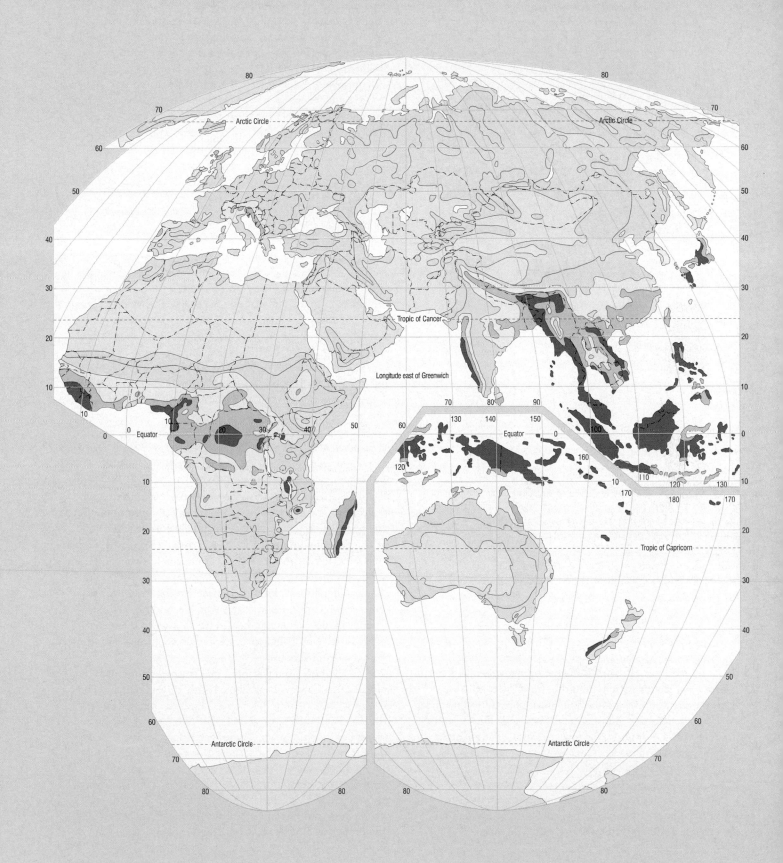

A Western Paragraphic Projection
developed at Western Illinois University

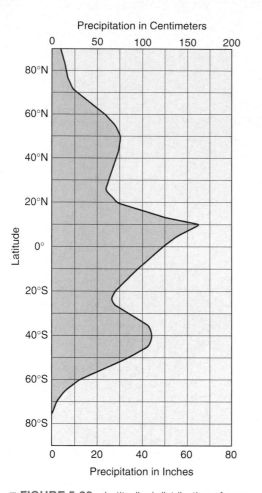

■ **FIGURE 5.23** Latitudinal distribution of average annual precipitation. Earth has distinctive precipitation zones: between the tropics, with high precipitation caused by air converging, the middle latitudes with precipitation associated with the polar front, and zones of low precipitation caused by subsiding air in the subtropical and polar regions.

Compare this graph with Figure 4.10. What is the relationship between world rainfall patterns and world pressure distribution?

In the United States, the interior lowlands are not as dry as we might expect within the prevailing westerlies. This is because of frontal activity resulting from the conflicting northward and southward movements of polar and subtropical air. If a high east–west mountain range extended from central Texas to northern Florida, the lowlands of the continental United States north of that range would be much drier because the mountains would block the moist air from the Gulf of Mexico.

Also characteristic of the westerlies are desert areas in the rain shadows of mountain ranges. This is one reason for the extreme aridity of California's Death Valley, as well as the deserts of eastern California and Nevada, the mountain-ringed deserts of eastern Asia, and Argentina's Patagonian Desert, which is in the rain shadow of the Andes. Note in Figure 5.23 that there is greater precipitation in the middle latitudes of the Southern Hemisphere, where the oceans cover more area than the continents, unlike the northern middle latitudes.

Moving poleward, low temperatures lead to low evaporation rates. In addition, the polar regions are generally areas of subsiding air and high pressure. These factors combine to cause low precipitation amounts in the polar zones.

Precipitation Variability

The rainfall amounts depicted in Figure 5.22 are annual averages. However, in many parts of the world there are significant variations in precipitation, both within any one year and between years. For example, areas like the Mediterranean region, California, Chile, South Africa, and Western Australia, located on the west sides of the continents roughly between 30° and 40° latitude, receive much more rain in the winter than in the summer. There are also areas between 10° and 20° latitude that receive much more of their precipitation in the summer (high-sun season) than in the winter (low-sun season).

Rainfall totals can change dramatically from one year to the next, and unfortunately for many of the world's people, the drier a place is on average, the greater will be the variability in its precipitation (compare ■ Fig. 5.24 with Fig. 5.22). To make matters worse for people in arid or semiarid regions, a year with a particularly high amount of rainfall may be balanced with several years of below-average precipitation. This situation has occurred recently in West Africa's Sahel, the Russian steppe, and the American Great Plains.

Thus, there can be years of drought and years of flood, each bringing its own kind of disaster. Farmers, resort owners, construction workers, and others whose economic well-being depends in one way or another on weather are at the mercy of highly variable probabilities of rainfall on an annual, monthly, or even a seasonal basis.

We all know from watching the weather reports that rainfall cannot always be predicted with 100% accuracy. This inaccuracy results from the interaction of the many factors involved in producing precipitation—temperature, moisture, atmospheric disturbances, landform barriers, fronts, air mass movement, upper air winds, and differential surface heating, among others. These weather factors, of course, also affect our daily lives. Using our knowledge of why various precipitation types develop and where they are likely to occur, we can begin to understand the reasons for the many patterns of environmental diversity that exist on our planet.

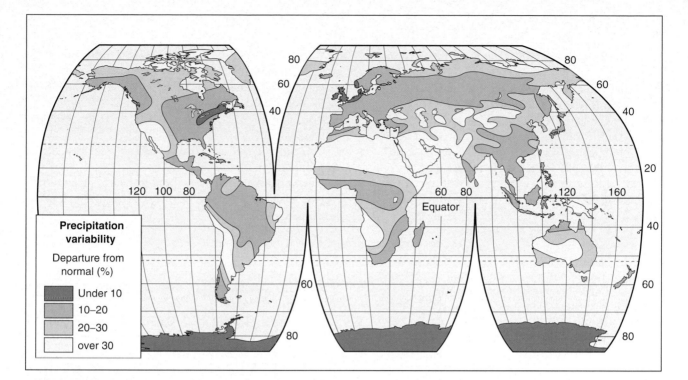

■ **FIGURE 5.24**   World map of precipitation variability. The greatest variability in annual precipitation totals occurs in the dry regions, accentuating the critical problem of moisture supply in those parts of the world. **Compare this map with Figure 5.22. What are some of the similarities and differences?**

:: Terms for Review

capillary action	frost	Bergeron (ice crystal) process
hydrologic cycle	precipitation	supercooled water
saturation	strato	rain
capacity	alto	drizzle
dew point	cirro	snow
humidity	cirrus	sleet
absolute humidity	cumulus	hail
specific humidity	stratus	freezing rain (glaze)
relative humidity	nimbus (nimbo)	uplift mechanism
transpiration	cumulonimbus	convectional precipitation
evapotranspiration	adiabatic cooling	front
potential evapotranspiration	adiabatic heating	frontal precipitation
condensation nuclei	dry adiabatic lapse rate	cyclonic (convergence) precipitation
radiation fog (*ground fog*)	wet adiabatic lapse rate	orographic precipitation
advection fog	instability	rain shadow
upslope fog	stability	
dew	collision–coalescence process	

:: Questions for Review

1. How is the hydrologic cycle related to Earth's water budget?
2. What is the difference between absolute and specific humidity? What is relative humidity?
3. Imagine that you are deciding when, in your daily schedule, to water the garden. What time of day would be best for conserving water? Why?
4. How is evapotranspiration related to the water budget of a region?
5. What factors affect the formation of temperature-inversion fogs?
6. What causes adiabatic cooling? How is it different from the environmental lapse rate?
7. How and why does the wet adiabatic lapse rate differ from the dry adiabatic lapse rate?
8. What atmospheric conditions are necessary for precipitation to occur?
9. Compare and contrast convectional, orographic, cyclonic, and frontal precipitation.
10. How is rainfall variability related to total annual rainfall? How might this relationship be considered a double problem for people?

:: Practical Application

1. Refer to Figure 5.5 to determine the following.
   a. What is the water vapor capacity of air at 0°C? 20°C? 30°C?
   b. If a parcel of air at 30°C has an absolute humidity (actual water vapor content) of 20.5 grams per cubic meter, what is the parcel's relative humidity?
   c. Low relative humidity indoors during the winter is a concern in cold climates. The problem results when cold air, which can hold little water vapor, is brought indoors and heated. Assume that the air outside is 5°C and has a relative humidity of 60%. What is the actual water vapor content of this air? If it comes indoors (through doors and windows) and is heated to 20°C, with no increase in water vapor content, what is the new relative humidity?
2. As air rises, it expands and cools. The cooling, at the dry adiabatic lapse rate, is 10°C per 1000 meters. (Descending air will always warm at this rate.) In addition, the dew point temperature decreases about 2°C per 1000 meters within a rising parcel of air. Above the height where the dew point temperature is reached, condensation occurs and the wet adiabatic lapse rate of 5°C per 1000 meters becomes operational. When the wet adiabatic lapse rate is in operation, the dew point temperature will be equal to the air temperature. When a parcel of air descends, its dew point temperature increases 2°C per 1000 meters. The height at which condensation begins, termed the lifting condensation level (LCL), can be determined by using the formula in this chapter's special section entitled The Lifting Condensation Level (LCL).
   a. An air parcel has a temperature of 25°C and a dew point temperature of 14°C. What is the height of the LCL? If that parcel rises to 4000 meters, what will be its temperature?
   b. A parcel of air at 6000 meters has a temperature of −5°C and a dew point of −10°C. If it descended to 2000 meters, what would be its temperature and dew point temperature?

3. Using the data set below, for each month of the year, calculate:
   a. A running total of precipitation month to month.
   b. The departure from the mean rainfall (in surplus or deficit) for each month.
   c. The departure from the mean (surplus or deficit) for the whole year.
   From that information, answer the following questions:
   d. How does the year begin with respect to surplus or deficit? How does the year end?
   e. Which month accumulated the greatest deficit for the year?
   f. Which month is the first to show a surplus?

Month	Recorded Rainfall (cm)	Mean Rainfall by Month (cm)
January	12.55	12.55
February	6.10	10.41
March	7.67	13.36
April	8.15	11.07
May	9.37	10.72
June	6.83	10.59
July	12.90	10.60
August	13.74	9.17
September	20.04	8.10
October	4.10	6.81
November	10.85	9.45
December	7.04	11.07

Air Masses and Weather Systems

6

:: Outline

Air Masses

Fronts

Atmospheric
Disturbances

Weather Forecasting

An enhanced satellite image
of Hurricane Katrina as
it swirls toward the New
Orleans area.

NASA

:: Objectives

When you complete this chapter you should be able to:

- Outline and explain the major air mass types, their characteristics, and their source regions.
- Describe all four types of fronts and the types of weather that occur with their passage or presence.
- Distinguish between the general atmospheric conditions associated with anticyclones and cyclones.
- Discuss the characteristics of a middle latitude cyclone, the factors that influence its movement, and its stages of development.

- Understand the potentially serious weather conditions, possible damage, and hazards associated with hurricanes, thunderstorms, and tornadoes.
- Explain the difference between weather and climate and be aware of the factors that make weather forecasting a complex process.
- Interpret a weather map and understand the symbols that are used to show atmospheric conditions, fronts, precipitation patterns, pressure cells, air masses, and winds.

In this chapter we will apply much that we have learned about insolation, heat energy, temperature, pressure, wind, and moisture conditions as we examine weather systems and the kinds of storms that accompany them. Understanding Earth's varied weather systems—their general characteristics, when they may occur, how they form, and how they may affect the regions they impact is of major importance in physical geography. In addition to involving temperature change and producing essential precipitation, weather systems are major means of energy exchange, and they can also present hazards, such as floods, damaging winds, lightning, and violent thunderstorms. We will begin the chapter with a detailed study of air masses and fronts, two atmospheric elements that not only affect weather systems, but also strongly influence regional climates and environmental diversity, two important topics to be considered in forthcoming chapters.

Air Masses

An **air mass** is a large body of air, sub-continental in size, which is relatively homogeneous in terms of temperature and humidity. However, the regional extent of an air mass may be 20 or 30 degrees of latitude, so temperature and humidity variations exist from its poleward to equatorward edges. The characteristics, locations, and movement of an air mass exert considerable impacts on the weather, whereas contact with differing land and ocean surfaces can modify air mass temperature and humidity.

The temperature and humidity characteristics of an air mass are determined by the nature of its **source region**—the area where an air mass originates. Only certain areas on Earth make good source regions because they require a nearly homogeneous surface. For example, a source region can be a desert, an ocean area, or a large landmass with relatively small elevational variations, but not a combination of surfaces.

On weather maps, air masses are identified by a two-letter code that refers to its source region. The first letter, always written in lowercase, will be either "m" or "c." The letter *m*, for *maritime*, means the air mass originated over water and is therefore relatively moist. The letter *c*, for *continental*, means the air mass originated over land and is

therefore relatively dry. The second letter, which is always capitalized, refers to the source region's latitudinal zone. *E* stands for *Equatorial*, and this air is very warm. The letter *T* identifies a *Tropical* origin and is also warm air. A *P* represents *Polar*, and this air can be quite cold; an *A* identifies *Arctic* air, which is *very* cold (*AA* is also used for *Antarctic air*). These letters give us the classification symbols for six air mass types: **Maritime Equatorial (*mE*), Maritime Tropical (*mT*), Continental Tropical (*cT*), Continental Polar (*cP*), Maritime Polar (*mP*), and Continental Arctic (*cA*).** Characteristics of these six air masses are described in Table 6.1. From now on, we will usually use the symbols rather than the full names in discussing each air mass type.

Air Mass Modification and Stability

Air masses require sufficient time to acquire the temperature and humidity characteristics of their source region. Once they have done so, they begin moving, typically driven by the divergent circulation of anticyclones. As an air mass travels over Earth's surface, in general, it retains its distinct characteristics. However, temperature and humidity modifications occur as the air mass gains or loses thermal energy or moisture by interacting with landmasses or water bodies. This gain or loss of thermal energy, humidity, or both, can either make an air mass more stable or cause it to become unstable.

If an air mass is colder than the surface that it passes over, heat will flow from the land or water surface below to the air mass. For example, an *mT* air mass originating over the Gulf of Mexico that moves onshore over a hot land surface during the summer will warm further, possibly becoming unstable and producing heavy convective precipitation. In contrast, a similar *mT* air mass moving onshore in winter would be warmer than the land surface and lose heat energy to the land. Consequently, the air mass would be cooled, possibly producing fogs, stratus clouds, or light precipitation.

Air mass modification can also involve a relatively dry air mass that picks up moisture by moving over a water body. During the early winter to midwinter seasons, cold, dry *cP* or *cA* air from Canada can move southeastward across the Great Lakes region. While passing over the lakes, this air mass can pick up moisture, increasing its humidity and it will rise

TABLE 6.1
Types of Air Masses

Source	Region	Usual Characteristics at Source	Accompanying Weather
Maritime Equatorial (*mE*)	Equatorial oceans	Ascending air, very high	High temperature and humidity, heavy moisture content rainfall; never reaches the United States
Maritime Tropical (*mT*)	Tropical and subtropical oceans	Subsiding air; fairly stable but some instability on western side of oceans; warm and humid	High temperatures and humidity, cumulus clouds, convectional rain in summer; mild temperatures, overcast skies, fog, drizzle, and occasional snowfall in winter; heavy precipitation along *mT/cP* fronts in all seasons
Continental Tropical (*cT*)	Deserts and dry plateaus of subtropical latitudes	Subsiding air aloft; generally stable but some local instability at the surface; hot and very dry	High temperatures, low humidity, clear skies, rare precipitation
Maritime Polar (*mP*)	Oceans between 40° and 60° latitude	Ascending air and general instability, especially in winter; mild and moist	Mild temperatures, high humidity; overcast skies and frequent fogs and precipitation, especially during winter; clear skies and fair weather common in summer; heavy orographic precipitation, including snow, in mountainous areas
Continental Polar (*cP*)	Plains and plateaus of subpolar and polar latitudes	Subsiding and stable air, especially in winter; cold and dry	Cool (summer) to very cold (winter) temperatures, low humidity; clear skies except along fronts; heavy precipitation, including winter snow, along *cP/mT* fronts
Continental Arctic (*cA*)	Arctic regions, Greenland, and Antarctica (cAA)	Subsiding very stable air; very cold and very dry	Seldom reaches United States, but when it does, bitter cold, subzero temperatures, clear skies, often calm conditions

slightly as it moves over the relatively warmer lake. When this modified *cP* or *cA* air reaches frigid land on the leeward shores of the Great Lakes, large amounts of *lake-effect snows* can accumulate. On satellite imagery these snowy areas can be clearly seen downwind from the lakes (■ Fig. 6.1). Lake-effect snows diminish in late winter as the lakes freeze, which cuts off the moisture supply to air masses that flow across them.

North American Air Masses

Most of us are familiar with the weather of the United States or Canada; therefore, this chapter will concentrate on the air masses of North America and their impacts on weather conditions. The processes involving North American air masses are also generally applicable worldwide, and are important to understanding the global climate regions that will be addressed in the following chapters.

Five types of air masses (*cA, cP, mP, mT,* and *cT*) influence the weather of North America. Because the middle-latitudes are located between several source regions, a great many storms and precipitation events in that latitudinal zone involve the collision of unlike air masses (■ Fig. 6.2). Further, as the source regions change with the seasons, primarily because of changing insolation, the air masses also will vary according to the season.

■ **FIGURE 6.1** This lake effect snow accumulated on the eastern shore as a storm moved from the west across Lake Michigan.
What two main factors contribute to increased precipitation caused by the lake effect?

Nasa/Johnson Space Center

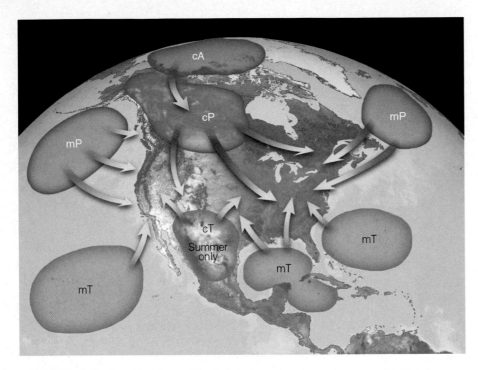

■ FIGURE 6.2 Source regions of North American air masses. Air mass movements import the temperature and moisture characteristics of these source regions into far distant areas.
Use Table 6.1 and this figure to determine which air masses affect your location. Are there seasonal variations?

Continental Arctic Air Masses (cA)

The frozen Arctic Ocean in winter and the frigid land surface of far northern Canada and Alaska serve as source regions for this air mass type. Extremely cold and very dry, during the winter, *cA* air masses can push south of the Canadian border. When continental Arctic air extends down into the Midwestern or even the southeastern United States, record-setting cold temperatures are typically experienced. If a *cA* air mass remains in regions that are not accustomed to extreme cold for extended periods of time, vegetation can be severely damaged or killed. Water pipes may also freeze and break, as they often are not as well insulated compared to those in the regions that expect hard freezes in the winter.

Continental Polar Air Masses (cP)

At their source in north-central North America, *cP* air masses are cold, dry, and stable, resulting in clear, cold, weather. Because North America has no east–west trending landform barriers, *cP* air can migrate into southern Canada, and in the United States as far south as the Gulf of Mexico or Florida. The movement of a continental polar air mass into the Midwest or South brings a cold wave characterized by clear, dry air and colder-than-average temperatures, sometimes with freezing conditions as far south as Florida and Texas. The westerly circulation in the middle latitudes rarely allows a *cP* air mass to move westward across the mountain ranges to the West Coast. When a *cP* air

mass does reach Washington, Oregon, or California, it brings with it unusual freezing temperatures that can do great damage to agriculture.

Maritime Polar Air Masses (mP)

During winter months, when oceans tend to be warmer than the land, *maritime polar* air masses, while damp and cool, tend to be warmer than their counterparts on land (*cP* air masses). Maritime polar air masses form in the northern Pacific Ocean, and move with the westerly circulation to affect the weather of the northwestern United States and southwestern Canada. When *mP* air meets an uplift mechanism (such as a mass of colder, denser air, or a mountain range), the result is usually cloudy weather and precipitation (snow in winter). Maritime polar air masses may also continue eastward, becoming the source of snowstorms after crossing the western mountain ranges.

Generally, the *mP* air masses that develop over the northern Atlantic Ocean do not affect the weather of the United States because the westerlies push those air masses toward Europe. On some occasions, however, a strong low pressure cell may stall off of the north Atlantic Coast. Cyclonic winds from the poleward side of the low pressure system cause a northeasterly onshore flow, with cool, damp winds, and rain, or heavy snows. Known as *nor'easters*, these weather systems can bring serious winter storms to the New England States.

Maritime Tropical Air Masses (mT)

The Gulf of Mexico and the subtropical areas of the Atlantic and Pacific Oceans are source regions for *mT* air masses. Long days and intense insolation during summer produce air masses in the *mT* source regions that are very warm and very humid. During the summer, however, the land is even warmer than the *mT* air masses. This difference in temperature results in convective precipitation and strong thunderstorms on hot, humid days. Maritime tropical air masses are responsible for much of the hot, humid summer weather of the southeastern and eastern United States.

In wintertime, the tropical and subtropical oceans and the Gulf of Mexico remain warm and the air above is warm and humid. As this warm, moist air moves northward into the south-central United States, it travels over increasingly cooler land surfaces. The lower layers of air are chilled, often resulting in fog. If a tropical air mass reaches *cP* air migrating southward from Canada, the warm *mT* air will be forced to rise over the colder, drier *cP* air, and frontal precipitation will occur.

Continental Tropical Air Masses (*cT*)

The *cT* air mass develops over large, homogeneous land surfaces in the arid subtropics, and affects only a few parts of North America. The weather typical of a *cT* air mass is usually very hot and dry, with clear skies and strong daytime solar heating. Continental tropical air masses form in summer over the deserts of the southwestern United States and northwestern Mexico. In its source region, a *cT* air mass provides hot, dry, clear weather. When these air masses move away from an arid region, however, they are usually modified by contact with air masses of lower temperature and higher humidity, or by passing over bodies of water.

Fronts

The middle latitudes are the global locations where clashes of unlike air masses are most common and frequent. When different air masses come together, they do not mix readily, but instead come in contact along sloping boundaries called *fronts*. The sloping surface of a *front* is created as a warmer, lighter air mass is lifted or pushed above a cooler, denser air mass. This rising of air, known as *frontal uplift,* is a major source of precipitation in middle latitude countries like the United States and Canada (as well as middle latitude European and Asian countries), where contrasting air masses are most likely to converge. The United States and southern Canada are located in a zone between the source regions for five different air masses, all of which migrate seasonally.

The steepness of the frontal surface is governed primarily by the degree of difference and the relative rate of advancement of the two converging air masses. When two strongly contrasting air masses converge—for example, when a warm and humid *mT* air mass meets a cold, dry *cP* air mass—the frontal surface tends to be steep, with strong frontal uplift. Given similar temperature and moisture content, a steep slope, with its greater frontal uplift, will produce heavier precipitation than will a gentler slope.

Fronts are differentiated based on whether a colder air mass is moving in on a warmer one, or vice versa. The weather that occurs along a front also depends on which air mass is the "aggressor." Clashing air masses form a frontal zone that can cover an area from 2 to 3 kilometers (1–2 mi) wide to as wide as 150 kilometers (90 mi). Although weather maps use a one-dimensional line symbol to separate two different air masses, a front is actually a three-dimensional surface with length, width, and height. Generally, then, it is more accurate to speak of a frontal *zone* rather than a frontal *line.*

Cold Fronts

A **cold front** occurs when a cold air mass actively moves in on a warmer air mass and pushes it upward. Because the colder air is denser and heavier than the warm air it displaces, it stays at the surface and forces the warmer air to rise. As we can see in ■ Figure 6.3, cold fronts usually have a relatively steep slope; for example, the warm air may rise 1 meter vertically for every 40–80 meters of horizontal distance. If the warm air mass is unstable and has a high moisture content, heavy precipitation can result, sometimes

■ **FIGURE 6.3** Cross section of a cold front. Cold fronts generally move rapidly, with a blunt forward edge that drives adjacent warmer air upward. This can produce violent precipitation from the warmer air.

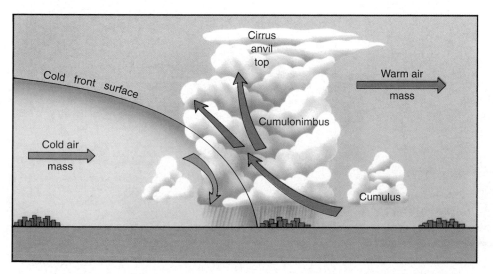

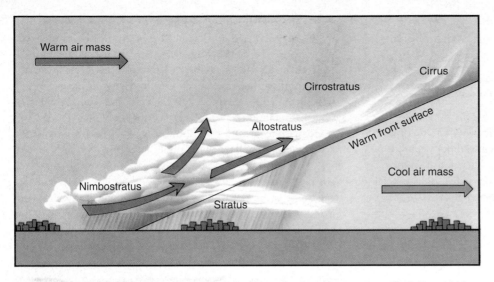

■ FIGURE 6.4 Cross section of a warm front. Warm fronts advance more slowly than cold fronts and replace rather than displace cold air by sliding upward over it. The gentle rise of the warm air produces stratus clouds and gentle rain.

Compare Figures 6.3 and 6.4. How are they different? How are they similar?

in the form of violent thunderstorms. **Squall lines** result when several storms align themselves along a cold front. Cold fronts are usually associated with strong weather disturbances and sharp changes in temperature, air pressure, and wind.

Warm Fronts

When a warmer air mass is the aggressor, invading a region occupied by a colder air mass, a **warm front** forms. At a warm front, warmer air slowly pushes against the cold air, and rises over the colder, denser air mass at the surface. The slope of a warm front is much more gentle than that of a cold front. For example, the warm air may rise only 1 meter vertically for every 100 or even 200 meters of horizontal distance. Thus, the uplift along a warm front will not be as strong as that occurring along a cold front. The result is that the weather associated with the passage of a warm front tends to be less violent and the changes less abrupt than those associated with cold fronts.

If we look at ■ Figure 6.4, we can see why the advancing warm front affects the weather of areas well in advance of the surface location of the frontal zone. The cloud types that precede a front typically indicate the weather changes that can be expected as a front approaches.

Stationary and Occluded Fronts

When a frontal boundary fails to move in any appreciable direction as air masses converge, this is a **stationary front.** Under the influence of a stationary front, locations may experience clouds, drizzle, and rain (or possible thunderstorms), sometimes for several days. A stationary front and its accompanying weather will remain until it dissipates as the contrasts

between the two air masses diminish, or the atmospheric circulation finally causes one of the air masses to move.

An **occluded front** occurs when a faster-moving cold front overtakes a warm front, pushing the warm air aloft. This frontal situation usually occurs in the latter stages of a storm and on the poleward side of a middle-latitude cyclone, a storm type that will be discussed next. Areas under an occluded front tend to experience gray, overcast skies and perhaps light precipitation. Map symbols for the four frontal types are shown in ■ Figure 6.5.

■ FIGURE 6.5 The four major frontal symbols used on weather maps.

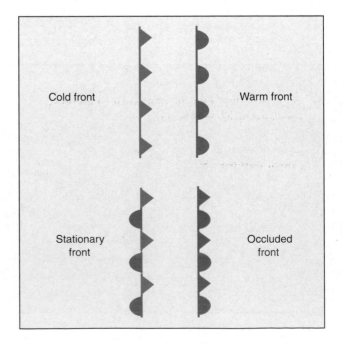

Atmospheric Disturbances

Anticyclones and Cyclones

We have previously distinguished anticyclones and cyclones according to differences in pressure and wind direction. We also have identified large areas of semi-permanent cyclonic and anticyclonic circulation in Earth's atmosphere (the subtropical high, for example). Secondary circulations, storms, and other atmospheric disturbances are embedded within the wind belts of the general atmospheric circulation. The term **atmospheric disturbance** is used because it is a more general term than "storm" and includes atmospheric conditions that cannot be classified as storms. Now, when examining middle-latitude atmospheric disturbances, we can use the terms *anticyclone* and *cyclone* to refer to moving cells of high and low pressure, which drift along the path of the prevailing westerly winds. It is important to remember that in a cyclone, pressure decreases toward the center, and in an anticyclone, pressure increases toward the center. The wind intensities involved in these systems will depend on the steepness of the *pressure gradients*, the change in pressure over a horizontal distance.

Anticyclone

An anticyclone is a high pressure area with subsiding air in the center, which displaces surface air with winds blowing outward, away from the center of the system. Hence, an anticyclone has diverging winds. An anticyclone tends to be a fair-weather system because the temperature and stability increase in subsiding air, reducing the possibility of condensation.

Air subsidence in the center of an anticyclone encourages stability because the air is warmed adiabatically, increasing its capacity to hold moisture. While the weather resulting from the influence of an anticyclone is often clear, with no rainfall, there are certain conditions under which some precipitation can occur within a high pressure system. When such a system passes near or crosses a large body of water, the resulting evaporation can cause variations in humidity significant enough to result in some precipitation.

There are two sources for the relatively high pressures that are associated with anticyclones in the middle latitudes of North America. Some anticyclones move into the middle latitudes from northern Canada and the Arctic Ocean in what are called **polar outbreaks** of frigid continental polar or even continental Arctic air. These outbreaks can be quite extensive, covering much of the midwestern and eastern United States, and they occasionally push into the subtropical Gulf Coast areas. The temperatures in an anticyclone that developed in a *cP* or *cA* air mass can be markedly lower than those expected for any given time of year, dipping far below freezing in the winter. Such an outbreak would typically be preceded by the squalls, clouds, and rain or snow associated with a cold front. A period of cold or cool, clear, and fair weather follows the frontal passage as the influences of modified polar air are felt.

Other anticyclones are generated in the subtropical high pressure regions. When they move across the United States toward the north and northeast, they bring waves of hot, clear weather in summer and unseasonably warm days in the winter.

Cyclone

A cyclone is a low pressure area that is typified by uplifting air and winds that tend to converge on the center of the low in an attempt to equalize pressure. As this air flows toward the center of a low pressure system, it feeds air into an upward spiral of air (convergence uplift, also known as cyclonic uplift), which results in clouds and precipitation. Compared to anticyclones, cyclones are much more varied and complex in the ways that they form, as well as in the weather that they generate. Low pressure systems generate storms and precipitation of all kinds through adiabatic cooling of rising air, and the resultant condensation that can occur. The types of cyclonic storms, their impacts, and their associated weather phenomena that will be discussed in detail in this chapter will focus on North American locations; but the discussion applies just as well to the middle latitude European locations.

Mapping Pressure Systems

Cells of high and low pressure are easy to visualize if we imagine these pressure systems as if they were land surfaces. A cyclone is shaped like a basin (■ Fig. 6.6). Winds converging toward the center of a cyclone will move rapidly if the pressure gradient is great, just as water will flow faster into a natural basin if the sides are steep and the depression is deep. If we visualize an anticyclone as a hill or mountain, then we can see that air diverging from an anticyclone will flow at speeds directly related to how high the pressure is at the center of the cell, in a manner similar to water moving down mountain slopes at speeds related to the landform's height.

On a surface weather map, cyclones and anticyclones are depicted by roughly concentric *isobars* of increasing pressure toward the center of a high and of decreasing pressure toward the center of a low. A high pressure cell will typically cover a larger area than a low, but both pressure systems are capable of covering and affecting extensive areas. There are times, for example, when nearly the entire midwestern United States is under the influence of the same system.

General Movement

The cyclones and anticyclones of the middle latitudes are steered, or guided, along paths influenced by the upper air westerlies (or the jet stream). Although the upper air flow can be quite variable and have large oscillations, a general west-to-east pattern prevails. As a result, people in the United States generally look at the weather occurring to the west to see what they might expect in the next few days. Most storms that develop in the Great Plains or on the West Coast move across the United States during a period of a few days at an average speed of 36 kilometers per hour (23 mph), and then travel into the North Atlantic.

Although neither cyclones nor anticyclones develop in exactly the same places at the same times each year, they do tend to arise in certain areas or regions more frequently than

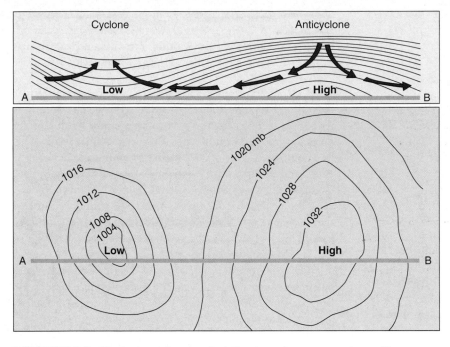

■ **FIGURE 6.6** The horizontal and vertical structure of pressure systems. Close spacing of isobars around a cyclone or anticyclone indicates a steep pressure gradient that will produce strong winds. Wide spacing of isobars indicates a weaker system.
Where would be the strongest winds in this figure? Where would be the weakest winds?

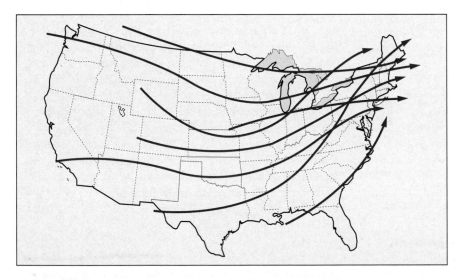

■ **FIGURE 6.7** Common storm tracks for the United States. Virtually all cyclonic storms move from west to east in the prevailing westerlies and swing northeastward across the Atlantic coast.
What storm tracks influence your location?

in others. Depending on the season, they also follow similar general paths, known as **storm tracks** (■ Fig. 6.7). In addition, atmospheric disturbances that develop in the middle latitudes during the winter are greater in number and intensity because the temperature variations between air masses are stronger during the winter months.

Middle-Latitude Cyclones

Because they are so important to the weather of North America, we will concentrate on examining **middle-latitude cyclones,** also known as **extratropical cyclones.** These migrating storms, with their opposing cold, dry polar air and warm, humid tropical air, can cause significant variation in the day-to-day weather of the locations over which they pass. Variability is common for middle-latitude weather, especially during fall and spring when conditions can change from a period of cold, clear, dry days to a period of snow, only to be followed by one or two more moderate but humid days.

The weather associated with middle-latitude cyclones can vary widely with the seasons as well as with air mass conditions, so no two storms are ever identical. Storms vary in intensity, longevity, speed of travel, wind strength, amount and type of cloud cover, the quantity and type of precipitation, and the area they affect. Yet, there is a useful model that describes and generalizes the typical characteristics of a middle-latitude cyclonic storm.

Shortly after World War I, Norwegian meteorologists Jacob Bjerknes and Halvor Solberg proposed the *polar front theory,* concerning the development, movement, and dissipation of middle-latitude storms. They recognized the middle latitudes as a region where different air masses, such as cold polar air and warm subtropical air, commonly meet at a boundary called the *polar front.* Though the polar front may be a continuous boundary that circles Earth, it is most often fragmented into several frontal segments. The polar front moves north and south with the seasons and is stronger in winter than in summer. The upper air westerlies, (see again Figures 4.14 and 4.15) also known as the *polar front jet stream,* develop and flow along the wavy track of the polar front.

Most middle latitude cyclones develop along the polar front where warm and cold air masses meet. These two contrasting air masses do not merge but may move in opposite directions along the frontal zone. Although there may be some slight uplift of the warmer air along the edge of the denser, colder air, uplift will not be significant. There may be some cloudiness and precipitation along such a frontal zone, but not conditions that we would call a storm.

The line of convergence along the polar front may develop a wave-form (in map view), for reasons that are related to wind flow in the upper troposphere, but not completely understood. These wave-forms represent an initial step in the development of a middle latitude cyclone (■ Fig. 6.8). At this bend in the polar front, warm air is pushing poleward (a warm front) and cold air is pushing equatorward (a cold front), with a low pressure center at the location where the two fronts are joined.

As the contrasting air masses compete for position, clouds and precipitation increase along the fronts, spreading over a larger area. Precipitation along the cold front will be less widespread than that along the warm front, but it will be more intense. One factor that affects the kind of precipitation that occurs at the warm front is the stability of the warm air mass. If it is relatively stable, then its movement over the cold air mass may cause only a fine drizzle or a light, powdery snow if the temperatures are cold enough. In contrast, if the warm

air mass is moist and unstable, uplift may set off heavier precipitation. As you can see by referring again to Figure 6.4, the precipitation that falls at a warm front may *appear* to be coming from the colder air. Though the weather may feel cool and damp, the precipitation is actually coming from the overriding warmer air mass above, and then falling through the colder air mass to reach Earth's surface.

Because a cold front typically moves faster, it will eventually overtake the warm front. This produces an *occluded front*. Once this occurs, the system will soon die because the temperature, pressure, and humidity differences that powered the storm diminish at the occluded front. Occlusions are usually accompanied by cloudy overcast conditions and rain (or snow), and are the major process by which middle-latitude cyclones dissipate.

Cyclones and Local Weather   The various sections of a middle-latitude cyclone generate different weather conditions. Therefore, the weather that a location experiences depends on which portion of the middle-latitude

■ **FIGURE 6.8**   Stages in the development of a middle-latitude cyclone. Each view represents the development somewhat eastward of the preceding view as the cyclone travels along its storm track. Note the occlusion in (e).

In (c), where would you expect rain to develop? Why?

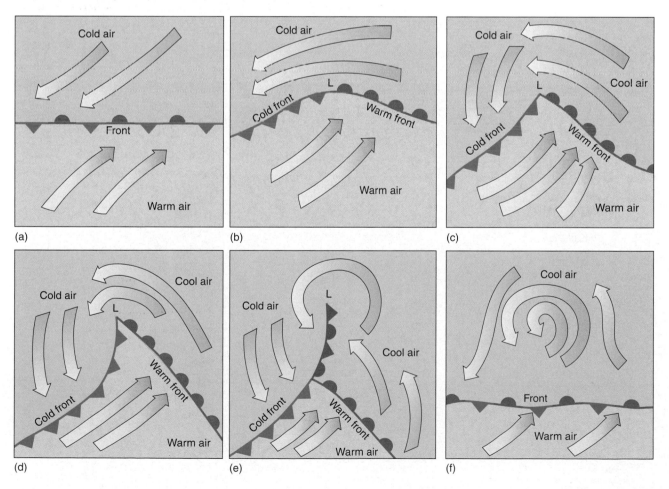

cyclone is over the location. Because the entire cyclonic system tends to travel as a unit from west to east, a specific sequence of weather can be expected at a given location as the cyclone passes.

Let's examine the typical passage of a middle latitude cyclone, following a track (see again Figure 6.7) that will take it across Illinois, Indiana, Ohio, and Pennsylvania, and finally out over the Atlantic Ocean. A view of this storm on a weather map, at a specific time in its journey, is presented in ■ Figure 6.9a. Figure 6.9b is a cross-sectional view north of the cyclone's center, and Figure 6.9c is a cross-sectional view south of the cyclone's center. As the storm continues eastward, the sequence of weather will be different for Detroit, where the warm and cold fronts will pass to the south, compared to Pittsburgh, which will experience the passage of both fronts. To illustrate this point, we will examine the changing weather in Pittsburgh, in contrast to that which Detroit experiences as the cyclonic system moves east. We will also examine the weather of other cities affected by passage of this storm system.

A cyclonic storm is composed of two dissimilar air masses. The sector of warm, humid *mT* air between the two fronts of the cyclone is usually considerably warmer than the cold *cP* air surrounding it. The temperature contrast is accentuated during the winter when the source region for *cP* air is the cold cell of high pressure in Canada. During the summer, the contrast between these air masses is greatly reduced.

Because of the temperature difference, atmospheric pressure in the warm sector is lower than the atmospheric pressure in the cold sector behind the cold front. In advance of the warm front, the pressure is also high, but as the warm front approaches Pittsburgh, the pressure will decrease. After the warm front passes through Pittsburgh, the pressure will stop falling, and the temperature will rise as *mT* air invades the area.

Indianapolis has already experienced the warm front's passage and is now awaiting the cold front. After the cold front passes, the pressure will rise rapidly and the temperature will drop. Detroit, which is to the north of the cyclone's center, will miss the warm air sector entirely and experience a

■ **FIGURE 6.9** Environmental Systems: Middle Latitude Cyclonic Systems. This diagram models a middle-latitude cyclone positioned over the Midwest as the system moves eastward: (a) a map view of the weather system; (b) a cross section along line AB north of the center of low pressure; (c) a cross section along line CD south of the center of low pressure.

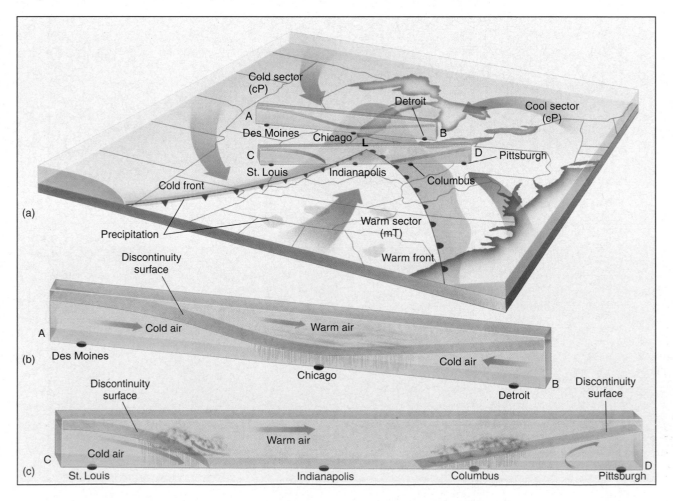

slight increase in pressure and a temperature change from cool to colder as the cyclone moves to the east.

Changes in wind direction are one signal of the approach and passing of a cyclonic storm and its associated fronts. Because a cyclone is a low pressure system, winds flow counterclockwise toward its center (in the northern hemisphere). The winds associated with a cyclonic storm are stronger in winter when the pressure and temperature differences between air masses are considerable.

Indianapolis, in the warm sector south of the center of the low, is receiving winds from the south. Pittsburgh, located to the east and ahead of the warm front, has southeast winds. As the cyclonic system moves eastward, and the warm front passes, the winds in Pittsburgh will shift to the south-southwest. After the cold front passes, the winds in Pittsburgh will be from the north-northwest. St. Louis has already experienced the passage of the cold front with cold weather and winds from the northwest.

Detroit's winds, now from the southeast, will shift to the northeast once the storm's center has passed to the south. Finally, after the storm has passed, the winds will blow from the northwest, as they are in Des Moines, to the west of the storm.

The type and intensity of precipitation and cloud cover also vary as a cyclonic storm moves through a location. In Pittsburgh, the first sign of the approaching warm front will be high cirrus clouds. As the warm front continues to approach, the clouds will thicken and lower and in Pittsburgh, light rain and drizzle may begin (or light snow in winter), and stratus clouds will blanket the sky. After the warm front has passed, precipitation will stop and the skies will clear.

As the cold front passes, warm air in its path will be forced to move aloft rapidly. This may mean that there will be a cold, hard rain (or heavy snowfall in winter), but the band of precipitation normally will not be very wide because of the steep angle of the surface along a cold front. In our example, the cold front and the band of precipitation have just passed St. Louis. During the winter, a cold front is likely to bring snow, followed by the cold, clear conditions of the cP air mass, and increasing atmospheric pressure.

Located in the latitudinal center of the cyclonic system, Pittsburgh can expect three zones of precipitation as it passes over its location: (1) a broad area of overcast and drizzle in advance of the warm front (or light snow in winter); (2) a zone within the warm sector where clearing occurs; and (3) a narrow band of heavy precipitation associated with the cold front (rain or snow, depending on the season and temperatures) (Fig. 6.9c). However, locations to the north of the center of the cyclonic storm, such as Detroit, will usually experience light precipitation and overcast cloudy conditions resulting from warm air being lifted above cold air from the north (Fig. 6.9b).

As you can see, the various portions of a middle-latitude cyclone are accompanied by different weather. If we know where the cyclone will pass relative to our location, we can make a fairly accurate forecast of what our weather will be like as the storm moves east along its track (see Map Interpretation: "Weather Maps", at the chapter end).

Cyclones and the Upper Air Flow

Steering surface storm systems is only one of the ways that the upper air winds influence our surface weather. A less obvious influence is related to the undulating, wave-like flow so often exhibited by the upper air winds. As the air passes through these waves, it undergoes divergence or convergence because of the atmospheric dynamics associated with curved flow of winds. These flow dynamics produce alternating pressure areas of ridges (highs/divergent winds) and troughs (lows/convergent winds).

The region between a ridge and an adjacent trough (A–B in ■ Fig. 6.10) is an area of upper-level convergence. Because any action taken in one part of the atmosphere is countered by an opposite reaction somewhere else, upper air convergence is compensated for by divergence at the surface. In this area, the air is pushed downward, promoting anticyclonic circulation. This pattern will either inhibit the formation of a middle-latitude cyclone altogether, or cause an existing storm to weaken or dissipate. In contrast, the region between a trough and the next downwind ridge (B–C in Fig. 6.10) is an area of upper-level divergence, which in turn is compensated for by surface convergence. This is an area where air is drawn upward, causing cyclonic circulation at the surface that will enhance the prospects for storm development or strengthen an existing storm.

In addition to storm development or dissipation, temperatures will also be affected by upper air flow. If we assume that our "average" flow of upper air is from west to east, then any deviation from that pattern will cause either colder air from the north or warmer air from the south to be advected into an area. For example, after the atmosphere has been in a wave-like pattern for a few days, the areas in the vicinity of a trough (area B in Fig. 6.10) will be colder than normal as polar air from higher latitudes dips into that area. Just the opposite will occur at locations near a ridge

■ **FIGURE 6.10** Waves in the jet stream. The upper air wind pattern, such as that depicted here, can have a significant influence on temperatures and precipitation on Earth's surface. **Where would you expect storms to develop?**

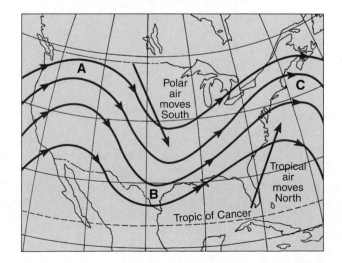

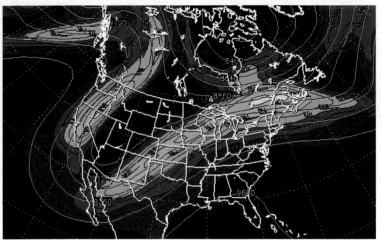

■ **FIGURE 6.11** Polar front jet stream analysis. This map shows winds at an altitude of 300 mb (approximately 10,000 m or 33,000 feet above sea level). At this height the long waves of the jet stream can more easily be seen. **Would Utah and Wyoming be clear or stormy?**

(area C in Fig. 6.10). Here, warmer air from more southerly latitudes will be drawn up toward the top of the ridge. As ■ Figure 6.11 shows, the jet stream actually curves with less regularity than in our model. Comparing Figures 6.10 and 6.11 you can see the difference between the theoretical and the real waves in the polar front jet.

Hurricanes

A **hurricane** is a circular, cyclonic system with wind speeds in excess of 118 kilometers per hour (74 mph) and a diameter of 160–640 kilometers (100–400 mi). These tropical storms form and develop in the tropical oceans, but steered by pressure and wind systems, they are often directed toward the middle-latitudes. Extending upward to heights of 12–14 kilometers (40,000–45,000 ft) or higher, the hurricane is a towering column of spiraling air (■ Fig. 6.12). Though its diameter may be less than that of a *middle latitude cyclone* with its extended *fronts*, a hurricane is essentially the largest storm on Earth. At its base, air is sucked in by low pressure surrounding the hurricane's center, and rises rapidly to the top where it spirals outward. The rapid upward movement of moisture-laden air produces enormous amounts of rain. Massive release of latent heat energy from condensation increases the power that drives the storm.

Hurricanes have extremely low pressure at their centers and very strong pressure gradients that produce powerful, high velocity winds. Unlike middle-latitude cyclones, hurricanes form from a single air mass and do not have the different sectors of strongly contrasting temperature that power a frontal system. Rather, at the surface, hurricanes have a circular distribution of warm temperatures.

■ **FIGURE 6.12** Cross-section of a hurricane, showing its circulation pattern: the inflow of air in the spiraling arms of the cyclonic system, rising air in the towering circular wall cloud, and outflow in the upper atmosphere. Subsidence of air in the storm's center produces the distinctive calm, cloudless "eye" of the hurricane. **Why is this so?**

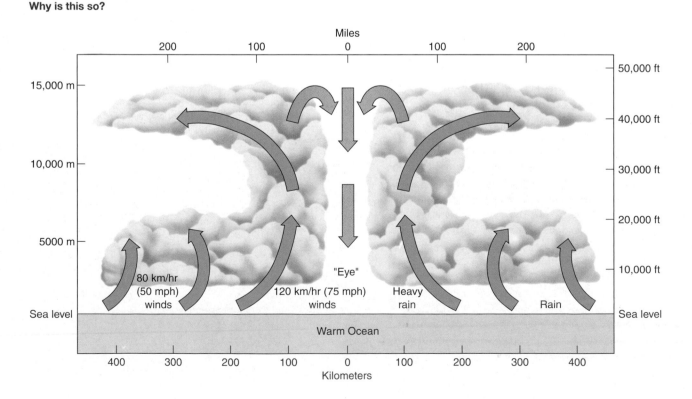

At the center is the eye of a hurricane, an area of calm, clear, usually warm and humid, but rainless air. Sailors traveling through the eye have been surprised to see birds flying there. Unable to leave the eye because of the strong winds surrounding it, these birds will often alight on the passing ship as a resting spot.

Hurricanes are severe *tropical cyclones* that receive a great deal of attention when they make landfall, primarily because of their tremendous destructive powers. Abundant, even torrential, rains and winds often exceeding 160 km/h (100 mph) characterize hurricanes.

Although a great deal of time, effort, and money has been spent on studying the development, growth, and paths of hurricanes, much is still not known. It is not yet possible to predict a hurricane's path with great certainty, even though it can be tracked with radar and studied by planes and weather satellites. As with tornadoes, there are also areas where hurricanes are most likely to develop and strike (Fig. 6.13). For North America, the most susceptible areas are the Atlantic and Gulf Coastal regions, but hurricanes can continue to do damage as they dissipate and move inland.

The development of a hurricane requires a warm ocean surface of about 27°C (80°F) or more and warm, moist overlying air. These factors explain why hurricanes occur most often in late summer and early fall when maritime air masses have maximum humidity and ocean surface temperatures are highest. Hurricane season in the Atlantic Basin officially runs from June 1st through November 30th.

Hurricanes begin as weak tropical disturbances over the ocean, called **easterly waves,** which are weak, trough-shaped, low pressure areas. Traveling slowly in the trade winds belt from east to west, an easterly wave is preceded by fair, dry weather and is followed by cloudy, showery weather. If the pressure becomes lower and the winds strengthen, a tropical storm will develop from the easterly wave.

Names are assigned to these storms once they reach *tropical storm* status, with wind speeds between 62–118 kilometers per hour (39–74 mph). Each year the names are selected from a different alphabetical list of alternating female and male names—one list for the North Atlantic and one for the North Pacific. If a hurricane is especially destructive and becomes a part of recorded history, its name is retired and never used again. Andrew, Carla, Hugo, and Katrina are just some of the nearly 70 names that have been retired since the naming of storms began in the 1950s.

Hurricanes do not last long over land because their source of moisture (and consequently their source of energy) is cut off, and friction with the land surface reduces wind speeds. North Atlantic hurricanes move first toward the west with the trade winds and then move north and northeast. Over land, they become simple cyclonic storms. Even then, however, with their power significantly reduced, hurricanes can still do great damage.

Hurricanes can occur over most subtropical and tropical oceans and seas. The South Atlantic was once the exception,

 FIGURE 6.13   A world map showing major "Hurricane Alleys" and their regional names. **Which coastlines seem unaffected by these tracks?**

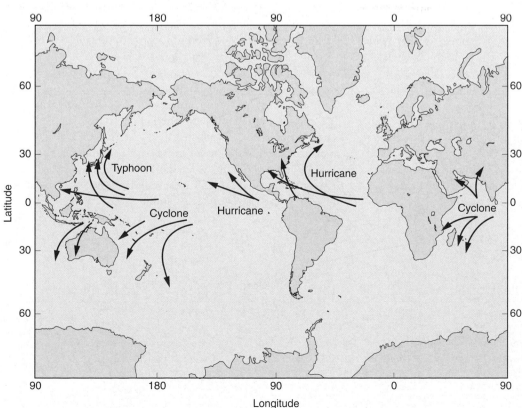

■ **FIGURE 6.14**   As a hurricane makes landfall, a storm surge may be generated. These surges are mounds of water, topped by battering waves, that can flow over low elevation coastal areas with tremendous destructive force.

What can people who live in such regions do to protect themselves when a serious storm surge is threatening?

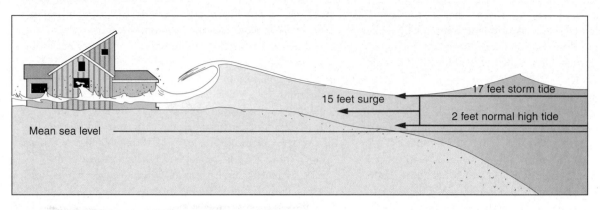

though it was not known why. However, in 2004, Catarina became the first known hurricane to strike the coast of Brazil, much to the amazement of atmospheric scientists. Referred to simply as *cyclones* in Australia and the South Pacific, as well as in the Indian Ocean, these storms are called **typhoons** in most of East Asia.

Hurricane Intensities and Impacts

The results of a hurricane landfall can be devastating destruction of property and loss of life. Among their most serious potential hazards, however are the high seas pushed onshore by the strong winds. These **storm surges** can flood, and sometimes destroy, entire coastal communities (■ Fig. 6.14). The **Saffir–Simpson Hurricane Scale** provides a means of classifying hurricane intensity and potential damage by assigning a category from 1 to 5 based on a combination of central pressure, wind speed, and the potential height of its storm surge (Table 6.2).

The year 2004 was a record-breaking year. Typhoon Tokage struck the Japanese Coast near Tokyo, with significant loss of life. In all, ten tropical cyclones pounded Japan in 2004. In the Caribbean and Gulf of Mexico, three hurricanes—Charley, Frances, and Jeanne—hit Florida directly. A fourth, Ivan, struck Gulf Shores, Mississippi, and caused devastation in Florida's western panhandle. The damages from these storms were estimated at $23 billion. This amount

exceeds even the $20 billion cost of damage by Hurricane Andrew, which, when it hit in 1992, was the costliest natural disaster in U.S. history.

Hurricane Katrina, with 225-kilometer per hour (140 mph) winds and storm surges of more than 16 feet, struck the Louisiana, Mississippi, and Alabama coasts in August, 2005. The storm surges breached the levee systems designed to protect the city of New Orleans, much of which is below sea level. The subsequent flooding caused massive destruction. Responsible for the deaths of more than 2000 people, the winds and floodwaters of Katrina caused damages estimated to be in excess of $125 billion. Hurricane Rita, even more powerful, followed in September with landfall near the Texas–Louisiana state line, devastating coastal areas in that region. Because Rita's eye missed the heavily populated Houston area, the estimated $10 billion damage was much less than for Katrina.

In 2006, the hurricane season was unusual because no hurricanes made landfall in the United States. The following year, two category 5 hurricanes struck Mexico and Central America, causing much destruction, but once again no powerful tropical storms affected the United States. In 2008, Hurricane Ike, a powerful storm with 230-kilometer per hour (145 mph) winds, made landfall on the Gulf Coast, with a storm surge that caused widespread damage to the coastal region near Galveston, Texas and in some cases complete

TABLE 6.2
Saffir–Simpson Hurricane Scale

Scale Number	Central Pressure	Wind Speed		Storm Surge		Damage
(CATEGORY)	(MILLIBARS)	(KPH)	(MPH)	(METERS)	(FEET)	
1	980	119–153	74–95	1.2–1.5	4–5	Minimal
2	965–979	154–177	96–110	1.6–2.4	6–8	Moderate
3	945–964	178–209	111–130	2.5–3.6	9–12	Extensive
4	920–944	210–250	131–155	3.7–5.4	13–18	Extreme
5	<920	>250	>155	>5.4	>18	Catastrophic

:: HURRICANE PATHS AND LANDFALL PROBABILITY MAPS

Hurricanes (also called typhoons or cyclones) are generated over tropical or subtropical oceans and build strength as they move over regions of warm ocean water. Ships and aircraft regularly avoid hurricane paths by navigating away from these huge violent storms. People living in the path of an oncoming hurricane try to prepare their belongings, homes, and other structures and may have to evacuate if the hazard potential of the impending storm is great enough. Landfall refers to the location where the eye of the storm encounters the coastline. Storm surges present the most dangerous hazard associated with hurricanes, where the ocean violently washes over and floods low-lying coastal areas. In 1900, 6000 residents of Galveston Island in Texas were killed when a hurricane pushed a 7-meter-high wall of water over the island. Much of the city was destroyed by this storm, the worst natural disaster ever

to occur in the United States in terms of lives lost.

Today we have sophisticated technology for tracking and evaluating tropical storms. Computer models, developed from maps of the behavior of past storms, are used to indicate a hurricane's most likely path and landfall location, as well as the chance that it may strike the coast at other locations. The nearer a storm is to the coast, the more accurate the predicted landfall site should be, but in some cases a hurricane may begin to move in a completely different direction. In general, hurricanes that originate in the North Atlantic Ocean tend to move westward toward North America and then turn northward along the Atlantic or Gulf Coasts.

Nature still remains unpredictable, so potential landfall sites are shown on probability maps, which show the degree of likelihood for the hurricane path. These maps help local authorities

and residents decide what course of action is best to take in preparing for the approach of a hurricane. A 90% probability means that nine times out of ten storms under similar regional weather conditions have moved onshore in the direction indicated by that level on the map. A 60% probability means that six of ten hurricanes moved as indicated, and so forth. Regions where the hurricane is considered likely to move next are represented on the map by color shadings that correspond to varying degrees of probability for the storm path. In recent years, the National Weather Service has worked to develop computer models that will yield better predictions of hurricane paths, intensities, and landfall areas. If you live in a coastal area affected by hurricanes and tropical disturbances, understanding these maps of landfall probability may be very important to your safety and your ability to prepare for a coming storm.

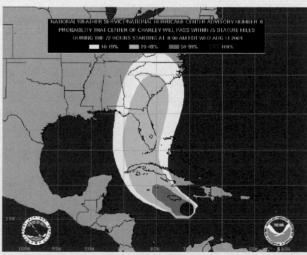

Landfall probability map for Hurricane Charley on August 11, 2004, showing the most likely place for landfall. This map was made 2 days before the hurricane struck Florida's coast.

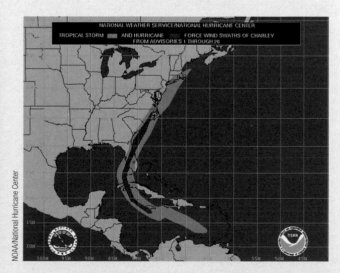

This was the actual path of Hurricane Charley. In this case, the landfall probability map proved fairly accurate.

National Weather Service Houston/Galveston and the Galveston County Office of Emergency Management

■ **FIGURE 6.15** Extensive destruction and damage to a community along the Texas Gulf Coast caused by the storm surge generated by Hurricane Ike in 2008.

destruction of coastal communities (■ Fig. 6.15). The storm's path continued directly to Houston, causing further damage. Also in 2008, major hurricanes struck Haiti, Cuba, Jamaica, and other Caribbean islands, causing many deaths and serious destruction. From one year to the next, the exact number and severity of *tropical cyclones* can vary dramatically.

In spite of the damage they cause, hurricanes also have beneficial impacts. They are an important source of much needed rainfall in regions of the southeastern United States such as Florida, and in other areas of the world as well. Hurricanes are a natural and important means of moderating latitudinal temperatures by transferring surplus heat energy away from the tropics and into the cooler latitudes.

Snowstorms and Blizzards

In the middle and higher latitudes, where freezing temperatures occur, *snowstorms* are common, due in part to seasonal changes in day length and solar heating, and in part to the influence of polar and arctic air masses. Snowfalls and snowstorms are triggered by the same uplift mechanisms that produce most other types of precipitation—orographic, frontal, and convergence (cyclonic). The exception is convectional precipitation, which is a warm weather phenomenon. In middle- to high-latitude winters, and also at high elevations, snowfall events vary greatly in severity. They can come as a *snow shower* or *snow flurry,* a brief period of snowfall in which intensity can be variable and may change rapidly. In contrast, the accumulation of snow during a **snowstorm** tends to be heavy. Often accompanied by strong winds, snowstorms can also create enough turbulence to produce lightning.

For a severe snowstorm to become a **blizzard,** the winds must be 55 kilometers per hour (35 mph) or greater. Falling and blowing snow can reduce visibility to zero, creating conditions that are known as a *whiteout.* In such a situation, visibility is so limited that all a person can see is white, making it easy to lose track of distance and direction. Airport closings and traffic accidents are common during blizzards (■ Fig. 6.16).

Thunderstorms

A thunderstorm is a storm accompanied by thunder and lightning, caused by an intense discharge of electricity. For lightning to occur, positive and negative electrical charges must be generated within a cloud. It is believed that the intense friction of air on moving ice particles within a cumulonimbus cloud generates these charges. Usually, but not always, a clustering of positive charges tends to occur in the upper portion of the cloud, with negative charges clustering in the lower portion. When the difference between these charges becomes large enough to overcome

■ **FIGURE 6.16** A weather station in central Illinois during a blizzard in February of 2000. With winds gusting to 45 miles per hour, a blizzard can greatly reduce visibility.
How far would you estimate the visibility to be in this area?

NOAA/NWS

Hail can be a product of thunderstorms when the vertical updrafts in the cells are sufficiently intense to carry water droplets repeatedly into a layer of air with sub-freezing temperatures. Fortunately, because thunderstorms are primarily associated with warm weather regions, only a very small percentage of storms around the world produce hail. In fact, hail seldom occurs in thunderstorms in the lower latitudes. In the United States, hail storms are unusual along the Gulf of Mexico where thunderstorms are most common.

Thunderstorms usually span an area of just a few miles, although a series of related storms may cover a larger region. The intensity of a storm will depend on the air's instability as well as the amount of water vapor it holds. Once most of this water vapor has condensed, removing the energy needed for continued uplift, a thunderstorm will die out, usually about an hour after it began.

Convective thunderstorms typically occur during the warmer months of the year and during the warmer hours of the day. The equatorial and humid tropics experience the world's highest thunderstorm frequency, but they can also occur in any region during hot, humid conditions. It is apparent, then, that the amount of solar heating affects the development of thunderstorms. This is true because intense heating of the surface steepens the environmental lapse rate, leading to increased instability of the air, allowing for greater moisture-holding capacity and adding to the buoyancy of the air.

Orographic thunderstorms occur when air is forced to rise over land barriers, providing the necessary initial trigger action leading to the development of thunderstorm cells. Thunderstorms of orographic origin play a large role in the tremendous precipitation of the monsoon regions of South and Southeast Asia. In North America, they occur over the mountains in the West (the Rockies and the Sierra Nevada), and the Appalachians in the East, especially during summer afternoons. For this reason, pilots of small planes try to avoid flying in the mountains during summer afternoons for fear of getting caught in the turbulence of a thunderstorm.

Frontal thunderstorms occur when a cooler air mass forces a warmer air mass to rise along a cold front. Frontal uplift can bring about the strong, vertical updrafts necessary for precipitation. At times a cold front is immediately preceded by a line of thunderstorms (a squall line) resulting from strong uplift along a front (see again Figure 6.3).

■ **FIGURE 6.17**   A cross-sectional view of a thunderstorm showing the distribution of electrical charges.
Where would you place a lightning bolt in this diagram?

the natural insulating effect of the air, a lightning flash, or discharge, takes place. These discharges, which often involve over 1 million volts, can occur within the cloud, between two clouds, or from cloud to ground. The air immediately around the discharge is momentarily heated to temperatures in excess of 25,000°C (45,000°F), and the heated air expands explosively, creating the shock wave that we call thunder (■ Fig. 6.17).

The intense precipitation that falls during a thunderstorm results from the rapid uplift of moist air. As is the case for other types of precipitation, the trigger mechanism causing that uplift can be thermal convection (warm unstable air rising on a warm afternoon, ■ Fig. 6.18a), orographic uplift (moist air ramping up a mountain side, Fig. 6.18b), or frontal uplift (see again Figures 6.3 and 6.4).

■ **FIGURE 6.18**   Thermal convection and orographic uplift.
What are the other mechanisms of uplift?

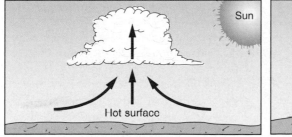

(a) Convectional uplift

(b) Orographic uplift

Tornadoes

A **tornado** is a small, intense cyclonic storm characterized by extremely low pressure, violent updrafts, and powerful converging winds. Tornadoes are the most violent storms on Earth (■ Figs. 6.19 and 6.20). Other than in the polar regions, they can occur almost anywhere, but are far more common in the interior of North America than in any other location in the world. In fact, Oklahoma and Kansas lie in the path of so many "twisters" that together they are sometimes referred to as the center of "Tornado Alley." Fortunately, tornados are small and short lived. Even in Tornado Alley, a tornado is likely to strike a given locale only once in 250 years.

Although only 1% of all thunderstorms produce a tornado, 80% of all tornadoes are associated with thunderstorms and middle-latitude cyclones. The remaining 20% of tornadoes are spawned by hurricanes that make landfall. In the past decade, more than 1000 tornadoes occurred each year in the United States, most of them from March to July in the late afternoon or early evening, and in the central part of the country.

A tornado first appears as a swirling, twisting funnel cloud, its narrow end as little as 100 meters (330 ft) across, which moves across the landscape at 35–51 kilometers per hour (22–32 mph). The funnel cloud becomes a tornado when its narrow end makes contact with the ground, where the greatest damage is done, often along a linear track (■ Fig. 6.21). Because of their small size and short life span, tornadoes are difficult to detect and forecast. However, a sophisticated radar technology, **Doppler radar,** improves tornado detection and forecasting significantly, allowing meteorologists to assess storms in greater detail (■ Fig. 6.22).

Doppler radar can measure wind speeds flowing toward and away from the radar site. When the energy emitted by radar strikes precipitation, a small portion is reflected back to the radar. Depending on whether the precipitation is moving toward or away from the radar site, the wavelength of the returned radar signal is either compressed or elongated. The faster the winds flow, the greater the wavelength change. Doppler can estimate the wind circulation and rotation within the storm. This technology allows meteorologists to see the formation of a tornado, thus increasing the warning time to the public.

Based on Doppler radar studies, most tornadoes (75%) are fairly weak, with wind speeds of 180 kilometers per hour (112 mph) or less. The remaining 25%, whose wind speeds reach up to 265 kph (165 mph) can be classified as strong. Nearly 70% of all tornado fatalities result from these more violent tornadoes. Although they are very rare, such killer storms can have wind speeds exceeding 320 kph (200 mph).

Before Doppler radar, tornado wind speeds could not be measured directly, and tornado intensity was estimated from the damage produced by the storm. The late Theodore Fujita developed a scale for comparing tornados, called the Fujita Intensity Scale or, more commonly, the F-scale (Table 6.3). In 2007, the National Weather Service adopted a refined and modified

■ **FIGURE 6.19** A powerful F4 tornado in central Illinois during July of 2004.

NOAA/NWS

■ **FIGURE 6.20** Terrible destruction caused by an F5 tornado at Greensburg, Kansas on May 16, 2007.

Greg Henshall/FEMA 7.23: NOAA

■ **FIGURE 6.21**   The destructive track of a powerful tornado is visible on this satellite image as a linear swath of damage across the landscape of La Plata, Maryland.

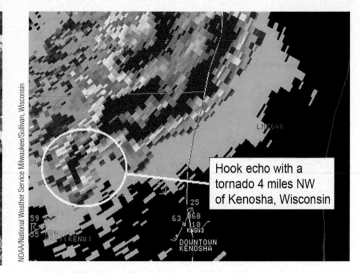

Hook echo with a tornado 4 miles NW of Kenosha, Wisconsin

■ **FIGURE 6.22**   This Doppler radar image near Kenosha, Wisconsin, shows a curving pattern in the precipitation patterns of the oncoming storm, called a hook echo, which means a tornado may be developing.

version of the original F-scale, based on new data and observations that were not available to Fujita. The result is the **Enhanced Fujita Scale,** the EF-scale, which is used today.

Although most tornado damage is caused by the violent winds, most injuries and deaths result from flying debris. The small size and short duration of a tornado greatly limit the

TABLE 6.3
The Fujita Tornado Intensity Scale and the Enhanced Fujita Scale

	Wind Speed			Wind Speed		
F-SCALE	**KPH**	**MPH**	**EF-SCALE**	**KPH**	**MPH**	**EXPECTED DAMAGE**
F-0	<116	<72	EF-0	<138	<86	**Light Damage** Damage to chimneys and billboards; broken branches; shallow-rooted trees pushed over
F-1	116–180	72–112	EF-1	138–177	86–110	**Moderate Damage** Surfaces peeled off roofs; mobile homes pushed off foundations or overturned; exterior doors blown off; windows broken; moving autos pushed off the road
F-2	181–253	113–157	EF-2	178–217	111–135	**Considerable Damage** Roofs torn off houses; mobile homes demolished; boxcars pushed over; large trees snapped or uprooted; light-object missiles generated
F-3	254–332	158–206	EF-3	218–265	136–165	**Severe Damage** Roofs and some walls torn off well-constructed houses; trains overturned; most trees in forest uprooted; heavy cars lifted off ground and thrown
F-4	333–419	207–260	EF-4	266–322	166–200	**Devastating Damage** Well-constructed houses leveled; structures with weak foundations blown some distance; cars thrown and large missiles generated
F-5	>419	>260	EF-5	>322	>200	**Incredible Damage** Strong frame houses lifted off foundations and carried considerable distance to disintegrate; automobile-sized missiles fly through the air farther than 100 meters; trees debarked; incredible phenomena occur

number of deaths caused by tornadoes. In fact, more people die from lightning strikes each year than from tornadoes. At times, however, severe storms may spawn a **tornado outbreak,** meaning multiple tornadoes are produced by the same system. The worst outbreak in recorded history occurred on 3–4 April 1974, when 148 twisters touched down in 13 states, injuring almost 5500 people and killing 330 others.

Weather Forecasting

Weather forecasting, at least in principle, is a fairly straightforward process. Meteorological observations are made, collected, and mapped to depict the current state of the atmosphere. From this information, the probable movements and the anticipated growth or decay of current weather systems are projected for a specific amount of time into the future.

When a forecast goes wrong—which we all know occurs—it is either because limited or incorrect information has been collected and processed, or because errors have been made in anticipating the path or growth of the storm systems. The further into the future one tries to forecast, the greater the uncertainty of a forecast.

Although forecasts are not perfect, they are much better today than they were in the past. Much of this improvement can be attributed to the development of sophisticated technology and equipment. Increased knowledge and surveillance of the upper atmosphere have improved the accuracy of weather prediction. Weather satellites provide meteorologists with images that lead to a better understanding of weather systems. Satellite imagery is extremely valuable to forecasters who track storms that develop over ocean areas and head toward landfall, like pacific storms on the West Coast and hurricanes in the Atlantic or Gulf of Mexico (■ Fig. 6.23). Before the advent of weather satellites, forecasters had to rely on information relayed from ships, leaving enormous ocean areas unobserved. Thus, forecasters were often caught off guard by unexpected weather events.

Computers can rapidly process weather data and map weather conditions, based on information and imagery downloaded from weather stations and satellites—all in what is termed *real time*, which means immediately, as changes occur. Complex digital processing of statistical forecast models, based on the physical processes that govern our atmosphere, are performed on computer systems. Some of the fastest and most powerful computers available

■ **FIGURE 6.23**   This weather satellite image shows two continents and the adjacent ocean areas.

Is there cloud cover over your state on this day?

NASA/NOAA GOES and NOAA AVHRR

today, are used to forecast the weather and model climate change.

Today, although weather forecasters possess a great deal of knowledge and apply highly sophisticated technologies, forecasting is still not a perfect process. Weather forecasters combine science and art, fact and interpretation, data and intuition, to provide their best judgments based on probabilities about future weather conditions. Realizing the atmospheric complexities that a weather forecaster must consider helps us comprehend the information presented in a weather forecast. The weather affects us on a daily basis, and having a basic knowledge of weather systems and atmospheric processes can enrich our lives, by helping us to know how to prepare for whatever weather conditions we may experience. This basic knowledge is also essential to understanding the global climates and environments, which are addressed in the following chapters.

:: Terms for Review

air mass	stationary front	Saffir–Simpson Hurricane Scale
source region	occluded front	snowstorm
Maritime Equatorial (*mE*)	atmospheric disturbance	blizzard
Maritime Tropical (*mT*)	polar outbreak	convective thunderstorm
Continental Tropical (*cT*)	storm track	orographic thunderstorm
Continental Polar (*cP*)	middle-latitude cyclone	frontal thunderstorm
Maritime Polar (*mP*)	extratropical cyclone	tornado
Continental Arctic (*cA*)	hurricane	Doppler radar
cold front	easterly wave	Enhanced Fujita Scale
squall line	typhoon	tornado outbreak
warm front	storm surge	

:: Questions for Review

1. Do all areas on Earth produce air masses? Why or why not?
2. What letter symbols are used to identify air masses on maps? How are these combined?
3. What air masses influence the weather of North America? Where and at what time of the year are they most effective?
4. Use Table 6.1 and Figure 6.2 to find out what kinds of air masses are most likely to affect your local area. How do they affect weather in your area?
5. Why are air masses classified by whether they develop over water or over land?
6. Why does *mP* air affect the United States, and what areas experience this air mass?
7. What is a front, and why do they occur? How can you tell which way it is moving on a weather map?
8. How do warm and cold fronts differ in duration and precipitation characteristics?
9. How does the configuration of upper air wind patterns affect surface weather conditions?
10. How can an understanding of cyclones be used to help forecast weather changes? For how long in advance? What atmospheric changes might occur to spoil your forecast?

:: Practical Application

1. Collect a 3-day series of weather maps from your local newspaper or the Internet. Based on the migration of high and low pressure systems during that period, predict their movement in the next few days.
2. Draw a diagram of a mature (fully developed) middle-latitude cyclone that includes the center of the low with several isobars, the warm front, the cold front, wind direction arrows, appropriate labeling of warm and cold air masses, and zones of precipitation.
3. Wind speeds are sometimes given in knots (nautical miles per hour) instead of statute miles per hour. A nautical mile (used mainly for air and sea navigation) equals 6,080 feet, slightly longer than a statute mile at 5,280 feet. Being longer than a statute mile, a knot is a little faster than a mile per hour. The conversion from miles per hour to knots is:

$$knots = mph \times 1.151$$

Using Table 6.2, convert the ranges of wind speeds for Hurricane Categories 1 through 5 from miles per hour to knots.

MAP INTERPRETATION

WEATHER MAPS

Weather maps that portray meteorological conditions over a large area at a given moment in time are important for current weather descriptions and forecasting. Simultaneous observations of meteorological data are recorded at weather stations across the United States (and worldwide). This information is electronically relayed to the National Centers for Environmental Prediction near Washington, D.C., where the data are analyzed and mapped.

Meteorologists then use the individual pieces of information and data to depict the general weather picture over a larger area. For example, isobars (lines of equal atmospheric pressure) are drawn to reveal the locations of cyclones (L) and anticyclones (H), and to indicate frontal boundaries. Areas that were receiving precipitation at the time the map depicts are shaded in green so that these areas are highlighted. The end result is a map of weather conditions that can be used to forecast changes in weather patterns. This map is accompanied by a satellite image that was taken on the same date.

1. Isobars are lines of equal atmospheric pressure expressed in millibars. What is the interval (in millibars) between adjacent isobars on this weather map?

2. What kind of front is passing through central Florida at this time?

3. Which Canadian high pressure system is stronger—the one located over British Columbia or the one near Newfoundland?

4. Which state is free of precipitation at this time: Nebraska, Connecticut, Mississippi, or Kentucky?

5. What kind of front is located over Nevada and Utah?

6. Does the surface map accurately depict the cloud cover indicated on the accompanying satellite image?

7. Can you find the low pressure systems on the satellite image?

8. Is there any cloud cover over West Virginia? How can you tell?

9. Do the locations of the fronts and areas of precipitation depicted on the map agree with the idealized relationship represented in Figure 6.9 (Middle-Latitude Cyclonic Systems)? Explain.

10. On the map, what kind of frontal symbol lies off the U.S. coast between New Jersey and Connecticut?

11. Looking at the map and the satellite image, comment about the relative strength of the fronts in Florida, in Illinois, and the one that is southeast of the New England coast.

12. What kind of storm is New England experiencing? (Hint: it is named after the wind direction.)

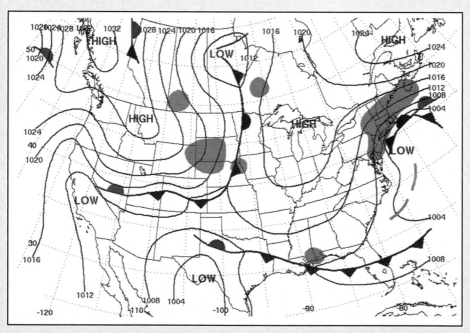

This weather map illustrates the spatial distribution of measurable weather elements (air pressure, temperatures, wind speed, wind directions) as well as the locations of fronts and areas of precipitation. Isobars define high and low pressure cells and the kinds of fronts are also identified. The weather conditions at meteorological stations are shown by numerical values and symbols.

Opposite:
A satellite image of the atmospheric conditions shown on the accompanying weather map.
©NOAA and Department of Geoscience, San Francisco State University

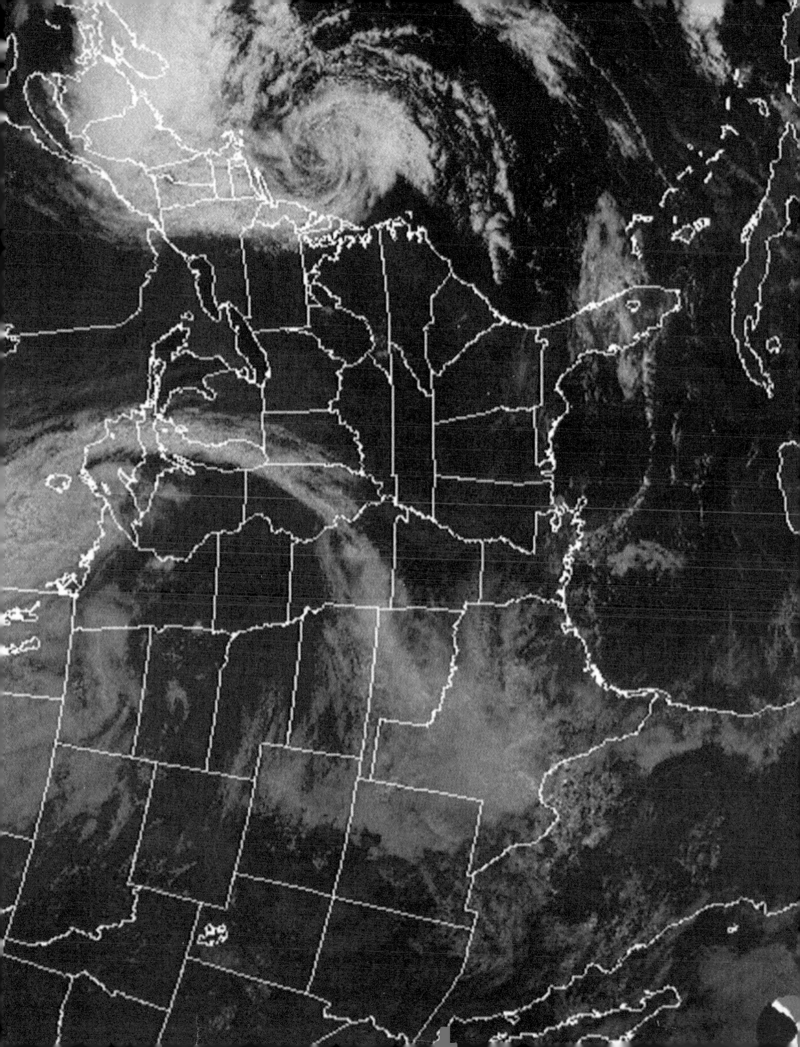

Climate Classification: Tropical, Arid, and Mesothermal Climate Regions

In tropical climate regions that experience a lengthy dry season, like this site in East Africa, waterholes are important resources for sustaining the wildlife population.

:: Objectives

When you complete this chapter you should be able to:

- Compare and contrast the advantages and limitations of the Thornthwaite and Köppen climate classification systems.
- Apply temperature and precipitation statistics to classify climates using the Köppen system.
- Recognize how the information on climographs is used to identify the climate of a place.
- Understand why vegetation types are so closely related to climate regions in the Köppen system.

- Outline the major characteristics of each humid tropical, arid, and mesothermal (moderate winter) climate.
- Describe the general locations of the humid tropical, arid, and mesothermal climates and the major factors that control their distributions.
- Understand the major vegetation types and human adaptations related to each humid tropical, arid, and mesothermal climate.

In the preceding chapters, we learned how the atmospheric elements combine to produce the weather. In this chapter and the following one we will examine how these elements interact over long periods of time to produce the climate, which gives character to the varied physical environments that cover our planet. As we noted earlier, climate is more than just average weather. Although the various climate types are primarily classified by statistical averages of elements like temperature and precipitation, the determination of a climate type also includes the likelihood of infrequent weather conditions such as storms, frost, and drought.

As a student seeking knowledge and understanding about the planet, this chapter and the next might prove to be particularly valuable to your life experience. In your lifetime you will likely observe a countless variety of natural landscapes and physical environments, either in person or in books, television,

and movies. These experiences will be enhanced when you understand and appreciate climatic characteristics and their impact on the landscape. The world's climates are ever-changing, and in this chapter and the next you will learn how and why this is so, and what impact a changing climate may have on interactions between humans and their physical environments.

Classifying Climates

Knowledge that climate varies from region to region dates to ancient times. The early Greeks (such as Aristotle, circa 350 B.C.) classified the known world into Torrid, Temperate, and Frigid zones based on their relative warmth. It was also recognized that these zones varied with latitude and that the flora and fauna also reflected these changes. With the further exploration of the world, naturalists noticed that the distribution of climates could be explained using factors such as sun angles, prevailing winds, elevation, and proximity to large water bodies.

The two weather variables used most often as indicators of climate are temperature and precipitation. To classify climates accurately, climatologists require a weather record of at least 30 years to describe the climate of an area. The invention of an instrument to reliably measure temperature—the thermometer—dates only to Galileo in the early 1600s. European settlement of distant colonies, and sporadic collection of temperature and precipitation data from those colonies, began in the 1700s, but was not routine until the mid-1800s or later. This was soon followed in the early 20th century by some of the first attempts to classify global climates using actual temperature and precipitation data.

Climatologists have worked to reduce the infinite number of worldwide variations in atmospheric elements to a comprehensible number of groups or varieties by combining elements with similar statistics (■ Fig. 7.1). Organizing and classifying the vast wealth of available climatic data into descriptions of major

■ **FIGURE 7.1** This map shows the diversity of climates possible in a relatively small area, including portions of Chile, Argentina, Uruguay, and Brazil. The climates range from dry to wet and from hot to cold, with many combinations of temperature and moisture characteristics.

What can you suggest as the causes for the major climate changes as you follow the 40°S latitude line from west to east across South America?

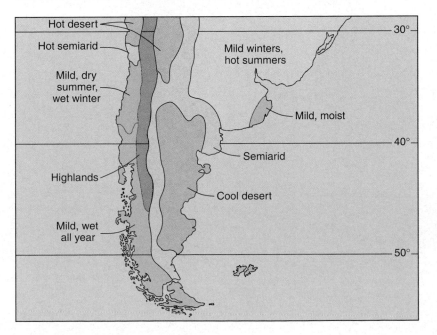

climatic groups enables geographers to concentrate on the large-scale causes of climatic differentiation. In addition, they can also examine exceptions to the general relationships, their distributions, and the processes related to those exceptions. Finally, differentiating climate types helps explain the geographic distribution of other climate-related phenomena of importance to global environments and humans.

Despite its usefulness, climate classification is not without its problems. Climate is a generalization about observed facts based on averages and probabilities of weather conditions. Each climate type is representative of a certain composite weather picture over the seasons. Within such a generalization, it is impossible to include the many variations that actually exist. On a global scale, generalizations, simplifications, and compromises are made to distinguish among climate types and regions.

The Thornthwaite System

One system used for classifying climates concentrates on characteristics at a local scale. This system is most useful for soil scientists, water resources specialists, and agriculturalists.

For example, for a farmer interested in growing a specific crop in a particular area, a global system for classifying large regions of Earth is too generalized. An important and locally-varying characteristic of climate concerns the amounts and timing of annual soil moisture surpluses or deficits. From an environmental or an agricultural perspective, it is very important to know if moisture will be available in the growing season, whether it comes directly in the form of precipitation or from the soil.

Developed by an American climatologist (and named in honor of him), the **Thornthwaite system** determines moisture availability (or shortages) at the subregional scale (■ Fig. 7.2). This system is often preferred when examining local climates. A locally detailed climate classification like the Thornthwaite system was only possible after temperature and precipitation data had been widely collected at numerous locations beginning in the latter half of the 19th century.

The Thornthwaite system is based on the concept of **potential evapotranspiration (potential ET),** which approximates the water use of plants and the loss through evaporation that would occur if an unlimited water supply were available. Evapotranspiration is a combination of *evaporation*

■ **FIGURE 7.2**   Thornthwaite climate regions in the contiguous United States are based on the relationship between precipitation (P) and potential evapotranspiration (ET). The moisture index (MI) for a region is determined by this simple equation:

$$MI = 100 \times \frac{P - \text{Potential ET}}{\text{Potential ET}}$$

Where precipitation exceeds potential ET, the index is positive; where potential ET exceeds precipitation, the index is negative.

What are the moisture index and Thornthwaite climate type for coastal California?

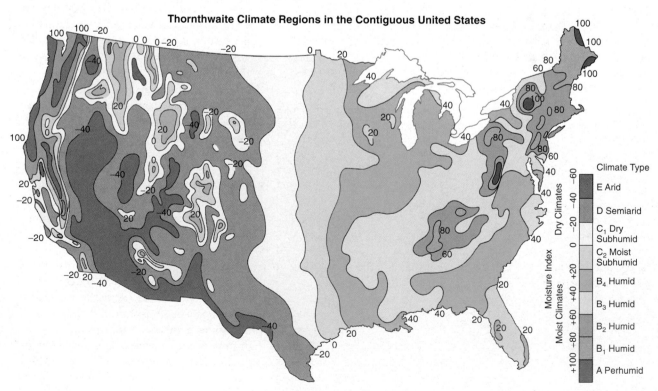

Thornthwaite Climate Regions in the Contiguous United States

Climate Type

Dry Climates	
E	Arid
D	Semiarid
C₁	Dry Subhumid

Moist Climates	
C₂	Moist Subhumid
B₄	Humid
B₃	Humid
B₂	Humid
B₁	Humid
A	Perhumid

Moisture Index

and *transpiration* (or water loss through vegetation). Potential ET is a theoretical value that increases with increasing temperature, winds, and length of daylight and decreases with increasing humidity. **Actual evapotranspiration (actual ET)** reflects the real (actual) evaporation loss and water use by plants at a location. Water for actual ET can be supplied during a dry season by soil moisture if the soil is not completely dry, and the available water supply may last through a dry season if the climate is relatively cool, and/or the day lengths are short. Measurements of actual ET relative to potential ET and available soil moisture are the determining factors for most vegetation and crop growth. The Thornthwaite system recognizes three climate zones based on potential ET values.

- Low-latitude climates, with potential ET greater than 130 centimeters (51 in).
- Middle-latitude climates, with potential ET less than 130 but more than 52.5 centimeters (20.5 in).
- High-latitude climates with potential ET less than 52.5 centimeters.

Climate zones are subdivided based on how long and by how much actual ET is below potential ET. Moist climates have either a surplus or a minor deficit of less than 15 centimeters (6 in). Dry climates are those with an annual deficit greater than 15 centimeters.

The Köppen System

The most widely used climate classification is based on temperature and precipitation patterns. It is referred to as the **Köppen system** after the German climatologist who developed it. Wladimir Köppen recognized that major vegetation associations reflect an area's climate. Hence, his climate regions were formulated to coincide with well-defined vegetation regions, and he named several after their most representative natural vegetation. Evidence of the strong influence of Köppen's system is seen in the wide usage of his climatic terminology, even in nonscientific literature (for example, tundra climate, rainforest climate).

Advantages and Limitations of the Köppen System   Temperature and precipitation are the two most commonly measured and recorded climatic variables. Variations caused by the atmospheric controls will show up most obviously in temperature and precipitation statistics. Further, temperature and precipitation are the weather elements that most directly affect humans, animals, vegetation, soils, and many elements of the natural landscape. Using temperature and precipitation statistics to define climate boundaries, Köppen derived precise numerical definitions for each climate region.

The climate boundaries in Köppen's classification were designed to correspond closely to global vegetation regions. Thus, many of Köppen's climate boundaries reflect "vegetation lines." For example, the Köppen classification uses the 10°C (50°F) monthly isotherm because of its relevance to the treeline—the line beyond which it is too cold and/or the growing season is too short for trees to thrive. For this reason, Köppen defined the treeless polar climates as those areas where the mean temperature of the warmest month is below 10°C (50°F). Clearly, if climates are divided according to associated vegetation types and if the division is based on the atmospheric elements of temperature and precipitation, then the result will be a visible association of vegetation with climate types. The relationship with the visible world in Köppen's climate classification system is one of its most appealing features to geographers.

There are of course limitations to Köppen's system. For example, Köppen considered only average monthly temperature and precipitation in making his climate classifications. These two elements permit estimates of precipitation effectiveness but do not measure it directly, or with enough precision to permit detailed comparison from one locality to another. For the purposes of generalization and because of very little available data concerning many other weather factors, Köppen did not consider winds, cloud cover, precipitation intensity, humidity, and daily temperature extremes, although these factors exert important influences on local weather and climate.

Simplified Köppen Classification   The Köppen system, as modified by later climatologists, divides the world into six major climate categories. The first four are based on temperature characteristics and adequate annual moisture: (*A*) **humid tropical climates,** (*C*) **humid mesothermal climates** (mild winter), (*D*) **humid microthermal climates** (severe winter), and (*E*) **polar climates** (■ Fig. 7.3). Another category, **arid climates** (*B*) includes *desert climates* (extremely arid) and *steppe climates* (semiarid), identifying characteristically dry regions by comparing the low precipitation they receive to their annual temperature characteristics (■ Fig. 7.4a). It should be noted, however, that the arid and semiarid climates include regions where the temperatures range from cold to very hot. The final category, (*H*) denotes the **highland climates,** the world's mountainous regions where vegetation and climate vary rapidly because of changes in elevation and exposure (Fig. 7.4b).

Within the first five major categories (all but the highland category), individual climate types and subtypes are differentiated from one another by specific parameters of temperature and precipitation. (See Table 1 in Appendix C, which outlines how to classify the climate types and subtypes of the Köppen classification system, and their designations by a set of letters.) We will refer to the climate types by their names as shown in Table 7.1. However, the letter symbols introduced in the Appendix will appear on graphs to identify specific climate types and subtypes and also on maps that indicate their locations and distribution. Familiarity with and frequent reference to the Appendix

■ **FIGURE 7.3**   The four humid climate categories of the Köppen climate classification are based on annual range of temperatures. (a) Tropical monsoon climate: Himalayan foothills, West Bengal, India; (b) Mediterranean mesothermal climate: village in southern Spain; (c) Microthermal, subarctic climate: these trees endure a long winter in Alaska; (d) Polar ice-sheet climate: glaciers near the southern coast of Greenland.

■ **FIGURE 7.4**   (a) Arid climates like this region of the Sonoran Desert, Arizona, are classified according to the deficiency of moisture they experience in a year. (b) Highland climates, as shown here in the Uncompahgre National Forest of Colorado, vary so widely according to their latitude, as well as local changes in elevation and exposure, that they are classified separately from the other Köppen climates.

TABLE 7.1
Simplified Köppen Climate Classes

Climates	Climograph Abbreviation
Humid Tropical Climates (*A*)	
Tropical Rainforest Climate	Tropical Rf.
Tropical Monsoon Climate	Tropical Mon.
Tropical Savanna Climate	Tropical Sav.
Arid Climates (*B*)	
Steppe Climate	Low-lat./Mid-lat. Steppe
Desert Climate	Low-lat./Mid-lat. Desert
Humid Mesothermal (Mild Winter) Climates (*C*)	
Mediterranean Climate	Medit.
Humid Subtropical Climate	Humid Subt.
Marine West Coast Climate	Marine W.C.
Humid Microthermal (Severe Winter) Climates (*D*)	
Humid Continental, Hot-Summer Climate	Humid Cont. H.S.
Humid Continental, Mild-Summer Climate	Humid Cont. M.S.
Subarctic Climate	Subarctic
Polar Climates (*E*)	
Tundra Climate	Tundra
Ice-sheet Climate	Ice-sheet
Highland Climates (*H*)	
Various climates based on elevation differences.	No single climograph can depict these varied (or various) climates

table and its contents are important. The major climate categories of the Köppen classification include enough differences in the ranges, total amounts, and seasonality of temperature and precipitation to produce the climate types listed in Table 7.1.

Climate Regions

Because each Köppen climate type is defined by specific numeric values for monthly averages of temperature and precipitation, it is possible to draw boundaries between these types on a world map. The areas within these boundaries are examples of one type of world region. The term **region,** as used by geographers, refers to an area that has recognizably similar internal characteristics that are distinct from those of other areas. A region may be described on any basis that unifies it and differentiates it from others.

As we examine the world's climate regions in this chapter and the one that follows, you should make frequent reference to the map of world climate regions (■ Fig. 7.5). It shows the patterns of Earth's climates as they are distributed over each continent. However, keep in mind that on a map of climate regions, distinct lines are used to separate one region from another. Obviously, the lines do not mark boundaries where there are abrupt changes in temperature or precipitation

conditions. Rather, the lines signify zones of transition between different climate regions. Furthermore, these zones or boundaries between regions are based on monthly and annual averages and may shift as temperature and moisture statistics change over the years.

The actual transition from one climate region to another is gradual, except in cases in which the change is brought about by an unusual climate control such as a mountain barrier. It would be more accurate to depict climate regions and their zones of transition on a map by showing one color fading into another. Keep in mind, as we describe Earth's climates, that it is the core areas of the regions that best exhibit the characteristics that distinguish one climate from another.

Climographs

The nature of the climate for anywhere on Earth can be summarized in graph form, as shown in ■ Figure 7.6, p. 154. Given data on mean (average) monthly temperature and rainfall, we can illustrate the changes in these two elements throughout the year by plotting their values on a graph. To make the pattern of the monthly temperature changes clearer, we can connect the monthly values with a continuous line, producing an annual temperature curve. To avoid

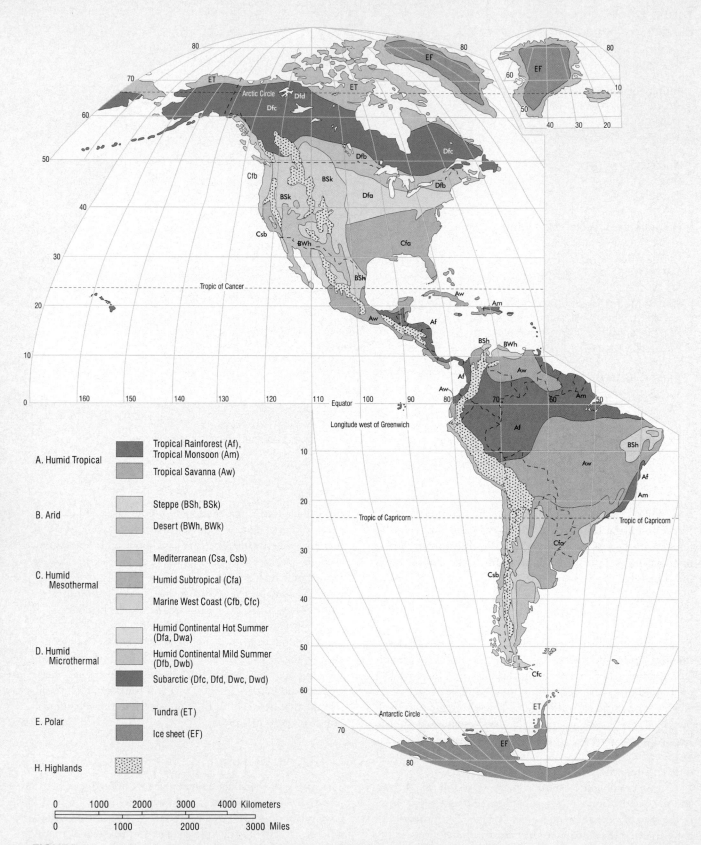

Legend:

A. Humid Tropical
- Tropical Rainforest (Af), Tropical Monsoon (Am)
- Tropical Savanna (Aw)

B. Arid
- Steppe (BSh, BSk)
- Desert (BWh, BWk)

C. Humid Mesothermal
- Mediterranean (Csa, Csb)
- Humid Subtropical (Cfa)
- Marine West Coast (Cfb, Cfc)

D. Humid Microthermal
- Humid Continental Hot Summer (Dfa, Dwa)
- Humid Continental Mild Summer (Dfb, Dwb)
- Subarctic (Dfc, Dfd, Dwc, Dwd)

E. Polar
- Tundra (ET)
- Ice sheet (EF)

H. Highlands

■ **FIGURE 7.5** World map of climates in the modified Köppen classification system.

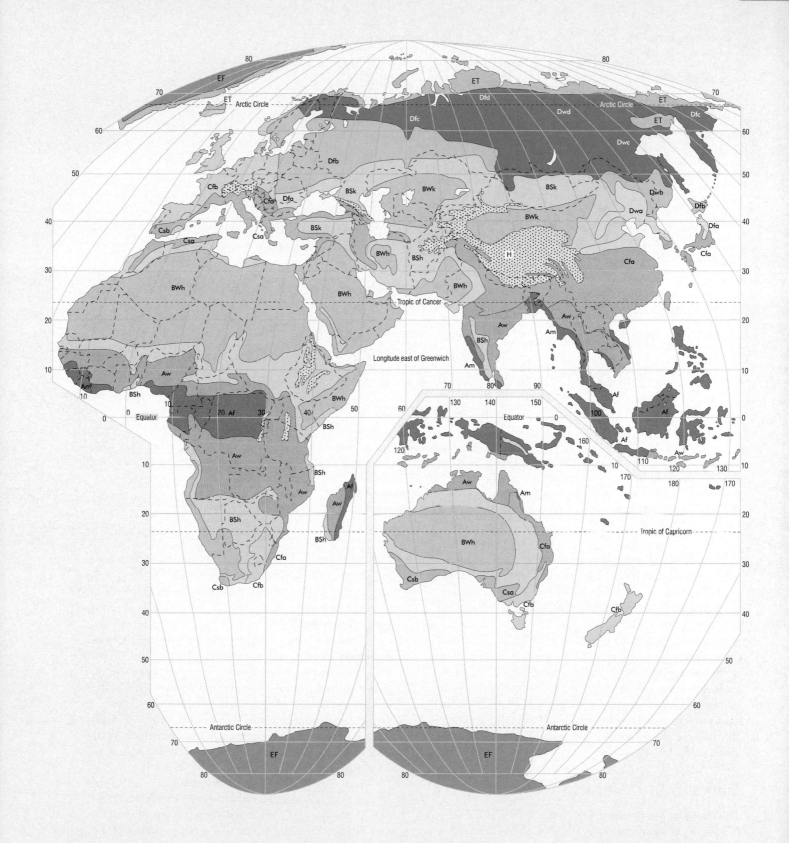

A Western Paragraphic Projection
Developed at Western Illinois University

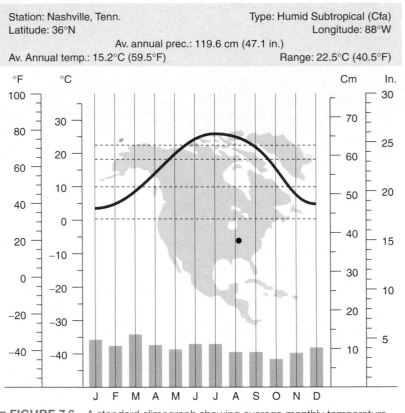

Station: Nashville, Tenn. Type: Humid Subtropical (Cfa)
Latitude: 36°N Longitude: 88°W
Av. annual prec.: 119.6 cm (47.1 in.)
Av. Annual temp.: 15.2°C (59.5°F) Range: 22.5°C (40.5°F)

■ **FIGURE 7.6** A standard climograph showing average monthly temperature (curve) and rainfall (bars). The horizontal index lines at 0°C (32°F), 10°C (50°F), 18°C (64.4°F), and 22°C (71.6°F) are the Köppen temperature parameters by which the station is classified.

What specific information can you read from the graph that identifies Nashville as a specific climate type (humid subtropical) in the Köppen classification?

confusion, average monthly precipitation amounts are usually shown on a graph as a bar instead of a curve. This kind of graphic display of a location's climate data is called a **climograph.** Other information may also be displayed, depending on the type of climograph. Figure 7.6 represents the type that we use in this chapter and in Chapter 8. To read these climographs, one must relate the temperature curve to the values given along the left side and the precipitation amounts to the scale on the right. A climograph can be used to determine the Köppen classification of a location as well as to show its specific temperature and rainfall regimes. The climate-type abbreviations relating to all climographs are found in Table 7.1.

Climate and Vegetation

The classification of plants at a global scale is as difficult as the classification of any other complex phenomenon that is influenced by a variety of factors. However, as Köppen recognized, plant communities are among the most highly visible of natural phenomena, so they can be categorized on the basis of form and structure or dominant physical characteristics. On Earth there are distinctive recurring

plant communities, which indicate a botanical response to systematic controls that are strongly related to climate. It is the dominant vegetation of these plant communities that we recognize when we classify Earth's major terrestrial ecosystems called **biomes.**

The major biome categories (forest, grasslands, desert, and tundra) are mapped in ■ Figure 7.7 (pp. 156–157) on the basis of the dominant associations of natural vegetation that gives each its distinctive character and appearance. The direct influence of climate on the distribution of biome types is apparent if you compare Figure 7.7 with Figure 7.5. Temperature (or latitudinal affect on temperature and insolation) and the availability of moisture are key factors in the location of biome types on a world regional scale (■ Fig. 7.8, p. 158).

As we consider the regions delineated by the Köppen climate system, we will emphasize both the characteristics of each climate and the natural vegetation associated with each climate. In addition, we could not provide a meaningful examination of each climate region if we did not also consider both the influences of climate on the humans that occupy the region and the vegetation that has to a significant extent displaced the natural vegetation as a result of human activity.

Humid Tropical Climate Regions

■ Figure 7.9 (p. 158) and a careful reading of Table 7.2 will provide the locations of humid tropical climates and a preview of the significant characteristics associated with them. The table also shows that, although all three humid tropical climates have high average temperatures throughout the year, they differ greatly in the amount and seasonal distribution of precipitation that they receive.

Tropical Rainforest and Tropical Monsoon Climates

The **tropical rainforest climate** probably comes readily to mind when someone says the word *tropical*. Hot and wet throughout the year, the tropical rainforest climate regions

TABLE 7.2
The Humid Tropical Climates

Name and Description	Controlling Factors	Geographic Distribution	Distinguishing Characteristics	Related Features
Tropical Rainforest Coolest month above 18°C (64.4°F); driest month with at least 6 cm (2.4 in.) of precipitation	High year-round insolation and precipitation of doldrums (ITCZ); rising air along trade wind coasts	Amazon R. Basin, Congo R. Basin, east coast of Central America, east coast of Brazil, east coast of Madagascar, Malaysia, Indonesia, Philippines	Constant high temperatures; equal length of day and night; lowest (2°C–3°C/3°F–5°F) annual temperature ranges; evenly distributed heavy precipitation; high amount of cloud cover and humidity	Tropical rainforest vegetation (selva); jungle where light penetrates; tropical iron-rich soils; climbing and flying animals, reptiles, and insects; slash-and-burn agriculture
Tropical Monsoon Coolest month above 18°C (64.4°F); one or more months with less than 6 cm (2.4 in.) of precipitation; excessively wet during rainy season	Summer onshore and winter offshore air movement related to shifting ITCZ and changing pressure conditions over large landmasses; also transitional between rainforest and savanna	Coastal areas of southwest India, Sri Lanka, Bangladesh, Myanmar, southwest Africa, Guyana, Surinam, French Guiana, northeast and southeast Brazil	Heavy high-sun rainfall (especially with orographic lifting), short low-sun drought; 2°C–6°C (3°F–10°F) annual temperature range, highest temperature just prior to rainy season	Forest vegetation with fewer species than tropical rainforest; grading to jungle and thorn forest in drier margins; iron-rich soils; rainforest animals with larger leaf-eaters and carnivores near savannas; paddy rice agriculture
Tropical Savanna Coolest month above 18°C (64.4°F); wet during high-sun season, dry during lower-sun season	Alternation between high-sun doldrums (ITCZ) and low-sun subtropical highs and trades caused by shifting winds and pressure belts	Northern and eastern India, interior Myanmar and Indo-Chinese Peninsula; northern Australia; borderlands of Congo R., south central Africa; llanos of Venezuela, campos of Brazil; western Central America, south Florida, and Caribbean Islands	Distinct high-sun wet and low-sun dry seasons; rainfall averaging 75–150 cm (30–60 in.); highest temperature ranges for humid tropical climates	Grasslands with scattered, drought-resistant trees, scrub, and thorn bushes; poor soils for farming, grazing more common; large herbivores, carnivores, and scavengers

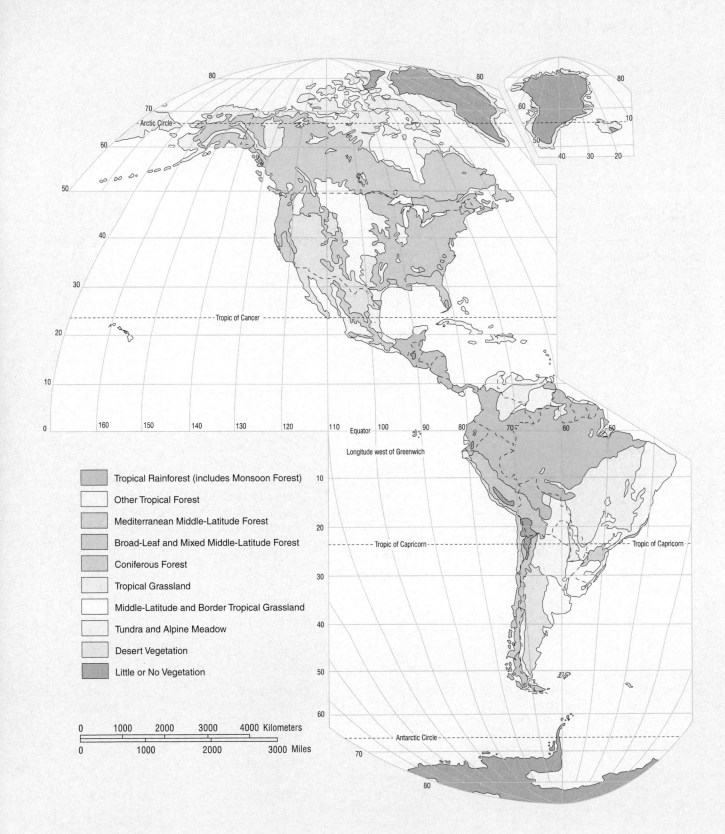

■ **FIGURE 7.7** World map of natural vegetation.

A Western Paragraphic Projection
Developed at Western Illinois University

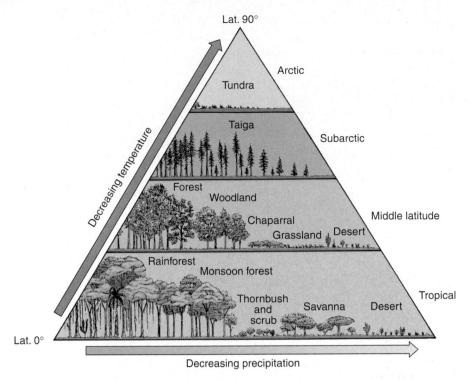

Influence of latitude and moisture on distribution of biomes

■ **FIGURE 7.8** This schematic diagram shows distribution of Earth's major biomes as they are related to temperature (latitude) and the availability of moisture. Within the tropics and middle latitudes, there are distinctly different biomes as total biomass decreases with decreasing precipitation.

What major biome dominates the wetter margins of all latitudes but the Arctic?

■ **FIGURE 7.9** Index map of humid tropical climates.

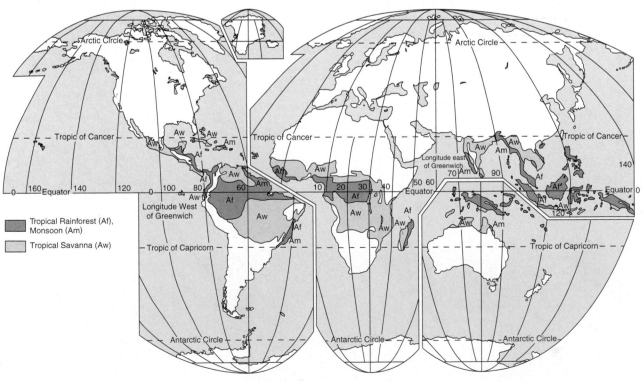

Humid Tropical Climates

remain among the most challenging environments for human occupation. Visiting this type of climate, an individual will experience high temperatures and humidity accompanied by frequent heavy rains that sustain the massive vegetative growth for which it is known. By comparison, although originally covered by tropical forests, the seasonal nature of precipitation in the tropical monsoon climate can support large populations in the coastal floodplain regions of Southeast Asia.

Contrasting Characteristics Comparing the climographs for Akassa, Nigeria, and Calicut, India, in ■ Figure 7.10 clearly indicates the major difference between the tropical rainforest and tropical monsoon climates. Although rainfall totals for the year are somewhat similar, the dry season–wet season nature of the **tropical monsoon climate** is in sharp contrast to the more consistent precipitation of the tropical rainforest regions. Heavy precipitation in both climates is associated with warm, humid air and unstable conditions along the ITCZ (intertropical convergence zone). In the tropical rainforest climate, both convection and convergence serve as uplift mechanisms, causing humid air to rise, condense, and produce heavy rains that are characteristic in this climate. Variations in rainfall can usually be traced to the ITCZ and its low pressure cells of varying strength. Many tropical rainforest locations (Akassa, for example) exhibit two maximum precipitation periods during the year, during each appearance of the ITCZ as it follows the migration of the sun's direct rays, which cross the equator on the March and September equinoxes. In addition, although no season can be called dry, during some months it may rain on only 15 or 20 days.

In the tropical monsoon climate, the alternating air circulation (from sea to land and from land to sea) is also related to the shifting of the ITCZ. During the northern hemisphere summer, the ITCZ moves north onto the Indian subcontinent and adjoining lands to latitudes of 20°–25°N. This is due in part to the attracting force of the strong low pressure system of the Asian continent in summer. Several months later, the moisture-laden summer monsoon is replaced by an outflow of dry air from the massive Siberian high pressure system that develops in the winter over central Asia. During the summer monsoon, the uplift mechanisms associated with the ITCZ are greatly enhanced by orographic uplift as the air is forced to rise over land barriers such as the windward side of India's Western Ghats and the south-facing slopes of the Himalayas.

As the climographs of Figure 7.10 indicate, the average monthly temperature of both rainforest and monsoon regions are consistently high. However, the annual march of temperature of monsoon climate regions differs from the monotony that exists in most rainforest regions. The heavy cloud cover during the rainy monsoon season reduces insolation and temperatures during that time of year. Higher temperatures are recorded just prior to the onslaught of the rainy season, when clear skies occur.

In addition to high rainfall totals and high average annual temperatures, an additional climatic characteristic distinguishes rainforest and monsoon climates from all other climate types. The **annual temperature range**—the difference between the average temperatures of the warmest and coolest months of the year—is very low, reflecting the consistently high sun angle in tropical latitudes. The annual range for tropical rainforest stations is seldom more than 2°C–3°C (4°F–5°F), and for tropical monsoon stations it is only slightly greater at 2°C–6°C (4°F–11°F). Another interesting distinction of the rainforest climate is that **daily (diurnal) temperature ranges**—the differences between the highest

■ **FIGURE 7.10** Climographs for a tropical rainforest and a tropical monsoon station.
What information is most important as you compare the two stations?

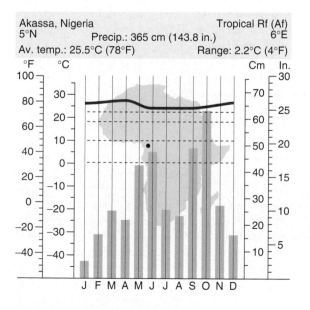

Akassa, Nigeria Tropical Rf (Af)
5°N Precip.: 365 cm (143.8 in.) 6°E
Av. temp.: 25.5°C (78°F) Range: 2.2°C (4°F)

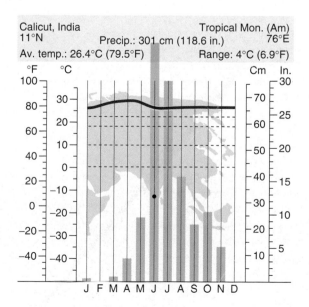

Calicut, India Tropical Mon. (Am)
11°N Precip.: 301 cm (118.6 in.) 76°E
Av. temp.: 26.4°C (79.5°F) Range: 4°C (6.9°F)

■ **FIGURE 7.11** (a) The typical vegetation in a tropical rainforest climate forms a cover of trees, growing at different heights to make a multilayered treetop canopy. This is the rainforest canopy in the Amazon region of Brazil. (b) Tall and massive hardwood trees with distinctive, buttressed trunks thrive in a climate that is hot and wet all year, but shady on the forest floor.
How many tree layers can you see in (a)?

and lowest temperatures during the day—are usually greater than the annual range. Highs of 30°C–35°C (86°F–95°F) and lows of 20°C–24°C (68°F–75°F) produce daily ranges of 10°C–15°C (18°F–27°F). However, high humidity causes even the cooler nights to seem oppressive.

Forest Biomes The forests of the world's tropics are far from uniform in appearance and composition. They grade poleward from the *tropical rainforests*, which support Earth's greatest biomass, to the last scattering of low trees that overlook the seemingly endless expanses of tall grasses and scattered trees that characterize the tropical savanna climate. In equatorial lowlands dominated by tropical rainforest climate, the only environmental limitation for vegetation growth is competition for light among adjacent species. Temperatures are warm enough all year to support constant growth, and water is always sufficient. The **tropical rainforests** consist of an amazing number of *broad-leaf evergreen* tree species. **Broad-leaf** refers to trees that do not have needles (like pines do, for example), and **evergreen** means that throughout the seasons they do not lose their leaves. A cross section of the rainforest often reveals concentrations of tree canopies at several different levels. The trees composing the distinctive individual

canopy tiers have similar light requirements—lower than those of the higher tiers but higher than those of the lower tiers (■ Fig. 7.11a). Little or no sunlight reaches the shady forest floor, which may support ferns but is often rather sparsely vegetated. Tropical rainforests are frequently traversed by woody vines, called **lianas,** which climb the tree trunks and intertwine toward the canopy in their own search for light. Aerial plants, called **epiphytes,** may grow on the limbs of the forest trees, deriving nutrients from the water and the plant debris that falls from higher levels.

The forest trees commonly depend on widely flared or buttressed bases for support because of their shallow root systems (Fig. 7.11b). This is a consequence of the richness of the surface soil and the poverty of its lower levels. The rainforest vegetation and soil are intimately associated. The forest litter is quickly decomposed, its nutrients released and almost immediately reabsorbed by root systems, which consequently remain near the surface. Tropical soils that maintain the incredible biomass of the rainforest are fertile only as long as the forest remains undisturbed. Clearing the forest interrupts the critical cycling of nutrients between the vegetation and the soil; the copious amounts of water percolating through the soil leach away its soluble constituents, leaving behind only inert

M. Trapasso

■ **FIGURE 7.12**   A jungle along the Usumacinta River on the Mexico–Guatemala border.
Why is the vegetation so dense here, when it is more open inside the forest at ground level?

iron and aluminum oxides that cannot support forest growth. The present rate of clearing threatens to wipe out the world-wide tropical rainforests within the foreseeable future. The largest remaining areas of unmodified rainforests are in the upper Amazon Basin where they cover hundreds of thousands of square kilometers.

Environmental conditions vary from place to place within climate regions; therefore, the typical rainforest situation that we have just described does not apply everywhere in the tropical rainforest climate. Some regions are covered by true jungle, a term often misused when describing the rainforest. **Jungle** is a dense tangle of vines and smaller trees that develops where direct sunlight does reach the ground, as in clearings and along streams (■ Fig. 7.12). Other regions have soils that remain fertile or have bedrock that is chemically basic and provides the soils above with a constant supply of soluble nutrients. Examples of the former regions are found along major river floodplains; examples of the latter are the volcanic regions of Indonesia and the limestone areas of Malaysia and Vietnam. Only in regions of continuous soil fertility can agriculture be intensive and continuous enough to support population centers in the tropical rainforest climate.

Toward the wetter margins of the tropical monsoon climate, the monsoon forest resembles the tropical rainforest but fewer plant species are present and certain ones become dominant. The seasonality of rainfall in the monsoon climate narrows the range of species that will prosper. Toward the drier margins of the climate, the trees grow farther apart, and the forest often gives way to jungle or a dwarfed thorn forest.

The composition of the animal population in monsoon forests also differs from that of rainforest regions. Because of the darkness and extensive root systems present on the forest floor, animals of the rainforest are primarily arboreal. A wide variety of species of tree-dwelling monkeys and lemurs, snakes, tree frogs, birds, and insects characterize the rainforest. Even the herbivorous and carnivorous mammals—such as sloths, ocelots, and jaguars—are primarily arboreal (living in trees). In monsoon regions, the climbing and flying species of rainforest regions are joined by larger, hoofed leaf eaters and by large carnivores such as the famous tigers of Bengal.

Human Activity

There are numerous challenges to human habitation of tropical rainforest and monsoon regions. In addition to the incessant heat and oppressive humidity of the rainforest and wet monsoon, humans must do constant battle with a host of insects. Mosquitoes, ants, termites, flies, beetles, grasshoppers, butterflies, and bees live everywhere in both climates. Insects can breed continuously without danger from cold or extended drought. A variety of parasites and disease-carrying insects even threaten human survival. Malaria, yellow fever, dengue fever, and sleeping sickness are all insect-borne (sometimes fatal) diseases of the tropics, but uncommon in the middle latitudes.

Whenever native populations have existed in the rainforest, subsistence hunting and gathering of fruits, berries, small animals, and fish have been important. Since the introduction of agriculture, land has been cleared, and crops such as manioc, yams, beans, maize (corn), bananas, and sugarcane have been grown. It has been the practice to cut down the

R. Gabler

■ **FIGURE 7.16** Tropical savanna climate: East African high plains.

the rainfall becomes. However, the rains are essential for human and animal survival in savanna regions. When rains are late or deficient, as they have been in West Africa in recent years, severe drought and famine result. Unfortunately, when the rains last longer than usual, or are excessive, they can cause major floods, often followed by outbreaks of disease.

The savannas of Africa have been veritable zoological gardens for the larger tropical animals, to such an extent that photo safaris have made the African savannas a major tourist destination. The grasslands support many different herbivores (plant eaters), such as the elephant, rhinoceros, giraffe, zebra, and wildebeest (■ Fig. 7.16). The herbivores in turn are eaten by carnivores (flesh eaters), such as the lion, leopard, and cheetah. Lastly, scavengers, such as hyenas, jackals, and vultures, devour what remains of the carnivore's kill. During the dry season, the herbivores find grasses and water along stream banks and forest margins and at isolated water holes. The carnivores follow the herbivores to the water, and, even in the game preserves, illegal hunters and poachers still follow them both.

Arid Climate Regions

There are two major types of locations where arid climates are found, and each illustrates an important climatic factor that causes dry conditions. The first type is centered on both the Tropics of Cancer and Capricorn (23½°N and S latitudes) and then extends 10°–15° poleward and equatorward. These regions contain the most extensive areas of arid climates in the world. The second type of location for arid climates is at higher latitudes and occupies continental interiors, particularly in the Northern Hemisphere. These two types of arid regions typically will have a central core of desert climate, bounded on the edges by transition zones of semiarid steppe climates.

The concentration of deserts in the vicinity of the two tropic lines is directly related to the subtropical high pressure systems. Although the boundaries of the subtropical highs migrate north and south with the direct rays of the sun, their

influence remains strong in these latitudes. The subsidence and divergence of air associated with these systems is strongest along the eastern portions of the oceans (recall that cold ocean currents off the western coasts of continents help stabilize the atmosphere). Hence, the clear weather and dry conditions of the subtropical high pressure extend inland from the western coasts of each landmass in the subtropics. The Atacama, Namib, and Kalahari Deserts, as well as the desert of Baja California, are restricted in their development by the small size of the landmass or by landform barriers toward the interior. However, the western portion of North Africa and the Middle East comprises the greatest stretch of desert in the world and includes the Sahara, Arabian, and Thar Deserts. Similarly, the Australian Desert occupies most of the interior of the Australian continent.

The second concentration of deserts is located within continental interiors that are remote from moisture-carrying winds. Such arid lands include the vast cold-winter deserts of inner Asia and the Great Basin of the western United States. The dry conditions of the latter region extend northward into the Columbia Plateau and southward into the Colorado Plateau, and are increased by the mountain barriers that restrict the movement of rain-bearing air masses from the Pacific. Similar rain-shadow conditions help explain the Patagonia Desert of Argentina and the arid lands of western China.

■ Figure 7.17 shows deserts of the world to be core areas of aridity, usually surrounded by the semiarid steppe regions. Hence, our explanations for the location of deserts hold true for the steppes as well. The steppe climates are transitional between humid climates and the deserts. As previously noted, we classify both steppe and desert on the basis of the relation between precipitation and potential evapotranspiration (ET). In the **desert climate,** the amount of precipitation received is less than half the potential ET (often much less—one-fifth to one-tenth or less is not uncommon). In the **steppe climate,** the precipitation is more than half but still significantly less than the total potential ET.

The criterion for determining whether a climate is desert, steppe, or humid is *precipitation effectiveness*. The amount of precipitation actually available for use by plants and animals is the *effective* precipitation. Precipitation effectiveness is related to temperature. At higher temperatures, it takes more precipitation to have the same effect on vegetation and soils than at lower temperatures. The result is that areas with higher temperatures that promote greater ET can receive more precipitation than cooler regions and yet have a more arid climate.

Desert Climates

The deserts of the world extend through such a wide range of latitudes that the Köppen system recognizes two major subdivisions. The first are low-latitude deserts where temperatures are relatively high year-round and frost is absent or infrequent even along poleward margins; the second are middle-latitude deserts, which have distinct seasons, including below-freezing temperatures during winter (Table 7.3). However, the significant characteristic of all deserts is their aridity.

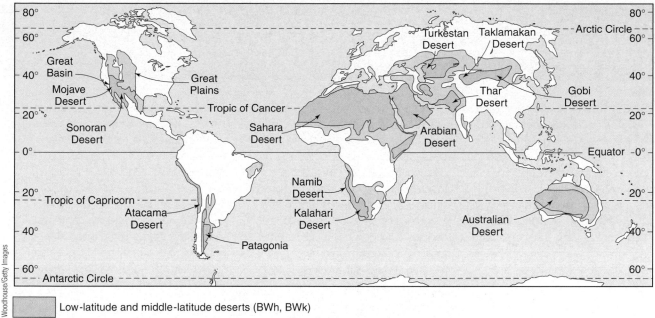

■ **FIGURE 7.17** A map of the world's arid lands.

What does a comparison of this map with the Map of World Population Density (inside back cover) suggest?

TABLE 7.3
The Arid Climates

Name and Description	Controlling Factors	Geographic Distribution	Distinguishing Characteristics	Related Features
Desert Precipitation less than half of potential evapotranspiration; mean annual temperature above 18°C (64.4°F) (low-lat.), below (mid-lat.)	Descending, diverging circulation of subtropical highs; continentality often linked with rainshadow location	Coastal Chile and Peru, southern Argentina, southwest Africa, central Australia, Baja California and interior Mexico, North Africa, Arabia, Iran, Pakistan and western India (low-lat.); inner Asia and western United States (mid-lat.)	Aridity; low relative humidity; irregular and unreliable rainfall; highest percentage of sunshine; highest diurnal temperature range; highest daytime temperatures; windy conditions	Xerophytic vegetation; often barren, rocky, or sandy surface; desert soils; excessive salinity; usually small, nocturnal burrowing animals; nomadic herding
Steppe Precipitation more than half but less than potential evapotranspiration mean annual temperature above 18°C (64.4°F) (low-lat.), below (mid-lat.)	Same as deserts; usually transitional between deserts and humid climates	Peripheral to deserts, especially in Argentina, northern and southern Africa, Australia, central and southwest Asia, and western United States	Semiarid conditions, annual rainfall distribution similar to nearest humid climate; temperatures vary with latitude, elevation, and continentality	Dry savanna (tropics) or short grass vegetation; highly fertile black and brown soils; grazing animals in vast herds; predators and small animals; ranching and dry farming

GEOGRAPHY'S ENVIRONMENTAL PERSPECTIVE

:: DESERTIFICATION

Desertification is the expansion of desert landscapes related to climate change but accelerated by human activities, and involves long-term environmental and human consequences. Desertification expands the margins of the desert when rare rains cause erosion and loss of soil, so eventually most vegetation cannot survive. It also increases wind erosion, causing dust storms and sand dune movement into grassland and farmland areas. Although climate change may be the trigger, the process is accelerated by deforestation, over-cultivation, accumulation of salt in the soil due to irrigation, and over-grazing by cattle, sheep, and goats.

Archeological evidence from the Middle East indicates that as far back as 4000 B.C. early farming communities may have destroyed the soil and deforested the hills, causing desertification. A similar pattern of denudation occurred in the hilly landscape of Greece as early as 3000 B.C. Along with the threat to the human population, desertification endangers habitats for wildlife.

It was not until the 1970s, however, that desertification became well known, as the media revealed starving and suffering people in the African Sahel. Television showed bone-thin

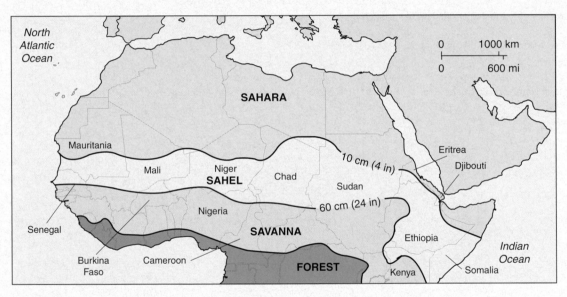

The Sahel region of Africa, shown here in a light-tan color, is the transition zone between the extreme aridity of the Sahara and the tropical humid areas of Africa (in green tones). In recent years, the Sahel has experienced desertification through climate change and overuse of this marginal land by human activities.

Land of Extremes By definition, deserts are associated with a minimum of precipitation, but they also represent extremes in other atmospheric conditions. With few clouds and low relative humidity in desert regions, as much as 90% of insolation reaches the surface. This is why the highest insolation and highest temperatures are recorded in low-latitude desert areas and not in the more humid tropical climates that are closer to the equator. Again, because cloud cover is light or absent and the air is clear and dry, much of the energy received during the day is radiated back to the atmosphere at night. Consequently, night temperatures in the desert drop far below their daytime highs.

The extremes of heating and cooling give low-latitude deserts the greatest diurnal temperature ranges in the world, and middle-latitude deserts are not far behind. In the spring and fall, these ranges may be as great as 40°C (72°F) in a day. More common diurnal temperature ranges in deserts are 22°C–28°C (40°F–50°F).

The sun's rays are so intense in the clear, dry desert air that temperatures in shade are much lower than those a few steps away in direct sunlight. (Note that all temperatures for meteorological statistics are recorded in the shade.) Khartoum, Sudan, in the Sahara, has an *average* annual temperature of 29.5°C (85°F), which is a *shade* temperature.

cattle trying to find a blade of grass in a barren landscape and villages being invaded by sand dunes. The Sahel is the semiarid zone bordering the southern margin of the Sahara. The term *desertification* was popularized at a United Nations conference that addressed problems like those of the Sahel, and many people associate the term with the continuing plight of the people in that region. Today, evidence of desertification is also visible in areas of Spain, in northwestern India, and throughout much of the Middle East, northern China, and northern Africa.

In 1994, 87 nations signed a treaty for budgeting funds to help protect the fertility of lands that are at the greatest risk of desertification. Only a major international effort can deal with a natural hazard that causes such large-scale environmental deterioration and human suffering.

UNEP Sudan

This pond in the Sudan, built to impound water, has dried up completely even after more rainfall had occurred than has been typical in recent years. Some vegetation exists around the margins of the pond, but the area in the background is seriously desertified.

During low-sun or winter months, deserts experience colder temperatures compared to more humid areas at the same latitude, and in summer they experience hotter temperatures. Just as with the high diurnal ranges in deserts, these high annual temperature ranges can be attributed to the lack of moisture in the air.

Annual temperature ranges are usually greater in middle-latitude deserts, such as the Gobi in Asia, than in low-latitude deserts, because of the colder winters experienced at higher latitudes deep within a continent. Compare, for example, the climograph for Aswan in south central Egypt—at 24°N, a low-latitude desert location—with the climograph for Turtkul,

Uzbekistan—at 41°N, a middle-latitude desert location (■ Fig. 7.18). The annual range for Aswan is 17°C (31°F); in Turtkul, it is 34°C (61°F).

Although the actual amount of water vapor in the air may be high in many desert regions, the relative humidity is low, so precipitation is irregular and unreliable. When rain does occur, it may arrive in an enormous cloudburst (■ Fig. 7.19). Because hot daytime temperatures increase the capacity of the atmosphere to hold moisture, desert nights are a different story. Radiation of energy is rapid in the clear air. As temperature drops, relative humidity increases and the formation of dew in the cool hours of early morning may occur. In some

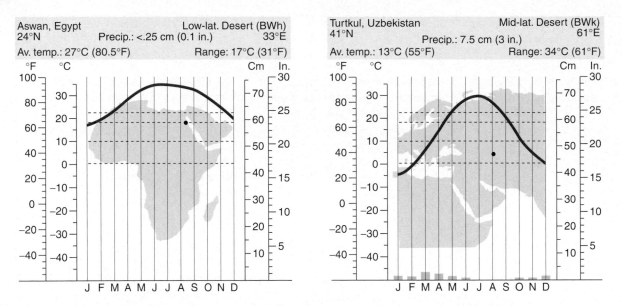

Aswan, Egypt	Low-lat. Desert (BWh)
24°N	33°E
Av. temp.: 27°C (80.5°F)	Range: 17°C (31°F)
Precip.: <.25 cm (0.1 in.)	

Turtkul, Uzbekistan	Mid-lat. Desert (BWk)
41°N	61°E
Av. temp.: 13°C (55°F)	Range: 34°C (61°F)
Precip.: 7.5 cm (3 in.)	

■ **FIGURE 7.18** Climograph for desert climate stations.
If you consider the serious limitations of desert climates, how do you explain why some people choose to live in desert regions?

©Johnathan Blair/CORBIS

■ **FIGURE 7.19** A rainstorm in the Mojave Desert of California produces a double rainbow.
What environmental clues suggest that rainfall is an infrequent event?

regions where measurements have been made, the amount of dew formed has sometimes considerably exceeded the annual rainfall for that location.

Adaptations by Plants and Animals
Deserts tend to have sparse vegetation, and large tracts may be barren bedrock, sand, or gravel. The plants that do exist are **xerophytic,** which means well adapted to extreme drought. They have some combination of thick bark, thorns, little foliage, a compact shape, and waxy leaves, all of which reduce

water loss by transpiration. Another characteristic adaptation is the storage of moisture in stem or leaf cells, as in the cactus (■ Fig. 7.20). Some plants, such as the creosote bush, mesquite, and acacias, have deep tap root systems to reach water; others, such as the Joshua tree, spread their roots widely near the surface for their moisture supply.

The non-xerophytic vegetation consists mainly of short-lived annual plants that germinate and hurry through a complete life cycle of growth—leaf production, flowering, and seed dispersal—in a matter of weeks when triggered by moisture availability. Like other species, these ephemeral plants also require days of a certain length, so they appear only in particular months; therefore, the month-to-month and year-to-year variation in form and appearance of desert vegetation is enormous. Animals of the deserts are primarily nocturnal so they can avoid the searing daytime heat, and many have evolved long ears, noses, legs, and tails that allow for greater blood circulation and cooling. The similar life forms and habits of different plant and animal species found in the deserts of widely separated continents are a remarkable display of evolutionary adaptation to ensure survival in similar climatic settings.

Humans also find the desert environment a lasting challenge. Inhabitants of deserts have generally been hunters and gatherers, nomadic herders, and subsistence farmers wherever there was a water supply from wells, oases, or exotic streams (streams bearing water from outside the region), such as the Nile, Tigris, Euphrates, Indus, and Colorado Rivers

(a)   (b)

■ **FIGURE 7.20** (a) Vegetation adapted to the arid conditions in the Atacama Desert in Chile. (b) After recent rainfall the desert landscape here in Organ Pipe National Park in Arizona becomes greener and comes into bloom.

What physical characteristics of cacti help them to survive the heat, drought, and evaporation rates of the desert?

■ **FIGURE 7.21** This oasis supplies enough water to support a small suburb of Ica, in southern Peru.

What can you see in the background that proves this is a desert environment?

(■ Fig. 7.21). Desert people have learned to adjust their habits to the environment. For example, they wear loose clothing to protect themselves from the burning rays of the sun and to prevent moisture loss by evaporation from the skin. At night, when the temperatures drop, the clothing keeps them warm by insulating and minimizing the loss of body heat.

Permanent agriculture has been established in desert regions all around the world wherever river or well water is available. Some produce mainly subsistence crops, but others have become significant producers of commercial crops for export.

Steppe Climates

Further study of Figure 7.17 and Table 7.3 provides a reminder that the distribution of the world's steppe lands is closely related to the location of deserts. Both moisture-deficient climate types share the controlling factors of continentality, rain-shadow location, subtropical high pressure systems, or some combination of these three. In both climates, the potential ET exceeds the precipitation. As in the deserts, precipitation in steppe regions is unpredictable and varies widely in total amount from year to year.

Steppes and Natural Vegetation Like deserts, steppe regions include both low-latitude and middle-latitude varieties, identified by mean annual temperature (see again Table 7.3). The climographs of ■ Figure 7.22 demonstrate that although summer temperatures are high in all steppe regions, the differences in winter temperatures can produce annual ranges in middle-latitude steppes that are two or three times as great as those in low-latitude steppes. Because of higher

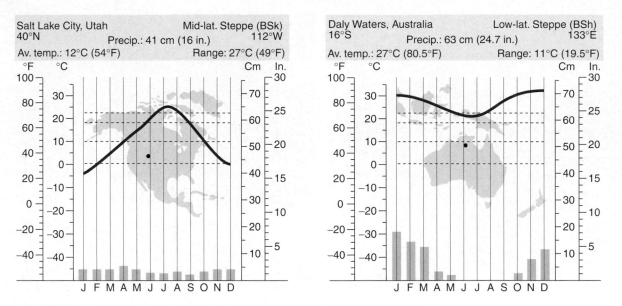

■ **FIGURE 7.22**   Climographs for steppe climate stations.
What are the major differences between the climograph data for Salt Lake City and Daly Waters? What are the causes of these differences?

evapotranspiration rates, annual rainfall totals for low-latitude steppe locations such as Daly Waters are usually much higher than those of middle-latitude stations such as Salt Lake City.

Where most desert locations receive fewer than 25 centimeters (10 in) of rain annually, steppe regions typically receive between 25 and 50 centimeters (10–20 in) of precipitation. The higher amounts in a steppe region support significant natural vegetation that is usually related to the ground cover of the more humid climate immediately adjacent to it. On areas of transition from tropical savanna to desert, the vegetation is the dry savanna type, including scrub tree and bush growth. Bordering the humid climates of the middle latitudes, most steppe areas were originally covered with grasslands like those found in the steppe region of Russia from which Köppen derived the climate's name. West of the 100th meridian in North America and extending across Eurasia from the Black Sea to northern China, there are vast, nearly level plains overgrown with a mixture of grasses, with shorter species becoming dominant in the direction of lower annual precipitation. In North America, the *short-grass prairie* usually coincides with a zone in which moisture rarely penetrates more than 60 centimeters (2 ft) into the soil, so the subsoil is permanently dry.

A Dangerous Appeal
Although the ground cover is often incomplete in low-latitude steppes, the grasslands in Africa were originally the domain of wildebeest, zebra, and a host of other grazing animals. These have been replaced today by herdsmen and their cattle or by villagers attempting to raise their crops despite meager and undependable precipitation. Overgrazing and ill-advised farming have often led to poverty, encroaching deserts, hunger, and the threat of starvation, especially in the Sahel region adjoining the Sahara.

■ **FIGURE 7.23**   Steppe climate: Sand Hills of Nebraska.
What might be different at this location 150 years ago?

In their natural conditions, the grassland steppes of North America and Eurasia supported high densities of grazing animals—bison and antelope in the former and wild horses in the latter. Today's animals are domesticated cattle, not the self-sufficient herds of wild species that formerly populated these regions, often in great numbers (■ Fig. 7.23).

Successful agriculture is possible in middle-latitude steppes but not without the use of irrigation or dry farming (alternate year) methods. Where such techniques have not been employed in higher precipitation steppe regions the results have been disastrous. During dry cycles, crops fail year after year, and with the land stripped of its natural sod, the soil is exposed to wind erosion. Even using the grasses for grazing domesticated animals can cause problems, for overgrazing can just as quickly create "Dust Bowl" conditions (as occurred during the extreme droughts of the 1930s in the southwestern United States).

The difficulties in making steppe regions more productive illustrate the sensitive ecological balance of Earth's systems. The natural rains in the steppe were usually sufficient to support a cover of short grasses that fed the roaming herbivores that grazed on them. The herbivore population in turn was kept in check by the carnivores that preyed on them. When people enter the scene, however, sending out more animals to graze, plowing the land, or merely killing off the predators, the ecological balance is tipped, and human suffering can be the tragic result.

Mesothermal Climate Regions

When we use the term *mesothermal* (from Greek: *mesos*, middle) in describing climates, we are referring to the moderate temperatures that characterize such regions. In the remainder of this chapter we include a detailed discussion of the mesothermal climates, which are outlined in Table 7.4. Mesothermal climates do not experience the high temperatures all year of the humid tropics or the cold winters of microthermal regions. However, two of the mesothermal climates may also be considered subtropical. The Mediterranean and the

TABLE 7.4
The Mesothermal Climates

Name and Description	Controlling Factors	Factors Geographic Distribution	Distinguishing Characteristics	Related Features
Mediterranean Warmest month above 10°C (50°F); coldest month between 18°C (64.4°F) and 0°C (32°F); summer drought; hot summers (inland), mild summers (coastal)	West coast location between 30° and 40°N and S latitudes; alternation between subtropical highs in summer and westerlies in winter	Central California; central Chile; Mediterranean Sea borderlands, Iranian highlands; Cape Town area of South Africa; southern and southwestern Australia	Mild, moist winters and hot, dry summers inland with cooler, often foggy coasts; high percentage of sunshine; high summer diurnal temperature range; frost danger	Sclerophyllous vegetation; low, tough brush (chaparral); scrub woodlands; varied soils, erosion in Old World regions; winter-sown grains, olives, grapes, vegetables, citrus, irrigation
Humid Subtropical Warmest month above 10°C (50°F); coldest month between 18°C (64.4°F) and 0°C (32°F); hot summers; generally year-round precipitation, winter drought (Asia)	East coast location between 20° and 40°N and S latitudes; humid onshore (monsoonal) air movement in summer, cyclonic storms in winter	Southeastern United States; southeastern South America; coastal southeast South Africa and eastern Australia; eastern Asia from northern India through south China to southern Japan	High humidity; summers like humid tropics; frost with polar air masses in winter; precipitation 62–250 cm (25–100 in.), decreasing inland; monsoon influence in Asia	Mixed forests, some grasslands, pines in sandy areas; soils productive with regular fertilization; rice, wheat, corn, cotton, tobacco, sugarcane, citrus
Marine West Coast Warmest month above 10°C (50°F); coldest month between 18°C (64.4°F) and 0°C (32°F); year-round precipitation; mild to cool summers	West coast location under the year-round influence of the westerlies; warm ocean currents along some coasts	Coastal Oregon, Washington, British Columbia, and southern Alaska; southern Chile; interior South Africa; southeast Australia and New Zealand; northwest Europe	Mild winters, mild summers, low annual temperature range; heavy cloud cover, high humidity; frequent cyclonic storms, with prolonged rain, drizzle, or fog; 3- to 4-month frost period	Naturally forested, green year-round; soils require fertilization; root crops, deciduous fruits, winter wheat, rye, pasture and grazing animals; coastal fisheries

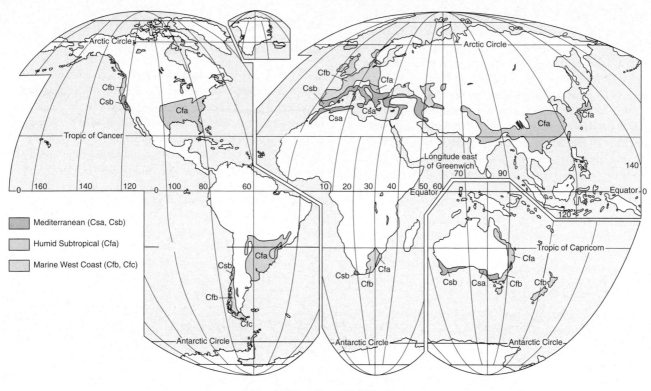

Humid Mesothermal Climates

■ **FIGURE 7.24** Index map of mesothermal climates.

humid subtropical climate regions have locations bordering the tropics and they possess certain tropical-like characteristics (■ Fig. 7.24). Summer temperatures in these two climates rival those of the tropics, and the low threat of frost permits nearly year-round growing seasons. Many plant species that originated in the tropics have been introduced successfully as cultivated vegetation in the subtropical regions.

The third mesothermal climate that we will examine is the marine west coast climate. The mesothermal marine west coast climate regions are not adjacent to the tropics and they lack the hot summers of the other mesothermal climates, thus they are not considered subtropical.

A brief review of Figure 7.5, the world map of climates, shows that marine coast regions, despite their moderate temperatures, extend well into (and sometimes beyond) the high middle latitudes. A moderate winter temperature with no months averaging below freezing is a common characteristic that links the mesothermal climate regions, although summer temperature and precipitation characteristics wary widely between them. Re-examining Figure 7.24 will reveal the locational character for each of the three mesothermal climates. On this map, it is obvious that the marine west coast climate always borders the Mediterranean climate in either poleward–coastal or inland locations. The humid subtropical and Mediterranean climates are found in regions centered between 30 and 40 degrees on opposite coasts of all major continents. As we further examine the humid mesothermal climates, we will be regularly reminded of what we have already learned about

the shifting of global winds and pressure cells and the differing nature of the subtropical highs on opposite sides of the oceans.

Mediterranean Climate

The **Mediterranean climate** is an especially good example for understanding world regions based on climate characteristics and classification. The alternating controls of subtropical high pressure in summer and the delivery of storms from westerly wind movement in winter are so predictable that the world's Mediterranean climate regions have notably similar and easily recognized temperature and precipitation characteristics (■ Fig. 7.25). The special appearance, combinations, and climatic adaptations of Mediterranean vegetation not only are unusual but also are clearly distinguishable from those of other climates. Agricultural practices, crops, recreational activities, and architectural styles all exhibit strong similarities within Mediterranean lands.

Warm, Dry Summers; Mild, Moist Winters
The major characteristics of the Mediterranean climate are dry summers; mild, wet winters; and abundant sunshine (90% of possible sunshine in summer and as much as 50–60% even during the rainy winter season). Summers are warm to hot, but there are enough differences between the monthly temperatures in coastal and interior locations to recognize two distinct subtypes. The moderate-summer subtype has lower summertime temperatures associated with a strong maritime influence. The hot-summer

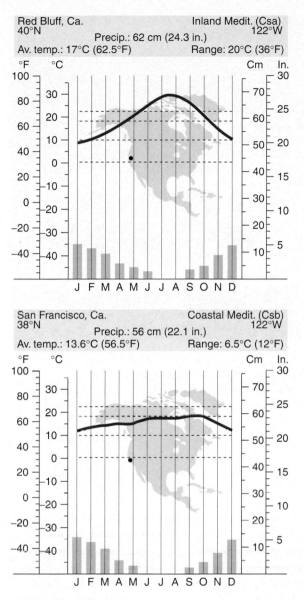

Red Bluff, Ca.   Inland Medit. (Csa)
40°N   122°W
Precip.: 62 cm (24.3 in.)
Av. temp.: 17°C (62.5°F)   Range: 20°C (36°F)

San Francisco, Ca.   Coastal Medit. (Csb)
38°N   122°W
Precip.: 56 cm (22.1 in.)
Av. temp.: 13.6°C (56.5°F)   Range: 6.5°C (12°F)

■ **FIGURE 7.25** Climographs for Mediterranean climate stations.

In what way do these climographs differ? What causes the differences?

subtype is located farther inland and reflects an increased influence of *continentality.*

Whichever the subtype, Mediterranean summers clearly show the influence of the subtropical highs. Weeks go by without a sign of rain, and evapotranspiration rates are high. Effective precipitation is lower than actual precipitation, and the summer drought is as intense as that of the desert. Days are warm to hot, skies are blue and clear, and sunshine is abundant. The high percentage of insolation coupled with nearly vertical rays of the noon sun may drive daytime temperatures as high as 30°C–38°C (86°F–100°F), except where moderated by a strong ocean breeze or coastal fog.

Fog is common throughout the year in coastal locations and is especially noticeable during the summer. As moist maritime

air moves onshore, it passes over the cold ocean currents that typically parallel west coasts in Mediterranean latitudes. The air is cooled, condensation takes place, and fog regularly creeps in during the late afternoon, remains through the night, but often "burns off" during the morning hours. As in the desert, radiation loss is rapid at night, and even in summer, nighttime temperatures are commonly only 10°C–15°C (50°F–60°F).

Winter is the rainy season in the Mediterranean climate. The average annual rainfall in these regions is usually between 35 and 75 centimeters (15–30 in), with 75% or more of the total rain falling during the winter months. The precipitation results primarily from cyclonic storms and frontal systems common with the westerlies. Annual amounts increase with elevation and decrease with increased distance from the ocean. Only because the rain comes during the cooler months, when evapotranspiration rates are lower, is there sufficient precipitation to make this a humid climate.

Despite the rain during the winter season, there are often many days of fine, mild weather. Insolation is still usually above 50%, and the average temperature of the coldest month rarely falls below 4°C–10°C (40°F–50°F). Frost is uncommon and, because of its rarity, many less hardy tropical varieties of fruits and vegetables are grown in these regions. However, when frost does occur, it can do great damage.

Special Adaptations The summer drought, not frost, is the challenge to vegetation in Mediterranean regions. The natural vegetation reflects the wet–dry seasonal pattern of the climate. During the rainy season, the land is covered with lush, green grasses that turn golden and then brown under the summer drought. Only with winter and the return of the rains does the landscape become green again. Much of the natural vegetation is **sclerophyllous** (hard-leafed) and drought resistant. Like xerophytes, these plants have tough surfaces, shiny, thick leaves that resist moisture loss, and deep roots to help combat aridity.

One of the most familiar plant communities is made up of many low, scrubby bushes that grow together in a thick tangle. In the western United States, this is called **chaparral** (■ Fig. 7.26). Most chaparral has or will experience the frequent fires that occur in this dry and dense brush. The fires help perpetuate the chaparral because the associated heat is required to open seedpods and allow many chaparral species to reproduce. People often remove chaparral as a preventive measure against fires, but the removal can have disastrous results because the chaparral acts as a check against erosion and mudslides during the rainy season.

Trees that appear in the Mediterranean climate are also well-suited to moisture conditions. Because of their drought-resistant qualities, the needle-leaf pines are among the more common species. Deciduous and evergreen oaks (eucalyptus trees in Australia) also appear as woodlands of grasses and scattered trees. Tree cover is denser in depressions where moisture collects and on the shady north sides of hills where evapotranspiration rates are lower. The coastal redwood forests of northern California probably could not survive without the heavy fogs and damp marine air that regularly invade the coastal lands in summer.

R. Gabler

■ **FIGURE 7.26** Typical remnant chaparral vegetation as found in southern Spain.
Why would you expect to find little "native" vegetation left in Old World Mediterranean lands?

In all Mediterranean regions, the most productive areas are the lowlands covered by stream deposits. Here, farmers have made special adaptations to climatic conditions. There is sufficient rainfall in the cool season to permit fall planting and spring harvesting of winter wheat and barley. These grasses originally grew wild in the eastern Mediterranean region. Grapevines, fig and olive trees, and the cork oaks, which were also native to Mediterranean lands, are well adapted to the dry summers because of their deep roots and thick, well-insulated stems or bark (■ Fig. 7.27). Where water for irrigation is available, an incredible diversity of crops may be seen. These include, in addition to those already mentioned, oranges, lemons, limes, melons, dates, rice, cotton, deciduous fruits, various types of nuts, and countless vegetables. California, blessed both with fertile valleys for growing fruits, vegetables, and flowers, and with snow meltwater for irrigation, is probably the most agriculturally productive of the Mediterranean climate regions.

Humid Subtropical Climate

The **humid subtropical climate** extends inland from continental east coasts between 15° and 20° and 40°N and S latitude (refer again to Table 7.4 and Fig. 7.24). Thus, it is located within approximately the same latitude range and in a

Lynn Betts NRCS

■ **FIGURE 7.27** Grapes thrive in a Mediterranean climate. This vineyard is in Sonoma County, California.
What landscape features suggest that this is not the dry summer season?

similar transitional position as the Mediterranean climate but on the eastern, instead of the western, continental margins. There is ample evidence of this climatic transition. Summers in the humid subtropics are similar to the humid tropical climates further equatorward. When the noon sun is nearly overhead, these regions are subject to the importation of moist tropical air masses. High temperatures, high relative humidity, and frequent convectional showers are all characteristics that they share with the tropical climates. In contrast, during the winter months, when the pressure and wind belts have shifted equatorward, the humid subtropical regions are more commonly under the influence of the cyclonic systems of the continental middle latitudes. Polar air masses can bring colder temperatures and occasional frost.

Comparison with the Mediterranean Climate
Like the inland version of the Mediterranean climate, the humid subtropical climate has mild winters and hot summers. But it has no dry season. Whereas the Mediterranean lands are under the drought-producing eastern flank of the subtropical high pressure systems, the humid subtropical regions are located on the weak western sides of the subtropical highs. Subsidence and stability are greatly reduced or absent, even during the summer months. Here again, ocean temperatures play a significant role. The warm ocean currents that are commonly found along continental east coasts in these latitudes also moderate the winter temperatures and warm the lower atmosphere, thus increasing lapse rates, which enhances instability. Furthermore, a modified monsoon effect (especially in Asia and to some extent in the southern

United States) increases summer precipitation as the moist, unstable tropical air is drawn in over the land.

Humid subtropical regions receive anywhere from 60 to 250 centimeters (25–100 in) of annual precipitation. Totals generally decrease inland toward continental interiors and away from the oceanic sources of moisture. It is not surprising that these regions are noticeably drier the closer they are to steppe regions inland, toward their western margins.

Both the Mediterranean and humid subtropical climates receive winter moisture from cyclonic storms, which travel with the westerlies along the polar front. As we have noted, the great contrast occurs in the summer when the humid subtropics receive substantial precipitation from convectional showers, supplemented in certain regions by a modified monsoon effect. In addition, because of the shift in the sun and wind belts during the summer months, the humid subtropical climates are subject to tropical storms, some of which develop into hurricanes (or typhoons), especially in later summer. These three factors—the modified monsoon effect, convectional activity, and tropical storms—combine in most of these regions to produce a precipitation maximum in late summer. The climographs for New Orleans, Louisiana, and Brisbane, Australia, illustrate these effects (■ Fig. 7.28).

Temperatures in the humid subtropics are much like those of the Mediterranean regions. Annual ranges are similar, despite a greater variation among climatic stations in the humid subtropical climate, primarily because the climate covers a far larger land area. Mediterranean stations record higher summer daytime temperatures, but summer months in both climates average around 25°C (77°F), increasing to as much as 32°C (90°F) as maritime influence decreases inland. Winter months in both climates average around 7°C–14°C (45°F–57°F). Frost is a similar problem. The long growing season in the warmest humid subtropical regions enables farmers to grow frost-sensitive crops such as oranges, grapefruit, and lemons, but, as in the Mediterranean climates, farmers must be prepared with various means to protect their more sensitive crops from the danger of freezing. The growers of citrus crops in Florida are concentrated in the Central Lake District to take advantage of the moderating influence of nearby bodies of water.

The summer temperatures in humid subtropical regions feel far warmer than they are because of the high humidity there. In fact, summers in this climate tend to be oppressively hot, sultry, and uncomfortable. There is little or no relief from lower night temperatures, as there is in the Mediterranean regions. The high humidity of the humid subtropical climate prevents much radiative loss of heat at night. Consequently, the air remains hot and sticky.

A Productive Climate

Vegetation generally thrives in humid subtropical regions, with their abundant rainfall, high temperatures, and long growing season. The wetter portions of less populated regions support forests of broad-leaf deciduous trees, pine forests on sandy soils, and mixed forests

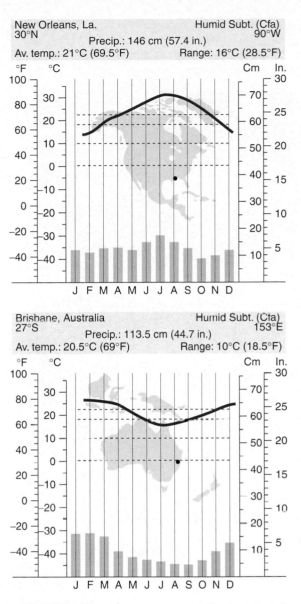

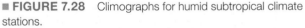

■ FIGURE 7.28 Climographs for humid subtropical climate stations.
What hemispheric characteristics are shown in these graphs?

(■ Fig. 7.29). In the drier interiors near the steppe regions, forests give way to grasslands, which require less moisture. There is an abundant and varied fauna. A few of the common species are deer, bears, foxes, rabbits, squirrels, opossums, raccoons, skunks, and birds of many sizes and species. Bird life in lake and marsh areas is incredibly rich. Alligators inhabit the swamps and other wetlands of the Atlantic coast from North Carolina to Florida, and along the Gulf of Mexico coast to Texas.

As in the tropics, soils tend to have limited fertility because of rapid removal of soluble nutrients. However, there are exceptions in drier grassland areas, such as the pampas of Argentina and Uruguay, which constitute South America's "bread basket." Whatever the soil resource, the humid subtropical regions are of enormous agricultural value because of their favorable

■ **FIGURE 7.29** Forest vegetation similar to this in Myakka State Park originally covered much of humid subtropical central Florida.
How has the physical landscape of central Florida been changed by human occupancy?

R. Gabler

temperature and moisture characteristics. They have been used intensively for both subsistence crops, such as rice and wheat in Asia, and commercial crops, such as cotton and tobacco in the United States. When we consider that this climate (with its monsoon phase) is characteristic of south China as well as of the most densely populated portions of both India and Japan, we realize that this climate regime contains and feeds far more human beings than any other type (■ Fig. 7.30).

Where forests still exist as a major form of vegetation, they can be important commercially (■ Fig. 7.31). The long-leaf and slash pines of the southeastern United States are a source of lumber, as well as the resinous products of the pine tree (pitch, tar, resin, and turpentine). The absence of temperature and moisture limitations strongly favors forest growth. In Georgia, for example, trees may grow two to four times faster than in colder regions such as New England. This means that trees can be planted and harvested in much less time than in cooler forested regions, offering distinct commercial advantages.

With the development of more nutritious grasses and new breeds of cattle that can stand the heat and humidity of the humid subtropics, livestock raising has increased in importance. Field corn (maize) often joins other commercial crops in subtropical regions as feed for cattle and more recently has joined sugar cane as a source of fuel for motor vehicles. However, despite the commercial advantages of this climate, people often find it an uncomfortable one in which to live. Fortunately, the development of air conditioning and its spread to urban areas and technologically advanced rural areas throughout humid subtropical regions have helped to mitigate this problem. Where the ocean offers relief from the summer heat, as in Florida, the humid subtropical climate is an attractive recreation and retirement region. The beauty of its more unusual features, such as its cypress swamps and forests draped with Spanish moss, has to be experienced to be fully appreciated.

■ **FIGURE 7.30** The terraced fields in this humid subtropical climate region near Wakayama, Japan, are ideally suited for rice production.
Why is rice a preferred crop in Japan?

©Robert Essel, NYC/CORBIS

CHINA
NORTH KOREA
Sea of Japan
SOUTH KOREA
JAPAN
Wakayama
Pacific Ocean

Marine West Coast Climate

Proximity to the sea and prevailing onshore winds make the **marine west coast climate** one of the most temperate climates in the world. Thus, this climate type is sometimes known as the temperate oceanic climate. Found in the middle-latitude regions (between 40° and 65°) that are continuously influenced by the westerlies, the marine west coast climate receives ample precipitation throughout the year (Table 7.4). However, unlike the subtropical mesothermal climates, it has mild to cool summers. The climograph for Bordeaux, France, in ■ Figure 7.32 is representative of a marine west coast climate.

USDA/NRCS Alabama

■ **FIGURE 7.31**   A commercial tree farm in Alabama. Note that the pine trees are planted in orderly rows to expedite cultivation and harvest.

Why are tree farms common in the U.S. Southeast and Pacific Northwest but not in New England or the upper Midwest?

■ **FIGURE 7.32**   Climographs for marine west coast climate stations.

What wind or pressure system would influence the decreases in precipitation during the summer months?

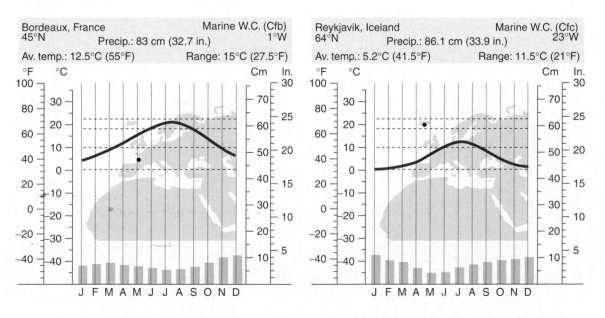

Oceanic Influences As the westerlies travel onshore, they carry with them the moderating marine influence on temperature, as well as much moisture. In addition, warm ocean currents, such as the North Atlantic Drift, bathe some of the coastal lands in the higher latitude locations of the marine west coast climate, further moderating climatic conditions and accentuating humidity. This latter influence is particularly noticeable in Europe where the marine west coast climate extends along the coast of Norway to beyond the Arctic Circle (see again Fig. 7.24). In these middle latitudes, the *maritime* influence is so strong that temperatures decrease little in a poleward direction. Thus, the influence of the ocean is stronger than latitude in determining these temperatures.

Another result of the ocean's moderating effect is that the annual temperature ranges in the marine west coast climates are relatively small, considering their latitudinal location. For an illustration of this, compare the monthly temperature graphs for Portland, Oregon, and Eau Claire, Wisconsin (■ Fig. 7.33). Though these two cities are at the same latitude, the annual range of temperature at Portland is 15.5°C (28°F), while at Eau Claire it is 31.5°C (57°F). The moderating effect of the ocean on Portland's temperatures is clearly contrasted to the effect of *continentality* on the temperatures in Eau Claire.

Despite the insulating effect of cloudy skies and high-moisture content of the air, slowing heat loss at night, frost can be a significant factor in the marine west coast climate. It occurs more often, may last longer, and is more intense than in other mesothermal regions. The growing season is limited to 8 months or less, but even during the months when freezing temperatures may occur, only half the nights or fewer may experience them. The possibility of frost and the frequency of its occurrence increase inland far more rapidly than they do poleward, once more illustrating the importance of the marine influence.

As final evidence of oceanic influences, study the distribution of the marine west coast climate by referring back to Figure 7.24. Where mountain barriers prevent the movement of maritime air inland, this climate is restricted to a narrow coastal strip, as in the Pacific Northwest of North America and in Chile. Where the land is surrounded by water, as in New Zealand, or where the air masses move across broad plains, as in much of northwestern Europe, the climate extends well into the interior of the landmass.

Clouds and Precipitation The marine west coast has a justly deserved reputation as one of the cloudiest, foggiest, rainiest, and stormiest climates in the world. This is particularly true during the winter season. Rain or drizzle may last off and on for days, though the amount of rain received is small for the number of rainy days recorded. Even when rain is not falling, the weather is apt to be cloudy or foggy. Advection fog may be especially common and long lasting in the winter months when maritime air masses pick up considerable moisture, which is then condensed as a fog when the air masses move over colder land. The cyclonic storms and frontal systems are also strongest in the winter when the subtropical highs have shifted equatorward. Conspicuous winter maximums in rainfall occur near the coasts and near boundaries with the Mediterranean climate. However, farther inland a summer maximum may occur.

■ **FIGURE 7.33** Effect of maritime influence on climates of two stations at the same latitude. Portland, Oregon, exemplifies the maritime influences dominating marine west coast climates. Eau Claire, Wisconsin, shows the effect of location in the continental interior. The difference in temperature ranges for the two stations is significant, but note also the interesting differences in precipitation distribution. **How do you explain these differences?**

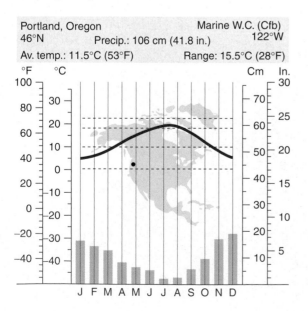

Portland, Oregon — Marine W.C. (Cfb)
46°N — Precip.: 106 cm (41.8 in.) — 122°W
Av. temp.: 11.5°C (53°F) — Range: 15.5°C (28°F)

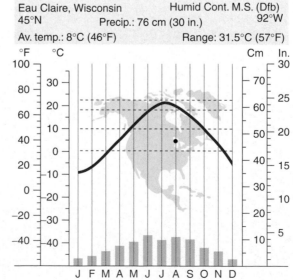

Eau Claire, Wisconsin — Humid Cont. M.S. (Dfb)
45°N — Precip.: 76 cm (30 in.) — 92°W
Av. temp.: 8°C (46°F) — Range: 31.5°C (57°F)

Though all places with this climate type receive ample precipitation, there is much greater site-to-site variation in precipitation averages than in temperature statistics. Precipitation tends to decrease very gradually as one moves inland, away from the oceanic source of moisture. It also decreases equatorward, especially during summer months, as the influence of the subtropical highs increases and the influence of the westerlies decreases. This can bring about periods of beautiful, clear weather, something rarely associated with this climate but not uncommon in our Pacific Northwest.

The most important factor in the amount of precipitation received is local topography. When a mountain barrier such as the Cascades in the Pacific Northwest or the Andes in Chile parallels the coast, abundant precipitation, both cyclonic and orographic, falls on the windward side of the mountains. On the windward side of the coast-facing ranges in Canada, Henderson Lake, British Columbia, averages 666 centimeters (262 in) of rain a year, the highest figure for the entire North American continent. During the last Ice Age, these high precipitation amounts, falling largely as winter snow, produced large mountain glaciers. In many cases, the glaciers flowed down to the sea, excavating deep troughs that now appear as elongated inlets or *fjords*. Fjord coasts are present in Norway, British Columbia, Chile, and New Zealand—all areas of marine west coast climates today (■ Fig. 7.34). In contrast, where there are lowlands and no major landforms of high elevation, precipitation is spread more evenly over a wide area, and the amount received at individual stations is more moderate, around 50–75 centimeters (20–30 in) annually. This is the situation in much of the Northern European Plain, extending from western France to eastern Poland.

Resource Potential   There is little doubt that the marine west coast climate offers certain advantages for agriculture. The small annual temperature ranges, mild winters, long growing seasons, and abundant precipitation all favor plant growth. Many crops, such as wheat, barley, and rye, can be grown farther poleward than in more continental regions. Although the soils common to these regions are not naturally rich in soluble nutrients, highly successful agriculture is possible with the application of natural or commercial fertilizers (■ Fig. 7.35a). Root crops (such as potatoes, beets, and turnips), deciduous fruits (such as apples and pears), berries, and grapes join the grains previously mentioned as important farm products. Grass, in particular, requires little sunshine, and pastures are always lush. The greenness of Ireland—the Emerald Isle—is evidence of these favorable conditions, as is the abundance of herds of beef and dairy cattle.

The magnificent **needle-leaf** evergreen trees that form the natural forests of marine west coast regions have been a readily available resource. Some of the finest stands of commercial timber in the world are found along the Pacific coast of North America where **conifers** (cone-bearing trees), like pines, firs, and spruces abound and commonly exceed 30 meters (100 ft) in height (7.35b). Europe and the British Isles were once heavily forested, but most of those forests (even the famous Sherwood Forest of Robin Hood fame) have been cut down for building material and have been replaced by agricultural lands and urbanization.

■ **FIGURE 7.34**   The scenic fjords of coastal Norway, shown here, were produced by glacial erosion during the Pleistocene ice advance.

In what other areas of the world are fjords common?

©Iconotec Royalty Free Photograph/Fotosearch

■ **FIGURE 7.35** (a) Reliable precipitation makes diversified agriculture possible in marine west coast climate regions, with emphasis on grain, orchard crops, vineyards, vegetables, and dairying. The village in the photograph is Iphofen, Germany. (b) A dense stand of needle-leaf coniferous trees in the Pacific Northwest.

Although the climate is similar, why would a photo taken in a marine west coast agricultural region of the United States depict a significantly different scene from the one in 7.35a?

:: Terms for Review

Thornthwaite system	humid tropical climate	highland climate
potential evapotranspiration (potential ET)	humid mesothermal climate	region
	humid microthermal climate	climograph
actual evapotranspiration (actual ET)	polar climate	biome
Köppen system	arid climate	tropical rainforest climate

tropical monsoon climate
annual temperature range
daily (diurnal) temperature range
tropical rainforests
broad-leaf trees
evergreen
liana
epiphyte

jungle
slash-and-burn (shifting cultivation)
tropical savanna climate
savanna
deciduous
desert climate
steppe climate
xerophytic

Mediterranean climate
sclerophyllous
chaparral
humid subtropical climate
marine west coast climate
needle-leaf trees
conifers

:: Questions for Review

1. Why are temperature and precipitation the two atmospheric elements most widely used as the sources of statistics for climate classification? How are these two elements used in the Köppen system to identify major climate categories?
2. Why are the Köppen climate boundaries often referred to as "vegetation lines?"
3. How does the Thornthwaite system of climate classification differ from the Köppen system? What are the advantages of the Thornthwaite system?
4. Describe the delicate balance between vegetative growth and soil fertility in a tropical rainforest environment.
5. What is the difference between a rainforest and a jungle? What factor differentiates these two forest types?

6. Explain the seasonal precipitation pattern of the tropical savanna climate. State some of the transitional features of this climate.
7. What conditions give rise to desert climates?
8. How do steppes differ from deserts? Why might human use of steppe regions in some ways be more hazardous than use of deserts?
9. Compare the humid subtropical and Mediterranean climates. What are their most obvious similarities and differences?
10. What factors combine to cause a precipitation maximum in late summer in most of the humid subtropical regions?
11. What climatic factors combine to produce the seasonal precipitation and temperature conditions that a marine west coast climate experiences?

:: Practical Applications

1. Based on the classification scheme presented in Appendix C, classify the following climate stations from the data provided.

		J	F	M	A	M	J	J	A	S	O	N	D	Yr.
a.	Temp (°C)	20	23	27	31	34	34	34	33	33	31	26	21	29
	Precip. (cm)	0.0	0.0	0.0	0.0	0.2	0.1	1.1	0.2	0.0	0.0	0.0	0.0	1.6
b.	Temp (°C)	21	24	28	31	30	28	26	25	26	27	25	22	26
	Precip. (cm)	0.0	0.0	0.2	0.8	7.1	11.9	20.9	31.1	13.7	1.4	0.0	0.0	87.2
c.	Temp (°C)	19	20	21	23	26	27	28	28	27	26	22	20	24
	Precip. (cm)	5.1	4.8	5.8	9.9	16.3	18.0	17.0	17.0	24.0	20.0	7.1	3.0	149.0
d.	Temp (°C)	23	23	22	19	16	14	13	13	14	17	19	22	18
	Precip. (cm)	0.8	1.0	2.0	4.3	13.0	18.0	17.0	14.5	8.6	5.6	2.0	1.3	88.1
e.	Temp (°C)	2	4	8	13	18	24	26	24	21	14	7	3	14
	Precip. (cm)	1.0	1.0	1.0	1.3	2.0	1.5	3.0	3.3	2.3	2.0	1.0	1.3	20.6
f.	Temp (°C)	27	26	27	27	27	27	27	27	27	27	27	27	27
	Precip. (cm)	31.8	35.8	35.8	32.0	25.9	17.0	15.0	11.2	8.9	8.4	6.6	15.5	243.8
g.	Temp (°C)	13	14	17	19	22	24	26	26	26	24	19	15	20
	Precip. (cm)	6.6	4.1	2.0	0.5	0.3	0.0	0.0	0.0	0.3	1.8	4.6	6.6	26.7
h.	Temp (°C)	9	9	9	10	12	13	14	14	14	12	11	9	11
	Precip. (cm)	17.0	14.8	13.3	6.8	5.5	1.9	0.3	0.3	1.6	8.1	11.7	17.0	97.6
i.	Temp. (°C)	3	3	5	8	10	13	15	14	13	10	7	5	9
	Precip. (cm)	4.8	3.6	3.3	3.3	4.8	4.6	8.9	9.1	4.8	5.1	6.1	7.4	65.8

2. The following nine locations are represented by the data in the previous table, although not in the order listed: Albuquerque, New Mexico; Edinburgh, Scotland; Belém, Brazil; Benghazi, Libya; Faya, Chad; Kano, Nigeria; Miami, Florida; Perth, Australia; Eureka, California. Use an atlas or Google Earth and your knowledge of climates to match the climatic data with the locations.

temperatures high enough during part of the year to have a recognizable summer with at least one month averaging over 10°C (50°F) and a distinct winter with the coldest month averaging less than 0°C (32°F). It is the cold winters that distinguish microthermal climates from the moderate winter mesothermal climates. The recognition of three separate microthermal climates is based mainly on latitude and the resulting differences in the length and severity of the seasons (Table 8.1).

Humid Microthermal Generalizations

The humid microthermal climates share the westerlies and the storms of the polar front with the marine west coast climates. However, their position in the continental interiors and in high latitudes prevents microthermal regions from experiencing the moderating influence of the oceans. In fact, the dominance of continentality in these climates is

TABLE 8.1
The Microthermal Climates

Name and Description	Controlling Factors	Geographic Distribution	Distinguishing Characteristics	Related Features
Humid Continental, Hot Summer				
Warmest month above 10°C (50°F); coldest month below 0°C (32°F); hot summers; usually year-round precipitation, winter drought (Asia)	Location in the lower middle latitudes (35°–45°); cyclonic storms along the polar front; prevailing westerlies; continentality; polar anticyclone in winter (Asia)	Eastern and midwestern U.S. from Atlantic coast to 100°W longitude; east central Europe; northern and northeast (Manchuria) China, northern Korea, and Honshu (Japan)	Hot, often humid summers; occasional winter cold waves; rather large annual temperature ranges; weather variability; precipitation 50–115 cm (20–45 in.), decreasing inland and poleward; 140- to 200-day growing season	Broad-leaf deciduous and mixed forest; moderately fertile soils with fertilization in wetter areas; highly fertile grassland and prairie soils in drier areas; "corn belt," soybeans, hay, oats, winter wheat
Humid Continental, Mild Summer				
Warmest month above 10°C (50°F); coldest month below 0°C (32°F); mild summers; usually year-round precipitation, winter drought (Asia)	Location in the middle latitudes (45°–55°); cyclonic storms along the polar front; prevailing westerlies; continentality; polar anticyclone in winter (Asia)	New England, the Great Lakes region, and south central Canada; southeastern Scandinavia; eastern Europe, west central Asia; eastern Manchuria (China) and Hokkaido (Japan)	Moderate summers; long winters with frequent spells of clear, cold weather; large annual temperature ranges; variable weather with less total precipitation than farther south; 90- to 130-day growing season	Mixed or coniferous forest; moderately fertile soils with fertilization in wetter areas; highly fertile grassland and prairie soils in drier areas; spring wheat, corn for fodder, root crops, hay, dairying
Subarctic				
Warmest month above 10°C (50°F); coldest month below 0°C (32°F); cool summers, cold winters poleward; usually year-round precipitation, winter drought (Asia)	Location in the higher middle latitudes (50°–70°); westerlies in summer, strong polar anticyclone in winter (Asia); occasional cyclonic storms; extreme continentality	Northern North America from Newfoundland to Alaska; northern Eurasia from Scandinavia through most of Siberia to the Bering Sea and the Sea of Okhotsk	Brief, cool summers; long, bitterly cold winters; largest annual temperature ranges; lowest temperatures outside Antarctica; low precipitation, 20–50 cm (10–20 in.); unreliable 50- to 80-day growing season; permafrost common	Northern Coniferous forest (taiga); strongly acidic soils; poor drainage and swampy conditions in warm season; experimental vegetables and rootcrops

best demonstrated by the fact that they do not exist in the Southern Hemisphere where there are no large landmasses in the appropriate latitudes.

In the microclimate regions, the winters tend to become longer and colder toward the poleward margins because of latitude and toward interiors because of the continental influence. Summers inland are also inclined to be hotter, but they become progressively shorter as the winter season lengthens poleward. Thus, the three microthermal climates can be defined as **humid continental, hot summer; humid continental, mild summer;** and **subarctic,** which has a cool summer and, in extreme cases, a long-bitterly cold winter.

All microthermal climates have several common characteristics. By definition, they all experience a surplus of precipitation over potential evapotranspiration, and they receive year-round precipitation. An exception to this rule lies in an area of Asia where the bitterly cold and dry Siberian High causes winter droughts. The greater frequency of maritime tropical air masses in summer and continental polar air masses in winter, combined with the monsoonal effect and strong summer convection, produce a precipitation maximum in the summer. Although the length of time that snow remains on the ground increases poleward and toward the continental interior (■ Fig. 8.2), all three microthermal climates experience significant snow cover. This decreases the effectiveness of insolation and helps explain their cold winter temperatures.

Finally, the unpredictable and variable nature of the weather is especially apparent in the humid microthermal climates.

Humid Continental Climates

The humid continental, hot-summer climate is relatively limited in its distribution on the Eurasian landmass (see again Fig. 8.1). This is unfortunate for the people of Europe and Asia because it has by far the greatest agricultural potential and is the most productive of the microthermal climates. In the United States, this climate is distributed over a wide area that begins with the eastern seaboard of New York, New Jersey, and southern New England and stretches continuously across the heartland of the eastern United States to encompass much of the American Midwest. By contrast, the humid continental mild-summer climate with less agricultural potential encompasses large areas poleward of hot summer regions on all Northern Hemisphere continents.

Hot-Summer, Mild-Summer Comparison

Both temperature and precipitation characteristics help to explain the differing productivity of the two humid continental climates. In hot-summer regions, the centers of migrating low pressure systems usually pass poleward during the summer season and the climate is dominated by tropical maritime air. So-called hot spells can go on day and night for a week or more,

■ **FIGURE 8.2** Map of the contiguous United States, showing average annual number of days of snow cover.

What areas of the United States average the greatest number of days of snow cover?

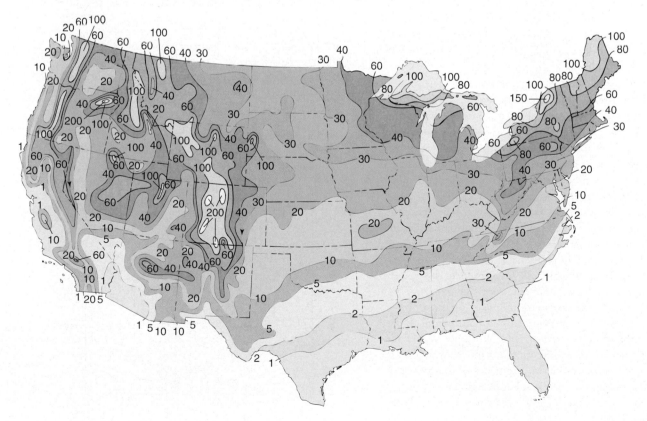

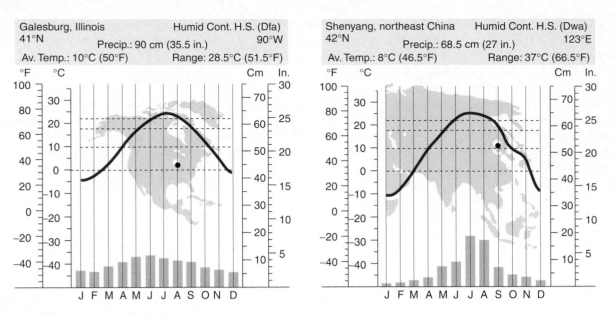

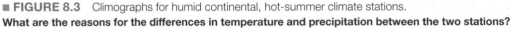

■ **FIGURE 8.3** Climographs for humid continental, hot-summer climate stations.
What are the reasons for the differences in temperature and precipitation between the two stations?

with only temporary relief from convectional thunderstorms or an occasional cold (cool) front. By the time humid tropical air migrates far enough inland to reach mild-summer regions, the air mass is modified and summers are not as long or as hot.

Winters in the mild-summer climate are more severe and longer than in its neighbor to the south. As a result, frost-free periods for natural and cultural vegetation are significantly different in the two humid continental climates. The combination of more severe winters and shorter summers in mild-summer regions makes for a growing season of between 90 and 130 days compared to growing seasons that vary from 130 days to as much as 200 days in equatorward margins of hot-summer regions.

The degree of continentality can have an effect on both summer and winter temperatures throughout humid continental regions and, as a result, on temperature range. Annual temperature ranges are consistently large but they become progressively larger poleward and toward continental interiors. Especially near coasts, temperatures may be modified by a slight marine influence so that temperatures are milder than those at comparable latitudes. Large lakes may cause a similar effect. Even the size of the continent exerts an influence. Galesburg, Illinois, a typical microthermal station in the United States, has a significantly lower temperature range than Shenyang, northeast China (Manchuria), which is located at almost the same latitude but which experiences the greater seasonal contrasts of the Eurasian landmass (■ Fig. 8.3).

The amount and distribution of precipitation vary from place to place throughout humid continental climates. Total precipitation received decreases both poleward and inland. A move in either direction is a move away from the source regions of warm maritime air masses that provide moisture for cyclonic storms and convectional showers. As a result, hot-summer regions are usually favored with

greater moisture totals and reliability than mild-summer regions. Decreases poleward are illustrated by comparing the annual precipitation in Duluth, Minnesota (■ Figure 8.4), with that of Galesburg, Illinois (see again Fig. 8.3). Duluth is poleward of Galesburg at nearly the same longitude. The decrease inland can be seen in the average annual precipitation figures for the following cities, all at a latitude of about 40°N: New York (longitude 74°W), 115 centimeters (45 in); Indianapolis, Indiana (86°W), 100 centimeters (40 in);

■ **FIGURE 8.4** Climograph for a humid continental mild-summer climate station.
How do you explain the differences in temperature and precipitation between Duluth and Galesburg, Illinois (see Fig. 8.3)?

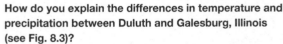

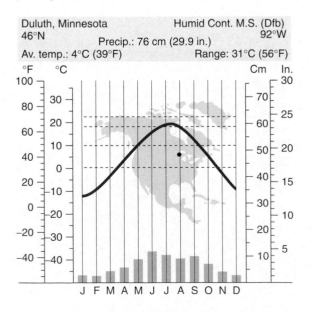

Hannibal, Missouri (92°W), 90 centimeters (35 in); and Grand Island, Nebraska (98°W), 60 centimeters (24 in). Most stations have a precipitation maximum in summer when the warm, moist air masses dominate. Asia, in particular, experiences heavy summer precipitation associated with the monsoonal effect. Not only does a summer maximum of precipitation exist, but also the monsoon circulation inhibits winter precipitation, and stations there experience winter drought (see again Shenyang, Fig. 8.3).

As might be expected, vegetation and soils vary with the differing climatic elements of the two humid continental climates. The wetter regions of both climates are associated with natural forest vegetation and forest soils. However, many trees common in the hot-summer climate, such as the broad-leaf deciduous species like oaks, hickories, and maples, find it difficult to compete with needle-leaf species like firs, pines, and spruces toward the colder, polar margins of the mild-summer climate. In areas of humid continental climates that are too dry to support trees (yet not semiarid), grasslands are the natural vegetation. The soils that developed under these grasslands are among the richest in the world.

Seasonal Changes

The four seasons are highly developed in the humid continental climates. Unlike in the subtropical climates, each season is distinct from the other three, with a character all its own. Winters are cold and snow is guaranteed, with amounts and accumulations rapidly increasing poleward. The springs are warm, with sudden changes in weather and showers that produce flowers, budding leaves, and green grasses. Average summer temperatures decrease slowly with movement from hot-summer to mild-summer regions, but both humid continental climates experience periods of hot, humid weather, with violent storms interspersed between extended periods of dry, sunny conditions. Fall is the colorful season, when the deciduous trees that are common throughout these climates enter an important phase of their annual growth cycles. To avoid frost damage during the colder winters and to survive periods of total moisture deprivation when the ground is frozen, deciduous trees whose leaves have large transpiring surfaces drop these leaves and become dormant, coming to life and producing new leaves only when the danger period is past. A large variety of trees have evolved this mechanism; certain oaks, hickory, chestnut, beech, and maples are common examples. The seasonal rhythms produce some beautiful scenes, particularly during the transitional period from activity to dormancy in the fall as brilliant leaf colors are associated with chemical substances drawing back into the trees for winter storage (■ Fig. 8.5).

Although seasonal differences are profound in humid continental climates, it is important to note that the atmospheric changes within seasons are just as significant as those between seasons. Humid continental climates serve as classic examples of variable middle-latitude weather. They are the domain of the polar front. Cyclonic storms are born as tropical air masses move northward and confront polar air masses migrating to the south. The daily weather in these regions is dominated by days of stormy frontal activity followed by the clear conditions of a following anticyclone. Above the land is a battlefield in which storms mark the struggles of air masses for dominance, and as in battle, the conflict occurs at the *front*. The general atmospheric circulation in these latitudes carries the cyclones and anticyclones toward the east along the polar front. When the polar front is most directly over these regions, as it is in winter and spring, one storm and its associated fronts seem to follow directly behind another with such speed and regularity that the only safe weather prediction is that the weather will change (see again Chapter 6).

Land Use in Humid Continental Regions

Perhaps the greatest contrast between the hot-summer and mild-summer humid continental regions is exhibited in agriculture. Despite the unpredictable weather, the humid continental hot-summer agricultural regions are among the finest in the world. The favorable combination of long-hot summers, ample rainfall, and highly fertile soils has made the American Midwest a leading producer of corn, beef cattle, and hogs. Soybeans, which are native to similar climate regions in northern China, are now second to corn throughout the Midwest, cultivated for animal feed and a raw material for the food-processing, plastics, and vegetable oil industries. Wheat, barley, and other grains are especially important in European and Asian regions, and winters are sufficiently mild that fall-sown varieties can be raised in the United States. In the mild-summer climate, on the other hand, a shorter growing season imposes certain limitations on agriculture and restricts the crops that can be grown. Farmers rely more on quick-ripening varieties, grazing animals, orchard products, and root crops. Dairy products milk, cheese, butter, cream—are mainstays in the economies of Wisconsin, New York, and northern New England. The moderating effect of the Great Lakes or other water bodies permits the growth of deciduous fruits, such as apples, plums, and cherries.

The length of the growing season is the most obvious reason for the differences in agriculture between the two humid continental climates, but there is another climate-related reason as well. The great ice sheets of the Pleistocene epoch have had significant but different effects on mild-summer and hot-summer regions, especially in North America. In the hot-summer regions, the ice sheets thinned and receded, releasing the enormous load of soil and solid rock debris they had stripped off the lands nearer to their centers of origin. The material was laid down in a blanket hundreds of feet thick in the areas of maximum glacial advance. As the ice retreated northward, less and less debris was deposited, much of it flushed away by meltwater streams. The more southerly, hot-summer region consequently has an undulating topography underlain by thick masses of glacial debris. The soils formed on this debris are well developed and fertile, and plant nutrients are more likely to be evenly distributed because steep slopes are lacking. The more northerly, mild-summer region, on the other hand, mainly shows the effects of glacial erosion. Rockbound lakes and marshy lowlands alternate with ice-scoured rock hills.

(a)

(b)

(c)

■ **FIGURE 8.5** The appearance of deciduous forests in humid microthermal regions changes dramatically with the seasons. The green leaves of summer (a) change to reds, golds, and browns in fall (b) and drop to the forest floor in winter (c). Leaf dropping in areas of cold winters, such as this example in western Illinois, is a means of minimizing transpiration and moisture loss when the soil water is frozen.

What length of growing season (frost-free period) is associated with the climate of western Illinois?

Soils are either thin and stony or waterlogged. Because of its lower agricultural potential, large sections of this area remain forested.

However, because of its wilderness character and the abundance of lakes in basins produced by glacial processes, recreational possibilities in a mild-summer region far exceed those of a more subdued hot-summer region. Minnesota calls itself the "Land of 10,000 Lakes," and in New York and New England, lakes, rough mountains, and forest combine to produce some of the most spectacular scenery east of the Rocky Mountains.

Subarctic Climate

The subarctic climate is the farthest poleward and most extreme of the microthermal climates. By definition, it has at least 1 month with an average temperature above 10°C (50°F), and its poleward limit roughly coincides with the 10°C isotherm for the warmest month of the year. As you may recall from our earlier discussion of the simplified Köppen system, forests cannot survive where at least 1 month does not have an average temperature over 10°C. Thus, the poleward boundary of the subarctic climate is the latitudinal limit of forest growth as well.

As Figure 8.1 indicates, the subarctic climate, like the other microthermal climates, is found exclusively in the Northern Hemisphere. It covers vast areas of subpolar Eurasia and North America. Conditions vary widely over such large areas. Extremely severe winter regions are located along the polar margins or deep in the interior of the Asian landmass. Climate subtypes with winter drought are found in association with the Siberian high and its clear skies, bitter cold, and strong subsidence of air over interior Asia. Other subarctic regions experience less severe winters or year-round precipitation.

High Latitude and Continentality Subarctic regions experience short, cool summers and long, bitterly cold winters (■ Fig. 8.6). The rapid heating and cooling associated with continental interiors in the high latitudes allow little time for the transitional seasons of spring and fall. At Eagle, Alaska,

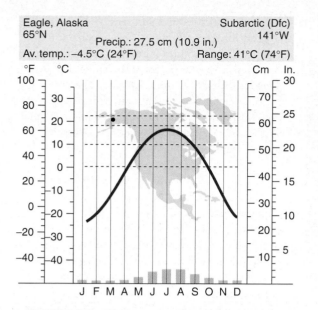

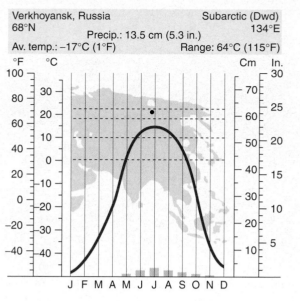

■ **FIGURE 8.6** Climographs of subarctic climate stations.
Why would people settle in such severe-winter climate regions?

a station in the Klondike region of the Yukon River Valley, the temperature climbs 8°C–10°C (15°F–20°F) per month as summer approaches and drops just as rapidly prior to the next winter season. At Verkhoyansk in Siberia, the change between the seasonal extremes is even more rapid, averaging 15°C–20°C (30°F–40°F) per month.

Because of the high latitudes of these regions, summer days are quite long, and nights are short. The noon sun is as high in the sky during a subarctic summer as during a subtropical winter. The combination of a moderately high sun angle and many hours of daylight means that some subarctic locations receive as much insolation during the summer solstice as the equator does. As a result, temperatures during the 1–3 months of the subarctic summer usually average 10°C–15°C (50°F–60°F), and on some days they may even approach 30°C (86°F). The brief summer in the subarctic climate can often be pleasantly warm, even hot, on some days.

The winter season in the subarctic is bitter, intense, and lasts for as long as 8 months. Eagle, Alaska, has 8 months with average temperatures below freezing. In the Siberian subarctic, the January temperatures regularly *average* −40°C to −50°C (−40°F to −60°F). The coldest temperatures in the Northern Hemisphere—officially, −68°C (−90°F) at both Verkhoyansk and Oymyakon; unofficially, −78°C (−108°F) at Oymyakon—have been recorded there. In addition, the winter nights, with an average 18–20 hours of darkness extending well into one's working hours, can be mentally depressing and can increase the impression of climatic severity.

As a direct result of the intense heating and cooling of the land, the subarctic has the largest annual temperature ranges of any climate. Average annual ranges in equatorward margins of the climate vary from near 40°C (72°F) to more than 45°C (80°F). The exceptions are near western coasts, where the maritime influence may significantly modify winter temperatures. Annual temperature ranges for poleward stations are great. The climograph for Verkhoyansk, which indicates a range of 64°C (115°F), is an extreme example.

Along with temperatures, latitude and continentality also influence subarctic precipitation. These climate controls combine to limit annual precipitation amounts to less than 50 centimeters (20 in) for most regions and to 25 centimeters (10 in) or less in northern and interior locations. Low temperatures in the subarctic reduce the moisture-holding capacity of the air, thus minimizing precipitation during the occasional passage of cyclonic storms. Location toward the center of large landmasses or near leeward coasts increases distance from oceanic sources of moisture. Finally, subarctic climates in the higher latitudes are dominated by the polar high, especially in the winter. The subsiding air in the polar anticyclone limits the opportunity for precipitation in the subarctic regions.

Subarctic precipitation is frontal, and because the polar anticyclone is weaker and farther north during the warmer summer months, more precipitation comes during that season. The meager winter precipitation falls as fine, dry snow. Though there is not as much snowfall as in less severe climates, the temperatures remain so cold for so long that the snow cover lasts for as long as 7 or 8 months. During this time, there is almost no melting of snow, especially in the dark shadows of the forest.

A Limiting Environment The climatic restrictions of subarctic regions place distinct limitations on plant and animal life and on human activity. The characteristic vegetation is coniferous forest, adapted to the severe temperatures, the physiologic drought associated with frozen soil water, and the infertile soils. Seemingly endless tracts of spruce, fir,

National Park Service—Wrangell-St. Elias

■ **FIGURE 8.7** Taiga (boreal forest) is typical of the vegetation throughout much of the American and Canadian subarctic. This photo was taken in Alaska's Wrangell-St. Elias National Park and Preserve.
Why are these kinds of virgin forests currently of little economic value?

The brief summers and long, cold winters severely limit vegetation growth in subarctic regions. Trees are shorter and more slender than comparable species in less severe climate regimes. There is little hope for agriculture. The growing season averages 50–75 days, and frost may occur even during June, July, or August. Thus, in some years, a subarctic location may have no truly frost-free season.

A difficult problem for people in subarctic (as well as *tundra*) regions is **permafrost,** a permanently frozen layer of subsoil and underlying rock that may extend to a depth of 300 meters (1000 ft) or more in the northernmost sections of the climate. Permafrost is present over much of the subarctic climate, but it varies greatly in thickness and is often discontinuous. Where it occurs, the land is frozen completely from the surface down in winter. The warm temperatures of spring and summer cause the upper few feet to thaw, but because the land beneath this thawed top layer remains frozen, water cannot percolate downward. The thawed soils become sodden with moisture, especially in spring, when there is an abundant supply of water from the melting snow. Landscapes called **patterned ground,** or **frost polygons,** are commonly found in regions of subarctic yearly freezing and thawing (■ Fig. 8.8).

and pine thrive over enormous areas, untouched by humans (■ Fig. 8.7). In Russia, the forest is called the **taiga** (or **boreal forests** in other regions), and this name is sometimes given to the subarctic climate type itself.

■ **FIGURE 8.8** Frost polygons, or patterned ground, as seen in the Arctic National Wildlife Refuge of Alaska. Repeated freezing and thawing cause the soil to produce polygonal shapes. When the ground freezes, it expands, and it shrinks when it thaws.
Why are the edges heaved upward?

Alaska Image Library/USFWS

With agriculture a questionable occupation at best, there is little economic incentive to draw humans to subarctic regions. Logging for timber is unimportant because of small and thin trees of the boreal forest. Even use of the vast forests for paper, pulp, and wood products is restricted by their interior location far from world markets. Miners occasionally exploit rich ore deposits; other people, many of them native to the subarctic regions, pursue hunting, trapping, and fishing of the relatively limited wildlife.

Polar Climate Regions

The polar climates are the last of Köppen's humid climate subdivisions to be differentiated on the basis of temperature. These climate regions are situated at the greatest distance from the equator, and they owe their existence primarily to the low annual amounts of insolation they receive. No polar station experiences a month with average temperatures as high as 10°C (50°F), and hence all are without a warm summer (Table 8.2). Trees cannot survive in such a regimen. In the regions where at least 1 month averages above 0°C (32°F), they are replaced by tundra vegetation. Elsewhere, the surface is covered by great expanses of frozen ice. Thus, there are two polar climate types, tundra and ice sheet.

An important characteristic of polar climates is the unique pattern of day and night. At the poles, 6 months of relative darkness, when the sun never rises above the horizon, alternate with 6 months of daylight during which the sun never sets. Even when the sun is above the horizon, however, the sun's rays are at a sharply oblique angle, and little effective insolation is received even with 24 hours of daylight.

Moving outward from the poles, the lengths of periods of continuous winter night and continuous summer day decrease rapidly from 6 months at the poles to 24 hours at the Arctic and Antarctic Circles (66½°N and S). Here, the 24-hour night or day occurs only at the winter and summer solstices, respectively.

Tundra Climate

Compare the location of the **tundra climate** with that of the subarctic climate in Figure 7.5. You can see that although the tundra climate is situated closer to the poles, it is also along the periphery of landmasses, and, with the exception of the Antarctic Peninsula, it is everywhere adjacent to the Arctic Ocean. Even though temperature ranges in the tundra are large, they are not as large as in the subarctic because of the maritime influence. Winter temperatures in the polar regions in particular are not as severe in the tundra as they are inland (■ Fig. 8.9).

It almost seems inappropriate to call the often unpleasantly chilly and damp conditions of the tundra's warmer season "summer." Temperatures average around 4°C (40°F) to 10°C (50°F) for the warmest month, and frosts occur regularly. The air does warm sufficiently to melt the thin snow cover and the ice on small bodies of water, but this only causes marshes, swamps, and bogs to form because drainage is blocked by permafrost (see again Fig. 8.8). Clouds of black flies, mosquitoes, and gnats swarm in this soggy landscape, known as **muskeg** in Canada and Alaska. A bright note in the landscape is provided by the enormous number of migratory birds that nest in the arctic regions in summer and feed on the insects. However, as soon as the

TABLE 8.2
The Polar Climates

Name and Description	Controlling Factors	Geographic Distribution	Distinguishing Characteristics
Tundra Warmest month between 0°C (32°F) and 10°C (50°F); precipitation exceeds potential evapotranspiration	Location in the high latitudes; subsidence and divergence of the polar anticyclone; proximity to coasts	Arctic Ocean borderlands of North America, Greenland, and Eurasia; Antarctic Peninsula; some polar islands	At least 9 months average below freezing; low evaporation; precipitation usually below 25.5 cm (10 in.); coastal fog; strong winds
Ice Sheet Warmest month below 0°C (32°F); precipitation exceeds potential evaporation	Location in the high latitudes and interior of landmasses; year-round influence of the polar anticyclone; ice cover; elevation	Antarctica; interior Greenland; permanently frozen portions of the Arctic Ocean and associated islands	Summerless; all months average below freezing; world's coldest temperature; extremely meager precipitation in the form of snow, evaporation even less; gale-force winds

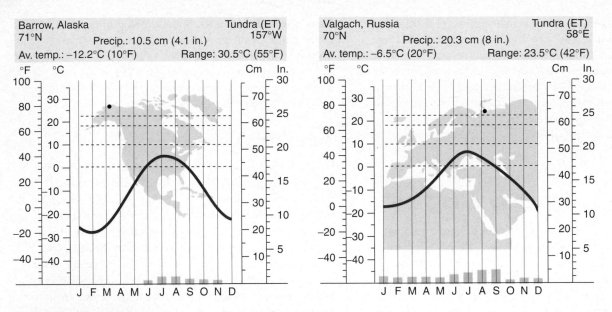

Barrow, Alaska   Tundra (ET)
71°N   157°W
        Precip.: 10.5 cm (4.1 in.)
Av. temp.: −12.2°C (10°F)   Range: 30.5°C (55°F)

Valgach, Russia   Tundra (ET)
70°N   58°E
        Precip.: 20.3 cm (8 in.)
Av. temp.: −6.5°C (20°F)   Range: 23.5°C (42°F)

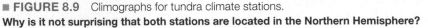

■ **FIGURE 8.9** Climographs for tundra climate stations.
Why is it not surprising that both stations are located in the Northern Hemisphere?

shrinking daylight hours of autumn approach, these birds depart for warmer climates.

Winters are cold and seem to last forever, especially in tundra locations where the sun may be below the horizon for days at a time. The climograph for Barrow, Alaska, illustrates the low temperatures of this climate. Note that average monthly temperatures are *below freezing* 9 months of the year. The average annual temperature is −12°C (10°F).

This climate is named for the treeless low-growing **tundra** vegetation that survives despite the forbidding environment. It consists of lichens, mosses, sedges, flowering herbaceous plants, small shrubs, and grasses. In particular, the plants have adjusted to the conditions associated with nearly universal permafrost (■ Fig. 8.10).

The tundra regions exhibit several other significant climatic characteristics. Diurnal temperature ranges are small because insolation is uniformly high during the long summer days and uniformly low during the long winter nights. Precipitation is generally low, except in eastern Canada and Greenland, because of exceedingly low absolute humidity and the influence of the polar anticyclone. Icy winds sweep across the open land surface and are an added factor in eliminating the trees that might impede their progress. Coastal fog is characteristic in marine locations, where cool polar maritime air drifts onshore and is chilled below the dew point by contact with the even colder land.

Ice-Sheet Climate

The **ice-sheet climate** is the most severe and restrictive climate on Earth. As Table 8.2 indicates, it covers large areas in both the Northern and Southern Hemispheres, a total of about 16 million square kilometers (6 million sq mi)—

nearly the same area as the United States and Canada combined. All average monthly temperatures are below freezing, and because most surfaces are covered with glacial ice, no vegetation can survive in this climate. It is a virtually lifeless region of perpetual frost.

Antarctica is the coldest place on Earth (although Siberia sometimes has longer and more severe periods of cold in winter). The world's coldest temperature, −88°C (−127°F), was recorded at Vostok, Antarctica. Consider the climographs for Little America, Antarctica, and Eismitte, Greenland, for a fuller picture of the cold ice-sheet temperatures (■ Fig. 8.11).

The primary reason for the low temperatures of ice-sheet climates is the minimal insolation received in these regions. Not only is little or no insolation received during half the year, but also the sun's radiant energy that is received arrives at sharply oblique angles. In addition, the perpetual snow and ice cover of this climate reflects nearly all incoming radiation. A further factor, in both Greenland and Antarctica, is elevation. The ice sheets covering both regions rise more than 3000 meters (10,000 ft) above sea level. Naturally, this elevation contributes to the cold temperatures.

The polar anticyclone severely limits precipitation in the ice-sheet climate to the fine, dry snow associated with occasional cyclonic storms. Precipitation is so meager in this climate that its regions are sometimes incorrectly referred to as "polar deserts." However, because of the exceedingly low evaporation rates associated with the severely cold temperatures, precipitation still exceeds potential evaporation, and the climate can be classified as humid.

The strong and persistent polar winds are another staple of the harsh ice-sheet climate. Mawson Base, Antarctica, for

■ FIGURE 8.10   One of the Hecho Islands off the Antarctic Peninsula displays thick, bright patches of moss as its most complex vegetation.

What climate controls help to form this stark landscape?

■ FIGURE 8.11   Climographs for ice-sheet climate stations.

If you were to accept an offer for an all-expense-paid trip to visit either Greenland or Antarctica, which would you choose, and why would you go?

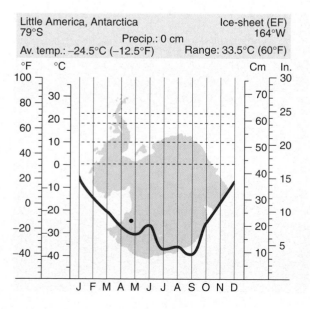

Little America, Antarctica   Ice-sheet (EF)
79°S   164°W
              Precip.: 0 cm
Av. temp.: −24.5°C (−12.5°F)   Range: 33.5°C (60°F)

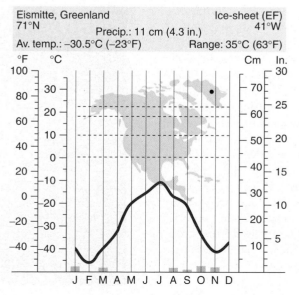

Eismitte, Greenland   Ice-sheet (EF)
71°N   41°W
              Precip.: 11 cm (4.3 in.)
Av. temp.: −30.5°C (−23°F)   Range: 35°C (63°F)

example, has approximately 340 days a year with gale-force winds of 54 kilometers per hour (33 mph) or more. The winds of these regions can result in whiteouts—periods of zero visibility due to blowing fine snow and ice crystals.

Human Activity in Polar Regions

The climatic severity that limits animal life in polar regions to a few scattered species in the tundra is just as restrictive on human settlement. The celebrated Lapps of northern Europe migrate with their reindeer to the tundra from the adjacent forest during warmer months. They join the musk ox, arctic hare, fox, wolf, and polar bear that manage to make a home there despite the prohibitive environment. Only the Inuit (Eskimos) of Alaska, northern Canada, and Greenland have, in the past, succeeded in developing a year-round lifestyle adapted to the tundra regime. Yet even this group relies less on the resources of the tundra than on the large variety of fish and sea mammals, such as cod, salmon, halibut, seals, walruses, and whales, which occupy the adjacent seas.

As their communication with the rest of the world has increased and they have become acquainted with alternative lifestyles, the permanent Inuit population living in the tundra has greatly diminished, and life for those remaining has changed drastically. Some have gained new economic security through employment at defense installations or at sites where they join other skilled workers from outside the region to exploit mineral or energy resources. However, the new population centers based on the construction and maintenance of radar and missile defense stations or, as in the case of Alaska's North Slope, on the production and transportation of oil, cannot be considered permanent (■ Fig. 8.12). Workers depend on other regions for support and often inhabit this region only temporarily.

The ice-sheet climate cannot serve as a home for humans or other animals. Even the penguins, gulls, leopard seals, and polar bears are coastal inhabitants. It is without question the harshest, most restrictive, most nearly lifeless climate

zone on Earth (■ Fig. 8.13). Yet, especially in Antarctica, it is of strategic importance and of great scientific interest. Antarctica's strategic value is so widely recognized that the world's nations have voluntarily given up claims to territorial rights on the continent in exchange for cooperative scientific exploration on behalf of all humankind.

■ **FIGURE 8.12**   Mile-marker zero on the Alaska pipeline, near Prudhoe Bay. This is a profitable venture for humans in the North Slope oil fields. However, hazards abound: in 1989, the *Exxon Valdez* spilled 11 million gallons of crude oil into Prince William Sound, Alaska.
Considering the vulnerability of Alaska's physical environment, should development of the North Slope oil fields have been permitted?

M. Trapasso

■ **FIGURE 8.13**   Greenland's ice sheet covers about 85 percent of its surface. Here we see ice swallowing up the landscape.
What kind of activities might bring individuals from other regions to an ice-sheet climate?

M. Trapasso

Highland Climate Regions

As we saw in Chapter 3, temperature decreases with increasing altitude at the rate of about 6.5°C per 1000 meters (3.6°F per 1000 ft). Thus, you might suspect that highland regions exhibit broad zones of climate based on changes in temperature with elevation that roughly correspond to Köppen's climate zones based on change of temperature with latitude (■ Fig. 8.14). This is indeed the case, with one important exception: Seasons only exist in highlands if they also exist in the nearby lowland regions. For example, although zones of increasingly cooler temperature occur at progressively higher elevations in the tropical climate regions, the seasonal changes of Köppen's middle-latitude climates are not present.

Elevation is only one of several controls of highland climates; **exposure** is another. Just as some coasts face the prevailing wind, so do some mountain slopes; others are leeward slopes or are sheltered behind higher topography. The nature of the wind, its temperature, and its moisture content depend on whether the mountain is (1) in a coastal location or deep in a continental interior, and (2) at a high or low latitude within or beyond the reaches of cyclonic storms and monsoon circulation. In the middle and high latitudes, mountain slopes and valley walls that face the equator receive the direct rays of the sun and are warm; poleward-facing slopes are shadowed and cool. West-facing slopes feel the hot afternoon sun, whereas east-facing slopes are sunlit only in the cool of the morning. This factor, known as **slope aspect,** affects where people live in the mountains and where particular crops will do best. The higher one rises in the mountains, the more important direct sunlight is as a source of warmth and energy for plant and animal life processes.

Complexity is the hallmark of highland climates. Every mountain range of significance is composed of a mosaic of climates far too intricate to differentiate on a world map or even on a map of a single continent. Highland climates are therefore given a separate designation, signifying climatic complexity. Highland climates are indicated on Figure 7.5 wherever there is considerable local variation in climate as a consequence of elevation, exposure, and slope aspect. We can see that these regions are distributed widely over Earth but are particularly concentrated in Asia, central Europe, and western North and South America.

The areas of highland climate on the world map are cool, moist islands in the midst of the climates that dominate the areas around them. Consequently, highland areas are also biotic islands, supporting a flora and fauna adapted to cooler and wetter conditions than those of the surrounding lowlands. This coolness is part of the highland charm, particularly where mountains rise cloaked with forests above arid or semiarid lands, as do the Canadian Rocky Mountains and California's Sierra Nevada.

Highlands stimulate moisture condensation and precipitation by forcing moving air masses to rise over them (■ Fig. 8.15). Where mountain slopes are rocky and forest free, their surfaces grow warm during the day, causing upward convection, which often produces afternoon thundershowers. Mountains receive abundant precipitation and are the source area for multitudes of streams that join to form the great rivers of all of the continents.

There are few streams of significance whose headwaters do not lie in rugged highlands. Much of the stream flow on all continents is produced by the summer melting of mountain snowfields. Thus, the mountains not only draw moisture from the atmosphere but also store much of it to be gradually released throughout summer droughts when water is most needed for irrigation, and for municipal and domestic use.

■ **FIGURE 8.14** Note the similarities between (a) Svolvar, Norway, 68°N latitude (an Arctic location), at sea level, and (b) Machu, Picchu, 13°S latitude (a tropical location), at 2590 meters (8500 feet) above sea level.
In what ways might high latitudes be similar to high elevations?

(a)

(b)

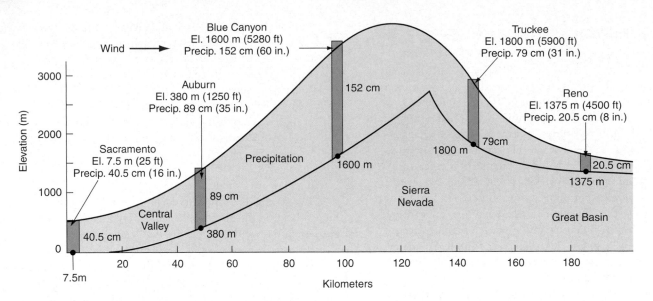

■ **FIGURE 8.15** Variation in precipitation caused by uplift of air crossing the Sierra Nevada range of California from west to east. The maximum precipitation occurs on the windward slope because air in the summit region is too cool to retain a large supply of moisture. Note the strong rain shadow to the lee that gives Reno a desert climate.

Taking into consideration the locations of the recording stations, during what season of the year does the maximum precipitation on the windward slope occur?

The Nature of Mountain Climates

A general characteristic of mountain weather is its variability from hour to hour as well as from place to place. Strong orographic flow over mountains often causes clouds to form very quickly, leading to thunderstorms and longer rains that do not affect surrounding cloud-free lowlands. Where the cloud cover is diminished, diurnal temperature ranges over mountains are far greater than those over lowlands. The thinner layer of low-density air above a mountain site does not greatly impede insolation, thus allowing surfaces to warm dramatically during the daytime. In addition, the atmosphere in these areas does little to impede longwave radiation loss at night. Consequently, air temperatures overnight are cooler than the elevation would indicate. Because the atmospheric shield is thinnest at high elevations, plants, animals, and humans receive proportionately more of the sun's shortwave radiation at high altitudes. Ultraviolet radiation is particularly noticeable; severe sunburn is one of the real hazards of a day in the high country.

In the middle and high latitudes, mountains rise from mesothermal and microthermal climates into tundra and snow-covered zones. The lower slopes of mountains are commonly forested with conifers, which become more stunted as one moves upward, until the last dwarfed tree is passed at the **tree line.** This is the line beyond which low winter temperatures and severe wind stress eliminate all forms of vegetation except those that grow low to the ground, where they can be protected by a blanket of snow (■ Fig. 8.16). Where mountains are high enough, snow or ice permanently covers the land

■ **FIGURE 8.16** As in this example in the Colorado Rocky Mountains, the last tree species found at the tree line are stunted, prostrate forms, which often produce an elfin forest. Where the trees are especially gnarled and misshapen by wind stress, the vegetation is called *krummholz* (crooked wood).

What do you see in the photograph that indicates prevailing wind direction?

R. Gabler

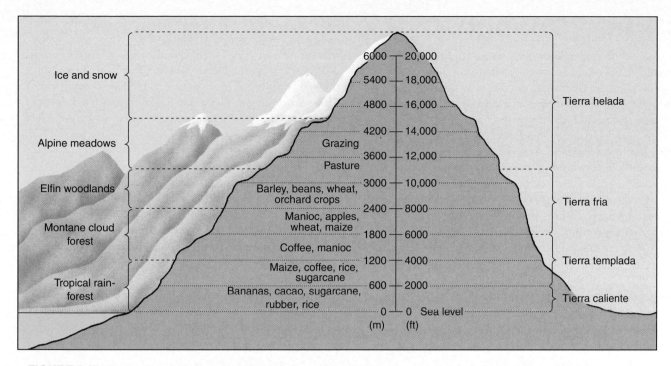

■ FIGURE 8.17 Natural vegetation, vertical climate zones, and agricultural products in tropical mountains. Note that this example extends from tropical life zones to the zone of permanent snow and ice. There is little seasonal temperature change in tropical mountains, which allows life forms sensitive to low temperatures to survive at relatively high elevations.

When Europeans first settled in the highlands of tropical South America, in which vertical climate zone did they prefer to live?

surface. The line above which summer melting is insufficient to remove all of the preceding winter's snowfall is called the **snow line.**

In tropical mountain regions, the vertical zonation of climate is even more pronounced. Both tree line and snow line occur at higher elevations than in middle latitudes. Any seasonal change is mainly restricted to rainfall; temperatures are stable year-round, regardless of elevation. Each climate zone has its own particular association of natural vegetation and has given rise to a distinctive crop combination where agriculture is practiced (■ Fig. 8.17). In South America, four vertical climate zones are recognized: *tierra caliente* (hot lands), *tierra templada* (temperate lands), *tierra fría* (cool lands) and *tierra helada* (frozen lands).

Adaptation to Highland Climates

In middle-latitude highlands, soils are poor, the growing season is short, and the winter snow cover is heavy in the conifer zone, which dominates the lower and middle mountain slopes. Therefore, little agriculture is practiced, and permanent settlements in the mountains are few. However, as the winter snow melts off the high ground just below the bare rocky peaks, grass springs into life, and humans drive herds of cattle and flocks of sheep and goats up from the warmer valleys. The high pastures are lush throughout the summer, but in early fall

they are once again vacated by the animals and their keepers, who return to the valleys. This seasonal movement of herds and herders between alpine pastures and villages in the valleys, termed *transhumance*, was once common in the European highlands and is still practiced on a reduced scale.

Otherwise, the middle-latitude highlands serve mainly as sources of timber and of minerals formed by the same geologic forces that elevated the mountains, and as arenas for both summer and winter recreation. Recreational use of the highlands is a relatively recent phenomenon, resulting both from new interest in mountain areas and from new access by road, rail, and air.

In contrast to mountain regions in higher latitudes, tropical highlands may actually experience more favorable climatic conditions and are often a greater attraction to human settlement (from ancient to modern times) than adjacent lowlands. In fact, large permanent populations are supported throughout the tropics where topography and soil favor agriculture in the vertical climate zones. Highland climates are at such a premium in many areas that steep mountain slopes have been extensively terraced to produce level land for agricultural use. Spectacular agricultural terraces can be seen in Peru, Yemen, the Philippines, and many other tropical highlands. Where the climate is appropriate and population pressure is high, people have created a topography to suit their needs, by carving it out of mountainsides.

Climate Change

Now that we have completed our detailed study of Earth's current climate regions it is appropriate to consider some questions that have challenged the minds and technologies of leading atmospheric scientists for decades as they seek answers to climate change. It is an accepted fact that all of Earth's systems are changing constantly, but how have climates changed in the distant past and what were the causes of major changes? Scientists study past climates, seeking clues to future climate change.

Undoubtedly the most important climatic questions today are related to the recent rise in atmospheric temperatures, termed **global warming.** Temperature increases over the last century or so are fully documented, but what are the causes? What is the relationship of current change to the most recent Ice Age that peaked about 18,000 years ago? What role have humans played in rising temperatures and how might temperature change affect other Earth systems? Let's look for answers to these questions by starting with the past.

Climates of the Past

Serious consideration by scientists of the premise that there have been major changes in global climates throughout Earth history did not begin until the 19th century. In 1837, Louis Agassiz, a European naturalist, was first to propose that temperatures colder than those experienced at the time had caused major periods of **glaciation,** periods commonly referred to as ice ages. He presented evidence, first in Europe and Asia, and later in North America, that large areas in all three continents had been covered by glaciers (flowing ice). The most striking evidence consisted of the rock debris carried forward by glaciers from distant sources and deposited by melting ice during warmer periods.

Today it is recognized that the most recent period of glaciation began as early as 2.4 million years ago. Prior to World War II, researchers widely believed that during this ice age (termed the Pleistocene Epoch by Earth scientists) Earth had experienced four major advances of the glaciers in North America, followed by warmer interglacial periods. Based on the southward limits of the glacial advances, these colder intervals were termed the Nebraskan (oldest), Kansan, Illinoian, and Wisconsinan glaciations in the United States. However, modern research has shown that glacial activity and climate change during the Pleistocene was far more complex than previously thought.

Modern Research

Two major developments in scientific knowledge about climate change occurred in the 1950s. First, radiometric techniques, such as radiocarbon dating, began to be widely used (organic material can be dated by measuring the extent to which radioactive elements in the material have decayed through time).

Radiocarbon dating of organic materials in deposits associated with the glaciers helped to provide a time sequence when the extensive ice sheets retreated for the last time. At their maximum extent, glaciers once covered nearly all of Canada and most of the northern United States, down to the Ohio and Missouri Rivers (■ Fig. 8.18). Glaciers flowed over the sites of modern-day cities such as Boston, New York, Indianapolis, and Des Moines.

The second major discovery was that evidence of detailed climate changes has been recorded in the sediments on the ocean floor. The slow, continuous deposition of deep-sea sediment records the history of climate changes during the past several million years. The most important discovery of the deep-sea record is that Earth experienced numerous major glacial advances during the Pleistocene, not just the four that had been identified previously. Today, the names of only two of the North American glacial periods, the Illinoian and Wisconsinan, have been retained.

■ **FIGURE 8.18**   This map identifies the extensive areas of Canada and the northern United States that were covered by moving sheets of ice as recently as 18,000 years ago. **Why does the ice move in various directions in different regions of the continent?**

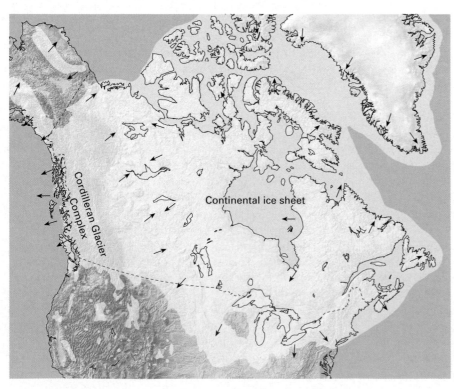

Cordilleran Glacier Complex

Continental ice sheet

:: CLIMATE CHANGE AND ITS IMPACT ON COASTLINES

When we look at a map or a globe, a part of geographic information that we see is so obvious and basic that we often take it for granted, perhaps failing to recognize that it is spatial information. This is the location of coastlines—the boundaries between land and ocean regions. The most basic geographic aspects of our planet that a world map shows us is where landmasses are and where the oceans are located, and the generally familiar shape of these major features. But maps of our planet today only show where the coastline is currently located. We know that sea level has changed over time and that it rose 20–30 centimeters (8–12 in) during the 20th century.

The hydrologic system on Earth is a closed system because the total amount of water (as a gas, liquid, and solid) on our planet is fixed. When the climate supports more glacial ice, sea level falls. When world climates experience a warming tendency, sea level rises. More ice in glaciers means less water in the oceans, and vice versa. If global warming trends continue at their present rate, the U.S. Environmental Protection Agency (EPA) predicts that sea level will rise 31 centimeters (1 ft) in the next 25–50 years. That amount of sea level rise will cause problems for low-lying coastal areas; the populations of some coral islands in the Pacific are already concerned, as their homelands are barely above the high-tide level.

A map of world population distribution shows a strong link between settlement density and coastal areas. For low-lying coastal regions, sea-level rise is a major concern, and the gentler the slope of the coast is, the farther inland the inundation would be with every increment of sea-level rise. Scientists at the United States Geological Survey (USGS) have determined that if all the glaciers on Earth were to melt, sea level would rise 80 meters (263 ft) and that a 10-meter (33-ft) rise would displace 25% of the U.S. population.

At the time when glaciers were most extensive, during the maximum advance of Pleistocene glaciers, sea level fell to about 100 meters (330 ft) below today's level. Maps that create the positions of coastlines and the shape of continents during times of major environmental change show how temporary and vulnerable coastal areas can be.

The accompanying map shows the present coastline (dark green), the pre-Pleistocene maximum rise of sea level (light green), the Pleistocene drop in sea level (light blue), and the impact on North American coastlines. It may seem odd to think of the coastlines shown on a world map as temporary, but because coasts can shift over time, a future world map could look quite different from the map we know today.

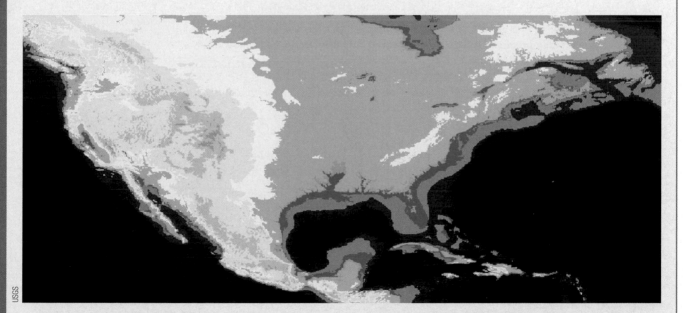

This map shows how changes in sea level would affect the coastline of much of North America. Dark green shows our coastlines as they appear today. Light green shows the coastline if sea level rose 35 meters (115 ft) in response to a massive glacial melting, similar to what occurred 3 million years ago (the dark green areas would be flooded also). Light blue shows the coastline if glaciers expanded to the level of maximum extent during the Pleistocene.

Because the deep-sea sedimentary record is so important to climate-change studies, we should understand how the record is deciphered. The deep-sea mud contains the microscopic remains of innumerable tiny marine animals that built shells for protection. When they died, these tiny shells sank to the seafloor, and were incorporated into the layers of mud. Different species thrive in different surface-water temperatures; therefore, the layers of sediment that contain the tiny fossils produce a detailed history of water-temperature fluctuations.

The tiny seashells of certain species are composed of calcium carbonate ($CaCO_3$); therefore, the analyses also record the oxygen composition of the seawater in which they were formed. One common measurement technique for determining oxygen composition is known as **oxygen-isotope analysis.** Modern seawater has a fixed ratio of the two oxygen isotopes. The O_{18}/O_{16} ratios will indicate changes in ocean temperatures relating to glacial cycles. A review of the oxygen-isotope record indicates that the last glacial advance about 18,000 years ago was only one of many major glacial advances during the past 2.4 million years. Evidence has suggested there may have been as many as 28 glacial-type climatic episodes.

Today, climatologists are aware that the present climate may be a short interval of relative stability in a time of major climate shifts. Moreover, the modern climate epoch, known as the Holocene (10,000 years ago to the present), is a time of stable, warm temperatures compared to much of the last 2.4 million years (■ Fig. 8.19). Based on the deep-sea record, it appears that global climates tend to rest at one of two extremes: a very cold interval characterized by a major glacial advance with lower sea levels and shorter intervals with a retreat of the glaciers with warm temperatures and high sea levels. Realizing that global climates have changed dramatically numerous times, two questions should now be considered: What causes global climate to change, and how quickly could global climate change from one extreme to the other?

Rapid Climate Change

As we previously noted, throughout most of Canada and into the United States, glaciers covered large areas north of the Missouri and Ohio Rivers 18,000 years ago. In the west, freshwater lakes more than 500 feet deep covered many areas of Utah and Nevada. However, the United States was mostly glacier free and most of the western lake basins were dry by about 9000 years ago. Abundant evidence has even been found that the climate about 7000 years ago (a time known as the **Altithermal**) was warmer than today (see again Fig. 8.19). For glaciers several thousand feet thick to melt completely and for deep lakes to evaporate, a substantial change of climate is required over a few thousand years. What were the causes?

To answer questions about such rapid rates of climate change requires a more detailed record of climate than the deep-sea sediments can provide. This is because the deep-sea sedimentary record is extraordinarily slow—a few centimeters of sea mud accumulates in a thousand years. Rapid shifts in climate during periods of a few hundred years are not recorded clearly in the seafloor sediments. Coring the glacial *ice sheets* of Antarctica and Greenland has solved this problem. Glacial ice records yearly amounts of snowfall in layers that can provide short-term evidence of climate changes. Oxygen-isotope analysis is also used, along with other techniques, to analyze the

■ **FIGURE 8.19** Analyses of oxygen-isotope ratios in ice cores taken from the glacial ice of Antarctica and Greenland provide evidence of surprising shifts of climate over short periods of time.
Has the general trend of temperatures on Earth been warmer or colder during the Holocene?

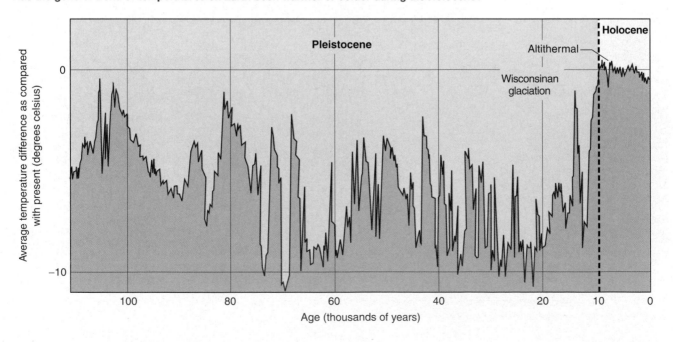

glacial ice of Antarctica and Greenland. These analyses have revealed a detailed record of climate changes during the past 250,000 years.

A surprising discovery of the ice-sheet analyses is how rapidly the climate can change. Rather than changing gradually from glacial to interglacial conditions over thousands of years, the ice record indicates that the shifts can occur in a few decades. Thus, whatever is most responsible for major climate changes can develop rapidly. This probably requires a positive feedback system, which means, as explained in Chapter 1, that a change in one variable will cause changes in other variables that magnify the amount of original change. For example, most glaciers have high albedos, reflecting significant amounts of sunlight back to space. However, if the ice sheets retreat for whatever reason, low-albedo land begins to absorb more insolation, increasing the amount of energy available to melt the ice. Thus, the more ice that melts, the more energy is available to melt the ice further, magnifying the initial glacial retreat.

Multiple Causes

Although theories about the causes of climate change are numerous, they can be organized into four broad categories: (1) astronomical variations in Earth's orbit; (2) changes in Earth's atmosphere; (3) changes in landmasses; and (4) asteroid and comet impacts.

Orbital Variations

Astronomers have detected slow changes in Earth's orbit that affect the distance between the sun and Earth as well as the deviation of Earth's axis on the plane of the ecliptic. These orbital cycles produce regular changes in the amount of solar energy that reaches Earth (or favors either the northern or southern hemisphere). The longest is known as the **eccentricity cycle,** which is a 100,000-year variation in the shape of Earth's orbit around the sun. In simple terms, Earth's orbit changes from an ellipse (oval), to a more circular orbit, and then back, affecting Earth–sun distance. More elliptical orbits seem to be associated with warm periods and more circular orbits may correspond to ice ages.

A second cycle, termed the **obliquity cycle,** represents a 41,000-year variation in the tilt of Earth's axis from a maximum 24.5° to a minimum of 22.0° and then back. The more Earth is tilted, the greater is the seasonality at middle and high latitudes. Therefore, less tilt should bring cooler summers to the polar regions and less melting of ice sheets, which may promote an ice age.

Finally, a **precession cycle** has been recognized with a periodicity of 21,000 years. The precession cycle determines the time of year that perihelion occurs. Today, Earth is closest to the sun about January 3 and, as a result, receives about 3.5% greater insolation than the average in January. When aphelion occurs on January 3 in about 10,500 years, the Northern Hemisphere winters should be somewhat colder (■ Fig. 8.20).

These cycles operate collectively, and the combined effect of the three cycles can be calculated. The person who examined all three of these cycles in detail was the mathematician

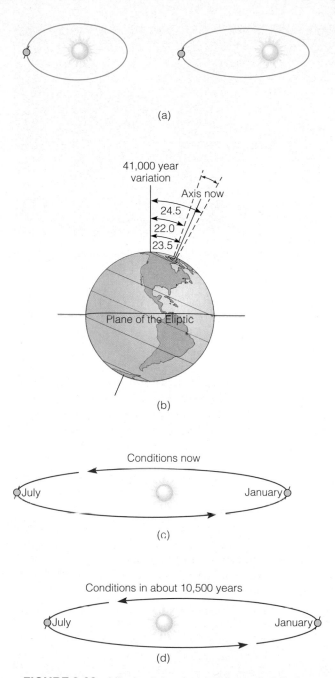

(a)

41,000 year variation

Axis now

24.5
22.0
23.5

Plane of the Eliptic

(b)

Conditions now

July   January

(c)

Conditions in about 10,500 years

July   January

(d)

■ **FIGURE 8.20** Milankovitch calculated the periodicity for (a) eccentricity, (b) obliquity, and (c) and (d) precession.
What effect should these changes in receipt of insolation have on global climates?

Milutin Milankovitch, who completed complex mathematical calculations to show how these changes in Earth's orbit and axis would affect insolation. Milankovitch's calculations indicated that numerous glacial cycles should occur during 1 million-year intervals.

By the late 1970s, most paleoclimate (ancient climate) scientists were convinced that a good correlation existed between the deep-sea record and Milankovitch's predictions. This suggests that an important driving force behind glacial cycles is regular orbital variations, and it suggests that long-term

USGS/Jim Vallance

■ **FIGURE 8.21**   Volcanic activity at Mount St. Helens in Washington State pumps gases and particulates into the atmosphere. The volcanic peak of Mount Rainier, a potentially active volcano, is in the background.
Besides affecting the climate, what other hazards result from volcanic explosions?

climate cycles are predictable. Unfortunately, the Milankovitch theory indicates that the warm Holocene interglacial will soon end and that Earth is destined to experience full glacial conditions (glacial ice possibly as far south as the Ohio and Missouri Rivers) in about 20,000 years.

Atmospheric Change
Many theories attribute climate changes to variations in atmospheric dust levels. The primary factor is volcanic activity, which pumps enormous quantities of particulates and aerosols (especially sulfur dioxide) into the stratosphere, where strong winds spread it around the world. The volcanic dust can reduce the amount of insolation reaching Earth's surface for periods of 1–4 years (■ Fig. 8.21 and ■ Fig. 8.22).

The climatic cooling effect of volcanic activity is unquestioned; all of the coldest years on record over the past two centuries have occurred in the year following a major eruption. Following the massive eruption of Tambora (1815 in Indonesia), 1816 was known as "the year without a summer." Killing frosts in July ruined crops in New England and Europe, resulting in famines. Several decades later, following the massive eruption of Krakatoa (also in Indonesia) in 1883, temperatures decreased significantly during 1884. Although no 20th-century eruptions have approached the magnitude of these two, the 1991 eruption of Mount Pinatubo (in the Philippine Islands) produced cool conditions in an otherwise continuous series of record warm years.

Another phenomenon closely associated with average global temperatures is the composition of atmospheric gases. Scientists have known for many years that carbon dioxide (CO_2) acts as a "**greenhouse gas.**" There is no question that CO_2 is transparent to incoming shortwave radiation and impedes outgoing longwave radiation, similar to the effect of the glass panes in a greenhouse or in your automobile on a sunny day (refer again to the greenhouse discussion in Chapter 3). Thus, as the atmospheric content of greenhouse gases rises, so will the amount of heat trapped in the lower atmosphere.

Captured in the glacial ice of Antarctica and Greenland are air bubbles containing minor samples of the atmosphere that existed at the time that the ice formed. One of the important discoveries of the ice-core projects is that prehistoric atmospheric CO_2 levels increased during interglacial periods and decreased during major glacial advances.

The fact that average global temperatures and CO_2 levels are so closely correlated suggests that Earth will experience record warmth as the atmospheric level of CO_2 increases. The present level of approximately 380 parts per million of CO_2 is already higher than at any time in the past million years.

Carbon dioxide is not the only greenhouse gas. Methane (CH_4) is more than 20 times more effective than CO_2 as a greenhouse gas but is considered less important because the atmospheric concentrations are lower. Also, the time the molecules of methane remain in the atmosphere (resident time) is much shorter. Garbage dump emissions and termite mounds both produce substantial quantities of CH_4. But a much more important source of atmospheric methane may come from the

■ **FIGURE 8.22**   Examination of global temperatures within 4 years before and after major volcanic eruptions provides compelling evidence that volcanic activity can have a direct effect upon the amounts of insolation reaching Earth's surface.
At what period after an eruption year does the effect seem the greatest?

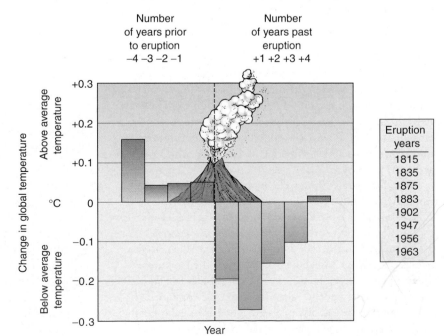

Eruption years
1815
1835
1875
1883
1902
1947
1956
1963

Greenhouse gases

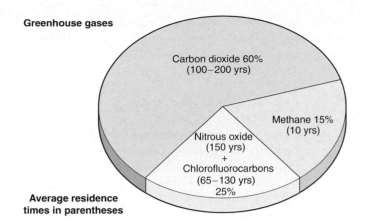

**Average residence
times in parentheses**

■ **FIGURE 8.23** Gases other than carbon dioxide released to the atmosphere by human activity contribute approximately 40% to the greenhouse effect. The figures in parentheses indicate the average number of years that the different gases remain in the atmosphere and contribute to temperature change.
Which gas has the longest residence time?

tundra regions or the deep sea. If warming the tundra or ocean water releases large amounts of methane as theorized, the positive feedback reinforcement of warming could be enormous.

Other greenhouse gases include CFCs (chlorofluorocarbons) and N_2 (nitrous oxide). The relative greenhouse contribution of common gases and their average residence times in the atmosphere are presented in ■ Figure 8.23.

Changes in Landmasses
The third category of climate change theories involves changes in Earth's surface to explain periods of cold or warm climates. A number of ice ages, some with multiple glacial advances have occurred during Earth's history. To explain some of the previous glacial periods, scientists have proposed several factors that might be responsible. For example, one characteristic that all of these glacial periods have in common with the Pleistocene is the presence of a continent in polar latitudes. Polar continents permit glaciers to accumulate on land, which results in lowered sea levels and consequent global effects.

Another geologic factor sometimes invoked as a cause of climate change is the formation, disappearance, or movement of a landmass that restricts oceanic or atmospheric circulation. For example, eruptions of volcanoes and the formation of the Isthmus of Panama severed the connection between the Atlantic and Pacific, greatly changing ocean circulation. The redirection of

ocean water created the Gulf Stream Current/North Atlantic Drift (see again Fig. 4.21). Another example is the uplift of the Himalayas, altering atmospheric flows, and monsoonal effects in Asia. Both of these events, and several other significant changes, immediately predate the onset of the modern series of glaciations. Which events caused climate changes and which are simply coincidences has yet to be determined.

Another group of theories involves changes in albedo, caused either by major snow accumulations on high-latitude landmasses or by large oceanic ice shelves drifting into lower latitudes. The increased reflection of sunlight from snow and ice starts a positive feedback cycle of cooling that may end when the polar oceans freeze, shutting off the primary moisture source for the polar ice sheets.

Impact Events
Rocky or metallic solar system bodies, called **asteroids,** usually less than 800 kilometers (500 mi) in diameter, may break into smaller pieces termed *meteoroids*. These objects orbit the sun, along with **comets** that are comprised of rocky or iron objects held together by ice. Through time, some of this material has struck Earth, occasionally with devastating impact.

Most of the objects with orbits intersecting that of Earth are so small that they burn up in our atmosphere before hitting the ground. At night they are seen as shooting stars. Objects that are smaller than 40 meters (131 ft) in diameter will incinerate because of the friction encountered in our atmosphere. Objects ranging from 40 meters to about 1 kilometer in diameter can do tremendous damage on a local scale when they reach Earth's surface (■ Fig. 8.24). An impact from an object this size can be expected on an average of every 100 years

■ **FIGURE 8.24** Barringer Crater, Arizona, shows the results of an impact with an iron-nickel meteorite of about 50 meters (165 ft) in diameter.
Can you conceive of the damage that would occur if a similar meteorite impacted an urban area?

or so. The last occurred in 1908 near Tunguska, Siberia, and devastated a huge area of the Siberian wilderness, with a blast estimated at 15 megatons (a megaton explosion is equal to 1 million tons of TNT).

On an average of very few hundred thousand years or so, an object with a diameter of greater than 1.6 kilometers (1 mi or greater) has struck Earth, producing severe environmental damage and climate change on a global scale. The blasts from such impacts could equal a million megatons of energy. The likely effect would be an "impact winter," characterized by skies darkened with particulates, blocking insolation and causing a drastic drop in temperatures. Firestorms would result from heated impact debris raining back down on Earth, and large amounts of acid rain would precipitate as well. A catastrophe of this kind would result in loss of crops worldwide, followed by starvation and disease. The largest known impacts on Earth in its history, like the one that may have contributed

to the extinction of the dinosaurs 65 million years ago, have been estimated to be about 15 kilometers (10 mi) in diameter, and may have exploded with a force of 100 million megatons.

Future Climates

With so many variables potentially responsible for climate change, reliably predicting future climate is an exceedingly difficult proposition. The primary problem in climate prediction is posed by natural variability. ■ Figure 8.25 displays the frequency and magnitudes of climate changes that have occurred naturally over the past 150,000 years. Although the Holocene has been the most stable interval of the whole period, a detailed examination of the Holocene record reveals a wide range of climates. For example, a long and warm interval, hotter than today's climate, occurred during

■ **FIGURE 8.25**   This figure shows the broad climate trends of the past 150,000 years, with significant details for the Holocene. Climatologists have been remarkably successful in dating recent climate change, but predicting future climates remains difficult. **Why is this so?**

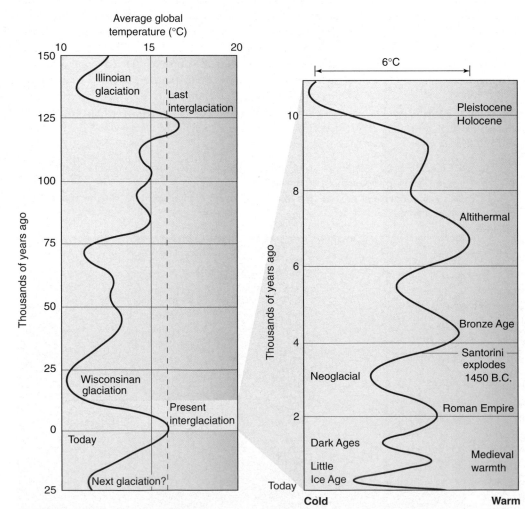

From Skinner & Porter, *Physical Geology.*

After Imbrie & Imbrie, *Ice Ages: Solving the Mystery,* Enslow Publishers, Short Hills, NJ, p. 179.

the **Altithermal**. This interval was characterized by the dominance of grasslands in the Sahara and severe droughts on the Great Plains. Other warm intervals occurred during the Bronze Age, during the second half of the Roman Empire, and in medieval times. An unusually cold interval began with the eruption of Santorini (which some believe was the basis for the Atlantis myth) in the Aegean Sea. Other cold periods occurred during the Dark Ages and again beginning about 1150 to 1460 in the North Atlantic and 1560 to 1850 in continental Europe and North America. These last episodes collectively have been termed the **Little Ice Age.** The Little Ice Age had major impacts on civilizations—from the demise of Viking settlements in Greenland that were established during the medieval warm period to the abandonment of the Colorado Plateau region by the Anasazi cultures. An important point to remember is that, with the exception of the cold interval that began with the eruption of Santorini, climatologists do not know what variables changed to cause each of these major climate fluctuations.

Predicting the Future

There have been numerous attempts to simulate the variables that affect climate. *General Circulation Models* (GCMs) are complex computer simulations based on the relationships among weather and climate variables discussed throughout this book—sun angles, temperature, evaporation rates, land versus water effects, energy transfers, and so on. The complexity and the usefulness of GCMs are both increasing rapidly. They appear to do a good job at predicting how conditions will change in specific regions as Earth's atmosphere warms or cools, and they have added new insights into how some climatic variables interact. However, GCMs are not infallible. Simulations from different computer programs often vary significantly and they can only provide painstakingly documented estimates of future climate.

Based on the record of climate changes during the past, only one thing can be concluded about future climates: they will change. Looking into the distant future, the Milankovitch cycles indicate that another glacial cycle is probably on the way. The most rapid cooling should occur between 3000 and 7000 years from now. In the near term, however, continued global warming is most likely. The rise of greenhouse gases such as carbon dioxide and methane, the widespread destruction of vegetation, and the feedback cycles that will most likely result are bound to increase the average global temperature for

Bruce Molnia, U.S. Geological Survey

■ **FIGURE 8.26**   With few exceptions, mountain glaciers worldwide are retreating. This glacier in Alaska's Glacier Bay National Park has continued to shrink for the last 200 years. In fact, in the last 70 years the glacier has retreated by 610 meters (2000 ft) and thinned by 244 meters (800 ft). Note the line on the valley wall just to the left, which marks the glacier extent, thickness, and height in the 1940s.
In what ways do melting glaciers become a problem?

the foreseeable future. An average increase of 1°C (nearly 2°F) globally would be equivalent to the change that has occurred since the end of the Little Ice Age in about 1850. A 2°C warming would be greater than anything that has happened in the Holocene, including the Altithermal. A 3°C warming would exceed anything that has happened in the past million years. Current estimates and the most reliable GCMs predict a 1°C–3.5°C (2°F–6°F) warming in the 21st century.

Not all areas on Earth will be affected equally. One of the most important effects is expected to be a more vigorous hydrologic cycle, fueled largely by increases in evaporation from the ocean. Intense rainfalls will be more likely in many regions, as will droughts in other regions such as the Great Plains. Temperatures are expected to rise the most in polar regions, mainly during the winter months. As a result of warming, sea levels will rise, because of melting ice sheets and the thermal expansion of ocean water. By 2100, sea levels are estimated to be between 15 and 95 centimeters (0.5–3.1 ft) higher than today. In addition, the ranges of tropical diseases could expand toward higher latitudes, tree lines will rise, and many mountain glaciers will continue to shrink or disappear (■ Fig. 8.26).

Global Warming

Concerns about climate change are voiced everywhere today, in television programs, movies, and the news media. Scientists, environmentalists, politicians, and celebrities are speaking out about global warming. Eleven of the 12 hottest years ever recorded have occurred since 1995, and subsequent years often set a new record. Average annual global temperatures have risen between 0.3°C and 0.6 C (0.5°F–1.1°F), and sea level has risen between 10 and 25 centimeters (4–10 in) during the past 100 years. Given the long residence times of many greenhouse gases (see again Fig. 8.23) and the heat capacity of the oceans, atmospheric scientists have concluded that additional global warming is inevitable and will continue for the near future.

What is often understated in discussions of global warming is the realization that climates are naturally subject to change in the future. Since the last major ice age ended 10,000 years or so ago, and the lesser "Little Ice Age" about a century ago, the climate has been warming. There is little controversy about that trend. According to extensive research and data sets produced by climate scientists from around the world, the most recent decades show global temperatures warming at an accelerated rate. However, some

controversy does arise when trying to establish the major causes of recent global warming and the degree to which humans are involved. Are human activities fully responsible or are these activities increasing the upward trend in global temperatures caused by natural processes? To what extent can humans slow the trend toward rising temperatures? Fortunately, international experts have been seeking answers to these questions.

In an effort to better understand global warming, including its causal factors, as well as its current and future impact, the United Nations and the Intergovernmental Panel on Climate Change (IPCC) have cooperated to gather as much relevant information and data as possible. The IPCC is a worldwide group of distinguished atmospheric scientists. In 2007, after years of research and study that involved more than 800 climate scientists from 130 countries, a series of

■ **FIGURE 8.27**   Using the best computer models available to evaluate global climate change, the International Panel on Climate Change has found that only the models that include increasing human releases of greenhouse gases (shown in pink) fit the temperature trends that have been observed in the last century (shown by the black lines). The blue tones estimate the ranges of what the temperatures would be without human impacts on global warming.

On what continent has the observed temperature fluctuated the most during this time period, and which one the least?

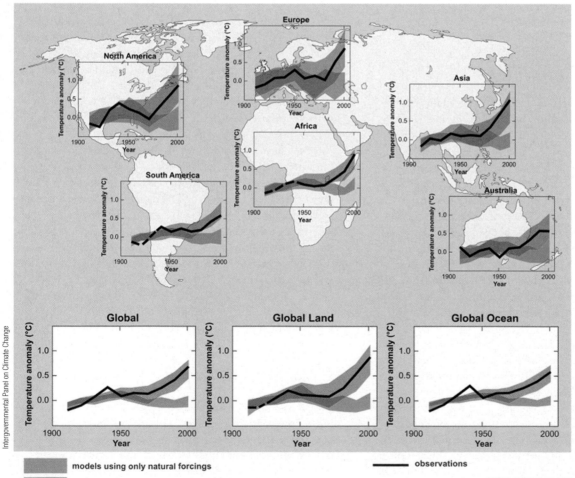

comprehensive reports on global warming was released. These scientists studied multiple lines of evidence worldwide, from tree-ring and ice-core data, to glacial retreat and sea-level rise, to changes in the atmosphere, to changes in weather phenomena. Further, they carefully considered the many potential influences of both natural processes and human activities. The conclusion of the IPCC was that it is "*very likely*" (>90% probability) that emissions of greenhouse gases from *anthropogenic* (human-induced) activities have caused ". . . most of the observed increase in globally averaged temperatures since the mid-20th century." They also state that in the last 50 years, the influence of Earth–sun relationships and volcanic activity would *likely* have caused a cooling trend. The results of computer models generated by the IPCC show how observed temperatures have increased in the last 100 years, compared to the predicted impact of natural influences alone, and compared to a combination of human and natural factors. The best fit is the one that includes human influence in global warming (■ Fig. 8.27).

The IPCC summarized their findings, stating: "Today, the time for doubt has passed. The IPCC has unequivocally affirmed the warming of our climate system, and linked it directly to human activities." The IPCC goes on to state that dealing with the environmental changes associated with global warming and/or working to minimize human impacts on climate change will be an important concern worldwide in the coming years.

Determining an appropriate course of action based on these findings could be complicated. If humans hope to halt or reduce the rate of global warming, and return to, or maintain, a more optimum climate, consensus must be built regarding some important questions. For example, what is the optimum climate, and who decides what levels of temperature and precipitation constitute the optimum climate? Further, the impact of global warming and in fact, of all major climate change, will always vary among different geographic locations and climate regions (see again Fig. 8.27). With the wide variety of environments on Earth, some geographic regions would benefit from a warmer climate and other areas will bear significant negative impacts (for example, the world's heavily populated coastal regions as sea level rises). We are not able to adjust our atmosphere as easily as we can set a thermostat in our homes.

We may not be able to find answers to all our questions or guarantee desirable climatic conditions in the future, but to do nothing only ignores our human responsibilities as stewards of planet Earth. There now is little question that the following recommendations should be given high priority. To the extent possible, the nations of the world should devote serious research and monetary resources to:

(1) Developing alternative sources of energy. Whatever the effects of burning fossil fuels on global temperatures, it also pollutes the air humans breathe, making it dangerous to human health. Energy from solar radiation, wind, geothermal heat, ocean tides, biofuels, hydroelectric generation, even nuclear reactors, helps keep the atmosphere cleaner for generations to come.

(2) Curtailing, or better managing, our energy usage. With Earth's growing human population, the developed nations' high rates of consumption, and the increasing industrialization in developing nations, energy demands are increasing and will continue to grow in the foreseeable future. How much energy we use and how we can conserve energy must be major considerations.

(3) Recycling our waste materials. Currently, human populations are consuming our nonrenewable resources at rates that cannot be sustained. Recycled resources will ease the strain on those that are vanishing at such a rapid rate, and will save the energy needed to create new resources.

(4) Curtailing deforestation. This destructive process should be restricted everywhere on Earth. Forest vegetation is a primary agent in the removal of CO_2 from our atmosphere through photosynthesis.

One of the few things all humans have in common, regardless of age, sex, ethnicity, religion, or nationality, is that we all occupy Earth together. It is our responsibility to care for the planet that sustains us. We must work toward the proper care of our world for our own descendants and for future generations.

:: Terms for Review

microthermal	tundra climate	oxygen-isotope analysis
humid continental, hot-summer climate	muskeg	eccentricity cycle
humid continental, mild-summer climate	tundra	obliquity cycle
	ice-sheet climate	precession cycle
subarctic climate	exposure	greenhouse gas
taiga	slope aspect	asteroid
boreal forest	tree line	comet
permafrost	snow line	Altithermal
patterned ground (frost polygons)	global warming	Little Ice Age
	glaciation	

:: Questions for Review

1. Explain why the microthermal climates are limited to the Northern Hemisphere.
2. List several features that all humid microthermal climates have in common.
3. What factors limit precipitation in the subarctic regions?
4. Identify and compare the climate factors that strongly influence the tundra and ice-sheet regions. How do these controlling factors affect the distribution of these climates?
5. What kind of plant and animal life can survive in the polar climates? What special adaptations must this life make to the harsh conditions of these regions?
6. How do elevation, exposure, and slope aspect affect the microclimates of highland regions? What are the major climatic differences between highland regions and nearby lowlands?
7. How have scientists been able to document the rapid shifts of climates that have occurred during the latter part of the Pleistocene?
8. What are the major possible causes of global climate change?
9. What effects can changes in the amounts of CO_2 and other greenhouse gases in the atmosphere have on global temperatures? How can past changes in amounts of CO_2 be determined?
10. What changes are likely to occur in Earth's major subsystems if global warming continues for the near term, as most scientists believe?

:: Practical Applications

1. Based on the classification scheme presented in Appendix C, classify the following climate stations from the data provided.

		J	F	M	A	M	J	J	A	S	O	N	D	Yr
a.	Temp. (°C)	−42	−27	−40	−31	−20	15	−11	−18	−22	−36	−43	−39	−30
	Precip. (cm)	0.3	0.3	0.5	0.3	0.5	0.8	2.0	1.8	0.8	0.3	0.8	0.5	8.6
b.	Temp. (°C)	−27	−28	−26	−18	−8	1	4	3	−1	−8	−18	−24	−12
	Precip. (cm)	0.5	0.5	0.3	0.3	0.3	1.0	2.0	2.3	1.5	1.3	0.5	0.5	10.9
c.	Temp. (°C)	−4	−2	5	14	20	24	26	25	20	13	3	−2	12
	Precip. (cm)	0.5	0.5	0.8	1.8	3.6	7.9	24.4	14.2	5.8	1.5	1.0	0.3	62.2
d.	Temp. (°C)	−3	−2	2	9	16	21	24	23	19	13	4	−2	11
	Precip. (cm)	4.8	4.1	6.9	7.6	9.4	10.4	8.6	8.1	6.9	7.1	5.6	4.8	84.8
e.	Temp. (°C)	0	0	4	9	16	21	24	23	20	14	8	2	12
	Precip. (cm)	8.1	7.4	10.7	8.9	9.4	8.6	10.2	12.7	10.7	8.1	8.9	8.1	111.5

2. The data in the previous table represent the following five locations, although not in this order: Beijing, China; Point Barrow, Alaska; Chicago, Illinois; Eismitte, Greenland; New York, New York. Use a world map or an atlas and your knowledge of climates to match the climatic data with the locations.
3. Chicago and New York City are located within a few degrees latitude of one another, yet they represent two different climates. Discuss their differences and identify the primary cause, or source, of the differences.
4. The precipitation recorded at Albuquerque, New Mexico (see Practical Applications, Chapter 7), is almost twice that recorded at Point Barrow, Alaska, yet Albuquerque is considered a dry climate and Point Barrow a humid climate. Why?

:: Outline

The living environment at the surface and the soils below are interdependent and intricately linked—the characteristics of one influences the characteristics of the other.

Natural Resources Conservation Service

:: Objectives

When you complete this chapter you should be able to:

- Define the four major components of an ecosystem, and explain their interdependence.
- Recognize that other environmental controls may be more important on a local scale, but that climate has the greatest influence over ecosystems on a worldwide basis.
- Explain how vegetation becomes established on barren or devastated areas, and cite an example of the steps in plant succession.
- Provide examples of how plants, animals, and the environments in which they live are interdependent, each affecting the others.
- Outline the climatic factors that have the greatest effect on plants and animals, and summarize the nature of those climatic impacts.

- Describe the major components of a soil and how they vary to produce different soil types.
- Discuss the role of water in soil processes and how different amounts of available soil water can affect the vegetation growing on a soil.
- Understand the factors that determine a soil's formation, development, and fertility, including the role of vegetation.
- Explain why soils are among the world's most critical resources, and the need for effective soil conservation practices.
- Cite a few reasons why humans affect ecosystems and soils more than all other life forms, and some examples of major impacts.

Biogeography is the study of how environmental factors affect the locations, distributions, and life processes of plants and animals. Basically, this discipline seeks explanations for the geography of life forms. Biogeographers delineate the spatial boundaries of ecosystems, and investigate how and why environmental characteristics change spatially and over time. Soils are intimately related to the factors that also influence the biogeography of an area. The characteristics of a soil reflect the interactions among the climate, vegetation, rocks, minerals, and fauna at its location. Soils are also environments that teem with organisms living on and beneath the surface. Relationships and interactions among the different climate regions, their associated vegetative biomes, and certain soils were introduced in Chapters 7 and 8. In this chapter, we take a closer look at biogeography and at the nature of soils.

Ecosystems

The term **ecosystem** refers to a community of organisms that occupy a given area, and the interdependent relationships—with each other and the environment—that allow the organisms to thrive (■ Fig. 9.1). Generally, ecosystems are studied on a local or regional basis, but the entire Earth system (the ecosphere) also functions as an ecosystem. When farmers plant crops, spread fertilizer, control weeds, and spray insecticides, new ecosystems (although artificial) are created. Despite environmental alterations by human activities, plants and animals that can adapt will still live together in interdependent relationships with soil, rainfall, temperatures, sunshine, and other characteristics of the physical environment.

Ecosystems are *open systems,* with movement of both energy and materials into and out of these systems. Ecosystems are not isolated in

nature, they are usually closely related to nearby ecosystems and integrated with the larger ecosystems of which they are a part. The ecosystem concept is a valuable model for examining the structure and function of life on Earth.

Major Components

Despite their great variety on Earth, the typical ecosystem has four basic components (■ Fig. 9.2). The first of these is the nonliving, or **abiotic,** part of the system. This is the physical environment in which the plants and animals of the system live. In a terrestrial ecosystem, the abiotic component provides life-supporting elements and compounds in the soil, groundwater, and atmosphere.

The second component of an ecosystem consists of the basic **producers,** or *autotrophs* (meaning "self-nourished"). Plants are important autotrophs, because they can use solar

■ **FIGURE 9.1**   This mountain ecosystem in Utah demonstrates the close relationship between living organisms and their nonliving environment.
Why might it be difficult for a biogeographer to determine boundaries for this ecosystem?

J. Petersen

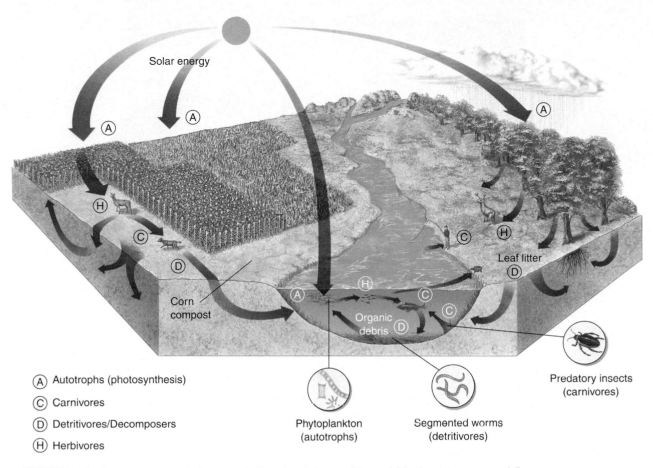

(A) Autotrophs (photosynthesis)

(C) Carnivores

(D) Detritivores/Decomposers

(H) Herbivores

Phytoplankton (autotrophs)

Segmented worms (detritivores)

Predatory insects (carnivores)

■ **FIGURE 9.2** Ecosystems clearly illustrate the interdependence of the variables in systems, especially the close relationships between living components of systems (biosphere) and the nonliving or abiotic components in systems (the atmosphere, hydrosphere, and lithosphere).

What examples of a producer, a consumer, and a decomposer are in this image?

energy to convert water and carbon dioxide into organic molecules through photosynthesis. The sugars, fats, and proteins produced by plants through photosynthesis supply the food that supports other forms of life. Some bacteria are also capable of photosynthesis, and sulfur-dependent organisms that dwell at thermal vents on the sea floor are also classified as autotrophs.

A third component of most ecosystems consists of **consumers,** or *heterotrophs* (meaning "other-nourished"). These are animals that survive by eating plants or other animals. **Herbivores** eat only plant material, **carnivores** eat other animals, and **omnivores** feed on both plants and animals. Animals contribute to the Earth ecosystem in many ways. They use oxygen in respiration and exhale carbon dioxide that is required for photosynthesis by plants. Animals also influence soil development through digging and trampling, and those activities also affect local plant distributions.

Without the fourth component of ecosystems, the decomposers, plant growth could soon come to a halt. The **decomposers,** or **detritivores,** feed on dead plants and animals, promoting decay and returning mineral nutrients that plants can use to the soil and bodies of water.

Trophic Structure

The living components of an ecosystem are organized in a sequence by their eating habits. Herbivores eat plants, carnivores may eat herbivores or other carnivores, and decomposers feed on dead plants and animals and their waste products. The sequence of feeding levels is referred to as a **food chain,** and organisms are identified by their **trophic level,** the number of steps they are removed from the producers (■ Fig. 9.3). Plants occupy the first trophic level, herbivores the second, carnivores feeding on herbivores the third, and so forth until the last level, the decomposers, is reached. Omnivores may belong to several trophic levels because they eat both plants and animals. The simplest food chain would include only plants and decomposers. More complex food chains may include six or more levels as carnivores feed on other carnivores—for example, zooplankton eat plants, small fish eat zooplankton, larger fish eat small fish, bears eat larger fish, and decomposers consume the bear after it dies.

In reality, most food chains do not operate in a simple linear sequence; they overlap and interact to form a feeding network within an ecosystem called a *food web.* Food chains

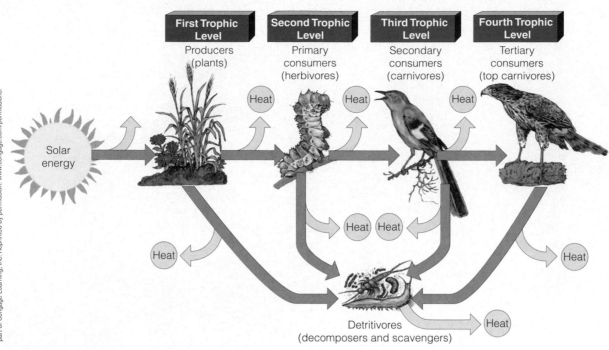

■ FIGURE 9.3   These trophic levels in an ecological example illustrate the dependence of higher trophic levels on all of the lower trophic levels in their food chain.

Can you outline, through four trophic levels, a trophic structure that exists in the area where you live?

and food webs can be used to trace the movement of food and energy from one level to another in an ecosystem.

Some biologists and ecologists find it helpful to separate the trophic structure into specific *nutrient cycles*. There are several such cycles that help explain the routing of nutrients through ecosystems. Particularly important cycles have been developed for water, carbon, nitrogen, and oxygen. Knowledge of chemical nutrient cycles is essential to an understanding of energy flow in ecosystems. Parts of the oxygen and carbon cycles as well as the water (hydrologic) cycle were discussed in previous chapters. ■ Figure 9.4 is a summary diagram that illustrates the major processes involved in these cycles.

Energy Flow and Biomass

Sunlight provides energy to an ecosystem, which is used by plants in photosynthesis, and the energy is stored in the organic materials of plants and animals. The total amount of living material in an ecosystem is referred to as the **biomass.** Because the energy of an ecosystem is stored in the biomass, scientists measure each trophic level's biomass to trace energy flow through the system. The second law of thermodynamics states that whenever energy is transformed from one state to another there will be a loss of energy through heat. When an organism at one trophic level feeds on an organism, not all of the food energy is used, some is lost from the system (see again Fig. 9.3). Additional energy is lost through respiration and movement. As energy flows from one trophic level to the next, the biomass successively decreases because of this loss of energy between trophic

levels (■ Fig. 9.5). At each higher trophic level, a greater amount of energy is required. A deer may graze in a limited area, but the wolf that preys on the deer must hunt over a much larger territory. As the flow of energy decreases with each successive trophic level, the biomass also decreases. This principle also applies to agriculture. A great deal more biomass (and food energy) is available in a field of corn than there is in the cattle that eat the corn.

Productivity

Productivity is defined as the rate at which new organic material is created at a particular trophic level. **Primary productivity** refers to the formation of new organic matter through photosynthesis by producers. **Secondary productivity** refers to the rate of formation of new organic material at the consumer level.

Primary Productivity   Photosynthesis requires sunlight, which varies widely by latitudinal effects on daylight hours and sun angles. Photosynthesis is also affected by soil moisture, temperature, the availability of nutrients, the atmosphere's carbon dioxide content, and the age and species of the individual plants.

Most studies of productivity in ecosystems have been concerned with measuring the net biomass at the producer level. The ecosphere's annual net primary productivity is enormous, estimated to be about 170 billion metric tonnes (a metric tonne is about 10% greater than a U.S. ton) of organic matter. Even though oceans cover approximately

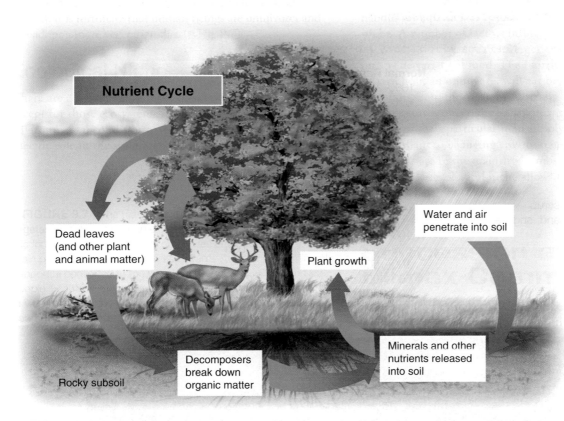

Nutrient Cycle

Dead leaves
(and other plant
and animal matter)

Water and air
penetrate into soil

Plant growth

Rocky subsoil

Decomposers
break down
organic matter

Minerals and other
nutrients released
into soil

■ **FIGURE 9.4** This simplified diagram displays the processes used by nutrient cycles to travel through an ecosystem.

What kinds of processes are taking place beneath the soil surface?

■ **FIGURE 9.5** Trophic pyramids showing biomass of organisms at various trophic levels in two contrasting ecosystems. Trophic levels get higher upward in the pyramids. Dry weight is used to measure biomass because the proportion of water to total mass differs from one organism to another.

How can you explain the exceptionally large loss of biomass between the first and second trophic levels of the tropical forest ecosystem?

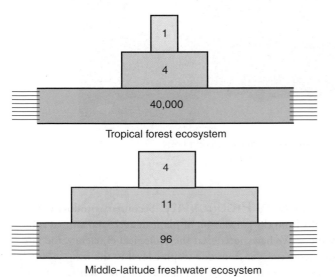

1

4

40,000

Tropical forest ecosystem

4

11

96

Middle-latitude freshwater ecosystem

Biomass expressed as dry weight (g/m²)

70% of Earth's surface, about two-thirds of net annual productivity is from terrestrial ecosystems and one-third is from marine ecosystems.

Latitudinal impacts on photosynthesis result in a noticeable decrease in terrestrial productivity from tropical ecosystems to those in middle and higher latitudes. Table 9.1 illustrates the wide range of net primary productivity displayed by various ecosystems. Today, satellites monitor Earth's biological productivity and give us a global perspective on our biosphere (■ Fig. 9.6).

The reasons for differences among aquatic, or water-controlled, ecosystems are not quite as apparent. Swamps and marshes are well supplied with plant nutrients and therefore have a relatively large biomass at the first trophic level. Water depth has a great impact on ocean ecosystems, because most nutrients in the open ocean sink to the bottom, to depths beyond the penetration of sunlight that can make photosynthesis possible. The most productive marine ecosystems are found in the sunlit, shallow waters of estuaries, continental shelves, and coral reefs, as well as in areas where ocean upwelling carries nutrients nearer to the surface.

Some agricultural (artificial) ecosystems can be fairly productive when compared with the natural ecosystems they have replaced. This is especially true in the warmer latitudes where farmers may raise two or more crops in a year or in arid lands where irrigation supplies the water for growth.

creation. Even after such damage, seeds lying dormant in the soil are ready to sprout and invade the newly opened gap. Compared to primary succession, secondary succession can occur more quickly.

A common form of secondary succession, associated with agriculture in the southeastern United States, is depicted in ■ Figure 9.8. After agriculture has ceased, pioneer plants such as weeds and grasses colonize the bare fields. These plants stabilize the topsoil, add organic matter, and produce favorable conditions for the growth of shrubs and brush such as sassafras, persimmon, and sweet gum. During this stage, the soil is enriched with nutrients and organic matter, and increases its ability to retain moisture. These conditions encourage the development of pine forests, the next stage in this vegetative succession. As pine forests thrive in this newly created environment, the pine trees eventually shade out and dominate the weeds, grasses, and brush.

Ironically, growth of a pine forest can also lead to its demise. Pine trees require much sunlight for their seeds to germinate. When competing with scattered brush, grasses, and weeds, there is adequate sunlight for germination, but once a pine forest develops, the shade and litter will not allow the pine seeds to germinate. Hardwood trees, such as oak and hickory, whose seeds can germinate in shady conditions, begin to grow as an understory, but eventually will replace the pines. In this southeastern United States example, a complete succession from field to oak–hickory forest will take about 100–200 years if it continues unimpeded. Through succession, the area can return to the natural oak–hickory forest that existed prior to agricultural clearing. In other ecosystems, such as tropical rainforests, succession from deforestation back to a natural forest may take many centuries.

The Climax Community

The concept of plant succession was introduced early in the 20th century, defined as a process of *predictable* steps ending with a vegetative cover that would remain in environmental balance unless affected by major climatic or environmental change. The final result in a succession is called a **climax community.** It was thought that climax communities would be self-perpetuating, and in a state of equilibrium or stability with the environment. In our illustration of plant succession in the southeastern United States, the oak–hickory forest would be considered the climax community. The tropical rainforests are also a good example of a climax community (■ Fig. 9.9).

Succession remains as a useful model for studying ecosystems, but some of the original ideas have been challenged. For one thing, early proponents

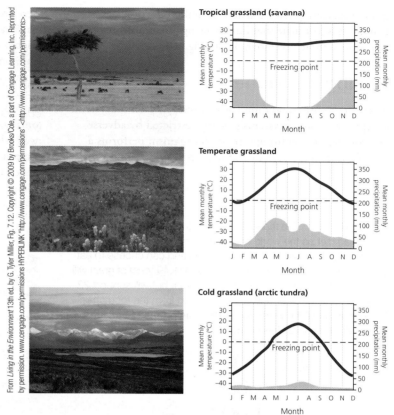

From *Living in the Environment* 13th ed. by G. Tyler Miller, Fig. 7.12. Copyright © 2009 by Brooks/Cole, a part of Cengage Learning, Inc. Reprinted by permission. www.cengage.com/permissions HYPERLINK "http://www.cengage.com/permissions" <http://www.cengage.com/permissions>.

■ **FIGURE 9.8**   A common plant succession in the southeastern United States. Each succeeding vegetation type alters the environment in such a way that species having more stringent environmental requirements can develop.

Why would plant succession be quite different in another region of the United States?

Sean Linehan/NOAA, NGS, Remote Sensing

■ **FIGURE 9.9**   A tropical rainforest on the island of St. Croix, U.S. Virgin Islands. Tropical rainforests are considered to be good examples of a climax community. The dense forest canopy conceals the vast numbers of other evergreen tree species and the relatively open forest floor.

How might this rainforest differ from the rainforests of the Pacific Northwest of the United States?

emphasized a *predictable* sequence of succession. One plant community would follow another in regular order as the ecosystem changed over time. But, many changes in ecosystems do not follow a rigid or completely predictable sequence.

Many scientists today no longer believe that only one type of climax vegetation is possible for each of the world's major climate regions. One of several different climax communities might develop within a given area, influenced not only by climate but also by local conditions of drainage, nutrients, soil, or topography. The dynamic nature of climate is also now better understood than it was when the original theories of succession and climax were developed. In the time it takes the species structure of a plant community to adjust to climatic conditions, the climate may change again. Also, because every habitat has a dynamic nature, no one climax community can exist in equilibrium with an environment indefinitely.

Today, many biogeographers and ecologists view plant communities and their ecosystems as a *landscape* that is the expression of all of its various environmental factors functioning together. They view the landscape of an area as a vegetative **mosaic** of interlocking parts, much like the tiles in a mosaic artwork. In a pine forest, for example, other plants also exist, and some areas may not support pine trees. The dominant area of the mosaic—in this case, the pine forest—is called the **matrix.** Gaps within the matrix, resulting from different soil conditions or from human or natural processes, are called **patches** within the matrix. Relatively linear features that cut across the mosaic, including natural features such as rivers and human-created structures such as roads, fence lines, and power lines, are termed **corridors** (■ Fig. 9.10). Every habitat is unique and constantly changing, and resultant plant and animal communities must constantly adjust to these changes.

Climate, a dominant environmental influence, is changing today, and has changed throughout Earth history. The world's climate has changed over relatively short time periods, such as over decades and centuries, and also over time frames measured in millennia. Climate change may be subtle, or may be sufficiently drastic to create ice ages or warm periods between ice ages. Plant and animal communities must be able to adapt to environmental changes, or they will not survive. Biogeographers work to reconstruct the vegetation communities of past climate periods by examining evidence such as tree rings, pollen, and fossils. By determining how past climate changes affected Earth's ecosystems, biogeographers hope to forecast future impacts that may develop as climate continues to change.

Environmental Controls

The plants and animals that exist in a particular ecosystem are those that have been successful in adjusting to their habitat's environmental conditions. Every living organism requires

■ **FIGURE 9.10** This aerial view of Taskinas Creek, Virginia, shows riparian (river) environments forming corridors that pass through the forest matrix on the Atlantic coastal plain.
How does a corridor differ from a patch?

certain environmental conditions to survive. Certain plants can exist under a wide range of temperature, whereas others have narrow temperature requirements. This refers to an organism's **range of tolerance** for certain environmental conditions. The ranges of tolerance will determine where a species may exist, and species with wide ranges of tolerance will be the most widely distributed. The *ecological optimum* refers to environmental conditions under which a species will thrive. The farther away a species is from its ecological optimum or from the geographic core of its plant or animal community, the conditions will become increasingly difficult for that species or the community to survive. However, those same conditions may be more amenable for another species or community. An **ecotone** is the overlap or the zone of transition between two plants or animal communities (■ Fig. 9.11). On a global basis, climate has the greatest influence over natural vegetation. The major types of terrestrial ecosystems, or *biomes,* are associated with certain temperature ranges as well as critical annual or seasonal precipitation and evaporation characteristics. Climate influences the sizes and shapes of tree leaves and determines whether trees can exist in a region, but, at the local scale, other environmental factors can be as important. A plant's range of tolerance for acidity, moisture, or salinity in the soil may also be a critical environmental determinant of whether a plant will grow, flourish, or die. The discussion that follows illustrates how major environmental factors influence the organization and structure of ecosystems.

GEOGRAPHY'S ENVIRONMENTAL PERSPECTIVE

:: THE THEORY OF ISLAND BIOGEOGRAPHY

Biogeographers are intrigued by the life forms and species diversity found on islands that are isolated from larger landmasses. How can land plants and animals be living on an island surrounded by a wide expanse of sea? How did this flora and fauna become established and flourish on these distant, often geologically recent, and originally barren terrains (volcanic islands, for example)? The farther an island is from the nearest landmass, the more difficult it is for species to migrate to the island and establish a viable population there. Winds, birds, or ocean currents can carry some seeds to islands, where they germinate to develop the vegetative environments on isolated landmasses. Humans have also introduced many species into island environments. But why did the species adapt and survive?

The theory of island biogeography offers an explanation for how natural factors interact to affect either the successful colonization or the extinction of species that come to live on an island. The theory considers an island's isolation (the distance from a mainland source of migrating species), its size, and the number of species living on it. Generally, the biological diversity on islands is low compared to mainland areas with similar climates and other environmental characteristics. Low species diversity typically means that the floral and faunal populations exist in an environmentally challenging location. Many extinctions have occurred on islands because of the introduction of some factor that made the habitat nonviable for that species to survive.

Several natural factors affect the species diversity on islands as long as

USCG

A beach on Palmyra Island in the Pacific Ocean, one of the world's most remote islands, illustrates how palm trees become established on tropical islands. Coconuts are the seeds for coconut palm trees. Washed away by surf from their original location, they float and drift hundreds of kilometers from one island to another. Waves deposit and bury coconuts on the beach, and palms sprout and grow. Note the coconuts, recently sprouted palms, and fully grown coconut palms. Very few other plant species grow on this very small island. Greater species diversity tends to exist on larger islands, and on islands that are nearer to major landmasses.

other environmental conditions such as climate are comparable:

1. The farther an island is from the area from which species must migrate, the lower the species diversity. Islands nearer to large landmasses tend to have higher diversity than those that are more distant.

2. The larger the island, the greater the species diversity. This is partly because larger islands tend to offer a wider variety of environments to colonizing organisms than smaller islands do. Larger islands also offer more space for species to occupy.

3. An island's species diversity results from an equilibrium between the rates of extinction of species on the island and the colonization rate of species. If the island's extinction rate is higher, only a few hardy species will live there; if the extinction rate is lower compared to the colonization rate, more species will thrive, and the diversity will be higher.

The theory of island biogeography has also been useful in understanding the ecology and biota of many other kinds of isolated environments, such as high mountain areas that stand, much like islands, above surrounding deserts. In those regions, plants and animals adapted to cool, wet environments live in isolation from similar populations on nearby mountains, separated by inhospitable arid environments.

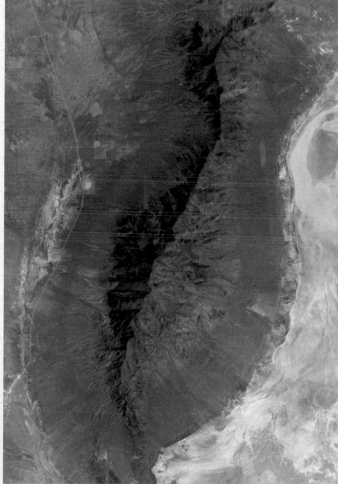

NASA Global Land Cover Facility

A mountain range along the Nevada—Utah state line supports green forests, alpine zones, and animal species associated with those environments. The blue and white areas to the east (right) are the edge of the Bonneville Salt Flats. Flora and fauna that could not survive in the surrounding desert environments flourish in cooler, wetter, higher elevation zones. Isolated mountains surrounded by arid environments tend to fit the concept of island biogeography.

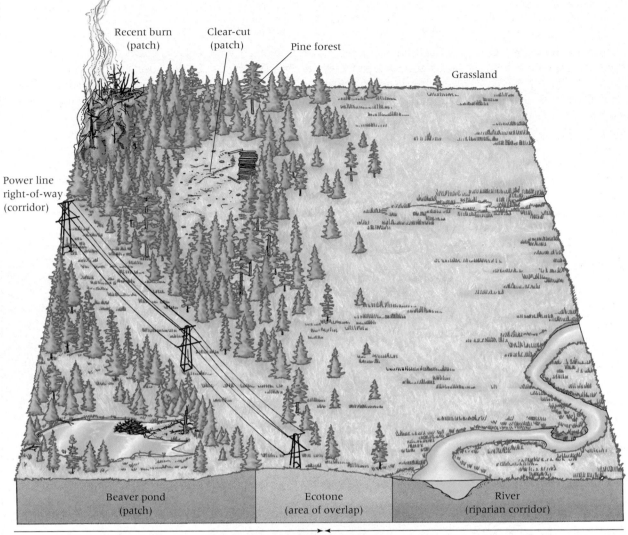

Recent burn
(patch)

Clear-cut
(patch)

Pine forest

Grassland

Power line
right-of-way
(corridor)

| Beaver pond (patch) | Ecotone (area of overlap) | River (riparian corridor) |

Conditions increasingly unfavorable
for pine forest

Conditions increasingly unfavorable
for grasslands

■ **FIGURE 9.11** The concepts of ecotone, *ecological optimum, range of tolerance, mosaic, matrix, patch,* and *corridor* are illustrated in the diagram.
What effect would a change to drier climate conditions throughout the area have on the relative sizes of the two ecosystems as well as the position of the ecotone?

Climatic Factors

Sunlight is one of the most critical climatic factors that influence an ecosystem. Sunlight is the energy source for plant photosynthesis, and it also strongly influences the behavior of both plants and animals. Competition for light can make forest trees grow taller, thereby limiting plant growth on the forest floor to shade tolerant species such as ferns. The sizes, shapes, and colors of leaves may result from variations in light reception, with large leaves developing in areas of limited light. The intense sunlight of the low latitudes produces a greater biomass in the tropical forests compared to the much lower light intensity that reaches the high latitude Arctic

regions. *Duration* of daylight, which varies seasonally and with latitude, has a profound effect on the flowering of plants as well as animal mating and migration.

Many plants can tolerate a wide range of temperatures, although every species has optimum conditions for growth. Vegetation, however, can be adversely affected by temperature extremes (unusual hot spells or cold temperatures) in climate regions where they rarely occur. Temperatures may also affect vegetation indirectly. For example, high temperatures lower the relative humidity, thus increasing transpiration. If a plant's root system cannot extract enough moisture from the soil to meet this increase in transpiration, the plant will wilt and eventually die.

Virtually all organisms require water. Plants need water for germination, growth, and reproduction, and most plant nutrients must be dissolved in soil water to be absorbed by plants. Marine and aquatic plants are adapted to living in water. Some trees, such as mangroves (■ Fig. 9.12a) and bald cypresses (Fig. 9.12b), rise from marshes and swamps. Certain tropical plants become dormant during dry seasons, dropping their leaves, and others store water received during wet periods

■ **FIGURE 9.12** (a) Mangrove thicket along the Gulf of Mexico coast of southern Florida. (b) Extensive cypress forests exist in swampy areas of the southeastern United States.
How might the vegetation and environments shown here have influenced the routes that were followed by the Spanish adventurers who first explored Florida?

(a)

(b)

for surviving the drought seasons. Desert plants, such as cacti, are well adapted to storing water when it is available while minimizing water loss from transpiration.

Luxuriant forests tower above the well-watered windward sides of mountain ranges such as the Sierra Nevada and Cascades, but semiarid grasslands, shrublands, and sparse forests cover the leeward sides. Orographic precipitation, rain shadow effects, and elevation changes produce variations in temperature, precipitation, drainage, and evapotranspiration that directly affect vegetation types and distributions. Life zones change progressively with elevation; upper zones are dominated by hardy plant communities that can tolerate the lower temperatures and the precipitation regimes found at higher elevations.

Animals, because of their mobility, are not as dependent as plants are on climatic conditions, although animals are subject to climatic stresses. In arid regions, animals employ adaptations to heat and aridity. Many become inactive during the hottest and driest seasons, and most leave their burrows or the shade only at night.

The geographic distribution of some animal groups reflects their degree of sensitivity to climate. Cold-blooded animals, for example, are more widespread in warm climates and more restricted in cold climates. Some warm-blooded animals develop layers of fat or fur for protection from the cold. During hot periods, they may sweat, shed fur, or lick their fur in an attempt to stay cool. Certain animals hibernate to survive in cold or arid regions. Cold-blooded animals such as the rattlesnake move into and out of shade in response to temperature changes. Seasonally, warm-blooded animals may migrate great distances out of environmentally harsh areas.

Some warm-blooded animals exhibit a link between body shape and size to variations in average environmental temperatures. The body size of a subspecies usually increases with the decreasing mean temperature of its habitat and, in warm-blooded species, the relative sizes of exposed body portions decrease as the mean temperature declines. Further, in cold climates, body sizes tend be larger to provide body heat needed for survival and for protecting vital organs in the trunk of the body (■ Fig. 9.13). Members of the same species living in colder climates eventually evolve shorter or smaller appendages (ears, noses, arms, legs, etc.) compared to their relatives in warmer climates. In cold climates, small appendages are advantageous because they reduce the body areas that are subjected to temperature loss and frostbite. In warm climates, long limbs, noses, and ears allow for heat dissipation in addition to that provided by panting or licking fur.

As a climatic control on vegetation, the wind is most significant in deserts, polar regions, coastal zones, and highlands. Wind may directly injure vegetation and can also have an indirect effect by increasing evapotranspiration. To prevent water loss in areas of severe wind stress, plants twist and grow close to the ground, minimizing their wind

■ **FIGURE 9.13** The Arctic polar bear provides an excellent example of a cold region subspecies having large body and small appendages (ears).
What other physical adaptations does the polar bear have to its Arctic environment?

■ **FIGURE 9.14** Krummholz (stunted) trees at the upper reaches of the subalpine zone in the Colorado Rockies. The healthy green vegetation has been covered by snow much of the year, protected from the bitterly cold temperatures. Note the flagged trees, which give a clear indication of wind direction.
What type of vegetation would be found at elevations higher than the one depicted in this photograph?

misshapen or swept bare of leaves and branches on their windward sides.

Soil and Topography

Soils supply much of the moisture and minerals for plant growth. Soil variations can strongly influence plant distribution and also can produce sharp boundaries between vegetation types. This is partly a consequence of varying chemical requirements of different plant species and partly a reflection of factors such as soil texture. Clay soils may retain too much moisture for certain plants, whereas sandy soils retain too little. Pines generally thrive in sandy soils, grasses in clays, cranberries in acid soils, and chili peppers in alkaline soils. The subject of soils will be explored in more detail later in this chapter.

Topography, particularly in highlands, influences ecosystems by providing diverse microclimates within a relatively small area. Plant communities vary from place to place in highland areas in response to the differing microclimatic conditions. *Slope aspect* has a direct effect on vegetation patterns in areas outside of the equatorial tropics. North-facing slopes in the middle and high latitudes of the Northern Hemisphere have microclimates that are cooler and wetter than those on south-facing exposures (■ Fig 9.15). Northern Hemisphere south-facing slopes tend to be warmer and drier because they receive more direct sunlight. The steepness and shape of a hillslope also affect how long water is present before draining downslope.

Natural Catastrophes

Plant and animal distributions are also affected by a variety of natural processes frequently termed *catastrophes*. It should be noted, however, that this term is applied from a strictly human perspective. What may be catastrophic to humans, such as a hurricane, wildfire, landslide, tsunami, or an avalanche are basically natural processes that can produce openings (gaps) in a region's vegetative mosaic (■ Fig. 9.16). The resulting succession, whether primary or secondary, produces a diverse set of patch habitats within the regional matrix of vegetation. Natural catastrophes and the resulting patch dynamics they create among an area's plant and animal residents are topics of much interest and research in modern landscape biogeography.

exposure (■ Fig. 9.14). During severe winters, they are better off being buried by snow than being exposed to bitterly cold gales. In some windy coastal regions, the shoreline may be devoid of trees or other tall plants. In windy coastal and mountain regions where the trees do grow, they are often

J. Petersen

■ **FIGURE 9.15**   The cooler, wetter north-facing slopes (facing to the left in the background) support coniferous evergreen forest, but the south-facing slopes receive more direct sunlight, and are warmer and drier, illustrating the strong impact of slope aspect here in Washington State.

Are there good examples of the influence of slope aspect on vegetation in the area where you live?

■ **FIGURE 9.16**   The vegetation mosaic of this area in Glacier National Park, Montana is coniferous forest, but frequent snow avalanches keep rigid-stemmed conifers from invading the patch of open, low shrubs and grasses.

Why are there so many broken tree stumps in the foreground?

USGS/P. Carrara

Biotic Factors

Although their influence on a particular species might tend to be overlooked, other plants and animals may also affect whether a given organism exists as part of an ecosystem. Some interactions between organisms are beneficial to both species involved; this is called a **symbiotic relationship.** However, other relationships may be directly *competitive* and have an adverse effect on one or both species. Because most ecosystems are suitable for a wide variety of plants and animals, there is always competition among species and also members of a given species to determine which organisms will survive. The greatest competition occurs between species that occupy the same ecological niche. Among plants, there is great competition for light. The dominant trees in the forest are those that grow tallest and partially shade the plants growing beneath them. Other competition occurs underground, where the roots compete for soil water and plant nutrients.

Interactions between animals and plants as well as competition both within and among animal species also can significantly affect an ecosystem. Many animals are helpful to plants through pollination or seed dispersal, and plants are the basic food supply for many animals. Grazing may also influence the species that make up a plant community. During dry periods, herbivores may be forced to graze an area very closely and the taller plants are grazed out. Plants that are unpalatable, that have thorns, or that have the strongest root development are the ones that survive. Grazing is a part of the natural selection process, yet serious overgrazing rarely results under natural conditions because wild animal populations increase or decrease with the available food supply. For most animals, predators are another control of population numbers.

Human Impact on Ecosystems

Throughout history, humans have modified ecosystems and their natural development. Except in regions too remote to be altered significantly by civilization, humans have eliminated or had a significant impact on much of Earth's natural vegetation. Farming, fire, domesticated animal grazing, deforestation, road building, urbanization, dam building and irrigation, impacts on water resources, mining, and the draining or infilling of wetlands are a few examples of how humans have modified plant communities. Overgrazing by domesticated animals can seriously harm marginal environments in arid and semiarid climates. Trampling and soil compaction by grazing herbivores may reduce the soil's ability to absorb moisture, leading to increased surface runoff of precipitation. In turn, the decreased absorption and increased runoff may lead to *land degradation* and gully erosion.

UNEP-Sudan

■ **FIGURE 9.17**   Overgrazing is a major cause of desertification here at this location in sub-Saharan Africa. The environment normally would have been a grassy savanna area.
What are some of the other causes of desertification?

As humans alter ecosystems and natural environments, the changes can often produce negative effects on humans themselves. The desertification of large semiarid sections of East Africa has resulted periodically in widespread famine (■ Fig. 9.17). Elsewhere, the continuing destruction of wetlands not only eliminates valuable plant and animal communities but also threatens the water supply quality and reliability for the people who drained the land.

Soils and Soil Development

Soil is a dynamic body of natural materials that is capable of supporting a vegetative cover (■ Fig. 9.18). It contains minerals, chemical solutions, gases, organic refuse, flora, and fauna. The interactions among the physical, chemical, and biological processes that take place demonstrate the dynamic character of soil. Soil responds to climatic conditions (especially temperature and moisture), the land surface configuration, its vegetation cover, and animal activity. The word *fertility,* so often associated with soils, has a meaning that takes into consideration the usefulness of a soil to humans. Soils are fertile in respect to their effectiveness in producing vegetation types (including crops) or plant communities. Soils are one of our most important and vulnerable resources.

Major Soil Components

What is soil actually made of? What soil characteristics support and influence variations in Earth's environments? Soil is an exceptional example of the interdependence and overlap

■ **FIGURE 9.18**   Scientists work to understand and control erosional losses and other problems that threaten soil, a precious natural resource.

USDA/ARS/Photo by Jack Dykinga

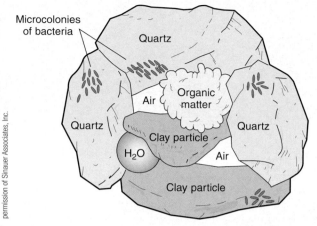

■ FIGURE 9.19   The four major components of soil. Soil contains a complex assemblage of inorganic minerals and rocks, along with water, air, and organic matter. The interaction among these components and the proportion of each are important factors in the development of a soil.

How does each soil component shown here contribute to making a soil suitable to support plant life?

among Earth's subsystems, because a soil develops through long-term interactions of atmospheric, hydrologic, lithologic, and biotic conditions. The nature of a soil reflects the ancient environments under which it formed as well as current environmental conditions. Soils contain four major components: inorganic materials, soil water, soil air, and organic matter (■ Fig. 9.19).

Inorganic Materials
Soils contain rock fragments and minerals that will not readily dissolve in water, and as well as soluble minerals, dissolved chemicals held in solution. Most soil minerals are composed of elements common in Earth's surface rocks, such as silicon, aluminum, oxygen, and iron. The chemical constituents of a soil typically come from many sources—the breakdown (*weathering*) of rocks, deposits of loose sediments, and solutions in water. As organic activities help to disintegrate rocks, they form new chemical compounds, and also release gases into the soil.

Soils sustain Earth's land ecosystems by providing vegetation with necessary chemical elements and compounds. Carbon, hydrogen, nitrogen, sodium, potassium, zinc, copper, iodine, and compounds of these elements are important in soils. Plants need many substances for growth, so the mineral and chemical content of a soil greatly affects its potential productivity. **Soil fertilization** is the process of adding nutrients or other constituents to meet the soil conditions that certain plants require.

Soil Water
When precipitation falls on the land, the water that does not run downslope or evaporate is absorbed into the rock or soil, or by vegetation. Moving through a soil, water dissolves certain materials and carries them through the soil. The water in a soil is not pure, as it contains dissolved nutrients in a liquid form that can be extracted by vegetation. Plants need air, water, and minerals from the soil to live and grow. Soil is a critical natural resource that functions as an *open system*. Matter and

energy flow in and out, and a soil also holds them in storage. Understanding these flows—inputs and outputs, the components and processes involved, and how they vary among different soils—is a key to appreciating the complexities of soil.

The water in a soil is found in several different circumstances (■ Fig. 9.20). Soil water adheres to soil particles and clumps by surface tension (the property that causes small water droplets to form rounded beads instead of spreading out in a thin film). This soil water, called **capillary water,** is a stored water supply that plants can use. Capillary water migrates through a soil from areas with more water to areas with less. During dry periods capillary water can move both upward or horizontally to supply plant roots with moisture and dissolved nutrients.

Capillary water moving upward moves minerals from the subsoil toward the surface. If this capillary water evaporates, the formerly dissolved minerals remain, generally as salt or lime (calcium carbonate) deposits in the topsoil. High concentrations of certain minerals like these can be detrimental to plants and animals that live in the soil. Lime deposited by evaporating soil water can build up to produce a cement-like layer, called *caliche,* which can prevent further downward percolation of water.

Soil water that percolates downward is called **gravitational water.** Gravitational water moves down through voids between soil particles and toward the *water table*—the level below which all available spaces are filled with water. The quantity of gravitational water a soil contains is related to several conditions, including the precipitation amounts, the time since it fell, evaporation rates, the space available for water storage and how easily water can move through the soil.

Gravitational water performs several functions. As gravitational water percolates downward, it dissolves soluble minerals and carries them to deeper levels of the soil, perhaps to the saturated zone. The depletion of soil nutrients by percolating water is called **leaching.** In regions of heavy rainfall, leaching can be intense, robbing a topsoil of all but the insoluble substances.

Gravitational water also can wash the finer solid particles (clay and silt) away from upper soil layers. This downward *removal* of solid components in a soil by water is called **eluviation** (■ Fig. 9.21). Eluviation tends to develop a coarse texture in the topsoil as the fine particles are removed, reducing the topsoil's ability to retain water. As gravitational water percolates downward, the fine materials transported from the topsoil are deposited at a lower level. This *deposition by water* in the subsoil is called **illuviation** (see again Fig. 9.21). Deposition of fine particles by illuviation may eventually form dense clay *hardpan* in the subsoil, which retards further downward percolation of gravitational water. Leaching, eluviation, and illuviation influence a soil by forming layered changes with depth, or **stratification**.

Soil Air
Much of a soil (sometimes nearly 50%) consists of spaces between soil particles and clumps (aggregates of soil particles). Voids that are not filled with water contain air or gases. For most microorganisms and plants that live in the ground, soil air supplies oxygen and carbon dioxide necessary for life. If all pore spaces are filled with water, there is no air supply and this lack of air is why many plants find it difficult to survive in water-saturated soils.

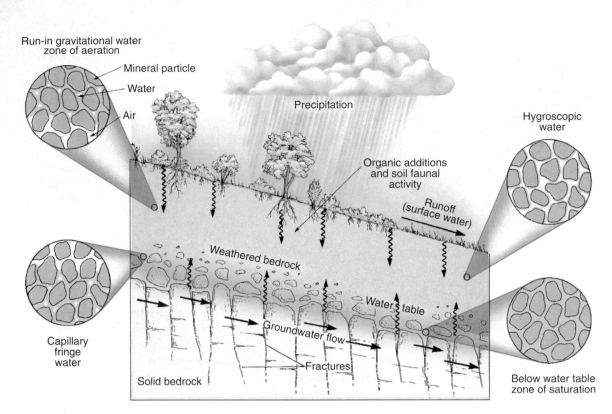

■ **FIGURE 9.20** The interrelationships between water and other environmental factors in the process of soil development. Soil is an example of an open system because it receives inputs of matter and energy, stores part of these inputs, and outputs matter and energy.
What are some examples of energy and matter that flow into and out of the soil system?

■ **FIGURE 9.21** Water is important in moving nutrients and particles vertically, both up and down, in a soil.
How does deposition by capillary water differ from deposition (illuviation) by gravitational water?

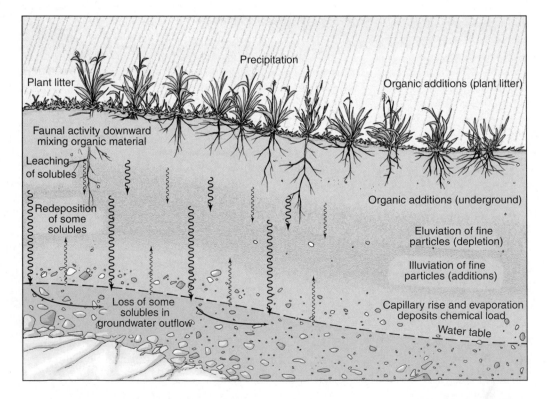

Organic Matter

Most soils contain decayed plant and animal materials, collectively called **humus.** Soils that are rich in humus are workable and have a good capacity for water retention. Humus supplies a soil with nutrients and minerals, and is important in chemical reactions that help plants extract nutrients. Humus also provides an abundant food source for microscopic soil organisms.

Most soils are actually microenvironments teeming with life that ranges from bacteria and fungi to earthworms, rodents, and other burrowers. Animals mix organic material deeper into the soil, and move inorganic fragments toward the surface. In addition, plants and their root systems are integral parts of the soil-forming system.

Soil Characteristics

Knowing a soil's water, mineral, and organic components and their proportions can help us determine its productivity and what the best use for that soil might be. Several soil properties that can be readily tested or examined are used to describe and differentiate soil types. The most important properties include color, texture, structure, acidity or alkalinity, and capacity to hold and transmit water and air.

Color

The color of a soil is immediately visible, but it might not be the most important characteristic. A soil's color is generally related to its physical and chemical characteristics. When describing soils in the field, or samples in the laboratory, soil scientists use a book of standard colors to identify this coloration (■ Fig. 9.22).

Decomposed organic matter is black or brown, so soils rich in humus tend to be dark. If the humus content is low because of limited organic activity or loss of organics through leaching, soil colors typically are light brown or gray. Soils with high humus contents are usually very fertile, so dark brown or black soils are often referred to as *rich*. However, this is not always true because some dark soils have little or no humus, but are dark because of other soil forming factors.

Red or yellow soils typically indicate the presence of iron. In moist climates, a light gray or white soil indicates that iron has been leached out, leaving oxides of silicon and aluminum, but in dry climates, the same color typically indicates an accumulation of calcium or salts.

Soil colors provide clues to the characteristics of soils and make the job of recognizing different soil types easier. But, color alone does not indicate a soil's qualities or its fertility.

Texture

Soil texture refers to the particle sizes (or distribution of sizes) in a soil (■ Fig. 9.23). In **clayey** soils, the dominant size is **clay,** particles defined as having diameters of less than 0.002 millimeter (soil scientists use the metric system). In **silty** soils, the dominant **silt** particles are defined as being between 0.002 and 0.05 millimeter. **Sandy** soils have mostly particles of **sand** size, with diameters between 0.05 and 2.0 millimeters.

■ **FIGURE 9.23**   Particle sizes in a soil. Sand, silt, and clay are terms that refer to the sizes of these particles for scientific and engineering purposes. Here, greatly enlarged, sand and silt sizes can be visually compared. Clays consist of tiny, flat, sheet-like particles that cannot be seen.

■ **FIGURE 9.22**   Scientists use a standardized classification system to determine precise color by comparing the soil to the color samples found in Munsell soil color books.
In general, how would you describe the color of the soils where you live?

Courtesy of James P. Shroyer, Kansas State University Research and Extension

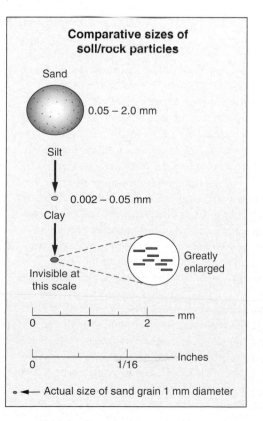

Comparative sizes of soil/rock particles

Sand

0.05 – 2.0 mm

Silt

0.002 – 0.05 mm

Clay

Invisible at this scale

Greatly enlarged

0   1   2   mm

0   1/16   Inches

●◄ Actual size of sand grain 1 mm diameter

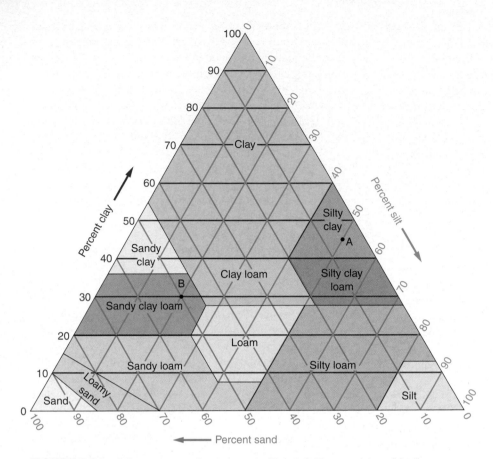

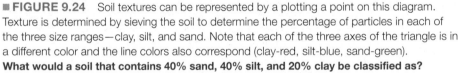

■ **FIGURE 9.24**   Soil textures can be represented by a plotting a point on this diagram. Texture is determined by sieving the soil to determine the percentage of particles in each of the three size ranges—clay, silt, and sand. Note that each of the three axes of the triangle is in a different color and the line colors also correspond (clay-red, silt-blue, sand-green). **What would a soil that contains 40% sand, 40% silt, and 20% clay be classified as?**

Rocks larger than 2.0 millimeters are regarded as pebbles, gravel, or rock fragments, and technically are not soil particles.

The proportion of particle sizes determines a soil's texture. For example, a soil composed of 50% silt-sized particles, 45% clay, and 5% sand would be identified as a silty clay soil. A triangular graph (■ Fig. 9.24) is used to discern different classes of soil texture based on the percentages of sand, silt, and clay within each class. Point A within the silty clay class represents the example just given. **Loam** soils, which occupy the central areas of the triangular graph, are soils with a good mix of the three grades (sizes) of soil particles without any size being greatly dominant. A second soil sample (B) that is 20% silt, 30% clay, and 50% sand would be a sandy clay loam. Loam soils are generally best suited for supporting vegetation growth.

Soil texture helps determine a soil's capacity to retain the moisture and air that are necessary for plant growth. Soils with a higher proportion of larger particles tend to be well aerated and allow water to **infiltrate** (seep through) the soil quickly—sometimes so rapidly that plants cannot use the water. Clayey soils retard water movement, becoming waterlogged and deficient in air.

Structure   Scientists classify soil structures according to their form. In most soils, particles clump into masses known as **soil peds,** which give a soil a distinctive structure. These range from columns, prisms, and angular blocks, to nutlike spheroids, laminated plates, crumbs, and granules (■ Fig. 9.25). Soils with massive or fine structures tend to be less useful than aggregates of intermediate size and stability, which permit good drainage and aeration.

Soil structure and texture both influence a soil's **porosity**—the amount of space that may contain fluids, and they also affect **permeability**—the rate at which water can pass through. Permeability is usually greatest in sandy soils, and poor in clayey soils.

Acidity and Alkalinity   An important aspect of soil chemistry is acidity, alkalinity (baseness), or neutrality. Levels of acidity or alkalinity are measured on the **pH scale** of 0 to 14. Low pH values indicate an acid soil, and high pH indicates alkaline conditions (■ Fig. 9.26a). Certain species tolerate alkaline soils, and others thrive under more acid conditions.

Most complex plants will grow only in soils with levels between pH 4 and pH 10, although the optimum pH varies

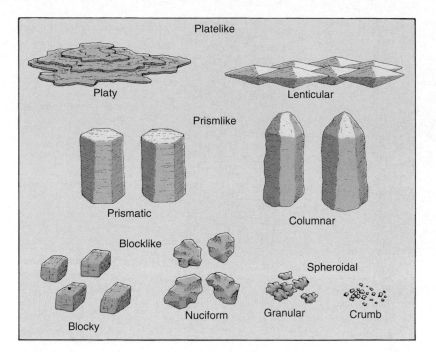

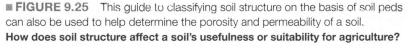

■ **FIGURE 9.25**   This guide to classifying soil structure on the basis of soil peds can also be used to help determine the porosity and permeability of a soil.
How does soil structure affect a soil's usefulness or suitability for agriculture?

with the plant species. In arid and semiarid regions, soils tend to be alkaline and soils in humid regions tend to be acidic (Fig. 9.26b). To correct soil alkalinity and to make the soil more productive, the soil can be flushed with irrigation water. Strongly acidic soils are also detrimental to plant growth, but soil acidity can generally be corrected by adding lime to the soil.

Development of Soil Horizons

Soil development begins when plants and animals colonize rocks, or deposits of rock fragments, the **parent material** on which soil will form. Once organic processes begin among mineral particles or rock fragments, chemical and physical differences begin to develop from the surface down through the parent material.

Initially, vertical differences result from surface accumulations of organic litter and the removal of fine particles and dissolved minerals by percolating water that deposits these materials at a lower level. A vertical section of a soil from the surface down to the parent material is called a **soil profile** (■ Fig. 9.27). Examining the vertical differences in a soil profile is important to recognizing different soil types and how a soil developed. Over time, as climate, vegetation, animal life, and the land surface affect soil development, this vertical differentiation becomes increasingly apparent.

■ **FIGURE 9.26**   (a) The degree of acidity or of alkalinity, called pH, can be easily understood when numbers on the scale are linked to common substances. Low pH means acidic, and high pH means alkaline; a reading of 7 is neutral. (b) The distribution of alkaline and acidic soils in the United States is generally related to climate. Soils in the East tend to be acidic and those in the West, alkaline.
Other than climate, what environmental factors might cause this east—west variation, and why are some places in the west acidic?

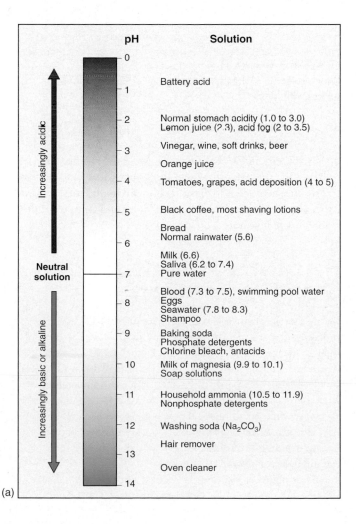

(a)

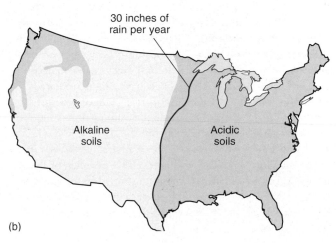

(b)

■ **FIGURE 9.27** A soil profile is examined by digging a pit with vertical walls to clearly show variations in color, structure, composition, and other characteristics that occur with depth. This soil is in a grassland region of northern Minnesota. **Why might you think that this is a fertile soil for vegetation growth?**

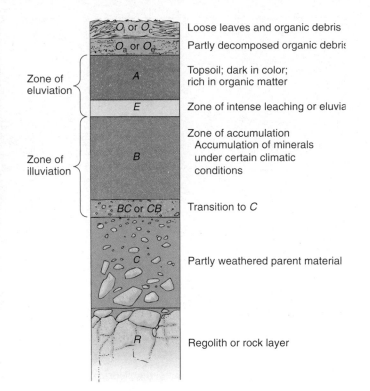

■ **FIGURE 9.28** The arrangement of horizons in a soil profile. Soils are categorized by their degree of development and the physical characteristics of their horizons. Many soils will not have all of these horizons, but horizons will appear in the vertical order shown here. Regolith is a generic term for broken bedrock fragments at or very near the surface. **What are some of the reasons why soils change color and texture with depth?**

Well-developed soils typically exhibit distinct layers in their soil profiles called **soil horizons** that are distinguished by their physical and chemical properties. Soils are classified largely on the differences in their horizons and the processes responsible for those differences. Soil horizons are designated by a set of letters that refer to their composition, dominant process, or position in the soil profile (■ Fig. 9.28).

At the surface, but only in locations where there is a cover of decomposed vegetation litter, there will be an O *horizon.* The "O" designation refers to this horizon's high content of organic debris and humus. The *A horizon,* immediately below, is commonly referred to as "topsoil." In general, *A* horizons are dark because they contain decomposed organic matter. Beneath the *A* horizon, certain soils have a lighter-colored *E horizon,* named for the action of strong eluvial processes. Below this is the *B horizon,* a zone of accumulation, where much of the materials removed from the *A* and *E* horizons are deposited. The *C horizon* is the weathered *parent material* from which the soil has developed—either bedrock, or deposits of rock materials that were transported to the site by a surface process such as running water, wind, or glacial activity. The lowest layer, sometimes called the *R horizon,* consists of unchanged parent material.

Certain horizons in some soils may not be as well developed as others, and some horizons may be absent. Because soils and the processes that form them vary widely and can be transitional between horizons, the horizon boundaries may be either sharp or gradual. Variations in color and texture within a horizon are not unusual.

Factors Affecting Soil Formation

Because of the great variety among parent materials and the processes that affected them, no two soils are identical in all of their characteristics. One important factor is rock *weathering,* which refers to the natural processes that break down rocks into smaller fragments (weathering will be discussed in detail in a later chapter). Chemical reactions can cause rocks and minerals to decompose and physical processes also cause the breakup of rocks. Just as statues, monuments, and buildings become "weather-beaten" over time, rocks exposed to the elements eventually break up and decompose.

Hans Jenny, a distinguished soil scientist, observed that soil development was a function of climate, organic matter, relief, parent material, and time—factors that are easy to remember by their initials: **Cl, O, R, P, T.** Among these factors, parent material is distinctive because it is the raw material. The other factors influence the type of soil that forms from a parent material.

Parent Material

All soil contains weathered rock fragments. If these weathered rock particles have accumulated in place—through the physical and chemical breakdown of bedrock directly beneath the soil—we refer to the fragments as **residual parent material.** If the rock fragments that form a soil have been carried to the site and deposited by streams, waves, winds, gravity, or glaciers, this mass of deposits is called **transported parent material.**

Parent materials influence soils to varying degrees. Soils that develop from weathering-resistant rocks tend to have a high level of similarity to their parent materials (■ Fig. 9.29). If the bedrock is easily weathered, the soils that develop tend to be more similar to soils in regions that have a similar climate. Soil differences related to variations in parent material are most visible on a local level.

In the long term, as a soil develops, the influence of parent material on its characteristics diminishes. Given the same soil-forming conditions, recently developed soils will show more similarity to their parent material, compared to soils that have developed over a long time.

The particle sizes that result from the breakdown of parent material are a prime determinant of a soil's texture and structure. Rock material such as sandstone, which contains little clay and weathers into relatively coarse fragments, will produce a soil of coarse texture.

Organic Activity

Plants and animals affect soil development in many ways. The life processes of plants growing in a soil are important, as are its microorganisms—the microscopic plants and animals that live in a soil.

Variations in vegetation species and density of cover can affect the evapotranspiration rates. Sparse vegetation allows greater soil moisture evaporation, but dense vegetation tends to maintain soil moisture. The characteristics of a plant community affect the nutrient cycles that are involved in soil development. Leaves, bark, branches, flowers, and root networks contribute to nutrients and the organic composition of soil, through litter and the remains of dead plants. Soils, however, can become impoverished by nutrient loss through leaching. The roots of plants help to break up the soil structure, making it more porous, and roots also absorb water and nutrients from the soil.

Bacteria are important to soil development, because they break down organic matter and humus, forming new organic compounds that promote plant growth. The number of bacteria, fungi, and other microscopic plants and animals living in a soil may be 1 billion per gram (a fifth of a teaspoon) of soil.

Earthworms, nematodes, ants, termites, wood lice, centipedes, and burrowing rodents stir up the soil, mixing mineral components from lower levels with organic components from the upper portion. Earthworms greatly contribute to soil development because they take soil in, pass it through their digestive tracts, and excrete, which mixes the soil and also changes the soil's texture, structure, and chemical qualities. In the late 1800s, Charles Darwin estimated that earthworm casts produced in a year would equal as much as 10–15 tons per acre.

Climate

On a world regional scale, climate is a very important factor in soil formation. Temperature directly affects soil microorganisms, which influences the decomposition of organic matter. In hot equatorial regions, intense soil microorganism activity precludes accumulations of organic debris or humus. The amounts of organic matter and humus in soils increase toward the middle latitudes and away from polar regions and the tropics. In the mesothermal and microthermal climates, microorganism activity is slow enough to allow decaying organic matter and humus to accumulate. Moving poleward into colder regions, retarded microorganism activity and limited plant growth result in thin accumulations of organic matter.

Chemical activity increases and decreases directly with temperature, given equal availability of moisture. Therefore,

■ **FIGURE 9.29**   Despite strong leaching in a wet tropical climate, these Hawaiian soils remain high in nutrients because they formed on volcanic parent materials of recent origin.
What other parent materials provide the basis for continuously fertile soils in wet tropical climates?

R. Gabler

parent materials of soils in hot, humid equatorial regions are chemically altered much more compared to the parent materials in colder zones.

Temperature affects soil indirectly through a climate's influence on vegetation associations. The combined effects of vegetative cover and the climate's regime tend to produce soils and soil profiles that share certain characteristics among different regions with similar climates and vegetation associations (■ Fig. 9.30).

Moisture conditions affect the development and character of soils more directly than any other climatic factor. Precipitation amounts affect plant growth, which directly influences a soils organic content and fertility. Extremely high rainfall will

■ **FIGURE 9.30**   Idealized diagrams of five different soil profiles illustrate the effects of climate and vegetation on the development of soils and their horizons.
Which two environments produce the most humus and which two produce the least?

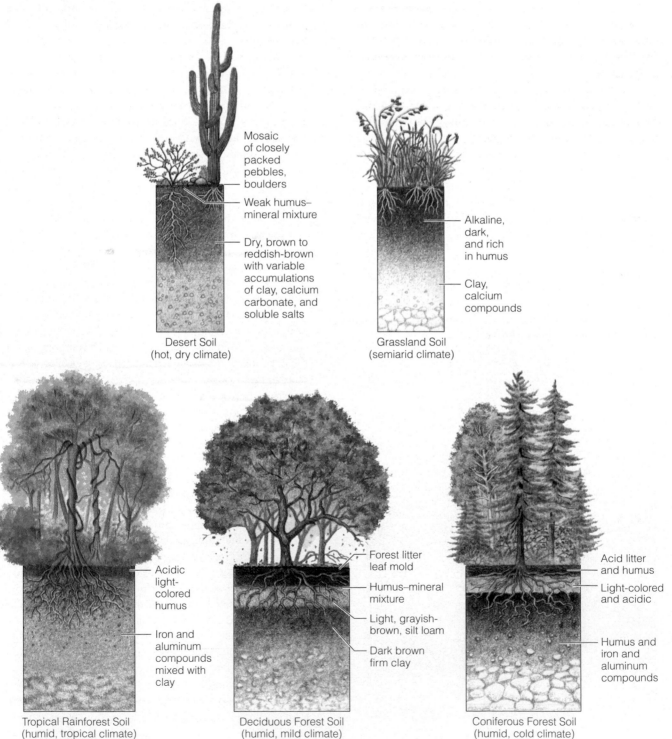

Mosaic of closely packed pebbles, boulders

Weak humus–mineral mixture

Dry, brown to reddish-brown with variable accumulations of clay, calcium carbonate, and soluble salts

Desert Soil
(hot, dry climate)

Alkaline, dark, and rich in humus

Clay, calcium compounds

Grassland Soil
(semiarid climate)

Acidic light-colored humus

Iron and aluminum compounds mixed with clay

Tropical Rainforest Soil
(humid, tropical climate)

Forest litter leaf mold

Humus–mineral mixture

Light, grayish-brown, silt loam

Dark brown firm clay

Deciduous Forest Soil
(humid, mild climate)

Acid litter and humus

Light-colored and acidic

Humus and iron and aluminum compounds

Coniferous Forest Soil
(humid, cold climate)

cause leaching of nutrients, and a relatively infertile soil. Extreme aridity may result in the absence of any soil development.

The amount of precipitation received affects leaching, eluviation, and illuviation, and thereby rates of soil formation and horizon development. Evaporation is also an important factor. Salt and gypsum deposits from the upward migration of capillary water are more extensive in hot, dry regions than in colder, dry regions (see again Fig. 9.30).

Land Surface

The slope of the land, its relief, and its aspect all influence soil development. Steep slopes are subject to rapid runoff of water, so, there is less infiltration on steeper slopes, which inhibits soil development. In addition, rapid runoff can erode steep slopes faster than soil can develop on them. On gentler slopes, more water tends to be available for soil development and vegetation growth. Erosion is also not as intense, so that well-developed soils typically form on flat or gently sloping land. Outside of the tropics in the northern hemisphere south-facing slopes receive the sun's rays at a steeper angle and are therefore warmer

and drier. This applies to north-facing slopes in the southern hemisphere. Local variations in soil depth, texture, and profile development result from these microclimate differences.

Time

Soils have a tendency to develop toward a state of equilibrium with their environment. A soil is called "mature" when it has reached such a condition of equilibrium. Mature soils have well-developed horizons that indicate the conditions under which they formed. Young or "immature" soils are in the early stages of the development process, and they typically have poorly developed horizons or perhaps none at all (■ Fig. 9.31). As soils develop over time, the influence of their parent material decreases and they increasingly reflect their climate and vegetative environments. On a global scale, climate typically has the greatest influence on soils, provided sufficient time has passed for the soils to become well developed.

The importance of time in soil formation is especially clear in soils developed on transported parent materials. Generally, these deposits have not been exposed to weathering long enough

■ **FIGURE 9.31** The time that a soil has been developing is important to its composition and physical character. Given enough time and the proper environmental conditions, soils will become more maturely developed with a deeper profile and stronger horizon development.

What major changes occur as the soil illustrated here becomes better developed over time?

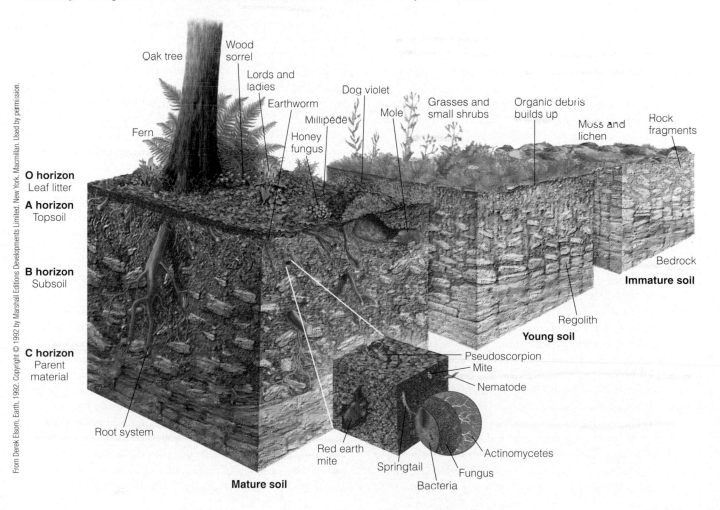

From Derek Elsom, Earth, 1992. Copyright © 1992 by Marshall Editions Developments Limited. New York. Macmillan. Used by permission.

for a mature soil to develop. Deposition occurs in a variety of settings: on river floodplains where the accumulating sediment is known as *alluvium*; downwind from dry areas where dust settles to form blankets of wind-deposited silts, called *loess*; and in volcanic regions showered by ash and covered by lava. Ten thousand years ago, glaciers withdrew from vast areas, leaving behind jumbled deposits of rocks, sand, silt, and clay.

Because of the great variability of materials and processes involved in soil formation, there is no fixed amount of time that it takes for a soil to become mature. It has been estimated that it takes about 500 years to develop 1 inch (2.54 cm) of soil in the agricultural regions of the United States. Generally, though, it takes thousands of years for a soil to reach maturity.

Soil-Forming Regimes and Classification

The characteristics that make major soil types distinctive and different from one another result from their **soil-forming regimes,** which vary mainly because of differences in climate and vegetation. At the broadest scale of generalization, climate differences produce three primary soil-forming regimes: laterization, podzolization, and calcification.

Laterization

Laterization is a soil-forming regime that occurs in humid tropical and subtropical climates as a result of high temperatures and abundant precipitation. These climatic environments encourage the rapid breakdown of rocks and decomposition of nearly all minerals. This soil type is known as **laterite**, and these soils are generally reddish in color from iron oxides (see again Fig. 9.29). Laterite, which means "brick-like," is quarried in tropical areas for building materials.

Despite the dense vegetation that is typical of these climate regions, little humus is incorporated into the soil because the plant litter decomposes so rapidly. Laterites do not have an *O* horizon, the *A* horizon loses fine soil particles, and most minerals are leached except for insoluble iron and aluminum compounds. As a result, the topsoil is reddish, coarse textured, and tends to be porous (■ Fig. 9.32). The *B* horizon in a lateritic soil has a heavy concentration of illuviated materials.

In the tropical forests, soluble nutrients released by weathering are quickly absorbed by vegetation, which eventually returns them to the soil where they are reabsorbed by plants. This rapid cycling of nutrients prevents them from being completely leached away of bases, leaving the soil only moderately acidic. Removal of vegetation permits the total leaching of bases, resulting in the formation of crusts of iron and aluminum compounds (laterites), as well as accelerated erosion of the *A* horizon.

Laterization is a year-round process because of the small seasonal variations in temperature or soil moisture in the humid tropics. This continuous activity and strong weathering of parent material cause some tropical soils to develop to depths of as much as 8 meters (25 ft) or more.

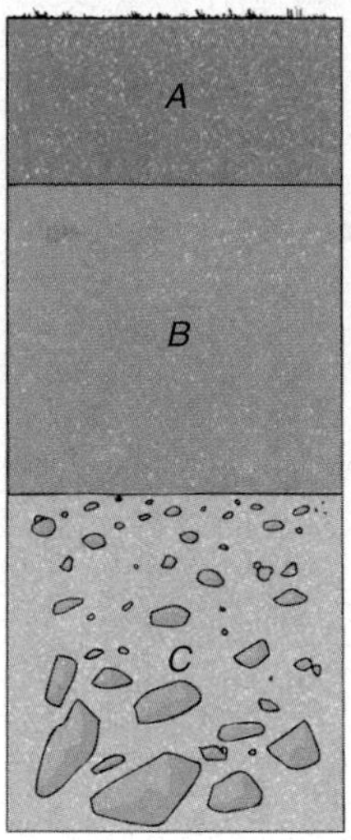

■ **FIGURE 9.32**   Soil profile horizons in a laterite. Laterization is a soil development process that occurs in wet tropical and equatorial climates that experience warm temperatures all year.

Little or no organic debris, little silica, much residual iron and aluminum, coarse texture

Some illuvial bases, much accumulated laterite

Much of the soluble material lost to drainage

Podzolization

Podzolization occurs mainly in the high middle latitudes where the climate is moist with short, cool summers and long, severe winters. The coniferous forests of these climate regions are an integral part of the podzolization process.

Where temperatures are low much of the year, microorganism activity is reduced enough that humus does accumulate; however, because of the small number of animals living in the soil, there is little mixing of humus below the surface. Leaching and eluviation by acidic solutions remove the soluble bases and aluminum and iron compounds from the *A* horizon (■ Fig. 9.33). The remaining silica gives a distinctive ash-gray color to the *E* horizon (*podzol* is derived from a Russian word meaning "ashy"). The needles that coniferous trees drop contribute to the soil acidity.

Podzolization can take place outside the typical cold, moist climate regions if the parent material is highly acidic— for example, on the sandy areas common along the East Coast of the United States. The pine forests that grow in such conditions return acids to the soil, promoting podzolization.

Calcification

A third distinctive soil-forming regime is called **calcification.** In contrast to both laterization and podzolization, which require humid climates, calcification occurs in regions where evapotranspiration significantly exceeds precipitation. Calcification is important in the climate regions where moisture penetration is shallow. The subsoil is typically too dry to support tree growth and shallow-rooted grass or shrubs are the primary forms of vegetation. Calcification is enhanced as grasses use calcium, drawing it up from lower soil layers and returning it to the soil when the annual grasses die. Grasses and their dense root networks provide large amounts of organic matter, which is mixed deep into the soil by burrowing animals. Middle-latitude grassland

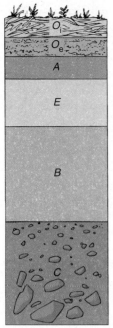

Well-developed organic horizons

Thin, dark

Badly leached, light in color, largely Si

Darker than E; often colorful; accumulations of humus; Fe, Al, N, Ca, Mg, Na, K

Some Ca, Mg, Na, and K leached down from B is lost to lateral movement of water below water table

■ **FIGURE 9.33** Soil profile horizons in a podzol. Podzolization occurs under cool, wet climates in regions of coniferous trees or in boggy environments, and forms very acidic soils.

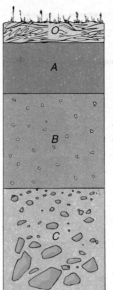

Dark color, granular structure, high content of residual bases

Lighter color, very high content of accumulated bases, caliche nodules

Relatively unaltered, rich in base supply, virtually no loss to drainage water

■ **FIGURE 9.34** Soil profile horizons in a calcified soil. Calcification is a soil development process that is most prominent in cool to hot subhumid or semiarid climate regions, particularly in grassland areas, but also occurs in deserts.

soils are rich in humus and are the world's most productive agricultural soils. The desert soils of the American West generally have no humus, and the rise of capillary water can leave deposits of calcium carbonate and salt at the surface.

In many dry regions, the air is often loaded with alkali dusts such as calcium carbonate ($CaCO_3$). When calm conditions prevail or when it rains, the dust settles and accumulates in the soil. The rainfall produces an amount of soil water that is just sufficient to translocate these materials to the B horizon (■ Fig. 9.34). Over hundreds to thousands of years, the $CaCO_3$-enriched dust concentrates in the B horizon, forming hard layers of *caliche*. Much thicker accumulations called *calcretes* form by the upward (capillary) movement of dissolved calcium in groundwater when the water table is near the surface.

Regimes of Local Importance

Two additional localized soil-forming regimes merit attention. Both characterize areas with poor drainage although they occur under very different climate conditions. **Salinization,** the concentration of salts in the soil, is often detrimental to plant growth (■ Fig. 9.35). Salinization occurs in stream valleys, interior basins, and other low-lying areas, particularly in arid regions with high groundwater tables. The high groundwater levels can be the result of water from adjacent mountain ranges, stream flow originating in humid regions, or a wet–dry seasonal precipitation regime. Salinization can also be a consequence of intensive irrigation under arid or semiarid conditions. Rapid evaporation leaves behind a high concentration of soluble salts and may destroy a soil's agricultural productivity.

Another localized soil regime, **gleization,** occurs in poorly drained areas in cold, wet environments. Gley soils, as they

■ **FIGURE 9.35** The white deposits on this field in Colorado were caused by salinization. Surface salinity resulted from upward capillary movement of water and evaporation at the surface causing deposits of salt. The soil cracks indicate shrinkage caused by evaporative drying of the soil.
What negative soil effects can result when humans practice irrigated agriculture in regions that experience great evaporation rates?

USDA/NRCS/Tim McCabe

are called, are typically associated with peat bogs where the soil has a humus accumulation overlying a blue-gray layer of gummy, water-saturated clay. In poorly drained regions that were formerly glaciated, such as Ireland, Scotland, and northern Europe, peat has long been harvested and used as a source of fuel.

Soil Classification

Soils, like climates, can be classified by their characteristics and mapped by their spatial distributions. In the United States, the Soil Survey Division of the Natural Resources Conservation Service (NRCS), a branch of the Department of Agriculture, is responsible for soil classification (termed *soil taxonomy*) and mapping.

Soil classifications are published in **soil surveys,** books that outline and describe the kinds of soils in a region and include maps that show the distribution of soil types, usually at the county level. These documents, available for most parts of the United States, are useful references for factors such as soil fertility, irrigation, and drainage.

The NRCS soil classification system is based on the development and composition of soil horizons. The largest division in the classification of soils is the **soil order,** of which 12 are recognized by the NRCS. To provide greater detail, soil orders can also be further divided into suborders, and four other increasingly localized subdivisions. The NRCS soil classification and illustrations of each soil order can be found in Appendix D.

The NRCS system uses names derived from root words from languages such as Latin, Arabic, and Greek to refer to the different soil categories. The names and the classification are precise in describing the distinguishing characteristics of each soil type. Some soil orders reflect regional climate conditions. Other soil orders, however, reflect their recency or type of parent material, and their distribution does not conform to climate regions. When examining a soil for classification under the NRCS system, particular attention is paid to characteristic horizons and textures.

Ecosystems and Soils: Critical Natural Resources

It is the responsibility of all of us to help protect our world's ecosystems and valuable soils. Soil erosion, degradation, depletion, and mismanagement of environments are of great global concern today (■ Fig. 9.36). The detrimental impacts of these factors on soils have negative consequences on the natural ecology and on the agricultural productivity that humankind depends upon. These problems, however, often have reasonable solutions (see again Fig. 9.18). Conserving soils and maintaining soil fertility are critical challenges, essential to natural environments, and to our planet's life-giving resources. The information and knowledge gained from studying biogeography and soils can help us learn to work in concert with nature to sustain and improve life on Earth.

■ **FIGURE 9.36**   Many areas of the world are experiencing the impact of soil degradation or loss through erosion, soil depletion, and many other factors. Compare this map to the population map in the back cover of the book.
Is there a general relationship between human population density and soil degradation?

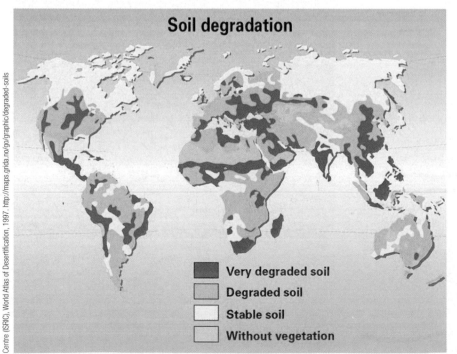

:: Terms for Review

ecosystem	corridor	soil grade
abiotic	range of tolerance	loam
producer	ecotone	infiltrate
consumer	symbiotic relationship	soil ped
herbivore	plankton	porosity
carnivore	phytoplankton	permeability
omnivore	zooplankton	pH scale
decomposer (detritivore)	soil	parent material
food chain	soil fertilization	soil profile
trophic level	capillary water	soil horizon
biomass	leaching	Cl, O, R, P, T
primary productivity	gravitational water	residual parent material
secondary productivity	eluviation	transported parent material
habitat	illuviation	soil-forming regime
ecological niche	stratification	laterization
generalist	humus	laterite
specialist	soil texture	podzolization
plant community	clayey	calcification
plant succession	clay	salinization
climax community	silty	gleization
mosaic	silt	soil survey
matrix	sandy	soil order
patch	sand	

:: Questions for Review

1. What are some of the reasons why the study of ecosystems is important in the world today?
2. What are the four basic components of an ecosystem?
3. What are the four main trophic levels?
4. How are productivity, energy flow, and biomass related to the sequence of trophic levels in a food chain?
5. What is plant succession, how do the two types differ, and in what ways has the original theory of succession been modified?
6. How do the terms *mosaic, matrix, patch, corridor,* and *ecotone* relate to each other and to a vegetation landscape?
7. Why is soil an outstanding example of integration and interaction among Earth's subsystems?

8. What factors are involved in the formation of soils? Which is most important on a global scale?
9. What are eluviation and illuviation, and what is the resulting impact on soil for each if these processes are carried to an extreme?
10. How is texture used to classify soils? Describe the ways scientists have classified soil structure.
11. What are the general characteristics of each horizon in a soil profile? How are soil profiles important to scientists?
12. Describe the three major soil-forming regimes.

:: Practical Applications

1. Kudzu is a climbing, woody, perennial vine that originates from Japan. From 1935 to 1950, farmers in the southeastern United States were encouraged to plant kudzu to help reduce soil erosion. Without natural enemies, this vine began to grow out of control. Since 1953, it was identified as a pest weed and efforts to eradicate this invader continue to this day. Today, kudzu inhabits about 30,000 km^2 of the southeastern U.S. From 1935 to present, what is the spread of this pest in square kilometers per year?
2. Refer to Figure 9.24. Using the texture triangle, determine the textures of the following soil samples.

	Sand	Silt	Clay
a.	35%	45%	20%
b.	75%	15%	10%
c.	10%	60%	30%
d.	5%	45%	50%

What are the percentages of sand, silt, and clay of the following soil textures? (*Note:* answers may vary, but they should total 100%.)
e. Sandy clay
f. Silty loam

Earth Materials and Plate Tectonics

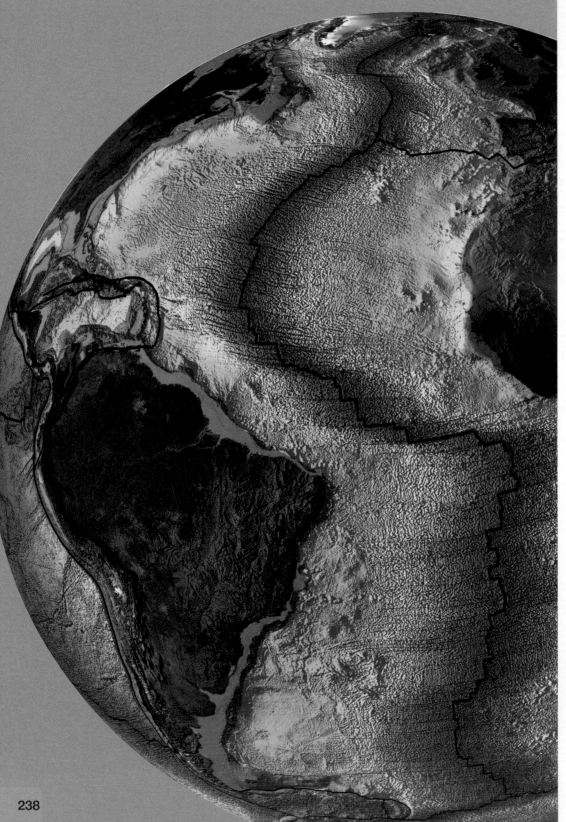

Relative age of rock material on the floor of the Atlantic Ocean. Rocks are youngest (red) along the extensive mid-Atlantic submarine mountain chain, and become progressively older with increasing distance from that midoceanic ridge. (Data by R.D. Muller, M. Sdrolias, C. Gaina, and W.R. Roest, 2008, doi 10.1029/2007GC001743)

E. Lim and J. Varner, CIRES & NOAA/ NGDC

:: Objectives

When you complete this chapter you should be able to:

- Compare the relative size and material properties of Earth's core, mantle, and crust.
- Recount the principal differences between oceanic and continental crust.
- Understand that the rigid lithosphere is dragged along with the flowing, plastic asthenosphere beneath it.
- Differentiate between minerals and rocks.
- Recall the definitions of the major categories of igneous, sedimentary, and metamorphic rocks.

- Explain the meaning of the rock cycle.
- Discuss the theory of plate tectonics.
- Provide evidence for the theory of plate tectonics.
- Describe major Earth features associated with plate convergence, divergence, and transform motion.
- Appreciate that the configuration of Earth's landmasses and environments has changed significantly over geologic time.

If we could travel back in time to view Earth as it was 90 million years ago, in addition to seeing now-extinct life forms, including dinosaurs, we would notice a very different spatial distribution of land and water than exists today. A vast inland sea cut across what is now the heartland of North America. Dinosaurs left their footprints in large trackways on floodplains of rivers that flowed out of the early Rocky Mountains. Forests grew above the present Arctic Circle. Grasses did not yet exist. The dramatic differences between then and now in the size, shape, and distribution of mountain ranges and water bodies as well as the differences in climate, soils, and organisms must be explained scientifically.

Like the atmosphere, hydrosphere, and biosphere, that part of the Earth system that lies beneath our feet—the lithosphere—undergoes change due to flows of energy and matter. Over long intervals of geologic time, the flows of energy and matter inside of Earth have significantly altered the size, shape, and location of major Earth surface features and environments. Internal Earth processes also help explain the present distribution of various rock types, mineral resources, and natural hazards. Processes originating within Earth create the structural foundation that surface-generated processes modify into the familiar landscapes in which we live.

Earth's Planetary Structure

Physical geography predominantly focuses on that part of the Earth system that lies at the interface of the atmosphere, hydrosphere, biosphere, and lithosphere, and these come together at Earth's surface. Still, basic knowledge of our planet's internal structure is needed to understand many aspects of Earth's natural surface characteristics.

From low-density gas molecules in the outermost layer of the atmosphere to high-density iron and nickel at the center of the planet, all of the gas, liquid, and solid matter comprising the Earth system is held within the system by gravitational attraction. Sir Isaac Newton taught us that the degree to which particles are drawn to each other by gravity depends on the mass of each particle, which is commonly expressed in units of grams or kilograms. The gravitational force of attraction is greater for objects that have a larger mass than for those with a smaller mass. Scientists commonly use density, which is mass per unit volume, to compare how the equal amounts of different materials vary in mass. Those types of Earth materials that have the greatest density have the greatest gravitational force of attraction, and as a result they have tended to concentrate close together at and near Earth's center.

Earth's interior is primarily composed of solids, the densest of the three states of matter. A less dense substance, liquid water, occupies most of Earth's surface thousands of kilometers above the densest substances that are deep inside the planet. Gases, with an even lower density, have the weakest gravitational attractive force and thus are held relatively loosely around Earth as the atmosphere, rather than within Earth or on its surface. Traveling outward from the center of the Earth system, there exists a density continuum (spectrum) that extends from the densest materials at the center of the planet to the least dense substances at the outer edge of the atmosphere. In previous chapters, we have learned a great deal about Earth's atmosphere as well as the hydrosphere and biosphere. In this chapter, we begin our study of the solid, or rock, portion of Earth—the lithosphere.

Earth has a radius of about 6400 kilometers (4000 mi). Through direct means by mining and drilling we have been able to penetrate and examine directly only an extremely small part of that distance. The lure of gold has taken people (specifically miners) to a depth of 3.5 kilometers (2.2 mi) in South Africa, while drilling for oil and gas has taken our machinery to a depth of about 12 kilometers (7.5 mi). These explorations have been helpful in providing information about the solid Earth's outermost layers, but they have just barely scratched the planet's surface. Scientists are continually working to understand the interior of Earth better. Extending scientific knowledge about the structure, composition, and processes operating within Earth helps us learn more about such lithospheric phenomena as earthquakes, volcanic eruptions, the formation of rocks and mineral deposits, and the origins of continents. It can even help us learn more about the origin of the planet itself.

Most of what we know about Earth's internal structure and composition has been deduced through indirect means by various forms of remote sensing. Thus far, the most important evidence that scientists have used to gain indirect knowledge of Earth's interior is the behavior of various shock waves, called **seismic waves,** as they travel through the planet (■ Fig. 10.1). Scientists generate some of these shock waves artificially with controlled explosions, but they mainly use evidence derived by tracking natural earthquake waves as they travel through Earth (■ Fig. 10.2a). By analyzing data collected over decades on worldwide travel patterns of earthquake waves, scientists have been able to develop a general model of Earth's interior. This information, supplemented by studies of Earth's magnetic field and gravitational pull, reveals a series of layers, or zones, in Earth's internal structure. These principal zones, from the center of Earth to the surface, are the core, mantle, and crust (Fig. 10.2b).

Earth's Core

Earth's innermost section, the **core,** contains one-third of Earth's mass and has a radius of about 3360 kilometers (2100 mi), which is larger than the planet Mars. Earth's core is under enormous pressure—several million times atmospheric pressure at sea level. Scientists have deduced that the core is composed primarily of iron and nickel, and consists of two distinct sections, the inner core and the outer core.

Earth's **inner core** has a radius of about 960 kilometers (600 mi). The speed of seismic waves traveling through the inner core shows that it is a solid with a very high material density of about 13 grams per cubic centimeter (0.5 lbs/in.³). The **outer core** forms a 2400 kilometer (1500 mi) thick band around the inner core. Rock matter at the top of the outer core has a density of about 10 grams per cubic centimeter (0.4 lbs/in.³). Because the outer core blocks the passage of a specific type of seismic wave, Earth scientists know that the outer core is molten (melted/liquid rock matter). The high density of both sections of Earth's core supports the notion that they are composed of iron and nickel.

Why is Earth's outer core molten while the inner core is solid? The answer involves the fact that the melting point of mineral matter depends not only on temperature but also on

■ **FIGURE 10.1**   Seismographs record earthquake waves for scientific study.

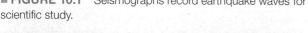

USGS Volcano Hazards Program

■ **FIGURE 10.2**   (a) Earth's internal structure is revealed by the refraction of P (primary) waves and the inability of S (secondary) waves to pass through the liquid outer core. (b) Cross section through Earth's internal structural zones.

How does the thickness of the crust compare with that of the mantle?

Earthquake focus

Reflected waves

P wave

S wave

Direct waves

A. Inner core
B. Outer core
C. Mantle
(a)   **D.** Crust

Refracted waves

Crust — 5–70 km (3–43 mi)

Mantle — 2885 km (1800 mi)

Outer core — 2400 km (1500 mi)

Inner core — 960 km (600 mi)

(b)

pressure. When rock matter is under higher pressure it melts at a higher temperature than when it is at a lower pressure. The material of the inner core is under higher pressure than the rock matter in the outer core. As a result of this great pressure, the material in the inner core remains solid despite its high temperature. Temperatures in the outer core are lower than in the inner core, but pressures are lower there as well, and this causes the outer core material to exist in the molten state. Internal temperatures are estimated to be 6900°C (12,400°F) at the very center of Earth, decreasing to 4800°C (8600°F) at the top of the outer core.

Earth's Mantle

With a thickness of approximately 2885 kilometers (1800 mi) and representing nearly two-thirds of Earth's mass, the **mantle** is the largest of Earth's interior zones. Earthquake waves that pass through the mantle indicate that it is composed of solid rock material, in contrast to the molten outer core that lies beneath it. It is also less dense than the core, with values ranging from 3.3 to 5.5 grams per cubic centimeter (0.12–0.20 lbs/in.^{3}). Although most of the mantle is solid, material near the top of the mantle especially displays characteristics of a *plastic solid,* meaning that the solid rock material can deform and flow very slowly, in this case at rates of a few centimeters per year. Scientists agree that the mantle consists of silicate rocks (high in silicon and oxygen) that also contain significant amounts of iron and magnesium.

The mantle is composed of various layers distinguished by different characteristics of strength and rigidity. Of special interest to us are the two uppermost layers. The outermost layer of the mantle, with an average thickness of about 100 kilometers (60 mi), is relatively cool, hard, and strong. This contrasts sharply with the hotter, weaker material in the next lower mantle layer that flows plastically in response to applied stress. The outermost layer of the mantle has a chemical composition like the rest of the mantle, but it responds to applied stress more like the overlying Earth layer, the crust. Together, the uppermost mantle layer and the crust form a structural unit called the **lithosphere.** The term lithosphere has traditionally been used to describe the entire solid Earth, as in Chapter 1 and earlier in this chapter. In recent decades, however, the term lithosphere has also been used in a separate, structural sense to refer to the more rigid outer shell of Earth, including the crust and the uppermost mantle layer (■ Fig. 10.3).

Extending down from the base of the lithosphere about 600 kilometers (375 mi) farther into the mantle is the **asthenosphere** (from Greek: *asthenias,* without strength), a thick layer of plastic mantle material. The material in the asthenosphere can flow both vertically and horizontally, dragging segments of the overlying, rigid lithosphere along with it. Earth scientists now believe that the energy for **tectonic forces,** large-scale forces that break and deform Earth's crust,

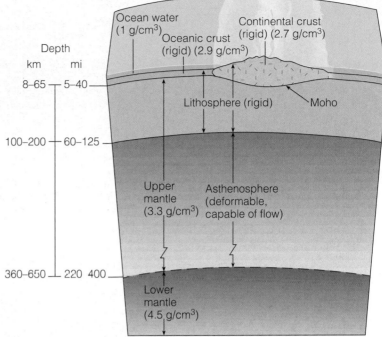

■ **FIGURE 10.3**   The lithosphere is the solid outer part of Earth, including the crust and the rigid, uppermost part of the mantle. Beneath the lithosphere is the plastic asthenosphere.

sometimes resulting in earthquakes and often responsible for mountain building, comes from movement within the plastic asthenosphere. Movement in the asthenosphere, in turn, is produced by thermal convection currents that occur in the rest of the mantle below the asthenosphere, and which are driven by heat from decaying radioactive materials in the planet's interior.

The interface between the mantle and the overlying crust is marked by a significant change of density, called a discontinuity, which is indicated by an abrupt increase in the velocity of seismic waves as they travel down through this internal boundary. Scientists call this zone the **Mohorovičić discontinuity,** or **Moho** for short, after the Croatian geophysicist who first detected it in 1909. The Moho does not lie at a constant depth but generally mirrors the surface topography, being deepest under mountain ranges where the crust is thick and rising to within 8 kilometers (5 mi) of the ocean floor (see again Fig. 10.3). No geologic drilling has yet penetrated through the Moho into the mantle, but an international scientific partnership, called the Integrated Ocean Drilling Program, is working on such a project. Rock samples eventually retrieved from cores drilled through the Moho will add greatly to our understanding of the composition and structure of Earth's lithosphere.

Earth's Crust

Earth's solid exterior is the **crust,** which is composed of a great variety of rock types that respond in diverse ways and at varying rates to surface processes. The crust is the only portion of the lithosphere of which Earth scientists have direct knowledge, yet it represents only about 1% of Earth's planetary

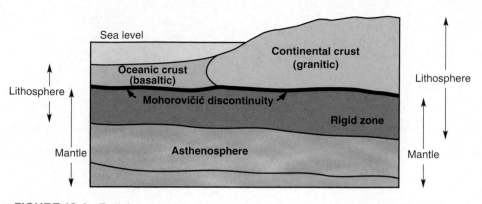

■ **FIGURE 10.4**   Earth has two distinct types of crust, oceanic and continental. The crust and rigid, uppermost mantle form the lithosphere. The plastic asthenosphere lies in the upper mantle beneath the lithosphere.

mass. As the outermost layer of the lithosphere, Earth's crust forms the ocean floor and the continents and is of primary importance in understanding surface processes and landforms.

The density of Earth's crust is significantly lower than that of the core and mantle, and ranges from 2.7 to 3.0 grams per cubic centimeter (0.10–0.11 lbs/in.^{3}). The crust is also extremely thin in comparison to the size of the planet. The two kinds of Earth crust, oceanic and continental, are distinguished by their location, composition, and thickness (■ Fig. 10.4). Crustal thickness varies from 3 to 5 kilometers (1.9–3 mi) in the ocean basins to as much as 70 kilometers (43 mi) under some continental mountain systems. The crust is relatively cold compared to the mantle, and behaves in a more rigid and brittle fashion, especially in its upper 10 to 15 kilometers (6–9 mi). The crust responds to stress by fracturing, crumpling, or warping.

Oceanic crust is composed of heavy, dark-colored, iron-rich rocks that are also high in silicon (Si) and magnesium (Mg). Its *basaltic* composition is described more fully in the next section. Compared to continental crust, oceanic crust is quite thin because its density (3.0 g/cm^3) is greater than that of continental crust (2.7 g/cm^3). Forming the vast, deep ocean floors as well as lava flows on all of the continents, basaltic rocks are the most common rocks on Earth.

Continental crust comprises the major landmasses on Earth that are exposed to the atmosphere. In addition to being less dense (2.7 g/cm^3) than oceanic crust, with an average thickness of 32 to 40 kilometers (20–25 mi) it is also much thicker than oceanic crust. At places where continental crust extends to high elevations in mountain ranges it also descends to great depths below the surface. Continental crust contains more light-colored rocks than oceanic crust does, and can be described as *granitic* in composition. The nature of granite, basalt, and other common rocks is discussed next.

Minerals and Rocks

Minerals are the building blocks of rocks. A **mineral** is an inorganic, naturally occurring substance represented by a distinct chemical formula and having a specific crystalline

TABLE 10.1
Most Common Elements in Earth's Crust

Element	Percentage of Earth's Crust by Weight
Oxygen (O)	46.60
Silicon (Si)	27.72
Aluminum (Al)	8.13
Iron (Fe)	5.00
Calcium (Ca)	3.63
Sodium (Na)	2.83
Potassium (K)	2.70
Magnesium (Mg)	2.09
Total	98.70

Source: J. Green, "Geotechnical Table of the Elements for 1953," *Bulletin of the Geological Society of America 64* (1953).

form. A **rock,** in contrast, is an aggregate of various types of minerals or an aggregate of multiple individual pieces (grains) of the same kind of mineral. In other words, a rock is not one single, uniform crystal. The most common elements found in Earth's crust, and therefore in the minerals and rocks that make up the crust, are oxygen and silicon, followed by aluminum and iron. As you can see in Table 10.1, the eight most common chemical elements in the crust, out of the more than 100 known, account for almost 99% of Earth's crust by weight. The most common minerals are combinations of these eight elements.

Minerals

Every mineral has distinctive and recognizable physical characteristics that aid in its identification. One of these characteristics is the nature of its crystalline form. Mineral crystals display consistent geometric shapes that express their molecular structure (■ Fig. 10.5). Halite, for example, which is used as table salt, is a soft mineral that has the specific chemical formula NaCl and a cubic crystalline shape. Quartz, calcite, fluorite, talc, topaz, and diamond are just a few examples of other minerals.

Jason Walz, National Park Service Photo

■ **FIGURE 10.5** Calcite mineral crystals.

Chemical bonds hold together the atoms and molecules that compose a mineral. The strength and nature of these chemical bonds affect the resistance and hardness of minerals and of the rocks that they form. Minerals with weak internal bonds undergo chemical alteration most easily. Charged particles, that is, ions, that form part of a molecule in a mineral may leave or be traded for other substances, generally weakening the mineral structure and forming the chemical basis of the breakdown of rocks at Earth's surface, called *rock weathering.*

Minerals can be categorized into groups based on their chemical composition. Certain elements, particularly silicon, oxygen, and carbon, combine readily with many other elements. As a result, the most common mineral groups are silicates, oxides, and carbonates. Calcite ($CaCO_3$), for example, is a relatively soft but widespread calcium-carbonate mineral that consists of one atom of calcium (Ca) linked together with a carbonate molecule (CO_3), which consists of one atom of carbon (C) plus three atoms of oxygen (O). The silicates, however, are by far the largest and most common mineral group, comprising 92% of Earth's crust.

The two most common elements in Earth's crust, oxygen and silicon (Si), frequently combine together to form SiO_2, which is called *silica. Silicate* minerals are compounds of oxygen and silicon that also include one or more metals and/or bases. Silica in its crystalline form is the mineral quartz, which has a distinctive prismatic crystalline shape. Quartz is one of the last silicate minerals to form from solidifying molten rock matter, and is a relatively hard and resistant mineral.

Rocks

Although a few rock types are composed of many particles of a single mineral, most rocks consist of several minerals (■ Fig. 10.6). Each constituent mineral in a rock remains separate and retains its own distinctive characteristics. The properties of the rock as a whole are a composite of those of its various mineral constituents. The number of rock-forming minerals that are common is limited, but they combine through a multitude of processes to produce an enormous variety of rock types (refer to Appendix E for information and pictures of common rocks mentioned in the text). Rocks are the fundamental building materials of the lithosphere. They are lifted, pushed down, and deformed by large-scale tectonic forces originating in the lower mantle and asthenosphere. At the surface, rocks are weathered and eroded, to be deposited as sediment elsewhere.

A mass of solid rock that has not been weathered is called *bedrock.* Bedrock may be exposed at the surface of Earth or it may be overlain by a cover of broken and decomposed rock fragments, called **regolith.** Soil may or may not have formed on the regolith. On steep slopes, regolith may be absent and the bedrock exposed if running water, gravity, or some other surface process removed the weathered rock fragments. A mass of exposed bedrock is often referred to as an *outcrop* (■ Fig. 10.7).

Geologists distinguish three major categories of rocks based on mode of formation. These rock types are igneous, sedimentary, and metamorphic.

■ **FIGURE 10.6** Granite contains intergrown mineral crystals of differing composition, color, and size that give the rock its distinctive appearance.

How does a rock differ from a mineral?

J. Petersen

■ **FIGURE 10.7**   Exposures of solid rock that are exposed (crop out) at the surface are often referred to as outcrops.
What physical characteristics of this rock outcrop have caused it to protrude above the general land surface?

Igneous Rocks   When molten rock material cools and solidifies it becomes an **igneous rock.** Molten rock matter below Earth's surface is called **magma,** whereas molten rock material at the surface is known specifically as **lava** (■ Fig. 10.8). Lava, therefore, is the only form of molten rock matter that we can see. Lava erupts from volcanoes or fissures in the crust at temperatures as high as 1090°C (2000°F). There are two major categories of igneous rocks: extrusive and intrusive.

Molten material that solidifies at Earth's surface creates **extrusive igneous rock,** also called *volcanic rock.* Extrusive igneous rock, therefore, is made from lava. Very explosive eruptions of molten rock material can cause the accumulation of fragments of volcanic rock, dust-sized or larger, that settle out of the air to form *pyroclastics* (from Greek: *pyros,* fire; *clastus,* broken), as a special category of extrusive rock (■ Fig. 10.9a). When molten rock beneath Earth's surface, that is, magma, changes to a solid (freezes), it forms **intrusive igneous rock,** also referred to as *plutonic rock* after Pluto, Roman god of the underworld. Igneous rocks are classified in terms of their mineral composition as well as the size of constituent minerals, which is referred to as *texture.* Igneous rocks vary in texture, chemical composition, crystalline structure, tendency to fracture, and presence or absence of layering.

Rocks composed of small-sized individual minerals not visible to the unaided eye are described as having a fine-grained texture, while those with large minerals that are visible without magnification are referred to as coarse-grained. Molten rock matter extruded on the surface cools very quickly—at Earth surface temperatures—and are fine-grained as a result of the brief time available for crystal growth prior to solidification. An extreme example is the extrusive rock *obsidian,* which cools so rapidly that it is essentially a glass (Fig. 10.9b). Large masses of intrusive rock matter solidifying deep inside Earth cool very slowly because surrounding rock

■ **FIGURE 10.8**   A flowing stream of molten lava appears reddish due to its very hot temperature. Adjacent, recently solidified lava looks black. These molten and solidified lava flows on the island of Hawaii are basaltic in composition.

(a)

(b)

■ **FIGURE 10.9** (a) Pyroclastic rocks are made of fragments ejected during a volcanic eruption. (b) Obsidian—volcanic glass—results when molten lava cools too quickly for crystals to form.

slows the loss of heat from the magma. Slow cooling allows more time for crystal formation prior to solidification. Exceptions include thin stringers of intrusive rocks and those that solidified close to the surface, which may cool rapidly and be fine-grained as a result.

The chemical composition of igneous rocks varies from *felsic*, which is rich in light-colored, lighter-weight minerals, especially silicon and aluminum (*fel* for the mineral feldspar;

si for silica), to *mafic*, which is lower in silica and rich in heavy minerals, such as compounds of magnesium and iron (*ma* for magnesium; *f* for *ferrum*, Latin for iron). Granite, a felsic, coarse-grained, intrusive rock, has the same chemical and mineral composition as *rhyolite*, a fine-grained, extrusive rock. Likewise, basalt is the dark-colored, mafic, fine-grained extrusive chemical and mineral equivalent to *gabbro*, a coarse-grained intrusive rock that cools at depth (■ Fig. 10.10).

■ **FIGURE 10.10** Igneous rocks are distinguished by texture (crystal size) and whether their mineral composition is mafic, felsic, or intermediate. Rocks with fine (small) crystals cooled rapidly at or near Earth's surface. Rocks that cooled slowly deep beneath the surface have coarse (large) crystalline texture. **What is the difference between granite and basalt?**

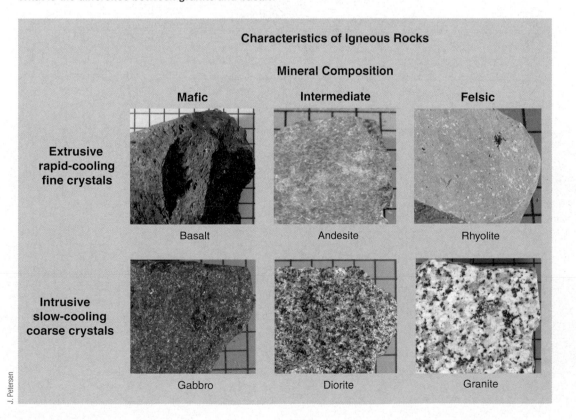

Characteristics of Igneous Rocks

Mineral Composition

	Mafic	Intermediate	Felsic
Extrusive rapid-cooling fine crystals	Basalt	Andesite	Rhyolite
Intrusive slow-cooling coarse crystals	Gabbro	Diorite	Granite

Igneous rocks also form with an intermediate composition, a rough balance between felsic and mafic minerals. The intrusive rock *diorite* and the extrusive rock *andesite* (named after the Andes where many volcanoes erupt lava of this composition) represent this intermediate composition (see again Fig. 10.10).

Many igneous rocks are fractured, often by multiple cracks that may be evenly spaced or arranged in regular geometric patterns. In the Earth sciences, simple fractures or cracks in bedrock are called **joints.** Although joints caused by regional stresses in the crust are common features in any type of rock, another way they develop in igneous rocks is by a molten mass shrinking in volume and fracturing as it cools and solidifies.

Sedimentary Rocks

As their name implies, **sedimentary rocks** are derived from accumulated sediment, that is, unconsolidated mineral materials that have been eroded, transported, and deposited. After the materials have accumulated, often in horizontal layers, pressure from the addition of material above compacts the sediment, expelling water and reducing pore space.

Cementation occurs when silica, calcium carbonate, or iron oxide precipitates between particles of sediment. The processes of compaction and cementation transform (lithify) sediments into solid, coherent layers of rock. There are three major categories of sedimentary rocks: clastic, organic, and chemical.

Broken fragments of solids are called *clasts* (from Latin: *clastus*, broken). In order of increasing size, clasts range from clay, silt, and sand to gravel, which is a general category for any fragment larger than sand (larger than 2.0 mm) and includes granules, pebbles, cobbles, and boulders. Most sediments consist of fragments of previously existing rocks, shell, or bone that were deposited on a river bed, beach, sand dune, lake bottom, the ocean floor, and other environments where clasts accumulate. Sedimentary rocks that form from fragments of preexisting rocks are called **clastic sedimentary rocks.**

Examples of clastic sedimentary rocks include *conglomerate, sandstone, siltstone,* and *shale* (■ Fig. 10.11). Conglomerate is a lithified mass of cemented, roughly rounded pebbles, cobbles, and boulders and may have clay, silt, or sand filling

■ **FIGURE 10.11**   Clastic sedimentary rocks are classified by the size and/or shape of the sediment particles they contain.
Why do the shapes and sizes of the sediments in sedimentary rocks differ?

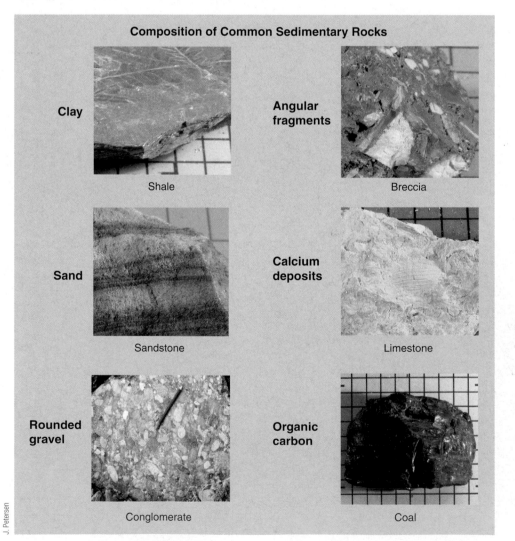

Composition of Common Sedimentary Rocks

Clay — Shale

Angular fragments — Breccia

Sand — Sandstone

Calcium deposits — Limestone

Rounded gravel — Conglomerate

Organic carbon — Coal

J. Petersen

in spaces between the larger particles. A somewhat similar rock composed of lithified fragments that are angular rather than rounded is called *breccia*. Sandstone consists of cemented sand-sized particles, most commonly grains of quartz. Sandstone is usually granular (has visible grains), porous, and resistant to weathering, but the cementing material influences its strength and hardness. If cemented by silica, sandstone tends to be more resistant to weathering than if it is cemented by calcium carbonate or iron oxide. Unlike sandstone, individual grains in siltstone, which is composed of silt-sized particles, are not easily visible with the unaided eye. Shale is produced by the compaction and cementation of very fine-grained sediments, primarily clays. Shale is often finely layered, smooth-textured, and has a low permeability. It is, however, easily cracked, broken apart, and eroded.

Sedimentary rocks may be further classified by their origin as either marine or terrestrial (continental). Marine sandstones typically form in nearshore coastal zones; terrestrial sandstones commonly originate in desert or floodplain environments on land. The nature and arrangement of sediments in a sedimentary rock provide a great deal of evidence for the kind of environment in which they were deposited, whether on a stream bed, a beach, or the deep-ocean floor.

Organic sedimentary rocks lithify from the remains of organisms, both plants and animals. *Coal* is created by the accumulation and compaction of partially decayed vegetation in acidic, swampy environments where water-saturated ground prevents oxidation and complete decay of the organic matter. The initial transformation of such organic material produces peat, which, when subjected to deeper burial and further compaction, is lithified to produce coal.

Other organic sedimentary rocks develop from the remains of organisms in lakes and seas. The remains of shellfish, corals, and microscopic drifting organisms called plankton sink to the bottom of such water bodies where they are compacted and cemented together. Rich in calcium carbonate ($CaCO_3$), they form a type of *limestone* that typically contains fossil shell and coral fragments (■ Fig. 10.12).

When the amounts of dissolved minerals in ocean and lake water reach saturation, they began to precipitate and build up as a deposit on the sea or lake bottom. These sediments may eventually lithify into **chemical sedimentary rocks.** Many fine-grained limestones form in this manner from chemical precipitates of calcium carbonate. Limestone, therefore, may vary from a jagged and cemented complex of visible shells or fossil skeletal material to a smooth-textured rock. Where magnesium is a major constituent along with calcium carbonate, the rock type is called *dolomite*. Because the calcium carbonate in limestone can slowly dissolve in water, limestone in arid or semiarid climates tends to be resistant, but in humid environments it tends to be weak.

Mineral salts that have reached saturation in evaporating seas or lakes will precipitate to form a variety of sedimentary deposits that are useful to humans. These include *gypsum* (used in wallboard), *halite* (common salt), and *borates*, which are important in hundreds of products such as fertilizer, fiberglass, detergents, and pharmaceuticals.

D. Sack

■ **FIGURE 10.12**   The White Cliffs of Dover, England. These striking, steep cliffs along the English Channel are made of chalky limestone from the skeletal remains of microscopic marine organisms.

Most sedimentary rocks display distinctive layering referred to as *stratification*. The many types of sedimentary deposits produce distinctive *strata* (layers) within the rocks. **Bedding planes,** the boundaries between sedimentary layers, indicate changes in energy in the depositional environment, but no real break in the sequence of deposition (■ Fig. 10.13). Where a marked mismatch and an irregular, eroded surface occur between beds, the contact between the rocks is called an **unconformity.** This indicates a gap in the section caused by erosion, rather than deposition, of sediment. One type of stratification, called *cross bedding*, is characterized by a pattern of thin layers that accumulated at an angle to the main strata, often reflecting shifts in direction of waves along a coast, currents in a stream, or winds over a sand dune (■ Fig. 10.14). All types of stratification provide evidence about the environment within which the sediments were deposited, and changes from

■ **FIGURE 10.13**   Bedding planes are boundaries between differing layers (strata) of sediment that mark some change in the nature of the deposited material. Numerous bedding planes, many represented by color changes, are visible in these rocks at Grand Canyon National Park, Arizona.
Where would the youngest strata is this photo be located?

■ **FIGURE 10.14**   Cross beds in sandstone at Zion National Park, Utah.
Under what circumstances might sand be deposited at a substantial angle, rather than as a more horizontal layer?

■ **FIGURE 10.15**   Vertical jointing of sandstone in Arches National Park, Utah, is responsible for creation of these vertical rock walls, called fins. Rock has been preferentially eroded along the joints. Only rock that was far from the locations of the joints remains standing.

one layer to the next reflect elements of the local geologic history. For example, a layer of sandstone representing an ancient beach may lie directly beneath shale layers that represent an offshore environment, suggesting that first this was a beach that the sea later covered.

Sedimentary rocks become jointed, or fractured, when they are subjected to crustal stresses after they lithify. The impressive "fins" of rock at Arches National Park, Utah, owe their vertical, tabular shape to joints in great beds of sandstone (■ Fig. 10.15).

Structures such as bedding planes and joints are important in the development of physical landscapes because they are weak points in the rock that weathering and erosion can attack with relative ease. Joints allow water to penetrate deeply into some rock masses, causing faster rock break down and removal than in the surrounding rock farther from these cracks.

Metamorphic Rocks   *Metamorphic* means "changed form." Enormous heat and pressure deep in Earth's crust can alter (metamorphose) an existing rock into a new rock type that is completely different from the original by recrystallizing the minerals without creating molten rock matter. Compared to the original rocks, the resulting **metamorphic rocks** are typically harder and more compact, have a reoriented crystalline structure,

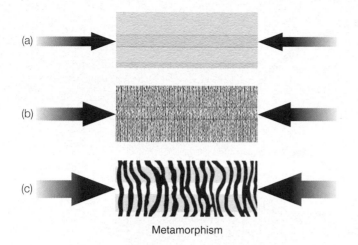

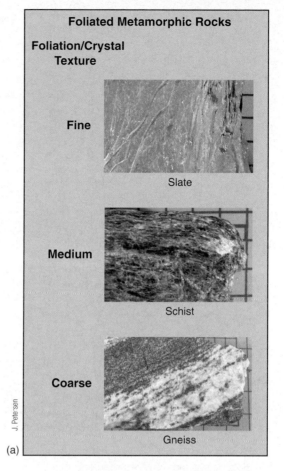

Foliated Metamorphic Rocks

Foliation/Crystal Texture

Fine — Slate

Medium — Schist

Coarse — Gneiss

(a)

J. Petersen

■ **FIGURE 10.16** During metamorphism, applied stress (arrows) can lead to an alignment of minerals, known as foliation. (a) Layered rocks under moderate pressure. (b) At greater pressure, metamorphism may realign minerals perpendicular to the applied stress, creating thin foliation layers and a platy structure. (c) Under even greater pressure, broader foliation layers may develop as wavy bands of light and dark minerals.
How does foliation differ from bedding planes?

and are more resistant to weathering. There are two major types of metamorphic rocks, based on the presence (foliated) or absence (nonfoliated) of platy surfaces or wavy alignments of light and dark minerals that form during metamorphism.

Metamorphism occurs most commonly where crustal rocks are subjected to great pressures by tectonic processes or deep burial, or where rising magma generates heat that modifies the nearby rock. Metamorphism causes minerals to recrystallize and, with enough heat and pressure, to reprecipitate perpendicular to the applied stress, forming platy surfaces (cleavage) or wavy bands known as **foliation** (■ Fig. 10.16). Some shales change to a hard metamorphic rock known as *slate*, which exhibits a tendency to break apart, or cleave, along smooth, flat surfaces that actually represent extremely thin foliation planes (■ Fig. 10.17a). Where foliation layers are moderately thin, individual minerals have a flattened but wavy, "platy" structure, and the rocks tend to flake apart along these bands. A common metamorphic rock with thin foliation layers is called *schist*.

Where foliation develops into broad mineral bands, the rock is extremely hard and is known as *gneiss* (pronounced "nice"). Coarse-grained rocks such as granite generally metamorphose into gneiss, whereas finer-grained rocks tend to produce schists.

Rocks that originally were composed of one dominant mineral are not foliated by metamorphism (Fig. 10.17b). Limestone is metamorphosed into much denser *marble*, and impurities in the rock can produce a beautiful variety of colors. Silica-rich sandstones fuse into *quartzite*. Quartzite is brittle, harder than steel, and almost inert chemically. It is virtually immune to chemical weathering and commonly forms cliffs or rugged mountain peaks after surrounding, less resistant rocks have been removed by erosion.

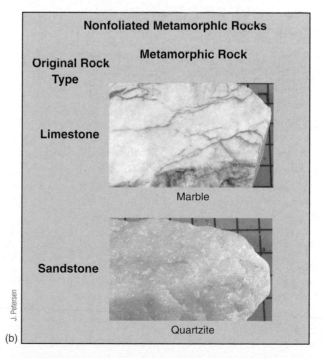

Nonfoliated Metamorphic Rocks

Metamorphic Rock

Original Rock Type

Limestone — Marble

Sandstone — Quartzite

(b)

J. Petersen

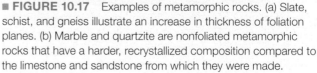

■ **FIGURE 10.17** Examples of metamorphic rocks. (a) Slate, schist, and gneiss illustrate an increase in thickness of foliation planes. (b) Marble and quartzite are nonfoliated metamorphic rocks that have a harder, recrystallized composition compared to the limestone and sandstone from which they were made.

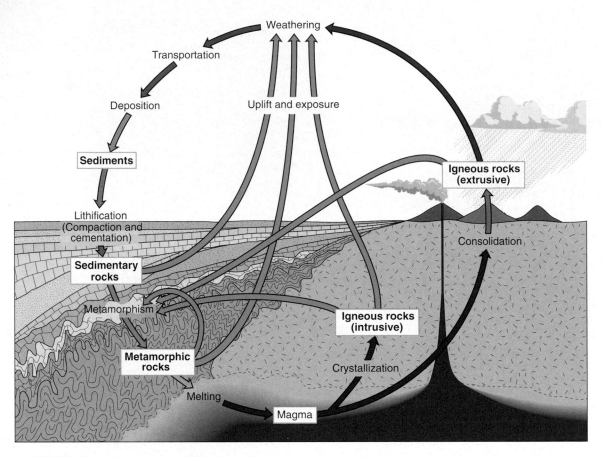

■ FIGURE 10.18 The rock cycle helps illustrate how igneous, sedimentary, and metamorphic rocks are formed. Note that some pathways bypass segments of the outer circle.
Can a metamorphic rock be metamorphosed?

The Rock Cycle Rock-forming materials do not necessarily remain in their initial form indefinitely but instead, over a long time, undergo processes of transformation. The *rock cycle* is a conceptual model for understanding processes that generate, alter, transport, and deposit mineral materials to produce different kinds of rocks (■ Fig. 10.18). The term *cycle* emphasizes that existing rocks supply the materials to make new and sometimes very different rocks. Whole existing rocks can be "recycled" to form new rocks. The geologic age of a rock is based on the time when it assumed its current state; metamorphism, melting, and lithification of sediments reset the age of origin.

A complete cycle is shown in the outer circle of Figure 10.18, but as indicated by the arrows that cut across the diagram, rock matter does not have to go through every step of the full rock cycle. For example, after igneous rocks are created by the cooling and crystallizing of magma or lava, they can weather into fragments that lithify into sedimentary rocks. Igneous rocks, however, could also be remelted and recrystallized to make new igneous rocks, or changed into metamorphic rocks by heat and pressure. Sedimentary rocks consist of particles and deposits derived from any of the three basic rock types. Metamorphic rocks can be created by means of heat and pressure changing any preexisting rock—igneous, sedimentary, or metamorphic—into a new rock type. In addition, with sufficient heat, metamorphic

rocks can melt completely into magma, eventually cooling to form igneous rocks. The rock cycle includes all the possible pathways for the recycling of rock matter over time.

Plate Tectonics

Scientists in all disciplines constantly search for broad explanations that shed light on the detailed facts, recurring patterns, and interrelated processes that they observe and analyze. Sometimes it requires years to develop, test, and refine a scientific concept to the point where it is more fully understood and broadly acceptable. As data and information are gathered and analyzed, new methods and technologies contribute to the process of testing hypotheses via the scientific method, and bit by bit an acceptable explanatory framework emerges. This was the case in the 20th century with the idea that segments of Earth's outer shell undergo changes in location and orientation over long periods of time.

Most of us have probably noticed on world maps that the Atlantic coasts of South America and Africa look as if they could fit together. In fact, if we could slide them together, several widely separated landmasses on Earth appear as though they would fit alongside each other without large gaps or

■ **FIGURE 10.19** The close fit of the edges of the continents that today border the Atlantic Ocean is a major basis for Wegener's continental drift hypothesis.

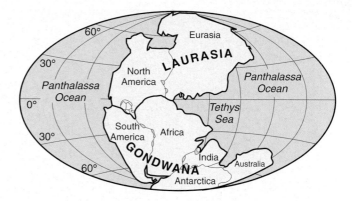

■ **FIGURE 10.20** The supercontinent of Pangaea included all of today's major landmasses joined together. Pangaea later split to make Laurasia and Gondwana. Further plate motion has produced the continents as they are today.

overlaps (■ Fig. 10.19). In the early 1900s, Alfred Wegener, a German climatologist, used his observations of the fit of the continents along with the spatial distribution of fossils, location of rock types, trends of mountain ranges, and glacial evidence, to propose the theory of **continental drift,** the idea that continents have shifted their positions during Earth history. Wegener hypothesized that all of the continents had once been part of a single supercontinent, which he called Pangaea, that later divided into two large landmasses, one in the Southern Hemisphere (Gondwana), and one in the Northern Hemisphere (Laurasia) (■ Fig. 10.20). He suggested that Gondwana and Laurasia later broke apart to produce the present continents, which eventually drifted to their current positions.

The reaction of most of the scientific community to Wegener's proposal ranged from skepticism to ridicule. A major objection to the notion of continental drift was that no one could provide an acceptable explanation for the energy that would be needed to break apart huge continental landmasses and slide them through the ocean over Earth's surface. It was almost a half century after Wegener first presented his ideas that Earth scientists began to seriously consider the notion of slowly moving landmasses. In the late 1950s and 1960s, new information appeared from research in oceanography, geophysics, and other Earth sciences, aided by sonar, radioactive dating of rocks, and improvements in equipment for measuring Earth's magnetism. These scientific efforts discovered

much new evidence that pointed to the horizontal movement of segments of the entire structural lithosphere, including the uppermost mantle, oceanic crust, and continental crust, rather than just the continents as Wegener had suggested.

Plate tectonics is the modern, comprehensive theory that explains the movement of the lithosphere. This rigid and brittle outer shell of Earth is broken into multiple sections called **lithospheric plates** that rest on, and are carried along with, the flowing plastic asthenosphere (■ Fig. 10.21). *Tectonics* involves large-scale forces originating within Earth that cause parts of the lithosphere to move around. In plate tectonics, the lithospheric plates move as distinct and discrete units. In some places they pull away from each other (diverge), in other places they push together (converge), and elsewhere they slide alongside each other (move laterally). To understand how and why plate tectonics operates, we must consider the scientific evidence that was gathered in the development and testing of this theory.

Seafloor Spreading and Convection Currents

In the 1960s, intensive study and mapping of the ocean floor yielded several key lines of evidence related to plate tectonics. First, detailed mapping conducted on extensive submarine mountain chains, called **midoceanic ridges,** revealed spatial trends remarkably similar to those of the continental coastlines. Second, it was discovered in the Atlantic and Pacific Oceans that basaltic seafloor displayed parallel bands of matching patterns of magnetic properties in rocks of the same age but on opposite sides of midoceanic ridges. Third, scientists made the surprising discovery that although some continental rocks are 3.6 billion years old, rocks on the ocean floor are all geologically young, having been in existence less than 250 million years. Fourth, the oldest rocks of the seafloor lie in deep trenches either beneath the deepest ocean waters or close to the continents, and rocks become progressively younger toward the midoceanic ridges where the youngest basaltic rocks are found (■ Fig. 10.22). Finally, temperatures

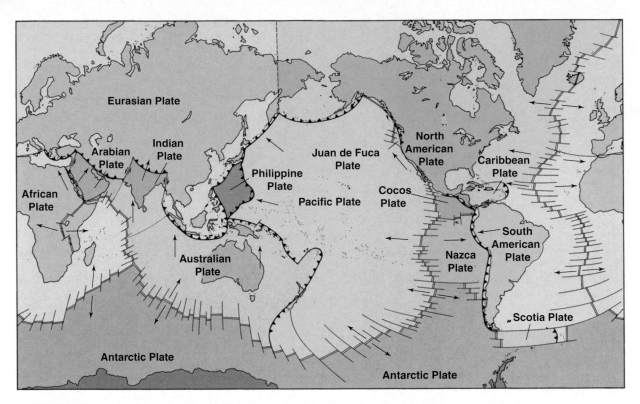

■ **FIGURE 10.21** · Earth's major lithospheric plates, also called tectonic plates, and their general directions of movement. Most tectonic and volcanic activity occurs along plate boundaries, where the large segments separate, collide, or slide past each other. Barbs indicate where the edge of one plate is moving downward (subducting) under another plate.

Does every lithospheric plate include a continent?

■ **FIGURE 10.22**   The global oceanic ridge system and the age of the seafloor, with red representing the youngest and blue the oldest seafloor. Detailed mapping and study of the ocean floors yielded much evidence to support the theory of plate tectonics by identifying the process of seafloor spreading.

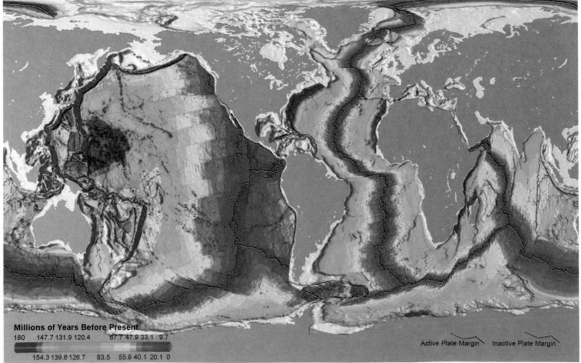

of rocks on the ocean floor vary significantly, being hottest near the ridges and becoming progressively cooler farther away.

Only one logical explanation emerged to fit this evidence. It became apparent that new oceanic crust is formed at the midoceanic ridges while older oceanic crust is destroyed in the deep trenches. The emergence of new oceanic crust is associated with the movement of large areas of seafloor in both directions away from the midoceanic ridges. This phenomenon is called **seafloor spreading** (■ Fig. 10.23). The young age of oceanic crust results from the creation of new basaltic rock at the midoceanic ridges and its movement with the lithospheric plates toward ocean basin margins where the older rock is remelted and destroyed. As the molten basaltic rock cools and crystallizes in the seafloor, the iron minerals that it contains become magnetized in a manner that records the orientation of Earth's magnetic field at the time. As a result, the iron-rich basaltic rocks of the seafloor have preserved symmetrically on both sides of a midoceanic ridge the historical record of Earth's magnetic field, including polarity reversals when the north and south magnetic poles exchange position.

By acknowledging that the entire rigid and brittle lithosphere, rather than just continental crust, is broken into multiple sections that move, the theory of plate tectonics includes a plausible explanation of the driving force for the movement, the explanation which had eluded Wegener. The mechanism is convection in the mantle. Hot mantle material travels upward toward Earth's surface and cooler material moves downward as huge subcrustal convection cells (■ Fig. 10.24). Mantle material in the convection cells rises to the asthenosphere where it spreads laterally and flows plastically in opposite directions, dragging the lithospheric plates with it. Pulling apart the brittle lithosphere breaks open a midoceanic ridge, which marks the boundary between two separate plates. Magma wells up into the fractures, cooling to form new crust. As the convective motion continues, the crust travels away from the ridges. Rigid lithospheric plates separate along midoceanic ridges at an average rate of 2 to 5 centimeters (1–2 in.) per year. In a time frame of up to 250 million years, older oceanic crust is consumed in the deep trenches near other plate boundaries where sections of the lithosphere meet and are recycled into Earth's interior.

Tectonic Plate Movement

Plate tectonics theory enables physical geographers to better understand not only our planet's ancient geography but also the modern global distributions and spatial relationships among such diverse, but often related, phenomena as earthquakes, volcanic activity, zones of crustal movement, and major landform features (■ Fig. 10.25). We will next examine the three ways in which lithospheric plates relate to one another along their boundaries as a result of tectonic movement: by pulling apart, pushing together, or sliding alongside each other.

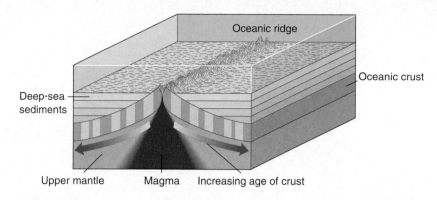

■ **FIGURE 10.23**   Seafloor spreading at a midoceanic ridge produces new seafloor.

■ **FIGURE 10.24**   Convection is the mechanism for plate tectonics. Heat causes convection currents of material in the mantle to rise toward the base of the solid lithosphere where the flow becomes more horizontal. As the asthenosphere undergoes its slow, lateral flow, the overlying lithospheric plates are carried along.
Why is plate tectonics a better name than continental drift for the lateral movement of Earth's solid outer shell?

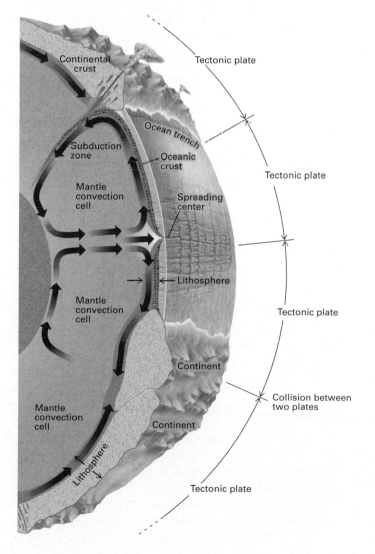

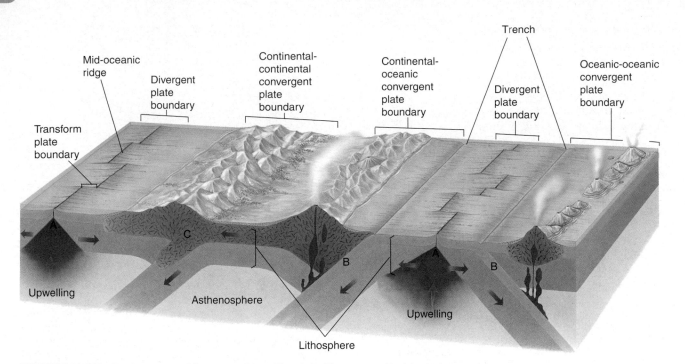

■ **FIGURE 10.25 Environmental Systems: Plate Tectonic Movement** Earth's plate tectonic system is powered by heat energy from inside Earth causing convection cells in the mantle. As lithospheric plates move, they interact with adjoining plates, forming different boundary types that display distinct landform features. Spreading centers (A) are divergent plate boundaries that have new crustal material emerging along active rift zones, eventually pushing older rock progressively away from the boundary in both directions. Subduction zones (B) occur where two plates converge, with the margin of at least one of them consisting of oceanic crust. The plate margin with the denser oceanic crust is subducted beneath the less dense plate, either continental or oceanic crust. Deep ocean trenches and either volcanic mountain ranges (continental crust) or island arcs (oceanic crust) lie along subduction zones. Continental collision zones (C) are places where two continental plates collide. Massive nonvolcanic mountains are built in those locations as the crust thickens because of compression.

Plate Divergence

The pulling apart of plates, as occurs with seafloor spreading, is tectonic **plate divergence** (see again Fig. 10.23). Tectonic forces that act to pull rock masses apart cause the crust to thin and weaken. Shallow earthquakes are often associated with this crustal stretching, and basaltic magma from the mantle wells up along crustal fractures. When oceanic crust is pulled apart the process creates new ocean floor as the plates are pulled away from each other along a spreading center. The formation of new crust in these spreading centers gives the label *constructive plate margins* to these zones. In some places, volcanoes, like those of Iceland, the Azores, and Tristan da Cunha, mark such boundaries (■ Fig. 10.26).

Most plate divergence occurs along oceanic ridges, but this process can also break apart continental crust, eventually reducing the size of the continents involved (■ Fig. 10.27a). The Atlantic Ocean floor formed as the continent that included South America and Africa broke up and moved apart 2 to 5 centimeters (1–2 in.) per year over millions of years. The Atlantic Ocean continues to grow today at about the same rate. The best modern example of divergence on a continent is the rift valley system of East Africa, stretching from the Red Sea south to Lake Malawi. Crustal blocks that have

moved downward with respect to the land on either side, with lakes occupying many of the depressions, characterize the entire system, including the Sinai Peninsula and the Dead Sea. Measurable widening of the Red Sea suggests that it may be the beginning of a future ocean that is forming between Africa and the Arabian Peninsula, similar to the young Atlantic between Africa and South America about 200 million years ago (Fig. 10.27b).

Plate Convergence

A wide variety of crustal activity occurs at areas of tectonic **plate convergence.** Despite the relatively slow rates of plate movement in terms of human perception, incredible energy is involved as two plates collide. Zones where plates are converging mark locations of major, and some of the tectonically more active, landforms on our planet. Deep trenches, volcanic activity, and mountain ranges may arise at convergent plate boundaries, depending on the type of crust involved in the plate collision. The distinctive spatial arrangement of these features worldwide can best be understood within the framework of plate tectonics.

If one or both margins of a convergent plate boundary consist of oceanic crust, the margin of one plate—always one composed of oceanic crust—is forced deep below the

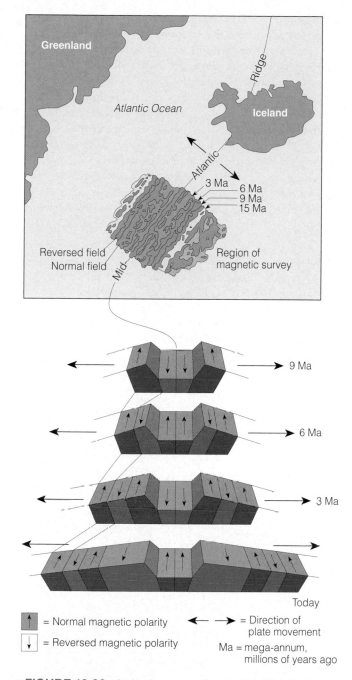

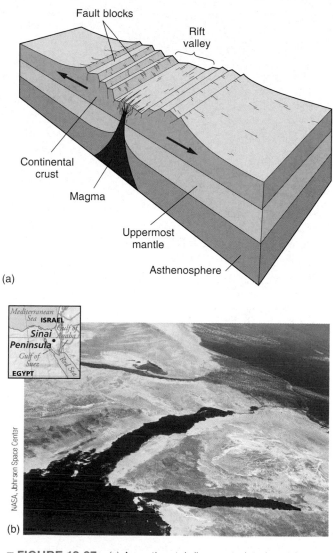

(a)

(b)

NASA, Johnson Space Center

■ **FIGURE 10.27** (a) A continental divergent plate boundary breaks continents into smaller landmasses. (b) The roughly triangular-shaped Sinai Peninsula, flanked by the Red Sea to the south (lower left), Gulf of Suez on the west (photo center), and Gulf of Aqaba toward the east (lower right), illustrates the breakup of a continental landmass. The Red Sea rift and the narrow Gulf of Aqaba are both zones of spreading.

■ **FIGURE 10.26** Iceland represents part of the Mid-Atlantic Ridge where it extends above sea level to form a volcanic island. The "striped" pattern of polarity reversals documented in the basaltic rocks along the Mid-Atlantic Ridge helped scientists understand the process of seafloor spreading.

surface in a process called **subduction.** Deep ocean trenches, such as the Peru–Chile trench and the Japanese trench, occur where oceanic crust is dragged downward in this way. The subducting plate is heated and rocks are melted as it plunges downward into the mantle. As the subducting plate grinds downward, enormous friction is produced, which explains the occurrence of major earthquakes in these regions.

Where oceanic crust collides with continental crust, the oceanic crust, which is denser, is subducted beneath the less dense continental crust (■ Fig. 10.28). This is the situation along South America's Pacific coast, where the Nazca plate subducts beneath the South American plate, and in Japan, where the Pacific plate dips under the Eurasian plate. As oceanic crust, and the lithospheric plate of which it forms a part, is subducted, it descends into the asthenosphere to be melted and recycled into Earth's interior. Frequently, hundreds of meters of sediments deposited at continental margins are carried along down into the deep trenches. As these melt, the resulting magma migrates upward into the overriding plate. Where molten rock reaches the surface, it produces a series of volcanic

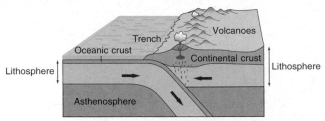

■ **FIGURE 10.28** An oceanic–continental convergent plate boundary where continent and seafloor collide. An example is the west coast of South America, where collision has formed the Andes and an offshore deep ocean trench.

peaks, as in the Cascade Range of the northwestern United States. Rocks can also be squeezed and contorted between colliding plates, becoming uplifted and greatly deformed or metamorphosed. The great mountain ranges, such as the Andes, are produced at convergent plate margins by these processes.

Where oceanic crust lies on either side of a convergent plate boundary, the plate with the denser oceanic crust will subduct below the other plate. Volcanoes may also develop at this type of boundary, creating major volcanic **island arcs** on the overriding plate. The Aleutians, the Kuriles, and the Marianas are all examples of island arcs lying near oceanic trenches that border the Pacific plate.

Continental crust converging with continental crust is termed **continental collision,** and causes two continents or major landmasses to fuse or join together, creating a new larger landmass (■ Fig. 10.29). This process closes an ocean basin that once separated the colliding landmasses, and therefore has also been called *continental suturing.* The crustal thickening that occurs along this type of plate boundary generally produces major mountain ranges due to massive folding and crustal block movement, rather than volcanic activity. The Himalayas, the Tibetan Plateau, and other high Eurasian ranges formed in this way as the plate containing the Indian subcontinent collided with Eurasia some 40 million years ago. India is still pushing into Asia today to produce the highest mountains in the world. In a similar fashion, the Alps were created as the African plate was thrust against the Eurasian plate.

■ **FIGURE 10.29** Continental collision along a convergent plate boundary fuses two landmasses together. The Himalayas, the world's highest mountains, were formed in this way when India drifted northward to collide with Asia.

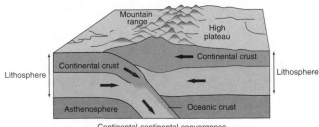

Transform Plate Movement Lateral sliding along plate boundaries, called **transform movement,** occurs where plates neither pull apart nor converge but instead slide past each other as they move in opposite directions. Such a boundary exists along the San Andreas Fault zone in California (■ Fig. 10.30). Mexico's Baja peninsula and Southern California are west of the fault on the Pacific plate. San Francisco and other parts of California east of the fault zone are on the North American plate. In the fault zone, the Pacific plate is moving laterally northwestward in relation to the North American plate at a rate of about 8 centimeters (3 in.) a year (80 km or about 50 mi per million years). If movement continues at this rate, Los Angeles will lie alongside

■ **FIGURE 10.30** Along this lateral plate boundary, marked by the San Andreas Fault in western North America, the Pacific plate moves northwestward relative to the North American plate. Note that north of San Francisco the boundary type changes.
What boundary type is found north of San Francisco and what types of surface features indicate this change?

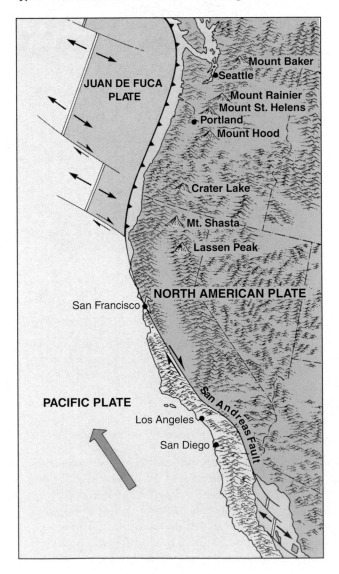

San Francisco (725 km northwest) in about 10 million years and eventually pass that city on its way to finally colliding with the Aleutian Islands at a subduction zone.

Another type of lateral plate movement occurs on ocean floors in areas of plate divergence. As plates pull apart, they usually do so along a series of fracture zones that tend to form at right angles to the major zone of plate contact. These cross-hatched plate boundaries along which lateral movement takes place are called *transform faults*. Transform faults, or fracture zones, are common along midoceanic ridges, but examples can also be seen elsewhere, as on the seafloor offshore from the Pacific Northwest coast between the Pacific and Juan de Fuca plates (see again Fig. 10.30). Transform faults result when adjacent plates travel at variable rates, causing lateral movement of one plate relative to the other. The most rapid plate motion is on the East Pacific rise where the rate of movement is more than 17 centimeters (7 in.) per year.

Hot Spots in the Mantle

The Hawaiian Islands, like many major landform features, owe their existence to processes associated with plate tectonics. As the Pacific plate in that region moves toward the northwest, it passes over a mass of molten rock in the mantle that does not move with the lithospheric plate. Called **hot spots,** these almost stationary molten masses occur in a few other places in both continental and oceanic locations. Melting of the upper mantle and oceanic crust causes undersea eruptions and the outpouring of basaltic lava on the seafloor, eventually constructing a volcanic island. This process is responsible for building the Hawaiian Islands, as well as the chain of islands and undersea volcanoes that extend for thousands of kilometers northwest of Hawaii. Today the hot spot causes active volcanic eruptions on the island of Hawaii. The other islands in the Hawaiian chain came from a similar origin, having formed over the hot spot as well, but these volcanoes have now drifted along with the Pacific plate away from their magmatic source. Evidence of the plate motion is indicated by the fact that the youngest islands of the Hawaiian chain, Hawaii and Maui, are to the southeast, and the older islands, such as Kauai and Oahu, are located to the northwest (■ Fig. 10.31). A newly forming undersea volcano, named Loihi, is now developing southeast of the island of Hawaii and will someday be the next member of the Hawaiian chain.

Growth of Continents

The origin of continents is still being debated. It is clear that the continents tend to have a core area of very old igneous and metamorphic rocks that may represent the deeply eroded roots of ancient mountains. These core regions have been worn down by hundreds of millions of years of erosion to create areas of relatively low relief that are located far

■ **FIGURE 10.31** Over the last few million years, a stationary zone of molten material in the mantle—a hot spot—has created each of the volcanic Hawaiian Islands in succession. Because they move to the northwest with the Pacific plate, the islands are progressively older toward the northwest (ages are in millions of years). The island of Hawaii, which is about 300 kilometers (185 mi) from Oahu, is currently located at the hot spot.

Approximately how long did it take the Pacific plate to move Oahu to its current position?

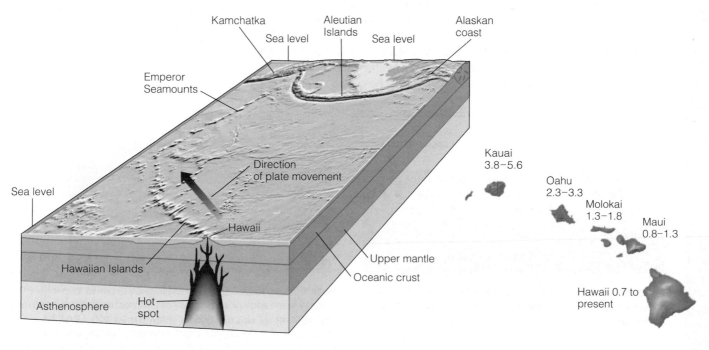

Paleogeography

The study of past geographical environments is known as *paleogeography*. The goal of paleogeography is to try to reconstruct the past environment of a geographical region based on geologic and climatic evidence. For students of physical geography, it generally seems that the present is complex enough without trying to know what the geography of ancient times was like. However, peering into the past helps us forecast and prepare for changes in the future.

The immensity of geologic time over which major events or processes (such as plate tectonics, ice ages, or the formation and erosion of mountain ranges) have taken place is difficult to picture in a human time frame of days, months, and years. The geologic timescale is a calendar of Earth history (Table 10.2). It is divided into *eras*, which are typically long units of time, such as the Mesozoic Era (which means "middle life"), and eras are divided into *periods*, such as the Cretaceous Period. *Epochs*, as for example the Pleistocene Epoch (recent ice ages), are shorter time units and are used to subdivide the periods of the Cenozoic Era ("recent life"), for which geologic evidence is more abundant. Today we are in the Holocene Epoch (last 10,000 years), of the Quaternary Period (last 2.6 million years), of the Cenozoic Era (last 65 million years). In a sense, these divisions are used like we would use days, months, and years to record time.

TABLE 10.2
The Geologic Timescale

Eon	Era	Period		Epoch	Millions of years ago	Major Geologic and Biologic Events
Phanerozoic	Cenozoic	Quaternary		Holocene (or Recent)	0.01	Ice Age ends
				Pleistocene	2.6	Ice Age begins / Earliest humans
		Tertiary	Neogene	Pliocene	5	
				Miocene	24	
			Paleogene	Oligocene	34	
				Eocene	56	Formation of Himalayas / Formation of Alps
				Paleocene	65	
	Mesozoic	Cretaceous				Extinction of dinosaurs / Formation of Rocky Mountains / First birds / Formation of Sierra Nevada
		Jurassic			144	
		Triassic			206	First mammals / Breakup of Pangaea / First dinosaurs / Formation of Pangaea / Formation of Appalachian Mountains
	Paleozoic	Permian			248	
					290	
		Pennnsylvanian				Abundant coal-forming swamps
		Mississippian			323	First reptiles
		Devonian			354	First amphibians
		Silurian			417	First land plants
		Ordovician			443	First fish
		Cambrian			490	
					543	Earliest shelled animals
Precambrian	Proterozoic Eon					
					2,500	
	Archean Eon					Earliest fossil record of life
					3,800	
	Hadeon Eon				~4,650	

If we took a 24-hour day to represent the approximately 4.6 billion-year history of Earth, the Precambrian, an era of which we know very little, would consume the first 21 hours. The current period, the Quaternary, which has lasted about 2.6 million years, would take less than 30 seconds, and human beginnings, over about the last 4 million years, about 1 minute.

Each era, period, and epoch in Earth's geologic history had a unique paleogeography with its own distribution of land and sea, climate regions, plants, and animal life. If we look at evidence for the paleogeography of the Mesozoic Era (248 million to 65 million years ago), for instance, we would find a much different physical geography than exists now. This was a time when the supercontinent Pangaea gradually split apart as new ocean floors widened, creating the continents that are familiar to us today. Global and local Mesozoic climates were very different from those of today but were changing as North America drifted to the northwest. During the Cretaceous Period, much of the present United States experienced warmer climates than now. Ferns and conifer forests were common. The Mesozoic was the "age of the dinosaurs,"

a class of large animals that ruled the land and the sea. Other life also thrived, including marine plants and invertebrates, insects, mammals, and the earliest birds.

The Mesozoic Era ended with an episode of great extinctions, including the end of the dinosaurs. Geologists, paleontologists, and paleogeographers are not in agreement as to what caused these great extinctions. Some of the strongest evidence points to a large meteorite striking Earth 65 million years ago, disrupting global climate and causing global environmental change. Other evidence points to plate tectonic changes in the distribution of oceans and continents or increased volcanic activity, either of which could cause rapid climate changes that might possibly trigger mass extinctions.

Available maps depicting Earth in early geologic times show only approximate and generalized patterns of mountains, plains, coasts, and oceans, with the addition of some environmental characteristics. These maps portray a general picture of how global geography has changed through geologic time (■ Fig. 10.33). Much of the evidence and the rocks that bear this information have been lost through metamorphism

■ **FIGURE 10.33**   Paleomaps showing Earth's tectonic history over the last 250 million years of geologic time.

How has the environment at the location where you live changed through geologic time?

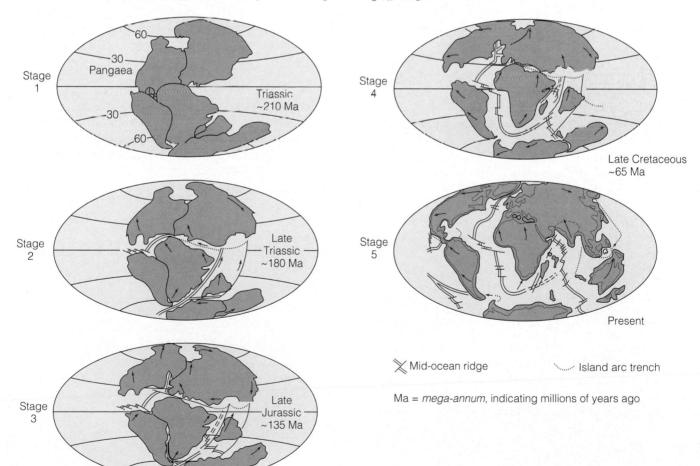

or erosion, buried under younger sediments or lava flows, or recycled into Earth's interior. The further back in time, the sketchier is the paleoenvironmental information presented on the map. Paleomaps, like other maps, are simplified models of the regions and times they represent.

As time passes and additional evidence is collected, paleogeographers may be able to fill in more of the empty spaces on those maps of the past that are so unfamiliar to us.

These paleogeographic studies aim not only at understanding the past but also at understanding today's environments and physical landscapes, how they have developed, and how processes act to change them. By applying the theory of plate tectonics to our knowledge of how the Earth system and its subsystems function, we can gain a better understanding of our planet's geologic past, as well as its present, and this will facilitate better forecasts of its potential future.

:: Terms for Review

seismic wave	regolith	foliation
core	igneous rock	continental drift
inner core	magma	plate tectonics
outer core	lava	lithospheric plate
mantle	extrusive igneous rock	midoceanic ridge
lithosphere	intrusive igneous rock	seafloor spreading
asthenosphere	joint	plate divergence
tectonic force	sedimentary rock	plate convergence
Mohorovičić discontinuity (Moho)	clastic sedimentary rock	subduction
crust	organic sedimentary rock	island arc
oceanic crust	chemical sedimentary rock	continental collision
continental crust	bedding plane	transform movement
mineral	unconformity	hotspot
rock	metamorphic rock	continental shield

:: Questions for Review

1. List the major zones of Earth's interior from the center to the surface and indicate how they differ from one another.
2. Define and distinguish (a) continental crust and oceanic crust, and (b) the lithosphere and the asthenosphere.
3. What is a mineral? What is a rock? Provide examples of each.
4. Describe the three major categories of rock and the principal means by which each is formed. Give an example of each.
5. What is the rock cycle?
6. What evidence has been found to support the theory that lithospheric plates move around Earth's surface?
7. What type of lithospheric plate boundary is found paralleling the Andes, at the San Andreas Fault, in Iceland, and near the Himalayas?
8. Explain why the eastern United States has relatively little tectonic activity compared to the western United States.
9. How does the formation of the Hawaiian Islands support plate tectonic theory?
10. Define paleogeography. Why are geographers interested in this topic?

:: Practical Applications

1. Two plates are diverging and both are moving at a rate of 3 centimeters per year. How long will it take for them to move 100 kilometers apart?
2. An area of oceanic crust has a density of 3.0 grams per cubic centimeter and a thickness of 4 kilometers. The same size area of continental crust has a density of 2.7 grams per cubic centimeter. How thick in kilometers would the continental crust have to be for its total mass to equal that of the oceanic crust?

:: Outline

Basaltic lava flows from
an erupting volcano on the
island of Hawaii. Lava that
has solidified and turned
dark in color at the surface is
carried along by the hot,
red molten lava that is
still flowing.

:: Objectives

When you complete this chapter you should be able to:

- Distinguish the endogenic from the exogenic landforming processes.
- Explain why some volcanic eruptions are explosive whereas others are effusive.
- Describe the principal differences among the six major types of volcanic landforms.
- Differentiate the various types of igneous intrusions.
- Demonstrate how compressional, tensional, and shearing forces each stress rock matter.
- Sketch examples of the different types of faults, indicating direction of motion.

- Associate the different types of faults with the type of tectonic force responsible for them.
- Draw a cross section of a fault that shows the relationship between an earthquake's focus and epicenter.
- Discuss the two ways in which the severity of an earthquake is measured.
- List various factors that help determine the devastation caused by an earthquake.

Our planet's surface **topography,** the distribution of landscape highs and lows, is intriguing and complex. Landscapes may consist of rugged mountains, gently sloping plains, rolling hills and valleys, or elevated plateaus cut by steep canyons. These are just a few examples of the varied types of surface terrain features, referred to as **landforms,** that contribute to the beauty and diversity of the environments on Earth. Landforms are one of the most appealing and impressive elements of Earth's surface. Local, state, and national parks attract millions of visitors annually seeking to observe and experience firsthand spectacular examples of landforms and associated environmental features. Landforms owe their development to processes and materials that originate within Earth's interior, at its surface, or, most typically, some combination of both. Understanding how landforms are made, why they vary, and their significance in a local, regional, or global context is the primary goal of **geomorphology,** a major subfield of physical geography devoted to the scientific study of landforms.

Landforming *igneous processes* (from Latin: *ignis,* fire), which are related to the eruption and solidification of rock matter, and *tectonic processes* (from Greek: *tekton,* carpenter, builder), which are movements of parts of the crust and upper mantle, constitute the primary geomorphic mechanisms that increase the topographic irregularities on Earth's surface. Areas of the crust can be built up by igneous processes, which include the ejection of volcanic rock matter from Earth's interior onto the surface, or can be uplifted or downdropped by tectonic processes. Igneous and tectonic processes build extensive mountain systems, but they also produce a great variety of other landforms. The geographical distribution of these terrain features, moreover, is not random. Volcanic landforms occur most commonly in association with lithospheric plate margins. Zones with the greatest tectonic forces and a concentration of dangerous earthquakes likewise lie along plate boundaries.

Igneous and tectonic processes have produced many impressively scenic landscapes, but they can also present serious natural hazards to people and their property. This chapter, and those that follow, focus on understanding how various landforms develop and on the potential hazards that are related to them. It is extremely important for us to understand how geomorphic processes work to shape Earth's surface landforms because they are active, ongoing, and often powerful processes that can impact human welfare. Landforms are a dynamic, beautiful, diverse, and sometimes dangerous aspect of the human habitat.

Landforms and Geomorphology

Landforms and landscapes are often described by their relative amount of **relief,** which is the difference in elevation between the highest and lowest points within a specified area or on a particular surface feature (■ Fig. 11.1). With no variations in relief our planet would be a smooth, featureless sphere and certainly much less interesting. It is hard to imagine Earth without dramatic terrain as seen in the high-relief mountainous regions of the Himalayas, Alps, Andes, Rockies, and Appalachians or in the huge chasm that we call the Grand Canyon. Interspersed with high-relief features, large expanses of low-relief features, like the Great Plains, can be equally impressive and inspiring.

Earth's landforms result from mechanisms that act to increase relief by raising or lowering the land surface and mechanisms that work to reduce relief by removing rock from high places and using it to fill in depressions. In general, geomorphic processes that originate within Earth, called **endogenic processes** (*endo,* within; *genic,* originating), result in an increase in surface relief, while the **exogenic processes** (*exo,* external), those that originate at Earth's surface, tend to decrease relief. Igneous and tectonic processes constitute the endogenic geomorphic processes. Exogenic processes consist of various means of rock breakdown, collectively known as *weathering,* and the removal, movement, and relocation of those weathered rock products in the continuum of processes known as *erosion, transportation,* and *deposition.* Erosion, transportation, and deposition occur through the force of gravity alone, as in the fall of a weathered clast from a cliff to the ground below, or operate with the help of a *geomorphic agent,* a medium that picks up, moves, and eventually lays

(a)   D. Sack

(b)   D. Sack

■ **FIGURE 11.1**   An area of (a) low relief in western Utah, and (b) high relief in Great Basin National Park of eastern Nevada.

down broken rock matter. The most common geomorphic agents are flowing water, wind, moving ice, and waves, but people and other organisms can also accomplish some erosion, transportation, and deposition of weathered pieces of Earth material. The exogenic processes decrease relief by eroding weathered rock material from highlands and depositing it in lowlands. High-relief features, including mountains, hills, and deep basins, exist where endogenic processes operate, or have operated, at faster rates than exogenic processes,

or where there has been insufficient time since creation of the relief for the exogenic processes to have made substantial progress toward leveling the terrain (■ Fig. 11.2).

In this chapter, we study the processes related to the buildup of relief through igneous and tectonic activity and examine the landforms and rock structures associated with these endogenic processes. Subsequent chapters focus on the exogenic processes and the landforms and landscapes formed by the various geomorphic agents.

■ **FIGURE 11.2**   The Teton range of Wyoming stands high above the valley floor because uplift rates due to endogenic processes exceed the rates of the exogenic processes of weathering and erosion.

National Park Service/Kimberly Finch

Igneous Processes and Landforms

Landforms resulting from igneous processes are related to eruptions of extrusive igneous rock material or emplacements of intrusive igneous rock. **Volcanism** refers to the extrusion of rock matter from Earth's subsurface to the exterior and the creation of surface terrain features as a result. *Volcanoes* are mountains or hills constructed in this way. **Plutonism** refers to igneous processes that occur below Earth's surface, including the cooling of magma to form intrusive igneous rocks and rock masses. Some masses of intrusive igneous rock are eventually exposed at Earth's surface where they comprise landforms of distinctive shapes and properties.

Volcanic Eruptions

Few spectacles in nature are as awesome as a volcanic eruption (■ Fig. 11.3). Although large, violent eruptions tend to be infrequent events, they can devastate the surrounding environment and completely change the nearby terrain. Yet volcanic eruptions are natural processes and should not be unexpected by people who live in the vicinity of active volcanoes.

Volcanic eruptions vary greatly in size and character, and the volcanic landforms that result are extremely diverse. *Explosive eruptions* violently blast pieces of molten and solid rock into the air, whereas molten rock pours less violently onto the surface as flowing streams of lava in *effusive eruptions*. Variations in eruptive style and in the landforms produced by volcanism stem mainly from temperature and chemical differences in the magma that feeds the eruption.

The mineral composition of the magma is the most important factor determining the nature of a volcanic eruption. Silica-rich *felsic* magmas tend to be relatively cool in temperature while molten and have a viscous (thick, resistant to flowing) consistency. *Mafic* magmas are more likely to be extremely hot and less viscous, and thus they flow readily in comparison to silica-rich magmas. Magmas contain large amounts of gases that remain dissolved when under high pressure at great depths. As molten rock rises closer to the surface, the pressure decreases, which tends to release expanding gases. If the gases trapped beneath the surface cannot be readily vented to the atmosphere or do not remain dissolved in the magma, explosive expansion of gases produces a violent, eruptive blast. Highly viscous, silica-rich magmas and lavas (rhyolitic in composition) tend to trap gases and have the potential to erupt with violent explosions. Mafic magmas, such as those with a basaltic composition, typically vent gases

more readily and therefore generally erupt with flowing lava rather than by exploding.

Explosive eruptions hurl into the air fragments of solidified lava, clots of molten lava that solidify in flight, or molten clots that solidify once they land. All of these represent *pyroclastic materials* (fire fragments), also referred to as **tephra.** These rock fragments erupt in a range of sizes, with **volcanic ash** denoting pyroclastic materials that are sand sized or smaller. In the most explosive eruptions, volcanic ash is thrown into the atmosphere to an altitude of 10,000 meters (32,800 ft) or more (■ Fig. 11.4). Fine-grained volcanic

USGS/VHP/B. Chouet

■ **FIGURE 11.3**   This spectacular eruption of Italy's Stromboli volcano, on an island off Sicily, lit up the night sky.

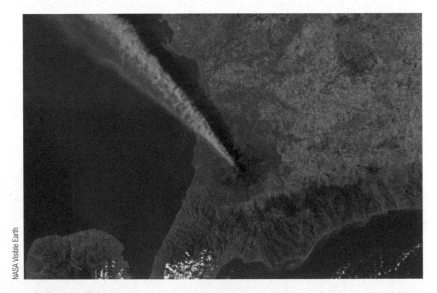

NASA Visible Earth

■ **FIGURE 11.4**   This photograph taken from the International Space Station in July of 2001 shows volcanic ash streaming from Mount Etna on the Italian island of Sicily. The ash cloud reportedly reached a height of about 5200 meters (17,000 ft) on that day.

What do you think conditions were like at the time of this eruption for settlements located under the ash cloud?

ash reaching such heights can eventually circle the globe, as happened in the 1991 eruptions of Mount Pinatubo in the Philippines. As a result of these eruptions, suspended material caused spectacular reddish orange sunsets due to enhanced scattering, and also lowered global temperatures slightly for 3 years by increasing reflection of solar energy back to space.

Volcanic Landforms

The type of landform that results from a volcanic eruption depends primarily on the explosiveness of that eruption. We will consider six major kinds of volcanic landforms, beginning with those associated with the most effusive (least explosive) eruptions. Four of the six major landforms are types of volcanoes.

Lava Flow
Lava flows are layers of erupted rock matter that when molten poured or oozed over the landscape. After cooling and solidifying, the rock retains the appearance of having flowed. Lava flows can be made from any lava type (see Appendix E), but basalt is by far the most common because its hot eruptive temperature and low viscosity allow gases to escape, greatly reducing the potential for an explosive eruption.

Solidified lava flows tend to have many fractures, known as *joints*. When basaltic lava cools and solidifies it shrinks, and this contraction can produce a network of vertical fractures that break the rock into numerous hexagonal columns. This creates columnar-jointed basalt flows (■ Fig. 11.5).

Lava flows commonly display distinctive surface characteristics. Extremely fluid lavas can flow rapidly and for long distances before solidifying. In this case, a thin surface layer of lava in contact with the atmosphere solidifies, while the

J. Petersen

■ **FIGURE 11.5** Columnar-jointed lava at Devil's Postpile National Monument, California.
Why are the cliffs shown in this photograph so steep?

molten lava beneath continues to move, carrying the thin, hardened crust along and wrinkling it into a ropy surface form called **pahoehoe.** Lavas of slightly greater viscosity flow more slowly, allowing a thicker surface layer to harden while the still-molten interior lava keeps on flowing. This causes the thick layer of hardened crust to break up into sharp-edged, jagged blocks, making a surface known as **aa.** The terms pahoehoe and aa both originated in Hawaii, where effusive eruptions of basalt are common (■ Fig. 11.6).

Lava flows do not have to emanate directly from volcanoes, but can pour out of deep fractures in the crust, called *fissures,* that can be independent of mountains or hills of

■ **FIGURE 11.6** Lava flow surfaces often consist of (a) the ropy-textured pahoehoe, where the molten lava was extremely fluid, and (b) the blocky and angular aa, where viscosity of the molten rock matter was slightly higher (less fluid).
In which direction relative to the photo did the pahoehoe flow?

D. Sack

(a)

D. Sack

(b)

volcanic origin. Very fluid basaltic lava erupting from fissures has traveled up to 150 kilometers (93 mi) before solidifying. In the geologic past, huge amounts of basalt have poured out of fissures, eventually burying existing landscapes under thousands of meters of lava flows. Multiple layers of basalt flows construct relatively flat-topped, but elevated, table-lands known as *basaltic plateaus*. The Columbia Plateau in Washington, Oregon, and Idaho, covering 520,000 square kilometers (200,000 sq mi), is a major example of a basaltic plateau (■ Fig. 11.7).

Shield Volcanoes

When numerous successive basaltic lava flows occur in a given region they can eventually pile up into the shape of a large mountain, called a **shield volcano**, which resembles a giant knight's shield resting on Earth's surface (■ Fig. 11.8a). The gently sloping, dome-shaped cones of Hawaii best illustrate this largest type of volcano (■ Fig. 11.9). Shield volcanoes erupt extremely hot, mafic lava at temperatures near 1090°C (2000°F). Escape of gases and steam occasionally hurl fountains of molten lava a few hundred meters into the air (■ Fig. 11.10), causing some accumulations of solidified pyroclastic materials, but the major feature is the outpouring of fluid basaltic lava flows. Compared to other volcano types, these eruptions are not very explosive, although still potentially dangerous and damaging. The extremely hot and fluid basalt can flow long distances before solidifying, and the accumulation of flow layers develops broad, dome-shaped volcanoes with very gentle slopes. On the island of Hawaii, active shield volcanoes also erupt lava from fissures on their flanks so that living on the island's edges, away from the summit craters, does not guarantee safety from volcanic hazards. Neighborhoods in Hawaii have

■ **FIGURE 11.7** River erosion has cut a deep canyon exposing the uppermost layers of basalt in the Columbia Plateau, southwestern Idaho.

©Jeff Gnass

■ **FIGURE 11.8** The four basic types of volcanoes are: (a) shield volcano, (b) cinder cone, (c) composite cone, also known as stratovolcano, and (d) plug dome.
What are the key differences in their shape and in their internal structure?

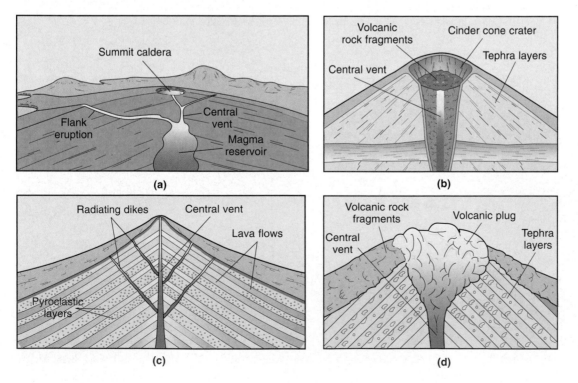

■ **FIGURE 11.9**   Mauna Loa, on the island of Hawaii, clearly displays the dome, or convex, shape of a classic shield volcano. From its base on the ocean floor to its summit at 4170 meters (13,681 ft) above sea level, Mauna Loa is almost 17 kilometers (56,000 ft) tall.

Why do Hawaiian volcanoes erupt less explosively than volcanoes of the Cascades or Andes?

■ **FIGURE 11.10**   A 300-meter (1000-ft) high fountain of lava in Hawaii.

been destroyed or threatened by lava flows. The Hawaiian shield volcanoes form the largest volcanoes on Earth in terms of both their height—beginning at the ocean floor—and diameter.

Cinder Cones

The smallest type of volcano, typically only a couple of hundred meters high, is known as a **cinder cone.** Cinder cones generally consist largely of gravel-sized pyroclastics. Gas-charged eruptions throw molten lava and solid pyroclastic fragments into the air. Falling under the influence of gravity, these particles accumulate around the almost pipelike conduit for the eruption, the *vent*, in a large pile of tephra (Fig. 11.8b). Each eruptive burst ejects more pyroclastics that fall and cascade down the sides to build an internally layered volcanic cone. Cinder cone volcanoes typically have a rhyolitic composition, but can be made of basalt if conditions of temperature and viscosity keep gases from escaping easily. The form of a cinder cone is very distinctive, with steep straight sides and a crater (depression) at the top of the hill (■ Fig. 11.11). The steep slopes of accumulated pyroclastics lie at or near the **angle of repose,** the steepest angle that a pile of loose material can maintain without rocks sliding or rolling downslope. The angle of repose for unconsolidated rock matter generally ranges between 30° and 34°. Cinder cone examples include several in the Craters of the Moon area in Idaho, Capulin Mountain in New Mexico, and Sunset Crater, Arizona. In 1943, a remarkable cinder cone called Paricutín grew from a fissure in a Mexican cornfield to a height of 92 meters (300 ft) in 5 days and to more than 360 meters (1200 ft) in a year. Eventually, the volcano began erupting basaltic lava flows, which buried a nearby village except for the top of a church steeple.

Composite Cones

A third kind of volcano, a **composite cone,** results when formative eruptions are sometimes effusive and sometimes explosive. Composite cones are therefore composed of a combination—that is, they represent a composite—of lava flows and pyroclastic materials (Fig. 11.8c). They are also called *stratovolcanoes* because they are constructed of layers (strata) of pyroclastics and lava. The topographic profile of a composite cone represents what many people might consider the classic volcano shape, with concave slopes that are gentle near the base and steep near the top (■ Fig. 11.12). Composite volcanoes form from andesite, which is a volcanic rock intermediate in silica content and explosiveness between basalt and rhyolite. Although andesite is only intermediate in these characteristics, composite cones are dangerous. As a composite cone grows larger, the vent eventually becomes plugged with unerupted andesitic rock. When this happens, the pressure driving an eruption can build to the point where either the plug is explosively forced out or the mountain side is pushed outward until it fails, allowing the great accumulation of pressure to be relieved in a lateral explosion. Such explosive eruptions may be

D. R. Crandell/USGS

■ **FIGURE 11.11** This cinder cone stands among lava flows in Lassen Volcanic National Park, California.

Why is the crater so prominent on this volcano?

USGS/CVO Oregon

■ **FIGURE 11.12** Composite cones, like Oregon's Mount Hood in the Cascade Range, are composed of both lava flows and pyroclastic material and have distinctive concave side slopes.

Along what type of lithospheric plate boundary is this volcano located?

that had been venting steam and ash for several weeks, exploded with incredible force on that day. A menacing bulge had been growing on the side of Mount St. Helens, and Earth scientists warned of a possible major eruption, but no one could forecast the magnitude or the exact timing of the blast. Within minutes, nearly 400 meters (1300 ft) of the mountain's north summit had disappeared by being blasted into the sky and down the mountainside (■ Fig. 11.13). Unlike most volcanic eruptions, in which the eruptive force is directed vertically, much of the explosion blew pyroclastic debris laterally outward from the site of the bulge. An eruptive blast composed of an intensely hot cloud of steam, noxious gases, and volcanic ash burst outward at more than 300 kilometers per hour (200 mph), obliterating forests, lakes, streams, and campsites for nearly 32 kilometers (20 mi). Volcanic ash and water from melted snow and ice formed huge mudflows that choked streams, buried valleys, and engulfed everything in their paths. More than 500 square kilometers (200 sq mi) of forests and recreational lands were destroyed. Hundreds of homes were buried or badly damaged. Choking ash several centimeters thick covered nearby cities, untold numbers of wildlife were killed, and 57 people lost their lives in the eruption. It was a minor event in Earth's history but a sharp reminder to the region's residents of the awesome power of natural forces.

Some of the worst natural disasters in history have occurred in the shadows of composite cones. Mount Vesuvius, in Italy, killed more than 20,000 people in the cities of Pompeii and Herculaneum in A.D. 79. Mount Etna, on the Italian island of Sicily, destroyed 14 cities in 1669, killing more than 20,000 people, and is still active much of the time. The greatest volcanic eruption in recent history was the 1883 explosion of Krakatoa in what is now Indonesia. Many of the casualties resulted from the subsequent *tsunamis,* large sets of ocean waves generated by a sudden offset of the water that swept the coasts of Java and Sumatra. The 1991 eruption of Mount Pinatubo in the Philippines killed more than 300 people, and airborne ash caused climatic effects for 3 years following the eruption. In 1997, a series of violent eruptions from the Soufriere Hills volcano destroyed more than half of the Caribbean island of Montserrat with volcanic ash and pyroclastic flows (■ Fig. 11.14). Mexico City, one of the world's most populous urban areas, continues to be threatened by ash falls from eruptions of a composite cone 70 kilometers (45 mi) away. Volcanic ash causes respiratory problems, stalls vehicles by clogging air intakes, and in large accumulations collapses roofs.

accompanied by *pyroclastic flows,* fast-moving density currents of airborne volcanic ash, hot gases, and steam that flow downslope close to the ground. The speed of a pyroclastic flow can reach 100 kilometers per hour (62 mph) or more.

Most of the world's best-known volcanoes are composite cones. Some examples include Fujiyama in Japan, Cotopaxi in Ecuador, Vesuvius and Etna in Italy, Mount Rainier in Washington, and Mount Shasta in California. The highest volcano on Earth, Nevados Ojos del Salado, is an andesitic composite cone that reaches an elevation of 6887 meters (22,595 ft) on the border between Chile and Argentina in the Andes, the mountain range after which andesite was named.

On May 18, 1980, residents of the American Pacific Northwest were stunned by the eruption of Mount St. Helens. Mount St. Helens, a composite cone in southwestern Washington

USDA Forest Service

(a)

R.P. Hoblitt/USGS Volcano Hazards Program

■ FIGURE 11.14 Beginning in 1995, the Caribbean island of Montserrat was struck by a series of volcanic eruptions, including pyroclastic flows, which devastated much of the island. Prior to the 1995 disaster, the volcano had not erupted for 400 years.

Plug Domes Where extremely viscous silica-rich magma has pushed up into the vent of a volcanic cone without flowing beyond it, it forms a **plug dome** (Fig. 11.8d). Solidified outer parts of the blockage create the dome-shaped summit, and jagged blocks that broke away from the plug, or preexisting parts of the cone, form the steep, sloping sides of the volcano. Great pressures can build up causing more blocks to break off, and creating the potential for extremely violent explosive eruptions, including pyroclastic flows. In 1903, Mount Pelée, a plug dome on the French West Indies island of Martinique, caused the deaths in a single blast of all but two people from a town of 30,000. Lassen Peak in California is a large plug dome that erupted with great violence less than 100 years ago (■ Fig. 11.15). Other plug domes exist in Japan, Guatemala, the Caribbean, and the Aleutian Islands.

USGS/J. Rosenbaum

(b)

■ FIGURE 11.15 Lassen Peak, in northern California, is a plug dome and the southernmost volcano in the Cascade Range. Silica-rich lava plugs are the darker areas protruding from the peak. Lassen was last active between 1914 and 1921.
Why are plug dome volcanoes considered dangerous?

USGS/Lyn Topinka

(c)

■ FIGURE 11.13 (a) Prior to the 1980 eruption, Mount St. Helens in the Cascade Range towered majestically over Spirit Lake. (b) On May 18, 1980, the violent eruption removed almost 3 cubic kilometers (1 cu mi) of material from the mountain's north slope. The blast cloud and mudflows decimated forests and took 57 human lives. (c) Two years later, the volcano continued to spew small amounts of gas, steam, and ash.
Could other volcanoes in the Cascade Range, such as Oregon's Mount Hood, erupt with the kind of violence that Mount St. Helens displayed in 1980?

USGS

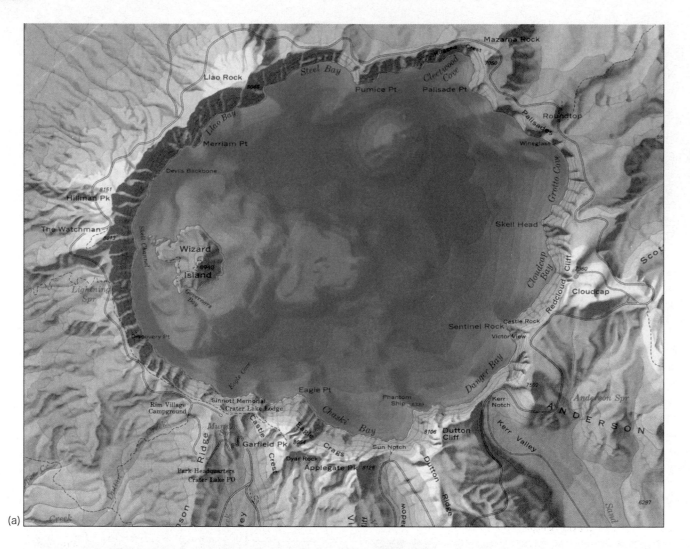

(a)

National Park Service

(b)

■ **FIGURE 11.16**   (a) Crater Lake formed about 7,700 years ago when a violent eruption of Mount Mazama blasted out solid and molten rock matter, leaving behind a deep crater, the caldera, that later accumulated water. (b) Wizard Island is a later, secondary volcano that has risen within the caldera.

Could other Cascade volcanoes erupt to the point of destroying the volcano summit and creating a caldera?

Calderas   Occasionally, the eruption of a volcano expels so much material and relieves so much pressure within the magma chamber that only a large and deep depression remains in the area that previously contained the volcano's summit. A large depression made in this way is termed a **caldera.** The best-known caldera in North America is the basin in south-central Oregon that contains Crater Lake, a circular body of water 10 kilometers (6 mi) across and almost 600 meters (2000 ft) deep, surrounded by near-vertical cliffs. The caldera that contains Crater Lake was formed by the prehistoric eruption and collapse of a composite volcano. A cinder cone, Wizard Island, has subsequently arisen from the caldera floor to a height above the lake's surface (■ Fig. 11.16). The area of Yellowstone National Park is the site of three ancient calderas, and the Valles Caldera in New Mexico is another excellent example. Krakatoa in Indonesia and Santorini (Thera) in Greece have left

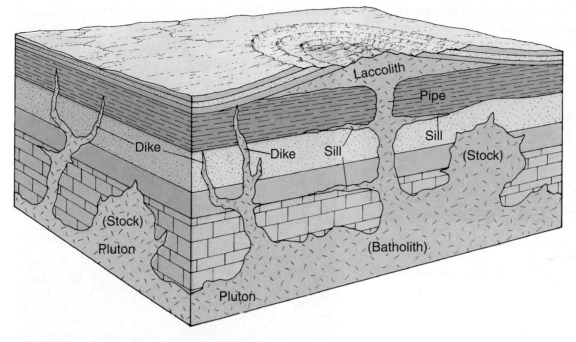

■ FIGURE 11.17   Because intrusive igneous rocks tend to be more resistant to erosion than sedimentary rocks, when they are eventually exposed at the surface sills, dikes, laccoliths, stocks, and batholiths generally stand higher than the surrounding rocks.

island remnants of their calderas. Calderas are also found in the Philippines, the Azores, Japan, Nicaragua, Tanzania, and Italy, many of them occupied by deep lakes.

Plutonism and Intrusions

Bodies of magma that exist beneath the surface of Earth or masses of intrusive igneous rock that cooled and solidified beneath the surface are called *igneous intrusions,* or *plutons.* A great variety of shapes and sizes of magma bodies can result from intrusive igneous activity, also called plutonism. When they are first formed, smaller plutons have little or no effect on the surface terrain. During their formation, larger plutons may be associated with uplift of the land surface under which they are intruded.

The many different kinds of intrusions are classified by their size, shape, and relationship to the surrounding rocks (■ Fig. 11.17). After millions of years of uplift and erosion of overlying rocks, intrusions may be exposed at the surface to become part of the landscape. Uplifted plutons composed of granite or other intrusive igneous rocks that crop out at the surface tend to stand higher than the landscape around them because their resistance to weathering and erosion exceeds that of many other kinds of rocks.

An irregularly shaped intrusion exposed at Earth's surface is a **stock** if its area is smaller than 100 square kilometers (40 sq mi); if larger, it is known as a **batholith.** Batholiths are complex masses of solidified magma, usually granite, that developed kilometers beneath Earth's surface. Because of the resistance of intrusive igneous rocks to weathering and erosion, batholiths form many major mountain ranges. The Sierra Nevada batholith, Idaho batholith, and Peninsular Ranges

batholith of Southern and Baja California cover hundreds of thousands of square kilometers of granite landscapes in western North America.

Magma creates other kinds of igneous intrusions by forcing its way into fractures and between rock layers without melting the surrounding rock. A **laccolith** develops where molten magma flows horizontally between rock layers, bulging the overlying layers upward, making a solidified mushroom-shaped structure. Laccoliths have a mushroomlike shape because the upper dome-like mass is usually connected to a magma source by a pipe or stem. Although laccoliths are smaller, like batholiths they form the core of mountains or hills after erosion has worn away the overlying less resistant rocks. The La Sal, Abajo, and Henry Mountains in southern Utah are exposed laccoliths, as are other mountains in the American West (■ Fig. 11.18).

Smaller but no less interesting landforms created by intrusive activity are also exposed at the surface by erosion of the overlying rocks. Magma sometimes intrudes between rock layers without bulging them upward, solidifying into a horizontal sheet of intrusive igneous rock called a **sill** (■ Fig. 11.19). Molten rock under pressure may also intrude into a nonhorizontal fracture that cuts across the surrounding rocks. The solidified magma in this case has a wall-like shape and is known as a **dike** (■ Fig. 11.20). At Shiprock, in New Mexico, resistant dikes many kilometers long rise vertically to more than 90 meters (300 ft) above the surrounding plateau (■ Fig. 11.21). Shiprock is a *volcanic neck,* a tall rock spire made of the exposed (formerly subsurface) pipe that fed a long-extinct volcano situated above it about 30 million years ago. Erosion has removed the volcanic cone, exposing the resistant dikes and neck that were once internal features of the volcano.

rock masses will move relative to each other along the fracture. **Faulting** is the slippage or displacement of rocks along a fracture surface, and the fracture along which movement has occurred is a **fault.** When compressional forces cause faulting either one mass of rock is pushed up along a steep-angled fault relative to the other, or one mass of rock slides along a shallow, low-angle

fault over the other. The steep, high-angle fault resulting from compressional forces is termed a *reverse fault* (■ Fig. 11.26a). Where compression pushes rocks along a low-angle fault so that they override rocks on the other side of the fault, the fracture surface is called a *thrust fault*, and the shallow displacement is an *overthrust* (Fig. 11.26b). In both reverse and thrust faults, one block of crustal rocks is wedged up relative to the other. Direction of motion along all faults is always given in relative terms because even though it may seem obvious that one block was pushed up along the fault, the other block may have slid down some distance as well, and it is not always possible to determine with certainty if one or both blocks moved. Reverse or thrust faulting also result from compressional forces that are applied rapidly and in some cases to rocks that have already responded to the force by folding. In the latter case, the upper part of a fold breaks, sliding over the lower rock layers along a thrust fault forming an overthrust. Major overthrusts occur along the northern Rocky Mountains and in the southern Appalachians.

Tensional Tectonic Forces

Tensional tectonic forces pull in opposite directions in a way that stretches and thins the impacted part of the crust. Rocks, however, typically respond by faulting, rather than

J. Petersen

■ **FIGURE 11.24** Compressional forces have made complex folds in these layers of sedimentary rock.

How can solid rock be folded without breaking?

■ **FIGURE 11.25** Folded rock structures become increasingly complex as the applied compressional forces become more unequal from the two directions.

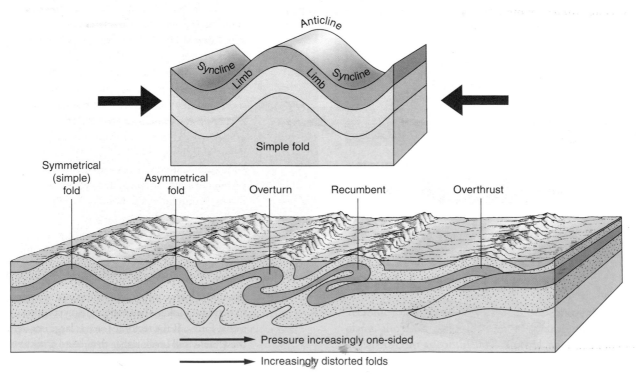

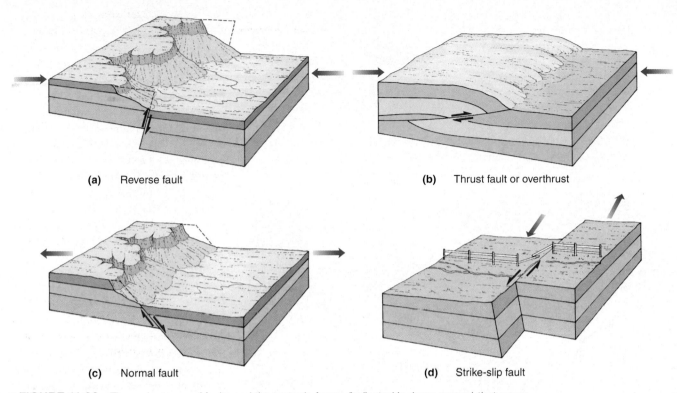

(a)   Reverse fault

(b)   Thrust fault or overthrust

(c)   Normal fault

(d)   Strike-slip fault

■ **FIGURE 11.26**   The major types of faults and the tectonic forces (indicated by large arrows) that cause them. Compressional forces form reverse (a) or thrust (b) faults, tensional forces result in normal faults (c), and shearing forces result in strike-slip faults (d).
How does motion along a normal fault differ from that along a reverse fault?

bending or stretching plastically, when subjected to tensional forces. Tensional forces commonly cause the crust to be broken into discrete blocks, called fault blocks, that are separated from each other by *normal faults* (Fig. 11.26c). In order to accommodate the extension of the crust, one crustal fault block slides downward along the normal fault relative to the adjacent fault block. Note that the direction of motion along a normal fault is opposite to that along a reverse or thrust fault.

In map view, tensional forces affecting a large region frequently cause a repeated pattern of roughly parallel normal faults, creating a series of alternating downdropped and upthrown fault blocks. Each block that slid downward between two normal faults, or that remained in place while blocks on either side slid upward along the faults, is called a **graben** (■ Fig. 11.27). A fault block that moved relatively upward between two normal faults—that is, it actually moved up or remained in place while adjacent blocks slid downward—is a **horst.** Horsts and grabens are rock structural features that can be identified by the nature of the offset of rock units along normal faults; topographically, horsts form mountain ranges and grabens form basins. The Basin and Range region of the western United States, which extends eastward

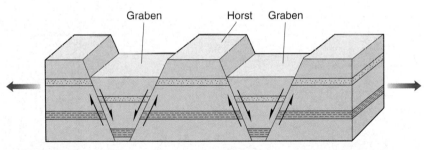

Graben   Horst   Graben

■ **FIGURE 11.27**   Horsts (upthrown blocks) and grabens (downdropped blocks) are bounded by normal faults.
What type of tectonic force causes these kinds of fault blocks?

from California to Utah and southward from Oregon to New Mexico, is an area undergoing tensional tectonic forces that are pulling the region apart to the west and east. A transect from west to east across that region, for example from Reno, Nevada, to Salt Lake City, Utah, encounters an extensive series of alternating downdropped and upthrown fault blocks comprising the basins and ranges for which the region is named. Some of the ranges and basins are simple horsts and grabens, but others are tilted fault blocks that result from the uplift of one side of a fault block while the other end of the same block rotates downward (■ Fig. 11.28). Death Valley, California, is a classic example of the down-tilted side of a tilted fault block (■ Fig. 11.29).

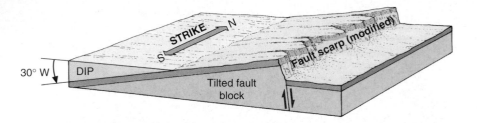
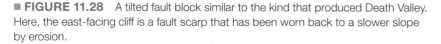

■ **FIGURE 11.28** A tilted fault block similar to the kind that produced Death Valley. Here, the east-facing cliff is a fault scarp that has been worn back to a slower slope by erosion.

Courtesy Sheila Brazier

■ **FIGURE 11.29** Death Valley, California, occupies the basin created by a tilted fault block.

D. Sack

■ **FIGURE 11.30** Movement along the normal fault that created this piedmont fault scarp in Nevada occurred about 30 years before the photograph was taken.
On which side of the fault does the horst lie?

Large-scale tensional tectonic forces can create **rift valleys,** which are composed of relatively narrow but long regions of crust downdropped along normal faults. Examples of rift valleys include the Rio Grande rift of New Mexico and Colorado, the Great Rift Valley of East Africa, and the Dead Sea rift valley.

An *escarpment,* often shortened to *scarp,* is a steep cliff, which may be tall or short. Scarps form on Earth surface terrain for many reasons and in many different settings. A cliff that results from movement along a fault is specifically a *fault scarp.* In areas of normal faults, unconsolidated sediments eroded from the uplifted block are deposited at the base of the slope near the fault zone and extending onto the downdropped block. If subsequent movement along the fault vertically offsets those unconsolidated sediments, it produces a *piedmont fault scarp* in the sediments (■ Figs. 11.30).

Shearing Tectonic Forces

Vertical displacement along a fault occurs when the rocks on one side move up or drop down relative to rocks on the other side. Faults with this kind of movement, up or down along the dip of the fault plane extending into Earth, are known as *dip-slip faults.* Normal and reverse faults have dip-slip motion. A completely different category of fault exhibits horizontal, rather than vertical, displacement of rock units. In this case, the direction of slippage is parallel to the surface trace, or strike, of the fault; thus it is called a *strike-slip fault* or, because of the horizontal motion, a *lateral fault* (Fig. 11.26d). Offset along strike-slip faults is most easily seen in map view (from above), rather than in cross-sectional view. Active strike-slip faults cause horizontal displacement of roads, railroad tracks, fences, streambeds, and other features that extend across the fault. The motion along a strike-slip fault is described as left lateral or right lateral, with left or right assessed by imagining yourself standing on one block looking across the fault to determine if the other block moved to your left or right. The San Andreas Fault, which runs through much of California, has right lateral strike-slip movement. A long and narrow, rather linear valley composed of rocks that have been crushed and weakened by faulting marks the trace of the San Andreas Fault zone (■ Fig. 11.31).

The amount that Earth's surface is offset during instantaneous movement along a fault varies from fractions of a centimeter to several meters. Faulting moves rocks laterally, vertically,

■ **FIGURE 11.31** The San Andreas Fault in California runs left to right across the center of this photo. The gullied background terrain is moving to the right relative to the smoother terrain in the foreground.

What type of fault is the San Andreas?

or both. The maximum horizontal displacement along the San Andreas Fault in California during the 1906 San Francisco earthquake was more than 6 meters (21 ft). A vertical displacement of more than 10 meters (33 ft) occurred during the Alaskan earthquake of 1964. Over millions of years, the cumulative displacement along a major fault may be tens of kilometers vertically or hundreds of kilometers horizontally, although the majority of faults have offsets that are much smaller.

Relationships between Rock Structure and Topography

Tectonic activity produces a variety of structural features that range from microscopic fractures to major folds and fault blocks. At Earth's surface, structural features comprise various landforms and are subject to modification by weathering, erosion, transportation, and deposition. It is important to distinguish between structural elements and topographic features because rock structure reflects endogenic factors while landforms reflect the balance between endogenic and exogenic factors. As a result, a specific type of structural element can assume a variety of topographic expressions (■ Fig. 11.32). For instance, an upfolded structural feature is an anticline even though geomorphically it may comprise a ridge, a valley, or a plain, depending on erosion of broken or weak rocks. Likewise, even though synclines are structural downfolds, topographically a syncline may contribute to the formation of a valley or a ridge. Some mountain tops in the Alps are the erosional remnants of synclines. Words like *mountain, ridge, valley, basin,* and *fault scarp* are geomorphic terms that describe the surface topography, while *anticline, syncline, horst, graben,* and *normal fault* are structural terms that describe the arrangement of rock layers. Elements of rock structure may or may not be directly represented in the surface topography. It is important to remember that the topographic variation on Earth's surface results from the interaction of three major factors: endogenic processes that create relief, exogenic processes that shape landforms and reduce relief, and the relative strength or resistance of different rock types to weathering and erosion.

Earthquakes

Earthquakes, evidence of ongoing tectonic activity, are ground motions of Earth caused when accumulating tectonic stress is suddenly relieved by displacement of rocks along a fault. The sudden, lurching movement of crustal blocks past one another to new positions represents a release of energy that moves through Earth as traveling *seismic waves.* Seismic waves can have a great impact on Earth's surface. It is primarily when these waves pass along the crustal exterior or emerge at Earth's surface from below that they cause the damage and subsequent loss of life that we associate with major tremors. The subsurface location where the rock displacement and resulting earthquake originated is the earthquake **focus,** which may be located anywhere from near the surface to a depth of 700 kilometers (435 mi). The earthquake **epicenter** is the point on Earth's surface that lies directly above the focus, and it is where the strongest shock is normally felt (■ Fig. 11.33).

■ **FIGURE 11.32** Structure, the rock response to applied tectonic forces, may or may not be directly represented in the surface topography, which depends on the nature and rate of exogenic as well as endogenic geomorphic processes. (a) A structural upfold (anticline) where the surface is a plain. (b) A topographic peak comprised of a structural downfold (syncline). (c) A topographic valley eroded from an anticline.

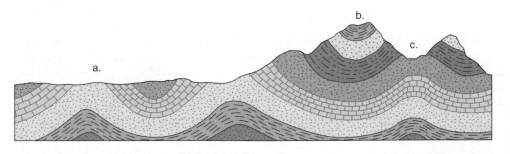

:: MAPPING THE DISTRIBUTION OF EARTHQUAKE INTENSITY

When an earthquake affects a populated area, one of the first pieces of scientific information reported is its magnitude, an expression of the energy released at the earthquake's focus. Because of their greater energy, earthquakes of larger magnitude have the potential to cause much more damage and human suffering than those of smaller magnitude, but magnitude is not the only important factor. A moderate earthquake in a densely populated area can cause much greater injury and damage than a very large earthquake in a sparsely populated region.

The modified Mercalli scale of earthquake intensity (I–XII) was devised to measure the impact of a tremor on people and their built environment. Although every earthquake has only one magnitude, intensity varies from place to place, and a single tremor typically generates a range of intensity values. Generally, the farther a location is from the earthquake epicenter, the lower the intensity, but this generalization does not always apply.

The spatial variation in Mercalli intensity is portrayed on maps by *isoseismals*, lines connecting points of equal shaking and earthquake damage expressed in Mercalli intensity values. Patterns of isoseismals are useful in assessing what local conditions contributed to the earthquake's impact.

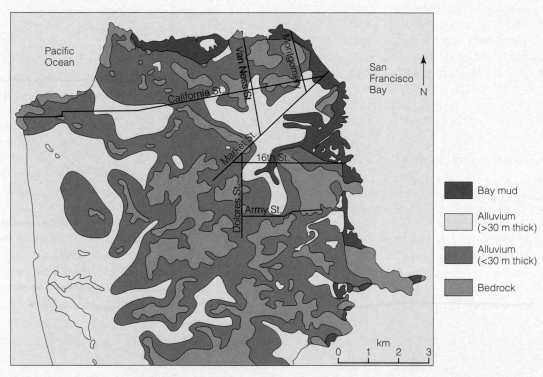

The location of different Earth materials during the 1906 San Francisco earthquake.

The vast majority of earthquakes are so slight that we cannot feel them and they produce no injuries or damage. Most earthquakes occur at a focus that is deep enough so that no displacement is visible at the surface. Others may cause mild shaking that rattles a few dishes, while a few are strong enough to topple buildings and break power lines, gas mains, and water pipes. Surface offset or ground shaking during an earthquake can also trigger rockfalls, landslides, avalanches, and tsunamis. Aftershocks commonly follow a large earthquake as crustal adjustments continue to occur, and foreshocks sometimes precede larger quakes.

Measuring Earthquake Size

Scientists express an earthquake's severity in two ways: (1) the size of the event as a physical Earth process, and (2) the degree of its impact on humans. These two

Earthquake intensity factors vary with the nature of the substrate, construction type and quality, topography, and population density. Areas of unconsolidated Earth materials, poor construction, or high population densities generally suffer more from shaking and experience greater damage.

The 1906 San Francisco earthquake and ensuing fire caused the destruction of many buildings, numerous injuries, and an estimated 3000 deaths. The fire resulted from earthquake-damaged utility lines. Neither the magnitude nor the intensity scales of measurement existed in 1906. Subsequent studies, however, have suggested that the earthquake magnitude was about 8.3, and cartographers have prepared maps of the distribution of the earthquake's Mercalli intensities. The geographic patterns of the isoseismals reveal that areas of bedrock experienced lower intensities (less damage) than areas of unconsolidated sediment. Much of the worst damage occurred on bayfront lands and in stream valleys that had been artificially filled with sediment to allow construction.

Analyzing the nature of sites where earthquake intensities were higher or lower than expected helps us understand the factors that contribute to earthquake hazards. The geographic patterns of Mercalli intensity that are generated even by small tremors help in planning for larger earthquakes in the same area. The overall pattern of ground shaking and isoseismals should be similar for a larger earthquake having the same epicenter as a smaller one, but the amount of ground shaking, Mercalli intensity, and size of the area affected would be greater for the larger magnitude tremor.

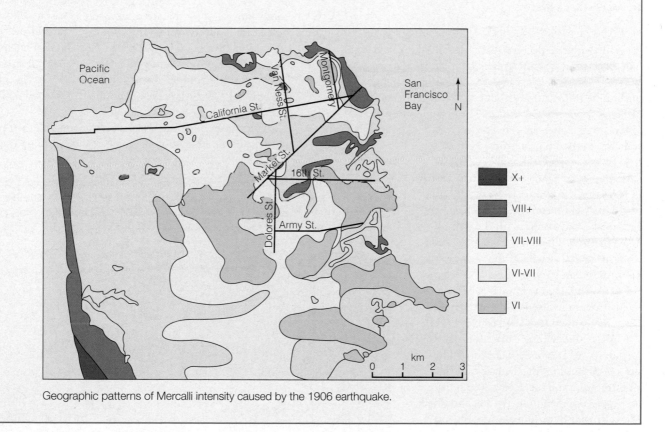

Geographic patterns of Mercalli intensity caused by the 1906 earthquake.

characteristics are sometimes related because, all other factors being equal, powerful earthquakes should have a greater effect on humans than smaller earthquakes. However, large tremors that strike in sparsely inhabited places will have limited human impact, while small earthquakes that strike densely populated areas can cause considerable damage and suffering. Many factors other than earthquake size affect the damage and loss of human life resulting from a tremor.

Measuring the physical size of earthquakes and, separately, their effects on people, help scientists and planners understand local and regional earthquake hazard potential.

The scale of *earthquake magnitude,* originally developed by Charles F. Richter in 1935, is based on the energy released in an earthquake as recorded by *seismographs*. Now measured in a different, more accurate way called *moment magnitude,* energy released in an earthquake is expressed in a number,

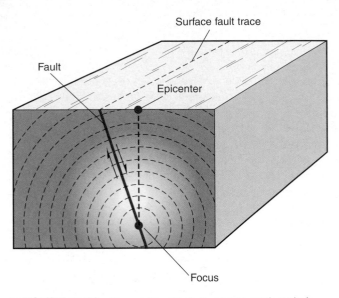

■ **FIGURE 11.33** The relationship between an earthquake's focus and epicenter.

Why is the epicenter in this example not located where the fault crosses Earth's surface?

generally with one decimal place. Each earthquake has only one magnitude, which represents the earthquake's size in terms of energy released. Every increase of one whole number in magnitude (for example, from 6.0 to 7.0) represents about 30 times more energy released. The extremely destructive 1906 San Francisco earthquake occurred before the magnitude scale was devised but is estimated to have been a magnitude 8.3. The strongest earthquake in North America to date, a magnitude 8.6, occurred in Alaska in 1964, and the strongest in the world in recent history was a 9.5 magnitude earthquake that occurred off the coast of Chile in 1960. The tragic December 2004 earthquake that struck off the west coast of northern Sumatra, Indonesia, called the Sumatra–Andaman earthquake, had a moment magnitude of 9.1 and generated the deadliest tsunami on record (■ Fig. 11.34).

A very different type of scale is that used to record and understand patterns of *earthquake intensity,* the damage caused by an earthquake and the degree of its impact on people and their property. The **modified Mercalli scale** of earthquake intensity utilizes categories numbered from I to XII (Table 11.1) to describe the effects of an earthquake on humans and the spatial variation of those impacts. The categories are represented with Roman numerals to avoid confusion with earthquake magnitude values. One earthquake typically produces a variety of intensity levels, depending on variable local conditions. These conditions include

distance from the epicenter, how long shaking lasted, the severity of shaking resulting from the local surface materials affected, population density, and building construction in the affected area. After an earthquake, observers gather information about Mercalli intensity levels by noting the damage and talking to residents about their experiences. The variation in intensity levels is then mapped so that geographic patterns of damage and shaking intensity can be analyzed. Understanding the spatial patterns of damage and ground response helps us plan and prepare so that we can reduce the hazards of future earthquakes.

Earthquake Hazards

Unfortunately, there is plenty of evidence concerning the deadliness of earthquakes. A magnitude 7.9 earthquake devastated extensive areas of eastern Sichuan, China, causing 68,800 fatalities in May of 2008. Together with the ensuing tsunami, the 2004 Sumatra–Andaman earthquake stunned the world by causing almost 300,000 fatalities. In Pakistan, 86,000 lives were lost due to a magnitude 7.6 earthquake in 2005. In 2003, 31,000 people perished from a magnitude 6.6 earthquake in southeastern Iran, and at least 40,000 died in that same country from a magnitude 7.4 earthquake in 1990. More than 5000 people were killed and more than 50,000 buildings were destroyed in Kobe, Japan, when a 7.2 magnitude earthquake struck there in 1995 (■ Fig. 11.35). A magnitude 7.8 earthquake in Peru in 1970 caused 65,000 deaths. These are just a few examples.

The impact of an earthquake on humans and their property depends on many factors. Some of the most important

■ **FIGURE 11.34** The 9.1 magnitude Sumatra–Andaman earthquake, and the tsunami that it generated, devastated the city of Banda Aceh on the Indonesian island of Sumatra. Here, above the lowest set of windows, the only piece of a building left standing displays dark scratches made by debris carried in the tsunami.

USGS/Dr. Guy Gelfenbaum

TABLE 11.1
Modified Mercalli Intensity Scale

I.	Not felt except by a very few under especially favorable conditions.
II.	Felt only by a few persons at rest, especially on upper floors of buildings.
III.	Felt quite noticeably by persons indoors, especially on upper floors of buildings. Many people do not recognize it as an earthquake. Standing motor cars may rock slightly. Vibrations similar to the passing of a truck. Duration estimated.
IV.	Felt indoors by many, outdoors by few during the day. At night, some awakened. Dishes, windows, doors disturbed; walls make cracking sound. Sensation like heavy truck striking building. Standing motor cars rocked noticeably.
V.	Felt by nearly everyone; many awakened. Some dishes, windows broken. Unstable objects overturned. Pendulum clocks may stop.
VI.	Felt by all, many frightened. Some heavy furniture moved; a few instances of fallen plaster. Damage slight.
VII.	Damage negligible in buildings of good design and construction; slight to moderate in well-built ordinary structures; considerable damage in poorly built or badly designed structures; some chimneys broken.
VIII.	Damage slight in specially designed structures; considerable damage in ordinary substantial buildings with partial collapse. Damage great in poorly built structures. Fall of chimneys, factory stacks, columns, monuments, walls. Heavy furniture overturned.
IX.	Damage considerable in specially designed structures; well-designed frame structures thrown out of plumb. Damage great in substantial buildings, with partial collapse. Buildings shifted off foundations.
X.	Some well-built wooden structures destroyed; most masonry and frame structures destroyed with foundations. Rails bent.
XI.	Few, if any (masonry) structures remain standing. Bridges destroyed. Rails bent greatly.
XII.	Damage total. Lines of sight and level are distorted. Objects thrown into the air.

Source: Abridged from *The Severity of an Earthquake: A U.S. Geological Survey General Interest Publication*. U.S. Government Printing Office: 1989, 288–913.

■ **FIGURE 11.35** One of the costliest earthquakes on record in terms of property value, the 7.2 magnitude earthquake that in 1995 struck Kobe, Japan, destroyed more than 50,000 buildings.
What factors might have contributed to the extensive damage?

Patrick Robert/Sygma/CORBIS

pertain to the likelihood of, and the potential number of people affected by structural collapse. Factors influencing structural collapse include the location of an earthquake epicenter relative to population centers, construction materials and methods, and the stability of Earth materials on which buildings were erected. A strong tremor of magnitude 7.5 that struck the Mojave Desert of Southern California in 1992 caused little loss of life because of sparse settlement. Structures made of brick, unreinforced masonry, or other inflexible materials or without adequate structural support do not withstand ground shaking well. Freeway spans collapsed in the San Francisco Bay area in the Loma Prieta earthquake of 1989 and in the 1994 6.7 magnitude Northridge earthquake in the San Fernando Valley area of Los Angeles (■ Fig. 11.36). Wood frame houses generally fare better than brick, adobe, or block structures because the wood frame has greater flexibility during ground shaking. Buildings located on loosely deposited Earth materials (unconsolidated sediment) tend to shake violently compared to areas of solid bedrock. In 1985, the effects of

USGS

■ **FIGURE 11.36** Elevated freeways collapsed in both the 1994 Northridge earthquake in the Los Angeles area (pictured here) and the 1989 Loma Prieta earthquake in the San Francisco Bay area. Each tremor killed close to 60 people, but lives were spared in 1994 because the Northridge earthquake occurred at 4:30 a.m. on a holiday.

an 8.1 magnitude earthquake with an epicenter 385 kilometers (240 mi) from Mexico City caused a death toll of more than 9000 and extensive damage to many buildings, including high-rise sections of the city. Mexico's densely populated capital city is built on an ancient lake bed made of soft sediments that shook greatly, causing much destruction even though the earthquake was centered in a distant mountain region.

In addition to structural collapse, loss of life in an earthquake is influenced by several other factors including time of day, time of year, weather conditions at the time of the quake, resulting fires from downed power lines and broken gas mains, and whether ground shaking triggers rockfalls, landslides, avalanches, or tsunamis. Much greater loss of life will result from freeways that collapse in an earthquake that occurs during rush hour than in one that strikes in the middle of the night. Contributing to the large death toll from the 1970 earthquake in Peru were enormous earthquake-induced avalanches that obliterated entire mountain villages. Even small slope failures that block roads can increase the death toll by hampering access for rescue operations.

Although most major earthquakes are related to known faults and have epicenters in or near mountainous regions, the one most widely felt in North America was not. It was one of a series of tremors that occurred during 1811 and 1812, centered near New Madrid, Missouri. It was felt from Canada to the Gulf of Mexico and from the Rocky Mountains to the Atlantic Ocean. Fortunately, the region was not densely settled at that time in history. That strong tremor accounts for the high earthquake hazard potential indicated in the St. Louis area today (■ Fig. 11.37). Although not common in these regions, recent earthquakes

■ **FIGURE 11.37** Earthquake hazard potential in the contiguous United States, expressed as greatest expectable earthquake intensities.
What is the earthquake hazard potential where you live, and what does that level of intensity mean according to the Mercalli scale in Table 11.1?

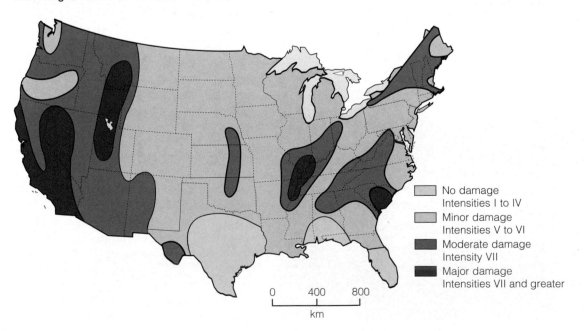

No damage
Intensities I to IV

Minor damage
Intensities V to VI

Moderate damage
Intensity VII

Major damage
Intensities VII and greater

0   400   800
km

have occurred in New England, New York, and the Mississippi Valley. Probably no area on Earth could be called entirely "earthquake safe."

Like volcanic processes, tectonic processes are natural-functioning parts of the Earth system. Many regions of active earthquake or volcanic activity are incredibly scenic and offer attractive environments in which to live, so it is not surprising that some of these hazard-prone areas are densely populated. It is essential, however, that residents and governmental agencies in areas where volcanic, tectonic, or other potentially hazardous natural processes are active make detailed preparations for coping with disasters before they occur.

:: Terms for Review

topography	shield volcano	folding
landform	cinder cone	anticline
geomorphology	angle of repose	syncline
relief	composite cone	faulting
endogenic process	plug dome	fault
exogenic process	caldera	graben
volcanism	stock	horst
plutonism	batholith	rift valley
tephra	laccolith	focus
volcanic ash	sill	epicenter
lava flow	dike	modified Mercalli scale
pahoehoe	strike	
aa	dip	

:: Questions for Review

1. How do endogenic processes differ from exogenic processes?
2. What is the main difference between volcanism and plutonism? How are they alike?
3. What are the four basic types of volcanoes and what are the distinguishing features of each?
4. How is a sill different from a dike?
5. Distinguish among compressional, tensional, and shearing forces.
6. Draw a diagram to illustrate folding, showing anticlines and synclines.
7. Why is a fault not the same thing as a joint?
8. How does a reverse fault differ from a normal fault?
9. What causes an earthquake?
10. What is the relationship between the focus and the epicenter of an earthquake?
11. How do earthquake magnitude and earthquake intensity differ? Why are there two systems for evaluating the severity of earthquakes?

:: Practical Applications

1. What potential hazards should a community located in a region with active faults and volcanoes plan for?
2. Everyone in the city felt the 6.5 magnitude earthquake, and it awakened those who were sleeping. People were frightened and stood under doorways for protection, but received only minor or no injuries. Some dishes broke, some plaster fell, and a few older chimneys were damaged. According to this information, what was the maximum intensity of the earthquake in this city?
3. Using Google Earth, identify the principal type of landform at the following locations (latitude, longitude). Provide a brief discussion of how the landform developed and why it is found at that location.
   a. 37.82°N, 117.64°W
   b. 40.43°N, 77.67°W
   c. 26.20°N, 122.10°W
   d. 39.98°N, 105.29°W

Weathering and Mass Wasting 12

Pressure release and
expansion have caused
the outer shell of this large
granite outcrop to break
parallel to the main mass,
and also into smaller
sections. Many broken
pieces of the outer shell have
already moved downslope off
the outcrop.

J. Petersen

Objectives

When you complete this chapter you should be able to:

- Appreciate that in disintegrating and decomposing rock, weathering prepares rock fragments for erosion, transportation, and deposition.
- Explain the principal differences among the various physical and chemical weathering processes.
- Discuss how variations in climate, rock type, and rock structure influence weathering rates.
- Provide examples of the topographic effects of differential weathering and erosion.

- Understand the role of gravity in mass wasting.
- Use appropriate terms to describe the types of material involved in slope failures.
- Categorize various mass wasting events as slow or fast.
- Describe the major ways by which materials can move downslope.
- Recognize some of the landforms and landscape features created by mass wasting.
- Identify ways in which people influence weathering and mass wasting, and ways in which weathering and mass wasting influence people.

In the previous two chapters we have examined the materials, processes, and structures associated with the construction of topographic relief at Earth's surface. We have also noted that the relief-building endogenic geomorphic processes of volcanism and tectonism, which arise from within Earth, are opposed by relief-reducing exogenic geomorphic processes, which originate at the surface. These exogenic processes break down rocks and erode rock fragments from higher energy sites, transporting them to locations of lower energy. The relocation of rock fragments can be accomplished by the force of gravity alone or with the help of one of the geomorphic agents—flowing water, wind, moving ice, or waves. This chapter focuses on the exogenic processes that cause rocks to decay and on the ways that erosion, transportation, and deposition of surficial Earth materials are accomplished when gravity, rather than a geomorphic agent, is the dominant factor in transporting them.

Gravity constantly pulls downward on all Earth surface materials. Weakened rock and broken rock fragments are especially susceptible to downslope movement by gravity, which may be slow and barely noticeable, or rapid and catastrophic. Slope instability causes costly damage to buildings, roadways, pipelines, and other types of construction, and is also responsible for injury and loss of life. Some gravity-induced slope movements are entirely natural in origin, but human actions contribute to the occurrence of others. Understanding the processes involved and circumstances that lead to slope movements can help people avoid these costly and often hazardous events.

Nature of Exogenic Processes

Most rocks originate under much higher temperatures and pressures and in very different chemical settings than those found at Earth's surface. Surface and near-surface conditions of comparatively low temperature, low pressure, and extensive contact with water cause rocks to undergo varying amounts of disintegration and decomposition (■ Fig. 12.1). This breakdown of rock material at and near Earth's surface is known as **weathering.** Rocks weakened and broken by weathering become susceptible to the other exogenic processes—erosion, transportation, and deposition. A rock fragment broken (weathered) from a larger mass will be removed from that mass (eroded), moved (transported), and set down (deposited) in a new location. Together, weathering, erosion, transportation, and deposition represent a chain or continuum of processes that begins with the breakdown of rock.

Erosion, transportation, and deposition of weathered rock often occurs with the assistance of a geomorphic agent, such as stream flow, wind, moving ice, and waves, but

■ **FIGURE 12.1**   This boulder, which was once hard and solid, has undergone disintegration and decomposition due to conditions at Earth's surface.

Why are some exposed parts of the boulder darker than others?

J. Petersen

sometimes the only factor involved is gravity. Gravity-induced downslope movement of rock material that occurs without the assistance of a geomorphic agent, as in the case of a rock falling from a cliff, is **mass wasting.** Although gravity plays a role in the redistribution of rock material by geomorphic agents, the term mass wasting is reserved for movement caused by gravity alone. Whether it is mass wasting or a geomorphic agent doing the work, fragments and ions of weathered rock are removed from high-energy locations and transported to positions of low energy, where they are deposited.

The variations in elevation at Earth's surface, as well as the shape of different landforms, reflect the opposing tendencies of endogenic and exogenic processes. Relief created by volcanism and tectonism will decrease over time if endogenic processes cease or operate at a slow rate compared to the exogenic processes (■ Fig. 12.2). Rates of the exogenic processes depend on such factors as rock resistance to weathering and erosion, amount of relief, and climate.

Different exogenic geomorphic processes impart visually distinctive features to a landform or landscape. Typically, weathering, mass wasting, or one of the geomorphic agents do not work alone in shaping and developing a landform. More often they work together to modify the landscape, and evidence of the multiple processes can be discerned in the appearance of the resulting landform

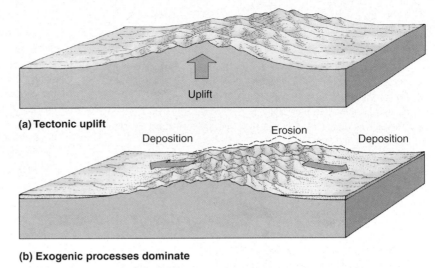

(a) Tectonic uplift

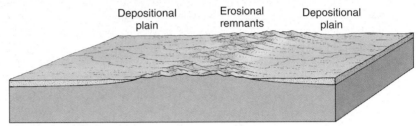

(b) Exogenic processes dominate

(c) Reduced relief

■ **FIGURE 12.2**   (a) Tectonic (endogenic) uplift is opposed by (b) the exogenic processes of weathering, mass wasting, erosion, transportation, and deposition that (c) eventually decrease relief at Earth's surface significantly if no additional uplift occurs.

(■ Fig. 12.3). For example, the northern Rockies were produced by tectonic uplift, but much of the spectacular terrain seen there today is the result of weathering, mass wasting, running water, and glacial activity that have

■ **FIGURE 12.3**   The landscape of most high mountain regions has been produced by at least three phases of landforming processes: stream cutting and valley formation during tectonic uplift; glacial enlargement of former stream valleys, and intense freeze–thaw weathering; and postglacial weathering and mass wasting with recent stream cutting.
How does the cross-sectional profile of the valley change at each phase?

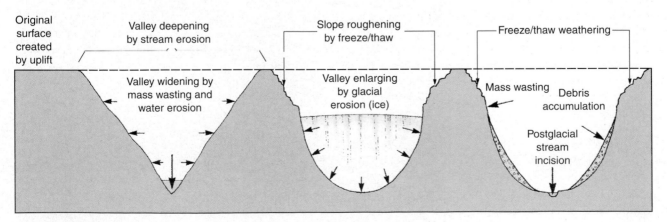

■ **FIGURE 12.4** Sculpted mountain summits in the Sawtooth Range of Idaho.
Can you identify evidence of the three phases shown in Figure 12.3?

USGS/Austin Post

sculpted the mountains and valleys in distinctive ways
(■ Fig. 12.4).

Weathering

Environmental conditions at and near Earth's surface subject rocks to temperatures, pressures, and substances, especially water, that contribute to physical and chemical breakdown of exposed rock. Broken fragments of rock, called *clasts*, that detach from the original rock mass continue to weather into smaller particles. Fragments may accumulate close to their source or be widely dispersed by mass wasting and the geomorphic agents. Many weathered rock fragments become sediments deposited in such landforms as floodplains, beaches, or sand dunes, while others blanket hillslopes as *regolith*, the inorganic portion of soils. Weathering is the principal source of inorganic soil constituents, without which most vegetation could not grow. Likewise, ions chemically removed from rocks during weathering are transported in surface or subsurface water to other locations. They are a major source of nutrients in terrestrial as well as aquatic ecosystems, including rivers, ponds, lakes, and the ocean.

Types of rock weathering fall into two basic categories. **Physical weathering,** also known as **mechanical weathering,** disintegrates rocks, breaking smaller fragments from larger blocks or outcrops of rock. **Chemical weathering** decomposes

rock through chemical reactions that change the original rock-forming minerals. Many different physical and chemical processes lead to rock weathering, and water plays an important role in almost all of them.

Physical Weathering

Mechanical disintegration of rocks by physical weathering is especially important to landscape modification in two ways. First, the resulting smaller clasts are more easily eroded and transported than the initial larger ones. Second, the breakup of a large rock into many smaller rocks encourages additional weathering because it increases the surface area exposed to weathering processes. There are several ways by which rocks can be physically weathered; even a person breaking a rock with a hammer is carrying out physical weathering. Although organisms are responsible for breaking some rocks (■ Fig. 12.5), most physical weathering occurs in other ways. Five principal types of physical weathering are considered here.

Unloading Most rocks form under much higher pressures (weight per unit area) than the 1013.2 millibars (29.92 in. Hg; 15 lbs/in.²) of average atmospheric pressure that exists at Earth's surface. Intrusive igneous rocks solidify slowly deep beneath the surface under great pressure from the weight of the overlying rocks. Sedimentary rocks solidify

J. Petersen

■ **FIGURE 12.5** A growing tree has broken a concrete sidewalk and a retaining wall, just as trees and other plants can break up natural rock into smaller fragments, contributing to physical weathering.
How might an animal cause physical weathering?

J. Petersen

■ **FIGURE 12.6** Outward expansion due to weathering by unloading has caused this granite in the Sierra Nevada, California, to break, forming joints and sheets of rock that parallel the surface.

J. Petersen

■ **FIGURE 12.7** Enchanted Rock in central Texas is a huge granite exfoliation dome.
Why is granite so susceptible to unloading and exfoliation?

partly due to compaction from the weight of overlying sediments. Metamorphic rocks originate when high pressure and temperature substantially change preexisting rocks. Many rocks that originated under conditions of high pressure from deep burial have been uplifted through mountain-building tectonic processes and eventually exposed at the surface. For example, a large pluton of the intrusive igneous rock granite can be uplifted in a fault block during tectonism. High elevation helps drive erosional stripping of the overlying rocks, and ultimately through this removal of overlying weight—the **unloading** process—the granite is exposed at the surface. As a result of the pressure differential between high pressure at depth and low atmospheric pressure at the surface, the outer few centimeters to meters of the rock mass expand outward toward the atmosphere. The expansion causes a crack to form that roughly parallels the exposed rock surface, creating a sheet of rock broken from the main granite mass (■ Fig. 12.6). The expansion crack is a joint, and the breakage is mechanical weathering that resulted from erosional removal of all the rocks originally overlying the pluton. Weathering by unloading is especially common on granite, but can affect other rock types as well.

As the outer sheet of an unloaded rock continues to weather, sections of it may slide off and slough away, further reducing the load on underlying rock and allowing additional concentric joints to form. Removal of successive outer rock sheets is **exfoliation,** and each concentric broken layer of rock is an *exfoliation sheet.* The term *exfoliation dome* designates an unloaded, exfoliating outcrop of rock with a dome-like surface form. Well-known examples of exfoliation domes are Stone Mountain in Georgia, Half Dome in Yosemite National Park, Sugar Loaf Mountain, which

overlooks Rio de Janeiro, Brazil, and Enchanted Rock in central Texas (■ Fig. 12.7).

Thermal Expansion and Contraction Many early Earth scientists believed that the extreme diurnal temperature changes common in deserts physically weather rocks by expansion and contraction as they heat and cool. Those scientists cited widespread existence of split rocks in arid regions as evidence of the effectiveness of this **thermal expansion and contraction** weathering (■ Fig. 12.8). Laboratory studies in the early 20th century seemed to refute the idea of rock weathering by this process, but recent desert field studies lend support to the concept that alternating heating and cooling can lead to the mechanical splitting of rocks. Less controversial has been the notion that differential thermal expansion and contraction of adjacent individual mineral grains in coarse crystalline rocks contributes to physical weathering that leads

■ FIGURE 12.8 This split rock may be the result of physical weathering by thermal expansion and contraction.

■ FIGURE 12.9 The numerous, individual mineral grains accumulated around the base of this intrusive igneous boulder show that the rock is undergoing granular disintegration.
What other evidence exists on the boulder to suggest that it has been subjected to considerable weathering?

to **granular disintegration,** the breaking free of individual mineral grains from a rock (■ Fig.12.9).

Freeze–Thaw Weathering

In areas subject to numerous diurnal cycles of freeze–thaw, water repeatedly freezing in fractures and small cracks in rocks contributes significantly to rock breakage by **freeze–thaw weathering,** sometimes referred to as *frost weathering,* or *ice wedging.* When water freezes, it expands in volume up to 9%, and this can cause large pressures to be exerted on the walls and bottom of the crack, widening it and eventually leading to a piece of rock breaking off (■ Fig. 12.10).

■ FIGURE 12.10 Within rock fractures, the force of expansion of water freezing into ice is great enough to cause some rock weathering.
How important is freeze–thaw weathering where you live?

The damaging effects of freezing water expanding is why vehicles driven in freezing temperatures must have antifreeze instead of water in their radiator systems. Likewise, water pipes to and within buildings will burst if they are not sufficiently insulated to protect the water inside them from freezing. Freeze–thaw weathering is particularly effective in the upper-middle and lower-high latitudes, and results of freeze–thaw weathering are especially noticeable in mountainous regions near tree line where angular blocks of rock attributed to freeze thaw weathering are common (■ Fig. 12.11). Freeze–thaw weathering is not significant at lower latitudes except in areas of high elevation.

Salt Crystal Growth The development of salt crystals in cracks, fractures, and other void spaces in rocks causes physical disintegration in a way that is similar to freeze–thaw weathering. With **salt crystal growth,** water with dissolved salts accumulates in these spaces and then evaporates, and the growing salt crystals wedge pieces of rock apart. This physical weathering process is most common in arid regions and in rocky coastal locations where salts are abundant, but people also contribute to salt weathering when they use salt to melt ice from roads and sidewalks in winter. Salt crystal growth leads to granular disintegration in coarse crystalline (intrusive igneous) rocks and to the removal of clastic particles from sedimentary rocks, especially sandstones.

Hydration In weathering by **hydration,** water molecules attach to the crystalline structure of a mineral without causing a permanent change in that mineral's composition. The water molecules are able to join and leave the "host" mineral during hydration and dehydration, respectively. The mineral

■ **FIGURE 12.11**   Angular blocks of rock attributed to freeze–thaw weathering, like these near tree line in Great Basin National Park, Nevada, are common in mountainous areas that experience numerous freeze–thaw cycles.

Why are these rocks angular in shape rather than rounded?

expands when hydrated and shrinks when dehydrated. As in freeze–thaw and salt crystal growth weathering, when hydrating materials expand in cracks or voids, pieces of the rock can wedge apart. Clasts, mineral grains, and thin flakes can be broken from a rock mass because of hydration weathering. Salts and *clay minerals,* which are clay-sized materials formed during chemical weathering, commonly occupy cracks and voids in rocks and are subject to hydration and dehydration.

Chemical Weathering

Chemical reactions between substances at Earth's surface and rock-forming minerals also work to break down rocks. In chemical weathering, ions from a rock are either released into water or recombine with other substances to form new materials, such as clay minerals. New materials made by chemical weathering are more stable at Earth's surface than the original rocks.

Oxidation   Water that has regular contact with the atmosphere contains plenty of oxygen. When oxygen in water comes in contact with certain elements in rock-forming minerals, a chemical reaction can occur. In this reaction, the element releases its bond with the mineral, leaving behind a substance with an altered chemical formula, and establishes a new bond with the oxygen. This chemical union of oxygen atoms with another substance to create a new product is **oxidation.** Metals, particularly iron and aluminum, are commonly oxidized in rock weathering and form iron and aluminum oxides as the new products. Compared to the original

rock, these oxides, including Fe_2O_3 and Al_2O_3, are chemically more stable, not as hard, larger in volume, and have a distinctive color. Iron oxides produced in this way often have a red or orange color, while oxidized aluminum from rock weathering frequently appears yellow (■ Fig. 12.12). Oxidation of iron is common; when it affects iron and steel objects we call it rust.

■ **FIGURE 12.12**   The reddish orange coloration on this boulder reveals that the rock contains iron-bearing minerals, and that it has been weathered by oxidation.

What is a likely chemical formula for the reddish orange substance?

Solution and Carbonation Under certain circumstances, some rock-forming minerals dissolve in water. In this **solution** process, a chemical reaction causes mineral-forming ions to dissociate, and the separated ions are carried away in the water. Rock salt, which contains the mineral halite (NaCl), is quite susceptible to solution in water. Most minerals that are insoluble or only slightly soluble in pure water will dissolve more readily if the water is acidic. Important sources of acidity in surface and subsurface waters include organic acids that are obtained from the soil, and carbon dioxide obtained from the air and soil.

The chemical weathering process of **carbonation** is a common type of solution that involves carbon dioxide and water molecules reacting with rock material to decompose it. Carbonation weathering is most effective on carbonate rocks (those containing CO_3), particularly limestone, which is an abundant sedimentary rock composed of calcium carbonate ($CaCO_3$). When water with sufficient carbon dioxide comes into contact with limestone, the chemical reaction decomposes some of the calcium carbonate into separate, detached ions of calcium (Ca^{2+}) and bicarbonate (HCO_3^-) that are carried away in the water (■ Fig. 12.13). The weathering event occurs as follows: $H_2O + CO_2 + CaCO_3 = Ca^{2+} + 2HCO_3^-$. Because of the role of water in carbonation, limestone in particular, but carbonate rocks overall, tend to weather extensively in humid regions, but are resistant and often form cliffs in arid climates. Because water can obtain carbon dioxide by moving through the soil, carbonation operates on rocks impacted by soil water or groundwater in the subsurface as well as on rocks exposed at the surface.

Hydrolysis In the weathering process of **hydrolysis,** water molecules alone, rather than oxygen or carbon dioxide in water, react with chemical components of rock-forming minerals to create new compounds, of which the H^+ and OH^- ions of water are a part. For example, hydrolysis of the feldspar mineral $KAlSi_3O_8$ can proceed as follows: $KAlSi_3O_8 + H_2O = HAlSi_3O_8 + KOH$. Many common minerals are susceptible to hydrolysis, particularly the silicate minerals that comprise igneous rocks. Hydrolysis of silicate minerals often produces clay minerals. Because water is the weathering agent, hydrolysis occurs in the subsurface through the action of soil and groundwater as well as on rocks exposed at the ground surface.

The chemical weathering process of hydrolysis differs from the physical weathering process of hydration. Both weathering processes, however, may involve clay minerals because those clay-sized substances are often produced by hydrolysis, and many clay minerals swell and shrink substantially during hydration and dehydration.

Variability in Weathering

How and why type and rates of weathering vary, at both the regional and local spatial scales, are of particular interest to physical geographers. Climate, rock type, and the nature and amount of fractures or other weaknesses in the rocks are major influences on the effectiveness of the various weathering processes. Understanding how specific rock types weather in natural settings is important for explaining the characteristics of regolith, soil, and relief, but it is also important in our cultural environment because weathering affects building stone as well as rock in its natural setting.

Climate

In almost all environments, physical and chemical weathering processes operate together, even though one of these categories usually dominates. Water plays a role in all but two of the physical weathering processes, but it is essential for all types of chemical weathering. Also, chemical weathering increases as more water comes into contact with rocks. Chemical weathering, then, is particularly effective and rapid in humid climates (■ Fig. 12.14). Most arid regions have enough moisture to allow some chemical weathering, but it is much more restricted than in humid climates. Arid regions typically receive sufficient moisture for physical weathering by salt crystal growth and the hydration of salts. Abundant salts, high humidity, and contact with seawater make salt weathering processes very effective in marine coastal locations.

■ FIGURE 12.13 Limestone weathered by carbonation often takes on a fluted, pitted, or even honeycombed appearance.
Why does the limestone near the bottom of the outcrop seem to be more weathered than that at the top?

J. Petersen

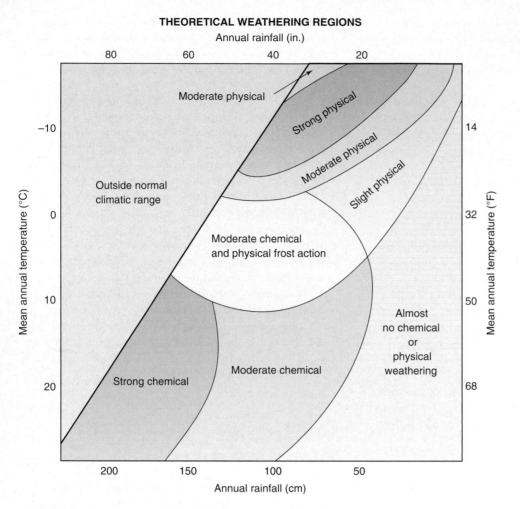

THEORETICAL WEATHERING REGIONS

■ **FIGURE 12.14** This generalized, theoretical diagram indicates that, although chemical and physical weathering processes both occur in any climatic environment, chemical weathering is more intense in regions of high temperature and rainfall. Physical weathering effects tend to be more pronounced where temperature and rainfall are both low.

The other principal climatic variable, temperature, also influences dominant types and rates of weathering. Most chemical reactions proceed faster at higher temperatures. Low-latitude regions with humid climates consequently experience the most intense chemical weathering. In the tropical rainforest, savanna, and monsoon climates, chemical weathering is more significant than physical weathering, soils are deep, and landforms appear rounded. Although chemical weathering is somewhat less extreme in the mid-latitude humid climates, its influence is apparent in the moderate soil depth and rounded forms of most landscapes in those regions. In contrast, the landforms and rocks of both arid and cold regions, where physical weathering dominates, tend to be sharper, angular, and jagged, but this depends to some extent on rock type, and rounded features may remain in an arid landscape as relics from wetter climates of the geologic past (■ Fig. 12.15). Comparatively low rates of chemical weathering are reflected in the thin soils found in arid, subarctic, and polar climate regimes. Daily temperature ranges are critical for thermal expansion and

contraction weathering in arid climates and freeze–thaw weathering in areas with cold winters.

Air pollution that contributes to the acidity of atmospheric moisture accelerates weathering rates. Extensive damage has already occurred in some regions to historic cultural artifacts made of limestone and marble (metamorphosed limestone), both containing calcium carbonate (■ Fig. 12.16). Because many of the world's great monuments and sculptures are made of these materials, there is growing concern about weathering damage to these treasures. The Parthenon in Greece, the Taj Mahal in India, and the Great Sphinx in Egypt are just a few examples of cultural artifacts undergoing pollution-induced solution, salt crystal growth, hydration of salts, and other destructive weathering processes.

Rock Type

Wherever many different rock types occupy a landscape, some will be more resistant and others will be less resistant to the weathering processes operating there. Because erosion

■ **FIGURE 12.15** (a) Due to the dominance of slow physical weathering and the sparseness of vegetation, slopes in arid and semiarid environments tend to be bare and angular. Slope angles reflect differences in component rock resistance to weathering and erosion. (b) Because chemical weathering dominates in humid regions, and abundant vegetation holds the regolith in place longer, slopes in wetter climates have a deep weathered mantle and a rounded appearance.

■ **FIGURE 12.16** These limestone tombstones have undergone extensive chemical weathering, accelerated by air pollution.

What kind of chemical weathering has impacted the iron fence?

removes small, weathered rock fragments more easily than large, intact rock masses, areas of diverse rock types undergo **differential weathering and erosion.**

A rock that is strong under certain environmental conditions may be easily weathered and eroded in a different environmental setting. Rocks that are resistant in a climate dominated by chemical weathering may be weak where physical weathering processes dominate, and vice versa. Quartzite is a good example. It is chemically nearly inert and harder than steel, but it is brittle and can be fractured by physical weathering. Shale is also chemically inert, but it is mechanically weak. In humid regions, limestone is highly susceptible to carbonation and solution, but under arid conditions, limestone is much more resistant. Granitic outcrops in an arid or semiarid region resist weathering. However, the minerals in granite are susceptible to alteration by oxidation, hydration, and hydrolysis, particularly in regions with warm, humid conditions. Accordingly, granitic areas are often covered by a deeply weathered regolith when they have been exposed to a tropical humid environment.

Structural Weaknesses

In addition to rock type, the relative resistance of a rock to weathering depends on other characteristics, such as the presence of joints, faults, folds, and bedding planes, that make rocks susceptible to enhanced weathering. In general, the more massive the rock, that is, the fewer joints and bedding planes it has, the more resistant it is to weathering.

The processes of volcanism, tectonism, and rock formation produce fractures in rocks that can be exploited by exogenic processes, including weathering. Joints can be found in any solid rock that has been subjected to crustal stresses, and some rocks are intensely jointed (■ Fig. 12.17). Joints and other fractures that commonly develop in igneous, sedimentary, and metamorphic rocks represent zones of weakness that expose more surface area of rock, provide space for flow or accumulation of water, collect salts and clay minerals, and offer a foothold for plants (■ Fig.12.18). Rock surfaces along fractures tend to experience pronounced weathering. Chemical and physical weathering both proceed faster along any kind of gap, crack, or fracture than in places without such voids.

Because joints are sites of concentrated weathering, the spatial pattern of joints strongly influences the appearance of the landforms and landscapes that develop. Multiple joints that parallel each other form a **joint set,** and two joint

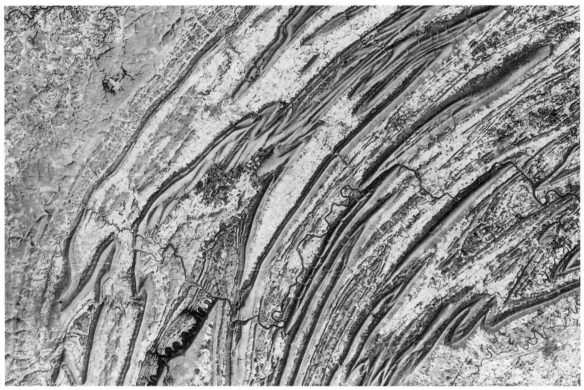

NASA

■ **FIGURE 12.22**   A satellite image of Pennsylvania's Ridge and Valley section of the Appalachians clearly shows the effects of weathering and erosion on folded rock layers of different resistance. Resistant rocks form ridges, and weaker rocks form valleys.
Can you see how the topography of the Ridge and Valley section influences human settlement patterns?

stand up to 700 meters (2000 ft) above agricultural lowlands that have been excavated by weathering and erosion out of weaker shales and soluble limestones.

Mass Wasting

As we previously noted, mass wasting, also called *mass movement,* is a collective term for the downslope transport of surface materials in direct response to gravity. Everywhere on the planet's surface, gravity pulls objects toward Earth's center. This gravitational force is represented by the weight of each object. Heavier objects have a greater downward pull from gravity than lighter objects. The force of gravity encourages rock, sediment, and soil to move downhill on sloping surfaces.

Mass wasting operates in a wide variety of ways and at many scales. A single rock rolling and tumbling downhill is a form of this gravity-driven transfer of materials (■ Fig. 12.23), as is an entire hillside sliding hundreds or thousands of meters downslope, burying homes, cars, and trees. Some mass movements act so slowly that they are imperceptible by direct observation and their effects appear gradually over long periods of time. Other types of mass wasting produce disastrous, instantaneous violence.

The cumulative impact of all forms of mass wasting rivals the work of running water as a modifier of physical landscapes

©AP/Wide World Photos

■ **FIGURE 12.23**   A massive boulder, which was loosened by heavy rains and pulled downhill by the force of gravity, blocks a road in Southern California.
What other kinds of problems on roads are related to mass wasting?

because gravitational force is always present. Wherever loose rock, regolith, or soil lies on a slope, gravity will cause some downslope movement. Friction and rock strength are factors that resist this downslope movement of materials. Friction increases with the roughness and angularity of a rock fragment and the roughness of the surface on which it rests. Rock strength depends on physical and chemical properties of the rock and is reduced by any kind of break or gap in the rock. Fractures, joints, faults, bedding planes, and spaces between mineral grains or clasts all weaken the rock. Furthermore, because all of these gaps invite the accumulation of water, their bonds to the outcrop continue to weaken further over time through weathering.

Slope angle also helps determine whether or not mass wasting will occur. Gravitational forces act to pull objects straight downward, toward the center of Earth. The closer a slope is to being parallel to that downward direction, that is, the steeper the slope angle, the easier it is for the gravitational forces to overcome the resistance of friction and rock strength. Gravity is more effective at pulling rock materials downslope on steep hillsides and cliffs than on gently sloping or level surfaces. The steeper the slope, the stronger the friction or rock strength must be to resist downslope motion (■ Fig. 12.24). Any surface materials on a slope that do not have the strength or stability to resist the force of gravity will respond by creeping, falling, sliding, or flowing downslope until stopping at the bottom of the slope or wherever there is enough friction to resist further movement. As a result, soil and regolith are thinner on steep slopes and thicker on gentle slopes, and intense mass wasting is one reason why bedrock tends to be exposed in areas of steep terrain.

Gravity is the principal force responsible for mass wasting, but water commonly plays a role and can do so in several ways. We have seen that water is involved in many weathering processes that break and weaken rocks, making them more susceptible to mass movement. Unconsolidated soil and regolith have a considerable volume of voids, or pore spaces, between particles. Usually, some of these voids contain air and some contain water, but storms, wet seasons, broken pipelines, irrigation, and other situations can cause the voids to fill with water. Saturated conditions encourage mass wasting because

water adds weight to the sediments, and an object's weight represents the amount of gravitational force pulling on it. As water replaces air in the voids, unconsolidated sediments and soil also experience decreasing strength because the rock fragments come into greater contact with the liquid, which tends to flow downslope. Finally, streams, and in coastal locations waves, undercut the base of slopes by erosion thereby increasing the slope angle and facilitating mass movement.

In some cases, factors besides water contribute to the occurrence of mass wasting. When people remove rock material from the base of a slope for construction projects, they steepen the slope angle, making mass movement more likely. Ground shaking, especially during earthquakes, often triggers mass wasting by briefly separating particles supported by other particles causing a decrease in material strength and friction, both of which resist downslope movement.

Understanding the conditions and processes that affect mass wasting is important because gravity-induced movements of Earth materials are common and frequently impact people and the built environments in which we live. Although this natural hazard cannot be eradicated, we can avoid actions that aggravate the hazard potential and pay close attention to evidence of impending failure in susceptible terrain.

Classification of Mass Wasting

Physical geographers categorize mass wasting events according to the kind of Earth materials involved and the manner in which they move. Mass wasting events are described with a specific term, for example *rockfall*, that summarizes the type of material and the type of motion. The several kinds of motion are also divided into two general groups according to the speed with which they occur.

Types of Earth Material
Anything on Earth's surface that exists in or on an unstable, or potentially unstable, land surface is susceptible to gravity-induced movement and can therefore be transported downslope as a result of mass wasting. Mass wasting involves almost all kinds of surface materials. Rock, snow and ice, soil, earth, debris, and mud commonly experience downslope movements. In a mass wasting sense, **soil** refers to a relatively thin unit of predominantly fine-grained, unconsolidated surface material. A thicker unit of the same type of material is referred to as **earth. Debris** specifies sediment with a wide range of grain sizes, including at least 20% gravel. **Mud** indicates saturated sediment composed mainly of clay and silt, which are the smallest particle sizes.

Speed of Motion
Surface materials move in response to gravity in many different ways depending on the material, its water content, and characteristics of the setting. Some types of mass movement happen so slowly that no one can watch the motion occurring. With these **slow mass wasting** types we can only measure the movement and observe its effects over long periods of time. The motion of **fast mass wasting** can be witnessed by people. The speed of downslope movement

■ **FIGURE 12.24** Besides slope angle and material strength, vegetation cover and soil moisture can influence the occurrence of mass wasting. G = total force of gravity (weight of the block); F = component of the block's weight resisting motion; f = frictional forces resisting motion; D = downslope component of gravity.

How might vegetative cover or moisture content affect the potential for downslope movement of soil?

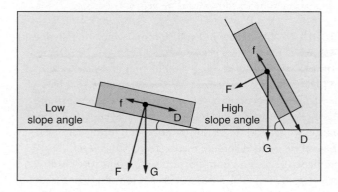

TABLE 12.1
Different Kinds of Mass Wasting Processes

Motion	Common Material	Typical Speed	Effect
Creep	Soil	Slow	
Solifluction	Soil	Slow	
Fall	Rock	Fast	
Avalanche	Ice and snow or debris or rock	Fast	
Slump (rotational slide)	Earth	Fast	
Slide (linear)	Rock or debris	Fast	
Flow	Debris or mud	Fast	

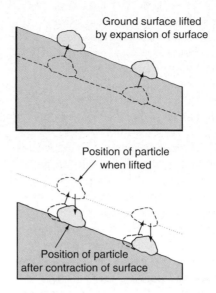

Ground surface lifted by expansion of surface

Position of particle when lifted

Position of particle after contraction of surface

■ **FIGURE 12.25** Repeated cycles of expansion and contraction cause soil particles to be lifted at right angles to the surface slope but to fall straight downward by the force of gravity, resulting in soil creep.
Are there places near where you live that show evidence of soil creep?

of material varies greatly according to details of the slope, the material, and if a triggering factor is involved. In addition, slow and fast mass movements often work in combination. Mass movement that is initially slow may be a precursor to more destructive rapid motion, and most materials that have undergone rapid mass wasting continue to shift with slow movements. Specific types of motion, and common Earth materials associated with them, are discussed next and are summarized in Table 12.1.

Slow Mass Wasting

Slow mass wasting has a significant, cumulative effect on Earth's surface. Landscapes dominated by slow mass wasting tend toward rounded hillcrests and an absence of sharp, angular features.

Creep   Most hillslopes covered with weathered rock material or soil undergo **creep,** the slow migration of particles to successively lower elevations. This gradual downslope motion often occurs as *soil creep*, primarily affecting a relatively thin layer of weathered rock material. Creep is so gradual that it is visually imperceptible; the rate of movement is usually less than a few centimeters per year. Yet creep is the most widespread and persistent form of mass wasting because it affects nearly all slopes that have weathered rock fragments at the surface.

Creep typically results from some kind of **heaving** process, which causes individual soil particles or rock fragments to be first pushed upward perpendicular to the slope, and then

eventually fall straight downward due to gravity. Freezing and thawing of soil water, as well as wetting and drying of soils or clays, can lead to soil heaving. For example, when soil water freezes, it expands, pushing overlying soil particles upward relative to the surface of the slope. When that ice thaws, the soil particles move back, but not to their original position because the force of gravity pulls downward on them, as shown in ■ Figure 12.25. Soil creep results from repeated cycles of expansion and contraction related to freezing and thawing, or wetting and drying, which cause lifting followed by the downslope movement of soil and rock particles. Additional slow downslope movement of soil is accomplished by organisms dislodging particles to a slightly lower position when they traverse, burrow, or extend roots into soil on a slope.

Rates of soil creep are greater near the surface and diminish into the subsurface of a slope because the factors instigating it are more frequent near the surface, and because frictional resistance to movement increases with depth. Telephone poles, fence posts, gravestones, retaining walls, other human structures, and even trees (■ Fig. 12.26) exhibit a downslope tilt when affected by the downward movement of creep.

Solifluction   The word **solifluction,** which literally means "soil flow," refers to the slow downslope movement of water-saturated soil and/or regolith. Solifluction is most common in high-latitude or high-elevation tundra regions that have *permafrost,* a subsurface layer of permanently frozen ground. Above the permafrost layer lies the **active layer,** which freezes during winter and thaws during summer. During summer thaw of the centimeters- to meters-thick active layer, the permanently frozen ground beneath it prevents downward percolation of melted soil water. As a result, the active layer

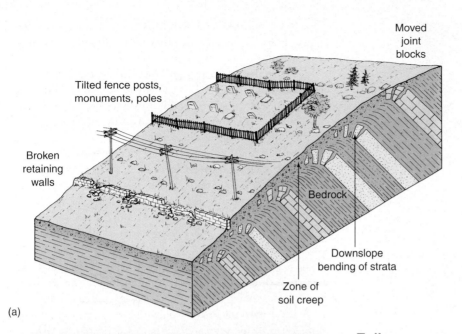

Tilted fence posts, monuments, poles

Broken retaining walls

Moved joint blocks

Bedrock

Downslope bending of strata

Zone of soil creep

(a)

J. Petersen

(b)

■ **FIGURE 12.26** (a) Effects of soil creep are visible in natural and cultural landscapes. (b) Trees attempt to grow vertically, but their trunks become bent if surface creep is occurring.
What other constructed features might be damaged by creep?

becomes a heavy, water-saturated soil mass that, even on a gentle incline, sags slowly downslope by the force of gravity until the next surface freeze arrives. Evidence of solifluction includes irregular lobes of soil that produce hummocky terrain or mounds (■ Fig. 12.27). Movement rates are typically only several centimeters per year.

Fast Mass Wasting

Four major kinds of mass wasting occur so quickly—from seconds to days—that people can watch the material move. The actual speed of movement varies with the situation and depends on the quantity and composition of the material, the steepness of the slope, the amount of water involved, the vegetative cover, and the triggering factor. The effects of fast mass wasting events on the land surface are more dramatic than those of slow mass wasting. Rapid mass movements usually leave a visible upslope scar on the landscape, revealing where material has been removed, and a definite deposit where transported Earth material has come to rest at a lower elevation.

Falls Mass wasting events that consist of Earth materials plummeting downward freely through the air are **falls. Rockfalls** are the most common type of fall. Rocks fall from steep bedrock cliffs, either one by one as weathering weakens the bonds between individual clasts and the rest of the cliff, or as large rock masses that fall from a cliff face or an overhanging

■ **FIGURE 12.27** Solifluction has formed these tongue-shaped masses of soil on a slope near Suslositna, Alaska.
How does solifluction differ from soil creep?

B. Bradley/NOAA, National Geophysical Data Center

■ FIGURE 12.28 Eventually this overhanging sandstone ledge will fail in a rockfall. As more individual rocks fall from the damper, shaded zone beneath the ledge, the size of the overhang will increase until its weight exceeds the strength of the bonds holding it in place.

What weathering processes might be acting on the sandstone beneath the overhang when it becomes wet?

ledge (■ Fig. 12.28). Large slabs that fall typically break into angular individual clasts when they hit the ground at the base of the cliff.

In steep mountainous areas, rockfalls are particularly common during spring when snowmelt, rain, and alternating freezing and thawing loosen and disturb rocks positioned on cliffs and slopes. Ground shaking caused by earthquakes is another common trigger for rockfalls.

Over time, a sloping accumulation of angular, broken clasts piles up at the base of a cliff that is subject to rockfall. This slope is known as **talus,** sometimes referred to as a *talus slope,* or, where cone-shaped, a *talus cone* (■ Fig. 12.29). The presence of talus is good evidence that the cliff above is undergoing rockfall. Like other accumulations of loose, unconsolidated sediment, such as the slopes of cinder cones or piles of gravel in a gravel pit, talus slopes typically lie at or near the steepest slope angle they can maintain before instability, their *angle of repose* (■ Fig. 12.30). The angle of repose commonly lies between 30° and 34° varying with the size and angularity of the clasts. Large angular clasts have a steeper angle of repose than small, more rounded rock fragments.

Falling rocks create hazardous conditions wherever bedrock cliffs are exposed, including mountainous regions and areas with steep roadcuts. Rockfall hazard mitigation remains a high priority along transportation routes in these locations. In July 1996, in Yosemite Valley, California, one hiker was killed and numerous others injured when a huge 180,000-kilogram (200-ton) mass of granite broke away from a cliff, initially slid along bedrock, and then fell through the air for 550 meters (1800 ft) before crashing on the ground with great force. The rockfall was so large that it generated a destructive blast of compressed air that destroyed trees hundreds of meters from the cliff.

■ FIGURE 12.29 Rockfall has constructed this talus, the steep slope of angular clasts, located along the base of the limestone cliff.

J. Petersen

Dave Saville/FEMA News Photos

■ **FIGURE 12.31** Snow avalanches like this one in Alaska can block roads, knock down trees, carry rocks and tree trunks downslope, and damage structures.

■ **FIGURE 12.30** (a) The angle of repose is the steepest natural slope angle that loose material can maintain. Particles are held at this angle by friction between clasts. (b) Nearly identical piles of sediment containing the same size and shape of particles lie at the same angle of repose.

How would the angle of repose of rounded particles differ from that of angular particles of the same size and shape?

Avalanches

An **avalanche** is mass movement in which much of the involved material is pulverized, that is, broken into small, powdery fragments, which then flow rapidly as a density current along Earth's surface. Although the word avalanche may bring to mind billowing torrents of snow and ice roaring down a steep mountainside, *snow avalanches* are not the only kind. Avalanches of pulverized bedrock, called *rock avalanches,* and those of a very poorly sorted mixture of gravel, sand, silt, and clay, called *debris avalanches,* are also common and have caused considerable loss of life and destruction to mountain communities. Many avalanches are triggered by falls of snow and ice, rock, or debris that pulverize when they impact a lower surface. Snow avalanches, the best-known type of avalanche to the public, present serious hazards in areas with steep slopes and deep accumulations of winter snow. Regardless of the specific type of Earth material

involved, avalanches are powerful and dangerous, traveling up to 100 kilometers per hour (60 mph). They easily knock down trees and demolish buildings, and have destroyed entire towns (■ Fig. 12.31).

Slides

In **slides,** a cohesive or semicohesive unit of Earth material slips downslope in continuous contact with the land surface. Water plays a somewhat greater role in most slides than it does in falls. Slides of all kinds threaten the lives and property of people who live in regions with considerable slope and such characteristics as tilted layers of alternating strong and weak rocks.

Slides of large units of bedrock, called *rockslides,* are frequent in mountainous terrain where originally horizontal sedimentary rock layers have been tilted by tectonism. The importance of water in reducing the resisting forces of rock strength and friction is seen in the fact that rockslides are most common in wet years, or after a rainstorm or snowmelt. As weathering and erosion by water weaken contacts between successive rock layers, the force of gravity can exceed the strength of the bonds between two rock layers. When this happens, a unit of rock, often of massive size, detaches and slides along the tilted planar surface of the contact (■ Fig. 12.32). Rockslides sometimes end as rockfalls if the topography is such that free fall is needed to transport the rock to a stable position on more level ground. Deposits from rockslides generally consist of larger blocks of rock than comprise rockfall deposits.

©AFP/Getty Images

■ **FIGURE 12.35** Very large slides that move a variety of materials are referred to as landslides. Landslides often cause much destruction when people build in susceptible areas.

Serious mudflow hazards exist in many active volcanic regions. Here, steep slopes may be covered with hundreds of meters of fine-grained volcanic ash. During eruptions, emitted steam, cooling and falling as rain, saturates the ash, sending down dangerous and fast-moving volcanic mudflows, known as **lahars.** Of particular concern are high volcanic peaks capped with glaciers and snowfields. Should an eruption melt the ice and snow, rapid and catastrophic lahars could rush down the mountains with little warning and bury entire valleys and towns. In the United States, there is concern that some of the high Cascade volcanoes in the Pacific Northwest pose a risk of eruptions and associated lahars. Lahars accompanied the 1980 eruption of Mount St. Helens, and Mounts Rainier, Baker, Hood, and Shasta all have the conditions in place, including nearby populated areas, for potentially disastrous mudflows to occur (■ Fig. 12.38).

which tend to move as cohesive units, flows involve considerable churning and mixing of the materials as they move.

When a relatively thick unit of predominantly fine-grained, unconsolidated hillside sediment or shale becomes saturated and mixes and tumbles as it moves, the mass movement is an *earthflow.* Earthflows occur as independent gravity-induced events or in association with slumps in a compound feature called a *slump-earthflow* (see again Fig. 12.34). A slump-earthflow moves as a cohesive unit along a concave surface in the middle and upper reaches of the failure. Downslope of the failure plane, the mass continues to move, but in the more fluid-like, less cohesive manner of an earthflow.

Debris flows and **mudflows** differ from each other primarily in grain size and sediment attributes. Both flow faster than earthflows, often move down gullies or canyon stream channels for at least part of their travel, create raised channel rims called *flow levees,* and leave lobate (tongue-shaped) deposits where they spill out of the channel. They result from torrential rainfall or rapid snowmelt on steep, poorly vegetated slopes, and are the most fluid of all mass movements. Debris flows transport more coarse-grained sediment than mudflows do.

Debris flows often originate on steep slopes, especially in arid or seasonally dry regions. In humid regions they may occur on steep slopes that have been deforested by human activity or wildfire. In both settings, rain and meltwater flush weathered rock material from the steep slopes into canyons where it acquires additional water from surface runoff. The result is a chaotic, saturated mixture of fine and coarse sediment, ranging in size from tiny clays to large boulders. As it flows down stream channels, some of the debris is piled along the sides as levees. Where a flow spills out of the channel, the unconfined mass spreads out and velocity decreases, resulting in deposition of a lobe of sediment (■ Fig. 12.36). Debris flows are powerful mass wasting events that can destroy bridges, buildings, and roads (■ Fig. 12.37).

■ **FIGURE 12.36** This small debris flow in western Utah left well-developed levees on either side of the channel and deposited a tongue-shaped mass (lobe) of sediment where the flow spread out at the base of the slope.
What evidence is there to indicate this is a site of repeated debris flows?

D. Sack

Weathering, Mass Wasting, and the Landscape

In this chapter, we have concentrated on the exogenic processes of weathering and mass movement. Although neither weathering nor the slower forms of mass movement usually attract much attention from the general public, they are critical to soil formation and, like faster forms of gravity-induced motion, they are significant factors in shaping the landscape. Weathering and mass wasting processes also significantly impact people and our built environment. The impact is reciprocal; our actions accelerate some weathering processes and help induce mass movements.

Every slope reflects the local weathering and mass wasting processes that have acted on it. These, in turn, are largely determined by the properties of the rocks and local climatic factors. Slow weathering of resistant rocks leaves steep hillslopes, while rapid weathering of weak rocks produces gentle hillslopes that are typically blanketed by a thick mantle of soil or regolith. Differential weathering and erosion in areas of multiple rock types or variations in structural weakness produce complex landscapes of variable slopes.

Weathering proceeds rapidly in warm, humid climates where heat and moisture accelerate the reactions that cause minerals and rocks to decompose chemically. Deep mantles of soil and regolith and rounded topographic forms subject to creep dominate these regions. In contrast, rocks in arid and cold climates weather more slowly, mainly by physical processes. Arid-region slopes typically have thin, discontinuous, sparsely vegetated accumulations of predominantly coarse-grained regolith that are easily mobilized during intense precipitation events. A tendency toward fast mass wasting is reflected in the angular slopes that are common in many arid regions of mountainous terrain.

In the following chapters, it will be important to remember the key role that weathering plays in preparing Earth materials for erosion, transportation, and deposition by the geomorphic agents. Because weathered rock fragments are often delivered to a geomorphic agent by gravity-induced movement from adjacent slopes, mass wasting is also important in preparing sediment for redistribution by streams, wind, ice, and waves.

Mark Reid/USGS LHP

■ **FIGURE 12.37**   A 1995 debris flow in La Conchita, California, destroyed several homes. Steep slopes consisting of weak unstable sediments failed during a period of heavy rainfall.
Why might a specific site experience repeated slope failures over time?

■ **FIGURE 12.38**   The violent 1980 eruption of Mount St. Helens in Washington generated lahars—mudflows consisting of volcanic ash. This house was half buried in lahar deposits associated with that volcanic eruption.

Lyn Topinka/USGS CVO

:: Objectives

When you complete this chapter you should be able to:

■ Recall the general distribution of Earth's freshwater resources.

■ Outline the principal differences among the subsurface water zone of aeration, the intermediate zone, and the zone of saturation.

■ Explain how water enters the subsurface and what facilitates its ability to move through the rock matter found there.

■ Describe the principal factors responsible for variations in groundwater supplies over space and time.

■ Provide an example in which the combined topographic and subsurface setting would lead to an artesian spring.

■ Appreciate why underground water is susceptible to pollution.

■ Determine why karst landforms are more common in warm and humid climates than in cold or dry climates.

■ Distinguish among the major landforms produced by solution of rock.

■ Understand what caverns are and how they are formed.

■ Explain how water dripping in caverns leads to the development of stalactites and stalagmites.

As part of our understanding of the distribution and effects of flowing water on Earth, we must consider the part of the hydrologic cycle that operates beneath the surface as well as surface water. Like water flowing at the surface, water beneath Earth's surface moves, carries other substances, influences the form and appearance of the landscape, and represents an important source of freshwater for human use.

Freshwater is a precious, but limited, natural resource. There is much concern today throughout the world about the quantity and quality of our freshwater resources. Many populated regions have limited supplies of freshwater, while ironically in some sparsely populated or uninhabited areas, such as tundra and tropical rainforest regions, potable water is plentiful. Ice and snow in the polar regions represent 70% of the freshwater on Earth, but it is generally unavailable for human use. When most people think of freshwater, they envision rivers and lakes, both of which are important sources; together, however, they represent less than 1% of this crucial resource. The remaining freshwater, nearly 30%, lies close to, but beneath, Earth's surface. These underground resources make up an impressive 90% of the freshwater that is readily available for human use (■ Fig. 13.1).

In the last 100 years, worldwide usage of freshwater has grown twice as fast as the population. At the same time, the quality of our freshwater resources has been declining due to pollution of both surface and subsurface water. Because of the severity of the global freshwater problem, and the crucial role of water for life on Earth, the United Nations has declared 2005–2015 the Decade for Action: Water for Life to promote development, conservation, and wise use of water resources. Understanding the nature and distribution of the largest source of our freshwater resources—underground water—is critical to maintaining enough water of suitable quality for domestic, agricultural, and industrial purposes, and to maintaining environmental quality in general. Thus, in this chapter we investigate the nature and distribution of underground water as well as its impact on the landscape.

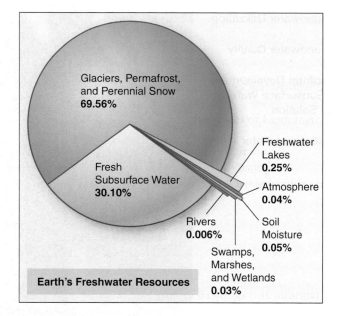

■ **FIGURE 13.1** With 97% of Earth's water existing as salt water in the oceans, Earth's freshwater resources are very limited. Of the freshwater, almost 70% is stored in glacial ice; much of the rest lies underground.

How can we work to conserve our freshwater resources?

The Nature of Underground Water

Subsurface water is a general term encompassing all water that lies beneath Earth's surface. It includes water contained within soil, sediments, and rock. Most of this water is delivered to the subsurface from the atmosphere as precipitation. Water from precipitation or meltwater from frozen precipitation that soaks into the ground, does so by the process of **infiltration.** During infiltration, water moves from the ground surface into void spaces in soil and loose sediments, and into cracks, joints, and other fractures in rock. Infiltration

recharges, that is, replenishes or adds to, the amount of water in subsurface locations. Water from underground sources, in turn, reaches the surface in seeps, springs, and wells, and it contributes substantially to water in streams and in standing water bodies, such as lakes and ponds.

Some underground water resources tapped and used by people today are irreplaceable because they accumulated during previous wetter times in geologic history. A small portion of subsurface water is so deep beneath Earth's surface that it may never have been part of the hydrologic cycle. Elsewhere, because of changes in Earth's surface, subsurface water has been cut off from the hydrologic cycle for a long period of time. This water is contained deep within sediment layers that were deposited by ancient rivers or seas. Future changes in the lithosphere could release these trapped waters and return them to the hydrologic cycle. Volcanic activity, for example, can release some of this water in the form of steam during eruptions and as steam or hot water in geysers and hot springs.

Subsurface Water Zones and the Water Table

Organized by their depth and water content, three distinct subsurface water zones exist in humid regions (■ Fig. 13.2). Under conditions of moderate precipitation and good drainage, water infiltrating into the ground first passes through a layer called the **zone of aeration** in which pore spaces in the soil and rocks almost always contain both air and water. This uppermost zone only rarely becomes saturated. If all the pore spaces do become filled with water because of a large rainfall or snowmelt event, it is temporary. Water will soon drain downward by gravity beyond the zone of aeration to lower levels by the process of **percolation.** Water in the zone of aeration is known as **soil water.**

In the lowest of the three layers of underground water, the openings in sediments and rocks are completely filled with water. The water in this **zone of saturation** is called

■ **FIGURE 13.2   Environmental Systems: Underground Water** The primary inputs to the underground subsystem of the hydrologic cycle are precipitation and snowmelt that infiltrate into the ground. From the surface down, the subsurface water system consists of the zone of aeration, the intermediate zone, and the zone of saturation. Air almost always occupies many of the void spaces in the zone of aeration; water occupies all of the void spaces in the zone of saturation, the top of which is the water table. Infiltrated water percolates down beyond the zone of aeration to the transitional intermediate zone and eventually into the zone of saturation, that is, the groundwater zone. Underground water exits the subsurface by direct evaporation from near-surface locations, through transpiration by plants, by seeping into streambeds and lakes, as natural springs, or by flowing into dug wells. Water table depth responds to changes in infiltration and outflow, falling during dry seasons or years and rising during wet seasons or years.

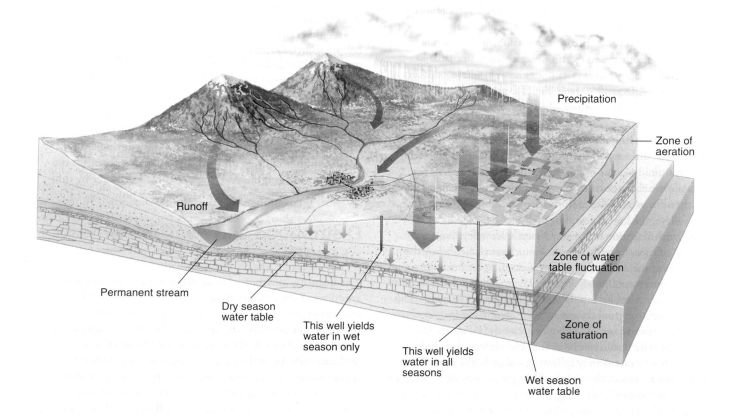

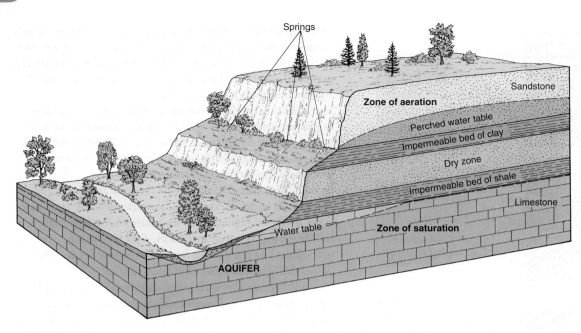

■ **FIGURE 13.5** An aquifer is a natural underground storage medium for groundwater. A perched water table can develop where impermeable rock lies between a permeable layer and the regional water table. Beneath the water table, water in the regional zone of saturation flows toward the nearby river.

Is a perched water table a reliable source of groundwater?

Groundwater Utilization

Groundwater is a vital resource to most of the world. Half of the population of the United States derives its drinking water from groundwater, and some states, such as Florida, draw almost all of their drinking water from this source. Irrigation, however, consumes the bulk of the groundwater—over two-thirds—used in the United States today. One of the largest aquifers supplying groundwater for irrigation is the Ogallala Aquifer, which underlies the Great Plains from west Texas northward to South Dakota. The Ogallala Aquifer alone supplies more than 30% of the groundwater used for irrigation in the United States (■ Fig. 13.6). Considerable concern exists about the future of this aquifer. Much of the water withdrawn from it accumulated thousands of years ago, and with the now-semiarid climate of the region, recharge is limited.

Groundwater also plays a major role in supporting many wetlands and in forming shallow lakes and ponds, all of which are ecologically invaluable resources. Water bodies and wetlands, fed by groundwater, are critical habitats for thousands of resident and migratory birds. Adequate groundwater flow is vital for the survival of the Everglades in southern Florida. This "river of grass," its great variety of birds, and many other animals are totally dependent on the continued southward movement of groundwater flow through the region.

Wells

Wells are artificial openings dug or drilled below the water table to extract water. Water is drawn from wells by lifting devices ranging from simple rope-drawn water buckets to pumps powered by gasoline, electricity, or wind. In many shallow wells, the supply of water varies with fluctuations of the water table. Deeper wells that penetrate aquifers beneath the zone of water table fluctuation provide more reliable sources of water and are less affected by seasonal periods of drought.

In areas where there are many wells or a limited groundwater supply, the rate of groundwater removal may exceed the rate of its natural replenishment through groundwater recharge. Many areas that are irrigated from wells have had their water table fall below the depth of the original wells (■ Fig. 13.7). Progressively deeper wells must be dug, or the old ones extended, in order to reach the supply of water. In the Ganges River Valley of northern India, the excavation of deep, modern wells to replace shallow hand-dug wells has increased the amount of groundwater being brought to the surface, but this greater usage has lowered the water table significantly. As previously mentioned, the Ogallala Aquifer is experiencing alarming water level declines, a condition that can be attributed to wells used mainly for agricultural irrigation.

In certain environments, particularly where high groundwater demand has led to extensive pumping, a sinking of the land, called *subsidence*, can occur due to compaction related to the water withdrawal. Mexico City, Venice (Italy), and the Central Valley of California, among many other places, have subsidence problems related to groundwater withdrawal. In parts of Southern California, withdrawn groundwater has been replaced artificially by diverting streams so that they flow over permeable deposits. This process is known as artificial recharge.

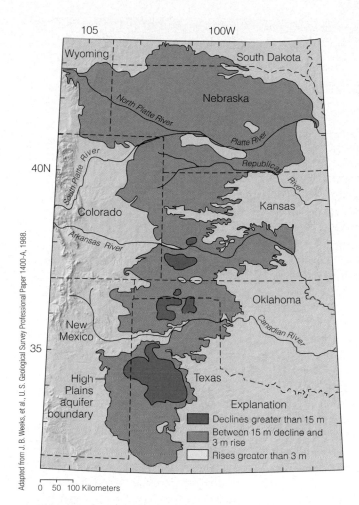

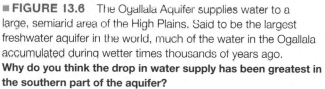

■ **FIGURE 13.6** The Ogallala Aquifer supplies water to a large, semiarid area of the High Plains. Said to be the largest freshwater aquifer in the world, much of the water in the Ogallala accumulated during wetter times thousands of years ago.
Why do you think the drop in water supply has been greatest in the southern part of the aquifer?

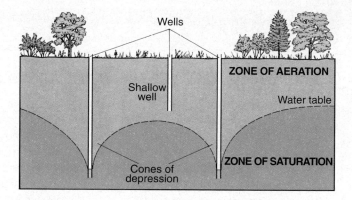

■ **FIGURE 13.7** Cones of depression may develop in the water table due to the pumping of water from wells. In areas with many wells, adjacent cones of depression intersect, lowering the regional water table and causing shallow wells to go dry.
What impact might this scenario have on some of the local natural vegetation?

Artesian Systems

In some cases, groundwater exists in **artesian** conditions, meaning the water is under so much pressure that if it finds an outlet it will flow upward to a level above the local water table. Where this water achieves outflow to the surface, it creates an **artesian spring** if natural; where the outlet to the surface is artificial, the result is an **artesian well.** In fact, the word artesian is derived from the Artois region of France, where the first known well of this kind was dug in the Middle Ages.

Certain conditions are required for artesian water flow (■ Fig. 13.8). First, a permeable aquifer, often sandstone or limestone, must be exposed at the surface in an area of high recharge by precipitation. This aquifer must receive water from the surface by infiltration, incline downward often hundreds of meters below the surface, and be confined

■ **FIGURE 13.8** Special conditions produce an artesian system. The Dakota Sandstone, an artesian aquifer system that averages 30 meters (100 ft) in thickness, transmits water under pressure from the Black Hills to locations more than 320 kilometers (200 mi) eastward beneath South Dakota.
What is unique about artesian wells?

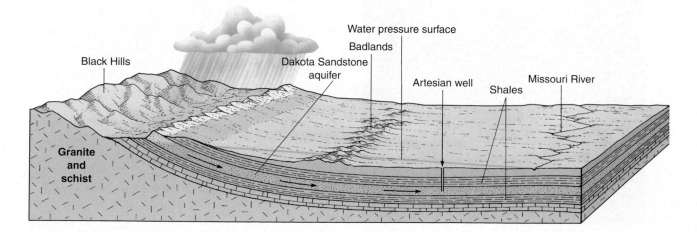

:: ACID MINE DRAINAGE

Acid mine drainage (AMD) typically occurs when subsurface water flowing through mines or mine tailings undergoes chemical reactions that leave the water highly acidic (low pH) and metal-rich. AMD is a serious environmental concern in some coal- and metal-mining regions, including much of the coal-mining area of the eastern United States, parts of Australia, South America, South Africa, and elsewhere. When sufficient quantities of the affected water flow onto the surface as springs or seeps and into streams, lakes, or ponds, the mineral content and low pH endanger aquatic organisms and make the water unsuitable for human consumption. Fish have completely disappeared from some streams with very low pH and high metal content due to AMD.

The chemical reactions involved in the formation of AMD occasionally happen under natural conditions, but at much slower rates and in much smaller amounts than on disturbed lands. Mining greatly increases the permeability of susceptible rocks, allowing greater amounts of water to undergo the chemical reactions. Susceptible rocks contain pyrite (FeS_2), a common substance in the extensive coals of Pennsylvanian age in the eastern United States. When water containing oxygen comes into contact with pyrite in an abandoned underground coal mine, the pyrite oxidizes readily. Iron, sulfate (SO_4), and hydrogen ions are released into the water as a result of the chemical weathering; it is the presence of hydrogen ions that increases the water's acidity. Additional chemical reactions lead to hydrolysis of the iron ions in the water, which releases more hydrogen ions, contributing to a further increase in acidity. Rate of hydrogen production is greatly accelerated if certain microorganisms that thrive in conditions of low pH are present.

Mines that lie in the zone of the fluctuating water table between the zone of aeration and the zone of saturation, like many mines in the Appalachian coal fields, are particularly susceptible to AMD because of the frequent introduction of new, moving, oxygenated water. Pumping removes water from these mines while they are operational. When they are abandoned, however, flowing underground water returns to the now highly permeable and chemically reactive environment producing AMD.

Controlling the flow of underground water represents a principal way to limit production of AMD. Approaches include locking susceptible mines out of subsurface water circulation by sealing water out, or flooding them with very slow moving, stagnant groundwater that has little opportunity to obtain new oxygen. AMD prevention and reduction remains an active field of research, and understanding the chemistry and circulation of underground water is critical to the ongoing investigation.

The orange-colored stream water and orange coating on some of the adjacent rocks result from acid mine drainage in this Appalachian coal-mining region. Acid mine drainage lowers the pH of local streams, which can lead to precipitation of orange-colored compounds of oxidized iron.

D. Sack

between impermeable layers that prevent escape of the water, except to the artesian springs or wells. These conditions cause the aquifer to act as a pipe that conducts water through the subsurface. Water at a given point in the "pipe" is under pressure from the water in the aquifer that lies upslope closer to the area of intake at the surface. As a consequence of this pressure, water flows toward any available outlet. If that outlet happens to be a well drilled through the overlying impervious layer and into the aquifer, the water will rise in the well. The height to which the water rises in a well depends on the amount of pressure exerted on the water. Pressure in turn depends on the quantity of water in the aquifer, the angle of incline for the aquifer, and the number of other outlets, usually wells, available to the water. The pressure, and the water height in the wells, increase with greater amounts of water, steeper inclines, and fewer outlets.

Sandstone exposed at the surface in Colorado and South Dakota transmits artesian water eastward to wells as far as 320 kilometers (200 mi) away (see again Fig. 13.8). Other well-known artesian systems are found in Olympia, Washington; the western Sahara; and eastern Australia's Great Artesian Basin, which is the largest artesian system in the world.

Groundwater Quality

Because most subsurface water percolates down through a considerable amount of soil and rock, by the time it reaches the zone of saturation it is mostly free of clastic sediment. However, it often carries a large amount of minerals and ions dissolved from the materials through which it passed. As a result of this large mineral content, groundwater is often described as "hard water," in comparison with "softer" (less mineralized) rainwater. Moreover, just as increases in population, urbanization, and industrialization have resulted in the pollution of some of our surface waters, they have also resulted in the pollution of some of our groundwater supplies. For example, subsurface water moving through underground mines in certain types of coal deposits that are widespread in the eastern United States becomes highly acidic. If this **acid mine drainage** reaches the surface, it can have very detrimental effects on the local aquatic organisms.

Other dangers to groundwater quality stem from the introduction of toxic substances or salt water into the zone of saturation. Excessive applications of pesticides and incompletely sealed surface or subsurface storage facilities for toxic substances, including gasoline and oil, are situations that can lead to groundwater pollution through downward percolation. In coastal regions with excessive groundwater pumping, denser salt water from the ocean seeps inland to replace freshwater withdrawn from the zone of saturation. This problem with saltwater replacement has occurred in many coastal localities, notably in southern Florida, New York's Long Island, and Israel.

Landform Development by Subsurface Water and Solution

In areas where the bedrock is soluble in water, subsurface water is an important agent in shaping landform features. Like surface water, subsurface water dissolves, removes, transports, and deposits rock-forming materials.

The principal mechanical role of subsurface water in landform development is to encourage mass movement by adding weight and reducing the strength of soil and sediments, thereby contributing to slumps, debris flows, mudflows, and landslides. Chemically, subsurface water also contributes to distinctive processes of landform development. Through the chemical breakdown of rock materials by carbonation and other forms of solution, and the deposition of those dissolved substances elsewhere, underground water is an effective landshaping agent, especially in areas where limestone is present. Subsurface as well as surface water can dissolve limestone through carbonation or simple solution in acidic water. In many of these areas, surface outcrops of limestone are pitted and pockmarked by chemical solution, especially along joints, in some cases forming large, furrowed limestone platforms (■ Fig. 13.9). Wherever water can act on any rock type that is significantly soluble in water, a distinctive landscape will develop.

■ **FIGURE 13.9**   Solution of limestone is most intense in fracture zones where the dissolved minerals from the rock are removed by surface water infiltrating to the subsurface. This landscape shows a limestone "platform" where intersecting joints have been widened by solution.

J. Petersen

Karst Landscapes and Landforms

Overwhelmingly, the most common soluble rock is limestone, the chemical precipitate sedimentary rock composed of calcium carbonate ($CaCO_3$). Landform features created by the solution and reprecipitation (redeposition) of calcium carbonate by surface or subsurface water are found in many parts of the world. The eastern Mediterranean region in particular exhibits large-scale limestone solution features. These are most clearly developed on the Karst Plateau along Croatia's scenic Dalmatian Coast. Landforms developed by solution are called **karst** landforms after this impressive locality. Other extensive karst regions are located in Mexico's Yucatán Peninsula, the larger Caribbean islands, central France, southern China, Laos, and many areas of the United States (■ Fig. 13.10).

The development of a classic karst landscape, in which solution has been the dominant process in creating and modifying landforms, requires several special circumstances. A warm, humid climate with ample precipitation is most conducive to karst development. In arid climates, karst features are typically absent or are not well developed, but some arid regions have karst features that originated during previous periods of geologic time when the climate was much wetter than it is today. Compared to colder humid climates, warmer humid climates have greater amounts of vegetation, which supplies carbon dioxide to subsurface water. Carbon dioxide is necessary for carbonation of limestone and it increases the acidity of the water, which encourages solution in general.

Another important factor in the development of karst landforms is movement of the subsurface water. This movement allows water that has become saturated with dissolved calcium carbonate to flow away and be replaced by water unsaturated with calcium carbonate, and therefore capable of dissolving additional limestone. Groundwater moves vigorously where it flows toward a low outlet, as might be provided along a deeply cut stream valley or a tectonic depression. In addition, everything else being equal, faster groundwater flow occurs in Earth material that has a higher permeability.

Because infiltration of surface water into the subsurface tends to be concentrated where cross-cutting sets of joints intersect, in limestone these intersections are subject to intense solution. Such concentrated solution may produce roughly circular surface depressions, called **sinkholes** or **dolines,** that are prominent features of many karst landscapes (■ Figs. 13.11a and b). Two types of sinkholes are distinguished by differences

■ **FIGURE 13.10** The distribution of limestone in the contiguous United States indicates where varying degrees of karst landform development exists, depending on climate and local bedrock conditions.
Where is the nearest karst area to where you live?

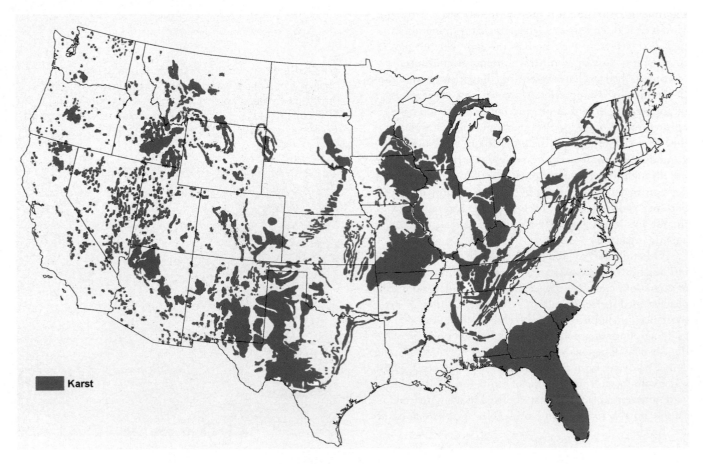

Karst

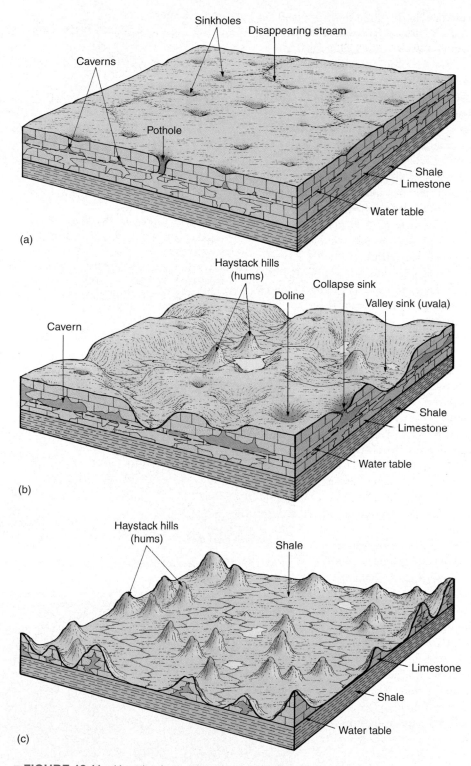

■ **FIGURE 13.11**   Karst landscapes can be quite varied. (a) Solution at joint intersections in limestone encourages sinkhole development. Caverns form by groundwater solution along fracture patterns and between bedding planes in the limestone. Cavern ceilings may collapse, causing larger and deeper sinkholes. Surface streams may disappear into sinkholes to join the groundwater flow. (b) Some limestone landscapes have more relief, such as this irregular terrain with merged sinkholes that make karst valleys, called uvalas, and some conical haystack hills (hums). (c) Areas that have experienced more intense solution may be dominated by limestone remnants, with haystack hills isolated above an exposed surface of insoluble shale.

Why are there no major depositional landforms created at the surface in areas of karst terrain?

in their mode of formation (■ Fig. 13.12). If the depressions are due primarily to the surface or near-surface solution of rock and the removal of dissolved materials by water infiltrating downward into the subsurface, the depressions are **solution sinkholes** (■ Fig. 13.13). **Collapse sinkholes** result when the land surface caves into large subsurface voids created by solution of bedrock some distance beneath the surface (■ Fig. 13.14).

The two processes, solution and collapse, cooperate to create most sinkholes in soluble rocks. Whether the depressions are termed solution or collapse sinkholes depends on which of these two processes was dominant in their formation. Collapse and solution sinkholes often occur together in a region, and they may be difficult to distinguish from one another based on their form. There is a tendency for solution sinkholes to be funnel-shaped, and collapse sinkholes to be steep-walled, but these shapes vary greatly.

Sudden collapse of sinkholes is a significant natural hazard that every year causes severe property damage and human injury. The rapid formation of sinkholes may result from excessive groundwater withdrawal for human use or from significant drought periods. Either of these conditions lowers the water table, causing a loss of buoyant support for the ground above, followed by collapse. Rapid sinkhole collapse has damaged roads and railroads and has even swallowed buildings.

Sinkholes may grow in area and merge over time to form larger karst depressions, called **uvalas,** or **valley sinks** (see again Fig. 13.11b). In many areas, uvalas are linearly arranged along former routes of subsurface water flow. Like the word *doline*, the term *uvala* is derived from Slavic languages used in the former Yugoslavia.

Despite their humid climates, many karst regions have few continually flowing surface streams. Surface water seeping into fractures in limestone widens the fractures by solution. These widened avenues for water flow increase the downward permeability, accelerate infiltration, and direct water rapidly toward the subsurface zone of saturation. In some cases, surface streams flowing on bedrock of low permeability upstream will encounter a highly permeable rock downstream, where it rapidly loses its surface flow to infiltration (see again Fig. 13.9). These are called **disappearing streams** because they "vanish" from the surface as the water flows into the subsurface (■ Fig. 13.15a).

Water moving along joints and bedding planes below the surface can also dissolve limestone, sometimes creating a system of connected passageways within the soluble bedrock (see again Figs. 13.11a and b). If the water table falls, leaving these passageways above the zone of saturation, they are called **caverns** or **caves.** In many well-developed karst landscapes with caverns, a complex underground drainage system all but replaces surface water flow. In these cases, the surface terrain consists of many

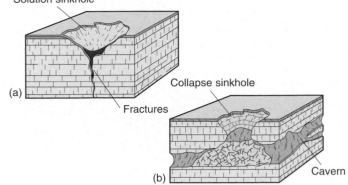

■ **FIGURE 13.12**   The two major types of sinkholes (dolines) based on principal mode of formation. (a) Solution sinkholes develop gradually where surface water funneling into the subsurface dissolves bedrock to create a closed surface depression. (b) Collapse sinkholes form when either the bedrock or the regolith above a large subsurface void fails, falling into the void.

D. Sack

■ **FIGURE 13.13**   A solution sinkhole in southern Indiana.

© St. Petersburg Times

■ **FIGURE 13.14**   This large collapse sinkhole appeared in Winter Park, Florida, when a drought-induced lowering of the water table caused the surface to collapse into an underground cavern system.
What human activities might contribute to such hazards?

large valleys that contain no streams. Surface streams originally existed and excavated the valleys, but the water flow was eventually diverted to underground conduits through the caverns. The site where a stream disappears into a cavern system is called a **swallow hole** (Fig. 13.15b). In some cases these underground-flowing "lost rivers" reemerge at the surface as springs where they encounter impervious beds below the limestone.

After intense and long-term karst development, especially in wet tropical conditions, only limestone remnants are left standing above insoluble rock below. These remnants usually take the form of small, steep-sided, and cave-riddled karst hills called **haystack hills, conical hills,** or **hums** (Figs. 13.11b and c). Multiple terms exist because similar features of slightly different form were independently named in different parts of the world. Excellent examples of these steep, karstic hills are found in Puerto Rico, Cuba, and Jamaica. Areas dominated by these features have been described as "egg box" landscapes because an aerial view of the numerous sinkholes and hums resembles the shape of an egg carton (■ Fig. 13.16). If the limestone hills are particularly high and steep-sided, the landscape is called **tower karst.** Spectacular examples of tower karst landscapes are found in southern China and Southeast Asia (■ Fig. 13.17).

Limestone Caverns and Cave Features

Of the various types of landforms created by limestone solution, caverns are the best-known and most spectacular. Caverns originate when subsurface water, sometimes flowing much like an underground stream, dissolves rock, leaving networks of passageways. Should the water table drop to the floor of the cavern or lower, the caves will become filled with air. Interaction between the cave air and mineral-saturated water percolating down to the cavern from above will then begin to precipitate minerals, especially calcium carbonate, on the cave ceiling, walls, and floor, decorating them with often-intricate depositional forms.

The nature of any fracturing that exists in soluble bedrock exerts a strong influence on cavern development in karst regions. Groundwater solution widens joints, faults, and bedding planes to produce passageways. The relationship between caverns and fracture distributions is evident on cave maps that show linear and parallel patterns of cave passageways (■ Fig. 13.18).

Limestone caverns vary greatly in size, shape, and interior character. All caves formed by solution, however, show evidence of previous water flow, such as deposits of clay and silt on the cavern floor; in some caves water still actively flows. Continued subterranean flow of water further deepens some caverns, and ceiling collapse enlarges them upward. Some cavern systems are extensive, with kilometers

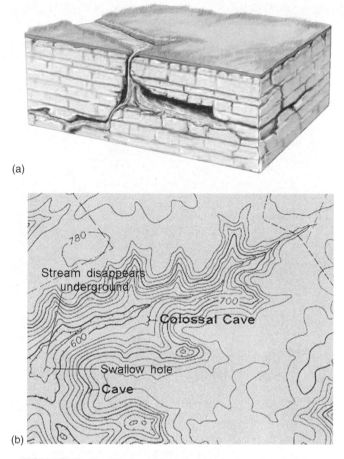

(a)

(b)

■ **FIGURE 13.15** Disappearing streams and swallow holes are common in some karst areas. (a) A surface stream disappears through a solution-widened hole in the ground, the swallow hole, to flow instead through a subsurface cavern system. The stream may emerge back onto the surface at a major spring. (b) This topographic map shows a disappearing stream and a swallow hole. The hachured contours indicate the closed depression into which the stream is flowing.

■ **FIGURE 13.16** Intense solution in wet tropical environments can produce a karst landscape consisting of a maze of haystack hills and intergrown sinkholes.

Courtesy Parris Lyew-Ayee

■ **FIGURE 13.17**   Guilin, a limestone region in southern China, is famous for its beautiful tower karst.
At one time in the geologic past, could this region have looked much like the landscape in Figure 13.16?

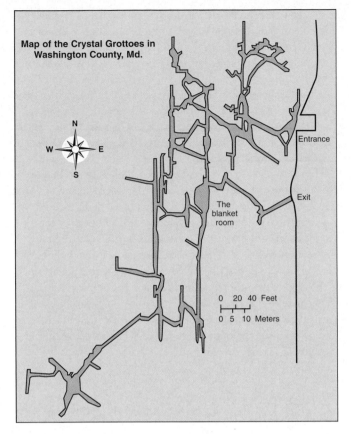

Map of the Crystal Grottoes in Washington County, Md.

■ **FIGURE 13.18**   This map of the Crystal Grottoes in Washington County, Maryland, illustrates the influence of fractures on the growth of the cavern system, as evidenced by the geometric arrangement and spacing of passageways. Groundwater flow widens fractures and other zones of weakness to develop a cavern system.

■ **FIGURE 13.19**   Stalactites form where mineralized water that has percolated down to the cave from above reaches the cave ceiling and deposits some of its dissolved minerals, typically calcium carbonate, as hanging features. These examples are found in Timpanogos Cave in Utah.
Why does evaporation of water tend to be only a minor process in the formation of stalactites?

of connecting passageways and rooms more than 30 meters (100 ft) high. Many cavern systems have several levels and are almost sponge-like in the pattern of their passageways, whereas others are linear in pattern. Some caves contain elaborate deposits of chemical precipitates, but others do not. The variations in cavern size and form may indicate differences in mode of origin. Some small caves might have formed above the water table by water percolating downward through the zone of aeration. The majority of caverns, however, developed just below the water table, where the rate of solution is most rapid. A subsequent drop in the water table level, caused by the incision of surface streams, climate change, or tectonic uplift, fills the cavern with air allowing calcium carbonate deposition to begin.

Speleothem is the generic term for any chemical precipitate feature deposited in caves. Speleothems develop in a great variety of textures and shapes, and many are delicate and ornate. Speleothems originate when previously dissolved substances, particularly calcium carbonate ($CaCO_3$), precipitate out of subsurface water and eventually accumulate into some of the most beautiful and intricate forms found in nature. The dripping water leaves behind a deposit of calcium carbonate called *travertine,* or *dripstone.* As these travertine deposits grow downward, they form icicle-like spikes called **stalactites** that hang from the ceiling (■ Fig. 13.19). Water

■ **FIGURE 13.20** Speleothems in Oregon Caves illustrate how stalactites and stalagmites join to create a column.

saturated with calcium carbonate dripping onto the floor of a cavern builds up similar but more massive structures called **stalagmites.** Stalactites and stalagmites often meet and continue growing to form pillar-like **columns** (■ Fig. 13.20).

People who have not studied caves often think that evaporation, leaving deposits of calcium carbonate behind, is the dominant process that produces the spectacular speleothems seen in caverns, but that is not the case. Caves that foster active development of speleothems typically have air that is fully saturated with water, having a relative humidity near 100%, so evaporation is minimal. Instead, much of the water that percolates into the cave from above first acquired carbon dioxide from the soil and then dissolved calcium carbonate by carbonation on its way to the cave. Cave air, in contrast, contains a comparatively low amount of carbon dioxide. When the dripping water contacts the cave air, it therefore releases carbon dioxide gas to the air. This degassing of carbon dioxide from the water essentially reverses the carbonation process, causing the water to precipitate calcium carbonate. Eventually, enough calcium carbonate is deposited in this way to form a stalactite, stalagmite, or other depositional cave feature.

Cavern development is a complex process involving rock structure, groundwater chemistry, and hydrology, as well as the regional tectonic and erosional history. As a result, the scientific study of caverns, **speleology,** is particularly challenging. Much of our knowledge of cavern systems has come from explorations as deep as hundreds of meters underground by individuals making scientific observations while crawling through mud, water, and bat droppings in dark, narrow passages (■ Fig. 13.21). Exploration and mapping of water-filled caves involve scuba diving. Cave diving

in fully submerged, totally dark, and confined passageways, which may contain dangerous currents, is an extremely risky operation.

Geothermal Water

Water heated by contact with hot rocks in the subsurface is referred to as **geothermal water.** Geothermal waters that flow out onto the surface fairly continuously form **hot springs.** When geothermal water flow is intermittent and somewhat eruptive it produces a **geyser,** a phenomenon that impresses with its sporadic bursts of expelled water and steam. The Old Faithful geyser in Yellowstone National Park is a well-known example (■ Fig. 13.22). The word *geyser* is an Icelandic term for the steam eruptions that are so common on that volcanic island. Geysers erupt when temperature and pressure of the water at depth reach a critical level, forcing a column of superheated water and steam out of the fissure in an explosive manner.

Hot springs and geysers are intriguing groundwater-related phenomena. Most hot springs and geysers contain significant amounts of minerals in solution. These minerals, typically rich in calcium carbonate or silica, precipitate out of the water often forming colorful terraces or mounds around the vent or spring (■ Fig. 13.23).

Geothermal activity is usually associated with areas of tectonic and volcanic activity, especially along plate boundaries and over hot spots. Geothermal energy has been used to produce electricity in California, Mexico, New Zealand, Italy, Iceland, and elsewhere. The best geothermal water for harnessing energy is not only very hot for generating steam, but also relatively free of dissolved minerals that can clog pipes and generating equipment.

■ **FIGURE 13.21** Cave exploration, called caving, or spelunking, can be exciting but also hazardous. This spelunker is exploring Spider Cave in Carlsbad Caverns, National Park.

National Park Service–Yellowstone

■ **FIGURE 13.22** One of the world's most famous geysers is Old Faithful in Yellowstone National Park, Wyoming.
How do geysers differ from hot springs?

■ **FIGURE 13.23** Because of the water's high mineral content, this hot spring in Yellowstone National Park, Wyoming, has left extensive calcareous deposits, known as travertine.

R. Gabler

:: Terms for Review

subsurface water
infiltration
recharge
zone of aeration
percolation
soil water
zone of saturation
groundwater
water table
intermediate zone
water mining
porosity
permeability
aquifer

aquiclude
perched water table
spring
well
artesian
artesian spring
artesian well
acid mine drainage
karst
sinkhole (doline)
solution sinkhole
collapse sinkhole
uvala (valley sink)
disappearing stream

cavern (cave)
swallow hole
haystack hill (conical hill, hum)
tower karst
spleleothem
stalactite
stalagmite
column
speleology
geothermal water
hot spring
geyser

:: Questions for Review

1. Why is it important to understand the basic nature and properties of underground water?
2. What is the name of the process by which water soaks into the subsurface, and what subsurface zones does water pass through on its way to the zone of saturation?
3. Define porosity and permeability and explain the importance of each for groundwater resources.
4. Distinguish between an aquifer and an aquiclude.
5. How do springs differ from wells, and under what conditions is a well or spring correctly described as artesian?

6. Under what circumstances might underground water decline in quantity and in quality?
7. Define karst. What conditions encourage the development of karst landscapes?
8. Describe a sinkhole and explain how sinkholes are formed.
9. How are caverns created?
10. Explain how calcium carbonate is deposited by subsurface water to form a stalactite in a humid cave.

:: Practical Applications

1. Drilling at seven sites spaced every 5 kilometers along a straight-line transect heading from west to east has provided the following data on depth to the water table and on rock type just below the water table: (a) 15 meters; sandstone, (b) 14.5 meters; sandstone, (c) 14 meters; sandstone, (d) 5 meters; limestone, (e) 13 meters; sandstone, (f) 12.5 meters; sandstone, (g) 12 meters; sandstone. What is the regional slope of the water table? Provide a reasonable explanation for the data obtained from site d.
2. Scientists studying an air-filled cavern located in an arid region determined that the speleothems present began forming 20,000 years ago and stopped growing 12,000 years ago. Suggest a reasonable history of climatic and hydrologic events that could have resulted in the observed cavern evidence.
3. Using Google Earth, identify the landforms at the following locations (latitude, longitude) and provide a brief explanation of how the landform developed.
   a. 18.40°N, 66.53°W
   b. 25.05°N, 110.37°E
   c. 29.66°N, 81.87°W
   d. 32.12°S, 125.29°E

MAP INTERPRETATION
KARST TOPOGRAPHY

The Map

The Interlachen area is in northeastern Florida. Florida's peninsula is the emerged portion of a gentle anticline called the Peninsular Arch. The region is underlain by thousands of meters of marine limestones and shales. This great thickness of marine sediments originated in the Mesozoic Era when Florida was a marine basin. As the arch rose, Florida became a shallow shelf and was eventually elevated above sea level.

Although nonresidents think of Florida mainly as a state with magnificent beaches and warm winter weather due to its humid subtropical climate, it is also a state with hundreds of lakes dotting its center. The lake region is formed on the Ocala Uplift, a gentle arch of limestone that reaches to 46 meters (150 ft) above sea level. Lake Okeechobee is the largest of these lakes and has an average depth of less than 4.5 meters (15 ft). Most of the lakes, such as those in the Interlachen map area, are much smaller.

Florida's central lake region is an ideal area for studying karst topography. Both the surface and the subsurface features express the geomorphic effects of subsurface water. Extensive cavern systems exist beneath the surface. Much of the state's runoff is channeled through huge aquifers, and springs are quite common.

You can also compare this topographic map to the Google Earth presentation of the area. Find the map area by zooming in on these latitude and longitude coordinates: 29.640425°N, 81.935683°W.

Interpreting the Map

1. What major type of landform (for example, mountain) is this map area? What is its average elevation?
2. What contour interval is used on this map? Why do you think the cartographers chose this interval?
3. On what type of bedrock is the map area situated? Do you think the climate has any influence on the landforms in the area? Explain.
4. What landform features on the map indicate that this is a karst region?
5. What are the round, steep depressions called? Why do lakes occupy some of the depressions?
6. Locate Clubhouse Lake on the full-page map (scale 1: 62,500) and the smaller map (1: 24,000). What is the elevation of Clubhouse Lake? What is its maximum width?
7. What type of feature is the area north of Lake Grandin?
8. What is the approximate elevation of the region's water table? (Note: You can determine this from the water surface elevation of the lakes.)
9. Underground, the water flows through an aquifer. Define an aquifer and list the characteristics an aquifer must have. What is the general direction of groundwater flow in the aquifer underlying the Interlachen area?
10. Because much of central Florida is rapidly urbanizing, what problems and hazards do you anticipate in this karst area?

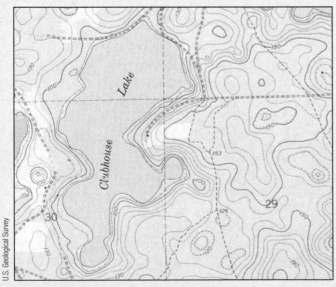

U.S. Geological Survey

Putnam Hall, Florida
Scale 1: 24,000
Contour interval = 10 ft

Opposite:
Interlachen, Florida
Scale 1:62,500
Contour interval = 10 ft
U.S. Geological Survey

(a)

(b)

L. Lynch/National Park Service

© NASA Earth Observatory/ Image courtesy NASA/GSFC/METI/ERSDAC/JAROS, and U.S./Japan ASTER Science Team

■ **FIGURE 14.1**   (a) Long-term fluvial erosion by the Gunnison River, with the aid of weathering and mass wasting, has carved the Black Canyon of the Gunnison through a part of the Rocky Mountains in Colorado. The narrow canyon is 829 meters (2722 ft) deep, with resistant rocks comprising the steep walls. (b) The Mississippi River has been building its delta outward into the Gulf of Mexico for thousands of years. Where the river enters the Gulf, the slowing current deposits large amounts of fine-grained sediment that came from the river's drainage basin. On this false-color image, muddy water appears light blue, and clearer, deeper water is dark blue.

■ **FIGURE 14.2**   The amount of runoff that occurs is a function of several factors, including the intensity and duration of a rainstorm. Surface features that increase infiltration and evapotranspiration will reduce the amount of water available to run off, and vice versa. Deep soil, dense vegetation, fractured bedrock, and gentle slopes tend to reduce runoff. Thin or absent soils, sparse vegetation, and steep slopes tend to increase runoff.

called **rills,** or somewhat larger channels, called **gullies.** Rills are typically a couple of centimeters deep and a couple of centimeters wide (■ Fig. 14.3), whereas gully depth and width may approach as much as a couple of meters.

Water does not flow in rills and gullies all the time but only during and shortly after a precipitation (or snowmelt) event. Channels that are empty of water much of the time like this are described as having **ephemeral flow.** As these small, ephemeral channels continue downslope, rills join to form slightly larger rills, which may join to make gullies. In humid climates, following these successively larger ephemeral channels downslope will eventually lead us to a point where we first encounter **perennial flow.** Perennial streams flow all year, but not always with the same volume or at the same velocity. Most arid region streams flow on an ephemeral basis. Because of this and other differences between arid and humid region streams, a full discussion of arid region stream systems appears in the separate chapter on arid region landforms, Chapter 15.

Perennial streams flow throughout the year even if it has been several weeks since a precipitation event. In most cases, this is possible only because the perennial streams continue to receive direct inflow of groundwater regardless of the date of most recent precipitation. Slow-moving groundwater that seeps directly into the stream through the channel bottom and sides at and below the level of the water surface is called

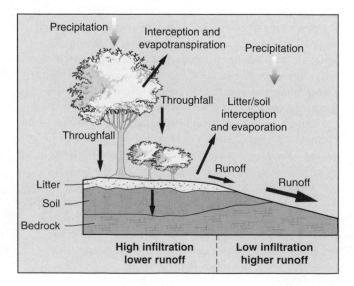

base flow. Except in rare instances, it takes a humid climate to generate sufficient base flow to maintain a perennial stream between rainstorms.

The Stream System

Most flowing water becomes quickly channelized into streams as it is pulled downhill by gravity. Continuing downslope, streams form organized channel systems in which small perennial channels join to make larger perennial channels, and larger perennial channels join to create even bigger streams. Smaller streams that contribute their water and sediment to a larger one in this way are *tributaries* of the larger channel, which is called the *trunk stream* (■ Fig. 14.4).

■ **FIGURE 14.3** Rills are the smallest type of channel. They are commonly visible on devegetated surfaces, like these mine tailings.

■ **FIGURE 14.4** **Environmental Systems: Streams** The stream system represents the surface runoff subsystem of the hydrologic cycle. Its major water input is from precipitation, but groundwater also contributes to the stream system, particularly in humid regions. The major water output for most stream systems is flow delivered to the ocean. Output or loss of water from a stream channel also occurs by evaporation back to the atmosphere and by infiltration into the groundwater system. Materials transported by streams, known as load, enter the stream system by erosion and mass movement, particularly in the headwaters of a drainage basin. As the number and size of tributaries increase downstream, the amount of load carried by the stream generally increases dramatically. Load leaves the stream system when deposited in the sea at the river mouth. Streams also deposit sediment adjacent to their channels as they overflow their banks during floods.

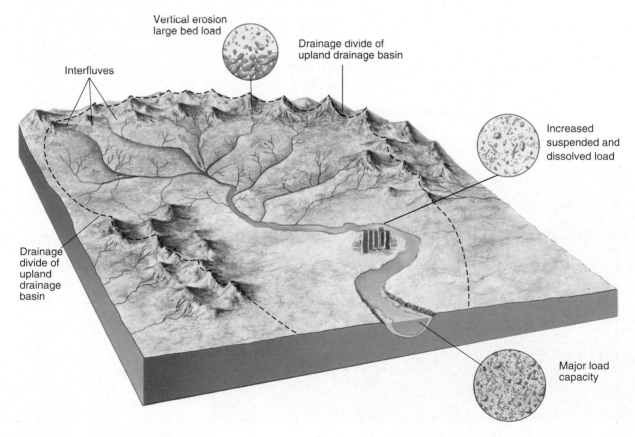

Drainage Basins

Each individual stream occupies its own **drainage basin** (also known as **watershed,** or **catchment**), the expanse of land from which it receives runoff. *Drainage area* refers to the measured extent of a drainage basin, and is typically expressed in square kilometers or square miles. Because the runoff from a tributary's drainage basin is delivered by the tributary to the trunk stream, the tributary's drainage basin also constitutes part of the drainage basin of the trunk stream. In this way, small tributary basins are nested within, or are *subbasins* of, a succession of larger and larger trunk stream drainage basins. Large river systems drain extensive watersheds that consist of numerous inset subbasins.

Drainage basins are open systems that involve inputs and outputs of water, sediment, and energy. To properly manage the water resources of a watershed, it is critical to know the boundaries of the drainage basin and its component subbasins. For example, pollution discovered in a river almost always comes from a source within its drainage basin, entering the stream system either at the point where the pollutant was first detected or at a location upstream from that site. Understanding the drainage basin system helps us track, detect, and correct sources of pollution.

The **drainage divide** represents the outside perimeter of a drainage basin and thus also the boundary between it and adjacent basins (■ Fig. 14.5). The drainage divide follows the crest of the interfluve between two adjacent drainage basins. In some places, this crest is a definite ridge, but the higher land that constitutes the divide is not always ridge-shaped, nor is it necessarily much higher than the rest of the interfluve. Surface runoff generated on one side of the divide flows toward a channel in one drainage basin, while runoff on the other side travels toward a channel in the adjacent drainage basin. The

Continental Divide separates North America into a western region, where most runoff flows to the Pacific Ocean, and an eastern region, where runoff flows to the Atlantic Ocean. The Continental Divide generally follows the crest line of high ridges in the Rocky Mountains, but in some locations the highest point between the two huge basins lies along the crest of gently sloping high plains.

A stream that has a very large number of tributaries and encompasses several levels of nested subbasins, like the Mississippi River, will differ in some major ways from a small creek that has no perennial tributaries and lies high up in the drainage basin near the divide. Knowing where each stream lies in the hierarchical order of tributaries helps Earth scientists make more meaningful comparisons among different streams. **Stream ordering** is the technique used to describe quantitatively the position of a stream and its drainage basin in the nested hierarchy of tributaries. *First-order streams* have no perennial tributaries. Even though they are the smallest perennial channels in the drainage basin, first-order channels are evident on large-scale topographic maps. Most first-order streams lie high up in the drainage basin near the drainage divide, the *source* area of the stream system. Two first-order streams must meet in order to form a *second-order stream*, which is larger than each of the first-order streams. It takes the intersection of two second-order channels to make a *third-order stream* regardless of how many first-order streams might independently join the second-order channels. The ordering system continues in this way, requiring two streams of a given order to combine to create a stream of the next higher order. The order of a drainage basin derives from the largest stream order found within it (■ Fig. 14.6). For example, the Mississippi is a tenth-order drainage basin

■ **FIGURE 14.5** An aerial photograph of a small stream channel network, basin, and divide. Every arm of the stream occupies its own subbasin (separated by divides) that combines with others to form the basin of the main stream channel.

Can you mentally trace the outline of the main drainage basin shown here?

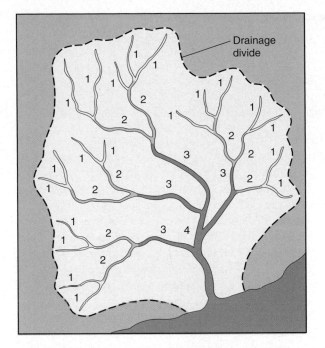

■ FIGURE 14.6   The concept of stream ordering is illustrated by the channels of this fourth-order drainage basin. Stream order changes only when two streams of equal order join, thereby creating a stream of the next higher order.

What is the highest-order stream in a selected drainage basin called?

because the Mississippi River is a tenth-order stream. Stream ordering allows us to compare streams and their various attributes quantitatively by relative size, which helps us better understand how stream systems work. Among other things, comparing streams on the basis of order has shown that as stream order increases, basin area, channel length, channel size, and amount of flow also increase.

Water in a stream system proceeds downslope through a succession of channels of ever-increasing order toward the downstream end, or *mouth*, of the trunk stream. The mouth of most humid-region, perennial stream systems lies at sea level, where the channel system finally ends and the stream water is delivered to the ocean. Drainage basins with channel systems that convey water to the ocean have **exterior drainage.** Many arid-region streams have **interior drainage** because they do not have enough flow to reach the ocean, but terminate instead in local or regional areas of low elevation.

Every stream has a **base level,** the elevation below which it cannot flow. Sea level is the *ultimate base level* for virtually all stream action. Streams with exterior drainage reach ultimate base level; the low point of flow for a stream with interior drainage is referred to as a *regional base level*. In some drainage basins, a resistant rock layer located somewhere upstream from the river mouth can act as *temporary base level* until the stream is finally able to cut down through it (■ Fig. 14.7).

■ FIGURE 14.7   The lowest point to which a stream can flow is its base level. Stream water travels downslope until it can flow no lower due to factors of topography, climate, or both. Sea level represents ultimate base level for all of Earth's streams, and most humid-region streams have sufficient flow to reach the ocean. Many arid-region streams lose so much water by evaporation to the air and infiltration into the channel bed that they cannot flow to the sea. The lowest point they can reach is a regional base level, a topographic basin on the continent. A temporary base level is formed when a rock unit lying in the pathway of a stream is significantly more resistant than the rock upstream from it. The stream will not be able to cut into the less resistant rock any faster than it can cut into the resistant rock of the temporary base level.

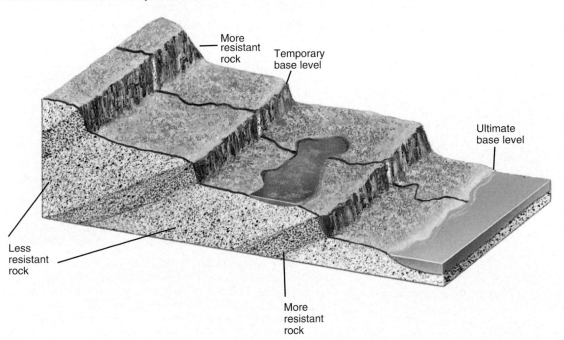

GEOGRAPHY'S SPATIAL PERSPECTIVE
:: DRAINAGE BASINS AS CRITICAL NATURAL REGIONS

Perhaps the most environmentally logical way to divide Earth's land surface into regions is by the drainage basins for stream systems. Virtually Earth's entire land surface comprises part of a drainage basin, or watershed. Like the channel systems that they contain, drainage basins are hierarchical and are conveniently subdivided into smaller subbasins for local studies and management. At the same time, higher-order watersheds of large river systems are subject to broad, integrated, regional-scale analyses.

The various components of a drainage basin are strongly interrelated. Problems in one part of a drainage basin system are likely to cause problems in other parts of the system. Along with the channel network that occupies each watershed, drainage basins consist of water, soil, rock, terrain, vegetation, people, wildlife, and domesticated animals that together form complex natural biotic habitats. Monitoring and managing these complex watershed systems requires an interdisciplinary effort that considers aspects of all four of the world's major spheres.

Surface water from watershed sources provides much of the potable water resources for the world's population. Increasing human populations and land-use intensities in previously natural drainage basins place pressures on these habitats and their water quality. Maintaining water quality within local subbasins contributes to maintaining water quality of larger river systems at the regional scale. Because streams comprise a network of flow in one direction (downstream), identifying sources of pollution involves upstream tracing of pollutants to the source and downstream spreading of pollution problems. Humans have a vital responsibility to monitor, maintain, and protect the quality of freshwater resources, and doing so at the local watershed scale has far-reaching as well as local benefits.

In recent years, many governmental agencies have designated watersheds as critical regions for environmental management. Some of these efforts involve the establishment of locally administered river conservation districts representing relatively small-order drainage basins in which community involvement plays a vital role. There are several

sound reasons, including the connectivity and hierarchical nature of watersheds, for this spatial management strategy.

Watersheds are clearly defined, well-integrated, natural regions of critical importance to life on Earth, and they make a logical spatial division for environmental management. However, a stream system may flow through many cities, counties, states, and countries, and management problems can arise where regional political and administrative boundaries do not coincide with the divide that defines the edge of a watershed. Each of these political jurisdictions may have very different needs, goals, and strategies for using and managing their part of a watershed and often a variety of strategies results in conflict. Still, most administrative units recognize that cooperative management of the watershed system as a whole is the best approach. Many cooperative river basin authorities have been established to encourage a united effort to protect a shared watershed. This watershed-oriented management strategy, based on a fundamental natural region, is an important step in protecting our freshwater resources.

The Mississippi River drainage basin covers an extensive region of the United States.

As a tributary of the Ohio River, the fourth-order Hocking River watershed in southeastern Ohio is a subbasin within the Mississippi River system. Community involvement in improving and maintaining the health of the Hocking River watershed, and other drainage basins, has regional as well as local benefits.

Drainage Density and Drainage Patterns

Each system of stream tributaries exhibits spatial characteristics that provide important information about the nature of the drainage basin. The extent of channelization can be represented by measuring **drainage density,** the total length of all channels in a drainage basin divided by its drainage area. Because drainage density indicates how dissected the landscape is by channels, it reflects both the tendency of the drainage basin to generate surface runoff and the erodibility of the surface materials (■ Fig. 14.8). Regions with high drainage densities have limited infiltration, considerable runoff, and at least moderately erodible surface materials.

The ideal climate for high drainage densities is semiarid. Humid climates encourage extensive vegetation cover, which promotes infiltration through interception and reduces channel formation by holding soils and surface sediment in place. In arid climates, although the vegetation cover is sparse, there is insufficient precipitation to create enough runoff to carve many channels. Semiarid climates

■ **FIGURE 14.8** Drainage density—the length of channels per unit area—varies according to several environmental factors. For example, everything else being equal, highly erodible and impermeable rocks tend to have higher drainage density than areas dominated by resistant or permeable rocks. Slope and vegetation cover can also affect drainage density.
What kind of drainage density would you expect in an area of steep slopes and sparse vegetation cover?

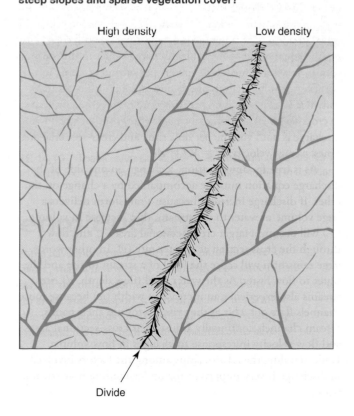

Divide

have enough precipitation input to generate overland flow, but not enough to support an extensive vegetative cover. Easily eroded surface materials in a semiarid climate can have a drainage density greater than 125 kilometers per square kilometer (75 mi/mi²), whereas very resistant granite hills in a humid climate may have a drainage density of only 5 kilometers per square kilometer (3 mi/mi²).

When viewed from an aerial perspective (map view), networks of stream tributaries display distinct spatial arrangements called **drainage patterns.** Two primary factors that influence drainage pattern are bedrock structure and surface topography. A *dendritic* (from Greek: *dendros,* tree) drainage pattern (■ Fig. 14.9a) resembles a branching tree with tributaries joining larger streams at acute angles (less than 90°). Dendritic patterns are common and develop where the rocks have a roughly equal resistance to weathering and erosion and are not intensely jointed. A *trellis* pattern consists of long, parallel streams with short tributaries entering at right angles (Fig. 14.9b). Trellis drainage indicates folded terrain where short streams flow down the sides of resistant rock ridges into a larger stream that occupies the adjacent valley of erodible rock, as in the Ridge and Valley region of the Appalachians. A *radial* pattern develops where streams flow away from a common high point on cone- or dome-shaped geologic structures, such as volcanoes (Fig. 14.9c). The opposite pattern is *centripetal,* with the streams converging on a central lowland as in an arid region basin of interior drainage (Fig. 14.9d). *Rectangular* patterns occur where streams follow sets of intersecting fractures to produce a blocky network of straight channels with right-angle bends (Fig. 14.9e). In some regions that were recently covered by extensive glacial ice, streams flow on low gradient terrain left by the receding glaciers, wandering between marshes and small lakes in a chaotic pattern called *deranged* drainage (Fig. 14.9f).

Stream Discharge

The amount of water flowing in a stream depends not only on the impact of recent weather patterns but also on such drainage basin factors as its size, relief, climate, vegetation, rock types, and land-use history. Stream flow varies considerably from time to time and place to place. Most streams experience occasional brief periods when the amount of flow exceeds the ability of the channel to contain it, resulting in the flooding of channel-adjacent land areas.

Just as it was important to develop the technique of stream ordering to indicate numerically a channel's place in the hierarchy of tributaries, it is also crucial to be able to describe quantitatively the amount of flow being conveyed through a stream channel. **Stream discharge** (Q) is the volume of water (V) flowing past a given channel cross section per unit time (t): $Q = V/t$. Discharge is most commonly expressed in units of cubic meters per second (m³/s) or cubic feet per second (ft³/s). A cross section is essentially a thin slice extending from one stream bank straight across the channel to the other stream

flow, has a great amount of channel roughness, which causes a considerable decrease in stream energy due to the resulting frictional effects. Smooth, bedrock channels without these irregularities have lower channel roughness and therefore a smaller loss in stream energy due to frictional effects. In all cases, however, friction along the bottom and sides of a channel slows down the flow, especially at the channel boundaries, so that the maximum velocity usually occurs somewhat below the stream surface at the deepest part of the cross section. In addition to the external friction at the channel boundaries, streams lose energy because of internal friction in the flow related to eddies, currents, and the interaction among water molecules.

Stream gradients are usually steepest at the headwaters and in new tributaries and diminish in the downstream direction (■ Fig. 14.11). Discharge, on the other hand, increases downstream in humid regions as the size of the contributing drainage basin increases. As the water originally conveyed in numerous small channels high in the drainage basin is collected in the downstream direction into a shrinking number of larger and larger channels, frictional resistance decreases and flow efficiency improves. As a result, flow velocity tends to be higher in downstream parts of a channel system than over the steep gradients in the headwaters. The ability of the stream to carry sediment, therefore, may be equivalent or even greater downstream than upstream.

Sediment being transported by a stream is called the **stream load.** Carrying sediment is a major part of the geomorphic work accomplished by a stream. From smallest to largest, the size of sediment that a stream may transport includes clay, silt, sand, granules, pebbles, cobbles, and boulders. Sand marks the boundary between fine-grained (small) clasts and coarse-grained (large) clasts. Gravel is a general term for any sediment larger in size than sand. The maximum size of rock particles that a stream is able to transport, referred to as the **stream competence** (measured as particle diameter in centimeters or inches), and the total amount of load a stream is able to move, termed **stream capacity** (measured as weight per unit time), depend on available stream energy and thus on the flow velocity. As a stream flows along, its velocity is always changing, reflecting the constant variations in stream discharge, stream gradient, and frictional resistance factors. As a result, the size and amount of load that the stream can carry is also changing constantly. If the material is available when flow velocity increases, the stream will pick up from its own channel bed larger clasts and more load. When its velocity decreases and the stream can no longer transport such large clasts and so much material, it deposits the larger clasts onto

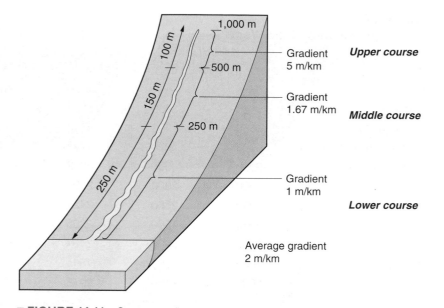

■ **FIGURE 14.11**   Stream gradient decreases from source to mouth for the ideal stream.

National Weather Service

■ **FIGURE 14.12**   It is not just the power of flowing water that causes damage in a flood. A flooding stream typically has a considerable load, can transport large and heavy materials, and may carry floating debris, all of which can ruin structures and belongings.

its bed until a new increase in velocity provides the energy to entrain and transport them. Thus, whether a stream is eroding or depositing at any particular moment, is determined by the complex of factors that control its energy.

Stream capacity and stream competence both increase in response to a relatively small increase in velocity. A stream that doubles its velocity during a flood may increase its amount of sediment load six to eight times. The boulders seen in many mountain streams arrived there during some past flood that greatly increased stream competence; they will be moved again when a flow of similar magnitude occurs. Rivers do most of their heavy sediment-moving work during short periods of flood (■ Fig. 14.12).

Streams have no influence over the amount of water entering the channel system, nor can they change the type of rocks over which they flow. Streams do, however, have some control over channel size, shape, and gradient. For example, when a stream undergoes a decrease in energy so that it deposits some of its sediment, the deposit raises the channel bottom and locally creates a steeper gradient downstream from the deposit. With continued deposition, the location will eventually attain a slope steep enough to cause a sufficient increase in velocity so that the flow will again entrain and carry that deposited sediment. Streams tend to adjust their channel properties so they can move the sediment supplied to them. *Dynamic equilibrium* is maintained by adjustments among channel slope, shape, and roughness; amount of load eroded, carried, or deposited; and the velocity and discharge of stream flow.

Fluvial Processes

Stream Erosion

Fluvial erosion is the removal of rock material by flowing water. Fluvial erosion consists of the chemical removal of ions from rocks and the physical removal of rock fragments (clasts). Physical removal of rock fragments includes breaking off new pieces of bedrock from the channel bed or sides and moving them as well as picking up and removing preexisting clasts that were temporarily resting on the channel bottom.

The removal of rock material by erosion does not necessarily mean that the landscape is undergoing long-term lowering. If sediments eroded from the bed of a stream channel are replaced by the deposition of other fragments transported in from upstream, there will be no net drop in the position of the channel bottom. Such lowering, known as channel incision, occurs only when there is net erosion compared to deposition. Net erosion results in the lowering of the affected part of the landscape and is termed **degradation.** Net deposition of sediments results in a building up, or **aggradation,** of the landscape.

One way that streams erode occurs when stream water chemically dissolves rock material and then transports the ions away in the flow. This fluvial erosion process, called **corrosion,** has only a limited effect on many rocks but can be significant in certain rock types, such as limestone.

Hydraulic action is the physical, as opposed to chemical, process of stream water alone removing pieces of rock. As stream water flows downslope it exerts stress on the streambed. Whether this stress results in entrainment and removal of a preexisting clast currently resting on the channel bottom, or even the breaking

off of a new piece of bedrock from the channel, depends on several factors, including the volume of water, flow velocity, flow depth, stream gradient, friction with the streambed, the strength and size of the rocks over which the stream flows, and the degree of stream turbulence. *Turbulence* is chaotic flow that mixes and churns the water, often with a significant upward component, that greatly increases the rate of erosion as well as the load-carrying capacity of the stream. Turbulent currents contribute to erosion by hydraulic action when they wedge under or pound away at rock slabs and loose fragments on the channel bed and sides, dislodging clasts that are then carried away in the current.

As soon as a stream begins carrying rock fragments as load, it can start to erode by **abrasion,** a process even more powerful than hydraulic action. As rock particles in the water bounce, scrape, and drag along the bottom and sides of a stream channel, they break off additional rock fragments. Because solid rock particles are denser than water, the impact of clastic load thrown against the channel bottom and sides by the current is much more effective than the impact of water alone. The wear and tear experienced by sediments as they tumble and bounce against one another and against the stream channel is called *attrition.* Attrition explains why gravels found in streambeds are rounded and why the load carried in the lower reaches of most large rivers is composed primarily of fine-grained sediments and dissolved minerals.

Stream Transportation

Streams transport their load in several ways (■ Fig. 14.13). Some minerals are dissolved in the water and are thus carried in the transportation process of **solution.** The finest solid particles are carried in **suspension,** buoyed by vertical turbulence. Such small grains can remain suspended in the water column for long periods, as long as the force of upward turbulence is stronger than the downward settling tendency of

■ **FIGURE 14.13** Transportation of solid load in a stream. Clay and silt particles are carried in suspension. Sand typically travels by suspension and saltation. The largest (heaviest) particles move by traction.

What is the difference between traction and saltation?

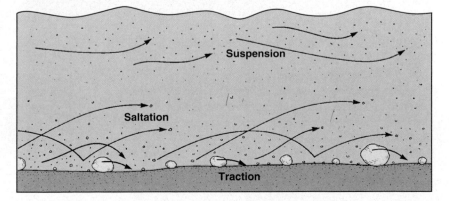

the particles. Some grains too large and heavy to be carried in suspension bounce along the channel bottom in a process known as **saltation** (from French: *sauter,* to jump). Particles that are too large and heavy to move by saltation may slide and roll along the channel bottom in the transportation process of **traction.**

There are three main types of stream load. Ions of rock material held in solution constitute the *dissolved load.* *Suspended load* consists of the small clastic particles being moved in suspension. Larger particles that saltate or move in traction along the streambed comprise the *bed load.* The total amount of load that a stream carries is expressed in terms of the weight of the transported material per unit time.

The relative proportion of each load type present in a given stream varies with such drainage basin characteristics as climate, vegetative cover, slope, rock type, and the infiltration capacities and permeabilities of the rock and soil types. Dissolved loads will be larger than average in basins with high amounts of infiltration and base flow because slow-moving groundwater that feeds base flow acquires ions from the rocks through which it moves. The intense weathering in humid regions produces much fine-grained sediment that is incorporated into streams as suspended load. Large amounts of suspended load give streams a characteristically muddy appearance (■ Fig. 14.14). The Huang He in northern China, known as the "Yellow River" because of the color of its silty suspended load, carries a huge amount of sediment in suspension, with often more than 1 billion tons of suspended load per year. Compared to the "muddy" Mississippi River, the Huang He transports five times the

suspended sediment load with only 15% of the discharge. Streams dominated by bed load tend to occur in arid regions because of the limited weathering rate in arid climates. Limited weathering leaves considerable coarse-grained sediment in the landscape available for transportation by the stream system.

Stream Deposition

Because the capacity and competence of a stream to carry material depend on flow velocity, a decrease in velocity will cause a stream to reduce its load through deposition. Velocity decreases over time when flow subsides—for example, after the impact of a storm—but it also varies from place to place along the stream. Shallow parts of a channel that in cross section lie far from the deepest and fastest flow typically experience low flow velocity and become sites of recurring deposition. The resulting accumulation of sediment, like what forms on the inside of a channel bend, is referred to as a *bar.* Sediment also collects in locations where velocity falls due to a reduction in stream gradient, where the river current meets the standing body of water at its mouth, and, during floods, on the land adjacent to the stream channel.

Alluvium is the general term for fluvial deposits, regardless of the type or size of material. Alluvium is recognized by the characteristic sorting and rounding of sediments that streams perform. A stream sorts particles by size, transporting the sizes that it can and depositing larger ones. As velocity fluctuates due to changes in discharge, channel gradient, and roughness, particle sizes that can be picked up, transported, and deposited vary accordingly (■ Fig. 14.15). The alluvium deposited by a stream with fluctuating velocity will exhibit alternating layers of coarser and finer sediment.

When streams leave the confines of their channels during floods, the channel cross-sectional width suddenly increases so much that flow velocity must slow down to counterbalance it *(Q = wdv).* The resulting decrease in stream competence and capacity causes the deposition of sediment on the flooded land adjacent to the channel. This sedimentation is greatest right next to the channel where aggradation constructs channel-bounding ridges known as *natural levees,* but some alluvium is left behind wherever load settles out of the receding flood waters.

Floodplains constitute the often extensive, low-gradient land areas composed of alluvium that lie adjacent to many stream channels (■ Fig. 14.16). Floodplains are aptly named because they are inundated during floods and because they are at least partially composed of the sediment that settles

■ **FIGURE 14.14**   Some rivers carry a tremendous load of sediment in suspension and show it in their muddy appearance, as in this aerial view of the Mississippi River in Louisiana. Much of the load carried by the Mississippi River is delivered to it by its major tributaries.
What are some of the major tributaries that enter the Mississippi River?

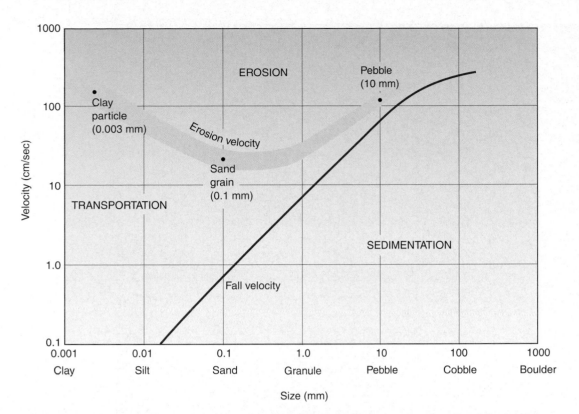

■ **FIGURE 14.15** This graph shows the relationship between stream flow velocity and the ability to erode or transport material of varying sizes (inability to transport particles of a particular size or larger will result in deposition). Note that small pebbles (particles with a diameter of 10 mm, for example) need a high stream flow velocity to be moved because of their size and weight. The fine silts and clays (smaller than 0.05 mm) also need high velocities for erosion because they stick together cohesively. Sand-sized particles (between 0.05 mm and 2.0 mm) are relatively easily eroded and transported, compared to gravel (particles larger than 2.0 mm) or clays.

■ **FIGURE 14.16** During floods, sediment-laden water that flows over the stream's banks inundates the adjacent, low floodplain areas and deposits alluvium, mainly silts and clays. This is the Missouri River floodplain at Jefferson City, Missouri, during the 1993 Midwest flood.
What would the river floodwaters leave behind in flooded homes after the water recedes?

out of slowing and standing floodwater. Most floodplains also contain channel bar deposits that get left behind as a stream gradually shifts its position in a sideways fashion (laterally) across the floodplain (■ Fig. 14.17).

Channel Patterns

Three principal types of stream channel have traditionally been recognized when considering the form of a given channel segment in map view. Although *straight channels* may exist for short distances under natural circumstances, especially along fault zones, joints, or steep gradients, most channels with parallel, linear banks are artificial features constructed by people.

If a stream has a high proportion of bed load in relation to its discharge, it deposits much of its load as sand and gravel bars in the streambed. The stream flows in multiple interweaving strands that split and rejoin around the bars to give a braided appearance, and indeed this

D. Sack

■ **FIGURE 14.17**   Sediment deposited in a bar on the inside of a channel bend becomes part of the floodplain alluvium if the stream laterally migrates away leaving the bar deposits behind. In this photo, winter ice fills swales between ridges that mark the crests of successive bar deposits.

■ **FIGURE 14.18**   The braided channel of the Brahmaputra River in Tibet, viewed from the International Space Station. Stream braiding results from an abundant bed load of coarse sediment that obstructs flow and separates the main stream into numerous strands.
What does the common occurrence of braided channels just downstream from glaciers tell you about the sediment transported and deposited by moving ice?

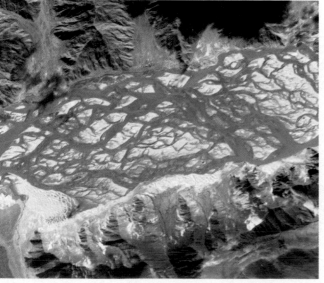

ASA/Earth Sciences and Image Analysis Laboratory at Johnson Space Center

is called a **braided channel** (■ Fig. 14.18). This channel pattern develops in settings that provide considerable amounts of loose sand and gravel bed load to the stream system, and are therefore common in arid regions and in glacial meltwater zones.

The most common channel pattern in humid climates is characterized by broad, sweeping bends. Over time, these sinuous, **meandering channels** also wander from side to side across their low-gradient floodplains, widening the valley by lateral erosion on the outside of meander bends and leaving behind bar deposits on the inside of meander bends (■ Fig. 14.19). These streams and their floodplains have a higher proportion of fine-grained sediment and greater bank cohesion than the typical braided stream.

Land Sculpture by Streams

One way to understand the variety of landform features resulting from fluvial processes is to examine the course of an idealized river as it flows from its headwaters in the mountains to its mouth at the ocean. The gradient of this river would diminish continually downstream as it flows from its source toward base level. In nature, exceptions

FIGURE 14.19 at left side credit: NASA/JPL TOPSAR; image generation performed at Washington University, St. Louis

■ **FIGURE 14.19** Over time, a meandering (sinuous) stream channel, like the Missouri River shown on this colorized radar image, may swing back and forth across its valley. Where the outside of a meander bend on the floodplain (purple and blue tones) impinges on the edge of higher terrain (orange areas), stream erosion can undercut the valley side wall and, with the assistance of mass wasting, contribute to floodplain widening.

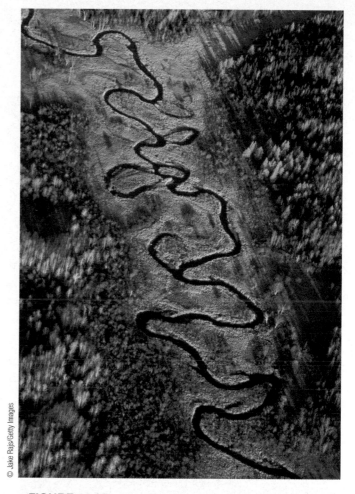

© Jake Rajs/Getty Images

■ **FIGURE 14.20** Not all streams fit the generalized pattern of characteristics for upper, middle, and lower stream segments. The Mississippi River, here a relatively small stream, meanders on a low gradient near its headwaters in Minnesota, far upstream from its mouth.

The following discussion subdivides the ideal stream course into upper, middle, and lower river sections flowing over steep, moderate, and low gradients, respectively. Fluvial erosion processes dominate the steep upper course, whereas deposition predominates in the lower course. The middle course displays important elements of both fluvial erosion and deposition.

Features of the Upper Course

At the headwaters in the ideal upper course, the stream primarily flows in contact with bedrock. Over the steep gradient high above its base level, the stream works to erode vertically downward by hydraulic action and abrasion. Erosion in the upper course creates a steep-sided valley, gorge, or ravine as the stream channel cuts deeply into the land. Little if any floodplain is present, and the valley walls typically slope directly to the edge of the stream channel. Steep valley sides encourage mass movement of rock material directly into the flowing stream. Valleys of this type, dominated by the

exist to this idealized profile because some streams flow entirely over a low gradient (■ Fig. 14.20), while other streams, particularly small ones on mountainous coasts, flow down a steep gradient all the way to the standing body of water at their mouth. Rather than having a smoothly decreasing slope from headwaters to mouth, we would expect real streams to have some irregularities in their *longitudinal profile*—the stream gradient from source to mouth (see again Fig. 14.11).

■ **FIGURE 14.21**   Where the upper course of a stream lies in a mountainous region, its valley typically has a characteristic "V" shape near the headwaters. Such a stream flows in a steep-walled valley, with rapids and waterfalls, as shown here in Yellowstone Canyon, Wyoming.

How does the gradient of the Yellowstone River compare with that of the stretch of the Mississippi River shown in FIGURE 14.20?

downcutting activity of the stream, are often called *V-shaped valleys* because in cross section their steep slopes attain the form of the letter V (■ Fig. 14.21).

The effects of *differential erosion* can be significant in the upper course where streams cut through rock layers of varying resistance. Streams flowing over resistant rock have a steeper gradient than where they encounter weaker rock. A steep gradient gives the stream flow more energy, which the stream needs to erode the resistant rock. Rapids and

waterfalls may mark the location of resistant materials in a stream's upper course. Where rocks are particularly resistant to weathering and erosion, valleys will be narrow, steep-sided gorges or canyons; where rocks are less resistant, valleys tend to be more spacious.

Features of the Middle Course

In the middle section of the ideal longitudinal profile, the stream flows over a moderate gradient and on a moderately smooth channel bed. Here the stream valley includes a floodplain, but remaining ridges beyond the floodplain still form definite valley walls. The stream lies closer to its base level, flows over a gentler gradient, and thus directs less energy toward vertical erosion than in its upper course. The stream still has considerable energy, however, due to the downstream increase in flow volume and reduction in bed friction. Much of its available energy is used for transporting the considerable load it has accumulated, and for lateral erosion of the channel sides. The stream displays a definite meandering channel pattern with its sinuous bends that wander over time across the valley floor. Erosion results in a steep **cut bank** on the outside of meander loops, where the channel is deep and centrifugal force accelerates stream velocity. In the low velocity and shallow flow on the inside of the meander bends, the stream deposits a **point bar** (■ Fig. 14.22). Erosion on the outside and deposition on the inside of meander bends result in the sideways displacement, or *lateral migration,* of meanders. This helps increase the area of the gently sloping floodplain when cut banks impact the confining valley walls.

Features of the Lower Course

The minimal gradient and close proximity to base level along the ideal lower river course make downcutting virtually impossible. Stream energy, now derived almost exclusively from the higher discharge rather than the downslope pull of gravity, leads to considerable lateral shifting of the channel and creation of a large depositional plain. The lower floodplain of a major river is much wider than the width of its meander belt and shows evidence of many changes in course (■ Fig. 14.23). The stream migrates laterally through its own previously deposited sediment in a channel composed exclusively of alluvium. During floods, these extensive floodplains, or **alluvial plains,** become inundated with sediment-laden water that contributes deposits to the large natural levees and to the already thick alluvial valley fill of the floodplain in general. Natural levees along the Mississippi River rise up to 5 meters (16 ft) above the rest of the floodplain.

A common landform in this deposition-dominated environment provides evidence of the meandering of a river over time. Especially during floods, **meander cut-offs** occur when a stream seeks a shorter, steeper, straighter path, and breaches

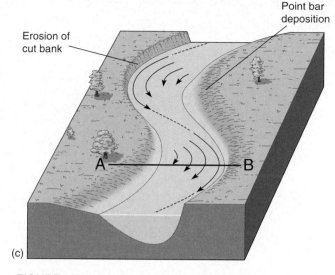

(a)

Original channel

Bank erosion (cut bank)

Sediment deposition (point bar)

Sediment deposition (point bar)

Bank erosion (cut bank)

(b)

Powerful downward velocity: undercutting, erosion

Sediment deposition

Bank erosion

Reduced upward velocity: deposition

Lower current

(c)

Erosion of cut bank

Point bar deposition

■ **FIGURE 14.22**   Characteristics of a meandering river channel. Note that water flowing in a channel has a tendency to flow downstream in a helical, or "corkscrew," fashion, which moves water against one side of the channel and then to the opposite side.

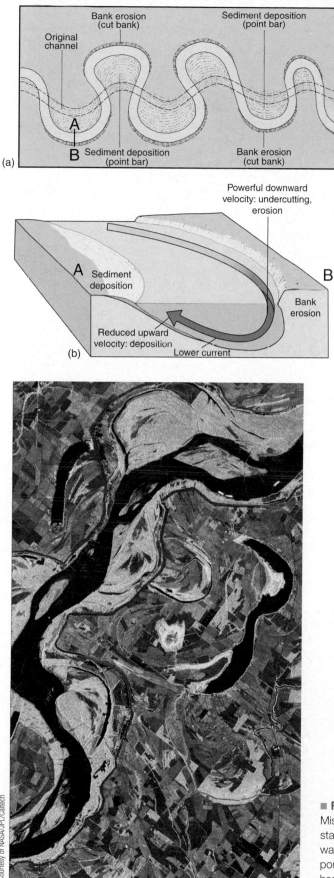

Courtesy of NASA/JPL/Caltech

through the levees, leaving behind a former meander bend now isolated from the new channel position. If the meander cut-off remains filled with water, which is common, it forms an **oxbow lake** (■ Fig. 14.24).

Sometimes people attempt to control streams by building up levees artificially in order to keep the river in its channel. During times of reduced discharge, however, when a river has less energy, deposition occurs in the channel. Thus, in an artificially constrained channel, a river may raise the level of its channel bed. In some instances, as in China's Huang He and the Yuba River in northern California, deposition has raised the streambed above the surrounding floodplain. Flooding presents a very serious danger in this situation with much of the floodplain lying below the level of the river. Unfortunately, when floodwaters eventually overtop or breach the levees, they can be even more extensive and destructive than they would have been in the natural case.

The presence of levees—both natural and artificial— can prevent tributaries in the lower course from joining the main stream. Smaller streams are forced to flow parallel to the main river until a convenient junction is found. These parallel tributaries are called *yazoo streams*, named after the Yazoo River, which parallels the Mississippi River for more than 160 kilometers (100 mi) until it finally joins the larger river near Vicksburg, Mississippi.

■ **FIGURE 14.23**   This colorized radar image shows part of the Mississippi River floodplain along the Arkansas–Louisiana–Mississippi state lines. The colors are used to enhance landscape features such as water bodies (dark), field patterns, and forested areas (green). Abandoned portions of former meanders visible on the floodplain show that the river has changed its channel position many times.

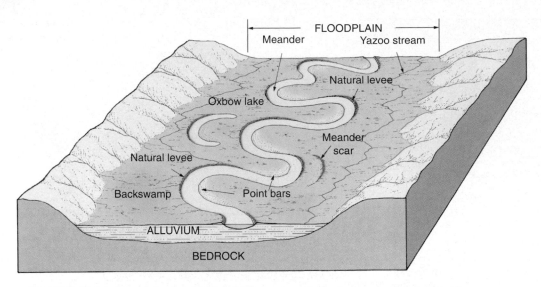

■ FIGURE 14.24 Features of a large floodplain common in the lower courses of major rivers. Backswamps are low, marshy or swampy parts of the floodplain, generally at the water table.
What is the origin of an oxbow lake?

Deltas

Where a stream flows into a standing body of water, such as a lake or the ocean, the flow is no longer confined in a channel. The current expands in width, causing a reduction in flow velocity and thus a decrease in load competence and capacity. If the stream is carrying much load, the sediment will begin to settle out, with larger particles deposited first, closer to the river mouth, and smaller particles deposited farther out in the water body. With continued aggradation, a distinctive landform,

called a **delta** because the map view shape resembles the Greek letter delta (Δ), may be constructed (■ Fig. 14.25a). Flowing over the low gradient of its delta, a river may divide into two channels, and may do so multiple times, to continue to convey its water and load to the lake or ocean. These multiple channels flowing out from the main stream are called **distributaries,** and help direct flow and sediment toward the lake or ocean.

Deltas develop only at those river mouths where the fluvial sediment supply is high, the underwater topography does not drop too sharply, and waves, currents, and tides cannot

■ FIGURE 14.25 Satellite views of two different types of deltas. (a) The Nile River delta at the edge of the Mediterranean Sea is an arcuate delta, displaying the classic triangular arc shape. (b) The unusual shape of the Mississippi River delta resembles a bird's foot.
Why are the shapes of some deltas controlled more by fluvial processes whereas the shapes of others are strongly influenced by coastal processes?

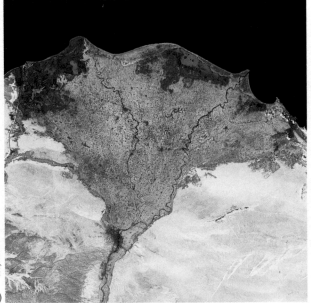

(a)

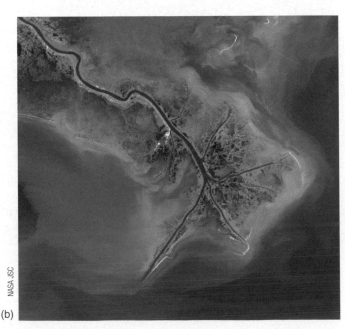

(b)

transport away all the sediments delivered by the river. These circumstances exist at the mouths of many, but not all, rivers. Furthermore, different types of deltas form in different settings. Where the Mississippi River flows into the Gulf of Mexico, the river has constructed a type of delta called a *bird's-foot delta* (Fig. 14.25b). Bird's-foot deltas are constructed in settings where the influence of the fluvial system far exceeds the ability of waves, currents, and tides of the standing water body to rework the deltaic sediment into coastal landforms or to transport it away. In this type of delta, numerous distributaries slightly above sea level extend far out into the receiving water body. Occasional changes in the distributary channel system occur when a major new distributary is cut that siphons flow away from a previous one, causing the center of deposition to switch to a new location far from its previous center. The appearance in map view of the distributaries extending toward present and former depositional centers leaves the delta resembling a bird's foot. An *arcuate delta*, on the other hand, like that of the Nile River, projects to a limited extent into the receiving water body, but the smoother, more regular seaward edge of this kind of delta shows greater reworking of the fluvial deposits by waves and currents than in the case of the bird's-foot delta. *Cuspate deltas,* like the São Francisco in Brazil, form where strong coastal processes push the sediments back toward the mainland and rework them into beach ridges on either side of the river mouth.

Base-Level Changes and Tectonism

A change in elevation along a stream's longitudinal profile will cause an increase or decrease in the stream's gradient, which in turn impacts the stream's energy. Elevation changes can occur anywhere along a stream due to tectonic uplift or depression; they can also occur due to rising or falling of base level at the river mouth. Changes of base level for basins of exterior drainage result principally from climate change. Sea level drops in response to large-scale growth of glaciers and rises with substantial glacier shrinking. Tectonic uplift or a fall in base level gives the stream a steeper gradient and increased energy for erosion and transportation. The landscape and its stream are then said to be *rejuvenated* because the stream uses its renewed energy to incise its channel to the new base level. Waterfalls and rapids may develop as rejuvenated channels are deepened by erosion. Tectonic depression of the drainage basin or a rise in sea level reduces the stream's gradient and energy, enhancing deposition.

If new uplift occurs gradually in an area where stream meanders have formed, these meanders may become *entrenched* as the stream deepens its valley (■ Fig. 14.26). Now, instead of eroding the land laterally, with meanders migrating across an alluvial plain, the rejuvenated stream's primary activity is vertical incision.

Virtually all rivers reaching the sea incised deep valleys near their mouths during the Pleistocene in response to the substantial lowering of their base level (sea level) associated with continental glaciation. Eventual melting of the glaciers returned water to the ocean, which raised sea level. Readjustment to this base-level rise caused the streams to deposit large amounts of their load, filling the Pleistocene valleys near their mouths with sediment.

While a drop in base level causes downcutting and a rise causes deposition, an upset of stream equilibrium resulting from sizable increases or decreases in discharge or load can have similar effects on the landscape. Variations in base level, tectonic movements, and changes in stream

■ **FIGURE 14.26**   A spectacular example of an entrenched meander from the San Juan River in the Colorado Plateau region of southern Utah.

© Rainer Duttmann

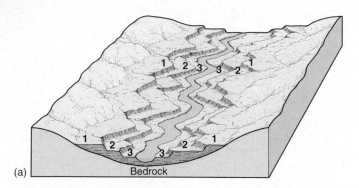

■ **FIGURE 14.27** (a) Diagram of a stream valley displaying two sets of alluvial stream terraces (labeled 1 and 2) and the present floodplain (labeled 3). The stream terraces are remnants of previous positions of the valley floor, with the higher (1) being the older. (b) River terraces in the Tien Shan Mountains of China.
How many terraces can you identify in this photo?

equilibrium can each cause downcutting in stream valleys, so the valley is slightly deepened and remnants of the older, higher valley floor are preserved in stair-stepped banks along the walls of the valley. These remnants of previous valley floors are **stream terraces.** Multiple terraces are a consequence of successive periods of downcutting and deposition (■ Fig. 14.27). Stream terraces provide a great deal of evidence about the geomorphic history of the river and its surrounding region.

Stream Hazards

Although there are many benefits to living near streams, doing so also has its risks, particularly in the form of floods. Variability of stream flow constitutes the greatest problem for life along rivers and is also an impediment to their use. Most stream channels can hold the maximum flows that occur once every year or two. Larger maximum flows that are probable over longer periods of 5, 10, 100, or 1000 years overflow the channel and inundate the surrounding land, sometimes with disastrous results (■ Fig. 14.28). Similarly, exceptionally low flows may produce crises in water supply and bring river transportation to a halt.

The U.S. Geological Survey maintains more than 6000 gaging stations for the measurement of stream discharge in the United States (■ Fig. 14.29). Many of these gaging stations measure stream discharge in an automated fashion and beam their data to a satellite that relays them to a receiving station. With this system, stream flow changes can be monitored at the time they occur, which is very beneficial in issuing flood

■ **FIGURE 14.28** Living on an active floodplain has its risks, as seen in this photo of the Coast Guard rescuing a man stranded on a rooftop in Olivehurst, California. A levee along the Feather River ruptured, sending acres of water into the Sutter County community.
What can be done to prepare for or to avoid flood problems?

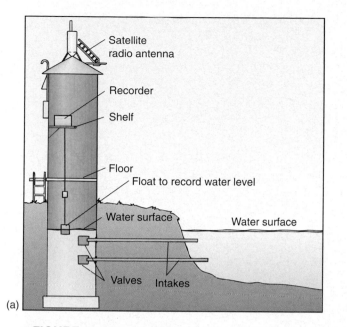

J. Petersen

(a) (b)

■ **FIGURE 14.29** U.S. Geological Survey (USGS) stream gaging station. (a) As illustrated here, when river level rises or falls, so does the water in the lower part of the station, connected to the channel by intake pipes. A gaging float moves up and down with the water level, and this motion is measured and recorded. Gaging stations electronically beam flow data to a satellite that transfers the information to a receiving station. (b) Gaging stations are commonly located where highway bridges cross rivers or streams.

warnings as well as in understanding how streams change in response to variations in discharge.

A **stream hydrograph** is the record of discharge changes in a stream over time (■ Fig. 14.30). Hydrographs cover a day, a few days, a month, or even a year depending on a scientist's purpose. Because of the relationship of discharge to water depth *(Q = wdv)*, hydrographs are often used to indicate how high and fast the water level rises in response to a precipitation event.

■ **FIGURE 14.30** A stream hydrograph depicts changes in discharge, which implies changes in flow depth, recorded over a period of time by a gaging station. This hydrograph example shows the rise in a river in response to flood runoff, the flood peak, and the recession of flow waters following the end of a storm. Note the lag between the time that precipitation starts and the rise in the river.

Why would such a time lag occur between the rainfall and rise in the river?

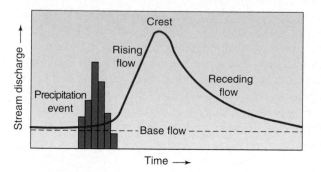

During and just after a runoff-producing rainfall, the stream level will rise in response to the increased discharge. The water level will peak, or crest, at the time of maximum discharge, and then fall as the river eventually returns to a more average amount of flow. At a gaging station, the rise, peak, and fall in discharge over time are recorded as a climbing, peaking, then descending curve on the hydrograph. The shape of the hydrograph curve can be used to understand a great deal about how a watershed and a stream channel respond to an increase in runoff, particularly during and after floods. If conditions in the watershed promote rapid runoff, the stream will rise quickly during a storm. That rapid increase in stream discharge will be represented on the hydrograph curve by a rising limb that climbs steeply to peak discharge. The greater the discharge at the flood crest, the higher the peak of the hydrograph curve. Compared to the rapid rise to peak discharge, the return to a more average flow typically takes longer because water continues to seep into the channel from the precipitation-saturated ground of the watershed. The falling limb of the hydrograph, therefore, usually plots as a comparatively gentle curve. How high a river rises, how fast it reaches peak flow, and how slowly the water recedes in response to a certain amount of precipitation are all important factors involved in preparing for future floods.

Studies have shown that urbanization of a watershed (particularly small drainage basins) tends to make the flood peak rise higher and faster than was the case before human development in the drainage basin. Reduction of vegetation and its replacement with impermeable surfaces, like roads, roofs,

D. Sack

■ **FIGURE 14.31**   Most aspects of urbanization and suburbanization, as in this area near Boston, increase the extent of impermeable cover within the drainage basin. As a result, the amount and rate of runoff from urbanized areas increase compared to their presettlement state.
What features of the urbanized landscape shown here enhance runoff?

and parking lots, contribute to higher rates of runoff and increased runoff, two conditions that directly affect the flow of the stream (■ Fig. 14.31). When a drainage basin becomes more highly developed and populated, the potential flood size, the rapidness of flood onset, and the flood hazard all tend to become greater.

To determine how often we can expect flows of a given magnitude to occur in a particular river, especially for those discharges that cause flooding, Earth scientists assess the river's previous flow history using maximum annual discharge data. These data are used to calculate *recurrence intervals,* the average number of years between past flow events that equaled or exceeded various discharges. For example, if flooding along a river resulted from a discharge of 7000 cubic meters per second (250,000 cu ft/sec) or larger in 2 out of 100 years of data, that discharge has a recurrence interval of 50 years—it is the 50-year flood—even if those flows occurred in two consecutive years. Because that discharge (and possibly greater) happened twice in the past 100 years, there is an estimated 2 out of 100, or 2%, probability that it will happen in any given year. Because

recurrence intervals are averages determined from historic data of limited duration, the size of a 10-, 50-, 100-year, or other flood is not absolute, but rather an estimate subject to change, and the length of time between 50-year floods, for example, will not necessarily be 50 years.

The Importance of Surface Waters

Historically, people have used rivers and smaller streams for a variety of purposes. The settlement and growth of the United States would have been very different without the Mississippi River and its far-reaching tributary system that drains most of the country between the Appalachians and the Rockies. The Mississippi, like many other rivers, has been used for exploration, migration, and settlement, and the number of major cities along it illustrates people's tendency to settle along rivers. Navigable rivers still compete successfully with railroads and trucks for transportation of bulk cargo, grain,

■ FIGURE 14.32   The multipurpose Lookout Point Dam on the Middle Fork of the Willamette River, Oregon.
What are some of the purposes that dams serve?

lumber, and mineral fuels, and streams have long been used to generate power. They supply irrigation water and agriculturally productive floodplain soils. We also use streams as sources of food and water, for recreation, and as a depository for waste.

Lakes are natural standing bodies of inland water. Lakes form wherever the water supply is adequate and geomorphic processes have created depressions on the land surface. The majority of the world's lake basins, such as those of the North American Great Lakes, are products of glaciation, but rivers, groundwater, tectonic activity, and volcanism also produce lake basins. Most of the world's lakes hold freshwater surface runoff in temporary storage along stream systems. However, some lakes, such as the Caspian Sea, Dead Sea, and Great Salt Lake in Utah, contain salty water because they exist in closed basins of interior drainage with no outflowing streams. Evaporation of water to the atmosphere from these lakes leaves behind the minerals that comprised the dissolved load of inflowing streams.

Lakes are important to humans for more than their scenic appeal and their value for fishing or recreational activities. Like oceans, lakes affect the nearby climates, particularly by reducing daily and seasonal temperature ranges and by increasing humidity. The moderating effect on temperature allows land adjacent to many large lakes to support agricultural activities that are not otherwise possible in the regional climate. Lakes can also cause a downwind increase in precipitation—snow or rain generated by the *lake effect*, in which storms either pick up more moisture from the lake or undergo uplift as they move over a relatively warm body of water.

Because stream flow can be so variable and in some cases unreliable, humans today regulate most rivers in some way. Many river systems now consist of a series of reservoirs—artificial lakes impounded behind dams. The dams hold back potentially devastating floodwaters and store the discharge of wet periods to make the water available during dry seasons or drought years (■ Fig. 14.32). Most great rivers in the United States, such as the Missouri, Columbia, and Colorado, have been transformed by dam construction into a string of reservoirs. The life of these reservoirs, however, will be only a few centuries at most because they will gradually fill with sediment carried by the inflowing streams. As part of a recent trend in the natural sciences aimed at mitigating effects of human disturbance on streams and other systems, increasing efforts are being directed toward assessing the impacts of dams and dam removal on stream geomorphology and ecology.

MAP INTERPRETATION

FLUVIAL LANDFORMS

The Map

Campti is in northwestern Louisiana on the Gulf Coastal Plain. The coastal plain region stretches westward from northern Florida to the Texas–Mexico border, and extends inland in some places for more than 320 kilometers (200 mi). Elevations on the Gulf Coastal Plain gradually increase from sea level at the shoreline to a few hundred meters far inland. The region is underlain by gently dipping sedimentary rock layers. The surface material includes marine sediments and alluvial deposits from rivers that cross the coastal plain, especially those of the Mississippi River drainage system. This is a landscape of meanders, natural levees, and bayous.

The Red River's headwaters are located in the semiarid plains of the Texas Panhandle, but it flows eastward toward an increasingly more humid climate. About midcourse, the Red River enters a humid subtropical climate region, which supports rich farmland and dense forests. The river flows into the Mississippi in southern Louisiana, about 160 kilometers (100 mi) downstream of the map area. The Red River is the southernmost major tributary of the Mississippi.

Louisiana has mild winters with hot, humid summers. Annual rainfall totals for Campti average about 127 centimeters (50 in.). The warm waters of the Gulf of Mexico supply vast amounts of atmospheric energy and moisture, producing a high frequency of thunderstorms, tornadoes, and, on occasion, hurricanes that strike the Gulf Coast.

You can also compare this topographic map to the Google Earth presentation of the area. Find the map area by zooming in on these latitude and longitude coordinates: 31.847019°N, 93.157614°W.

Interpreting the Map

1. How would you describe the general topography of the Campti map area? What is the local relief? What is the elevation of the banks of the Red River at the town of Campti?
2. What kind of landform is the low-relief surface that the river is flowing on? Is this mainly an erosional or depositional landform?
3. In what general direction does the Red River flow? Is the direction of flow easy or hard to determine just from the map area? Why?
4. Does the Red River have a gentle or steep gradient? Why is it difficult to determine the river gradient from the map area?
5. What is the origin of Smith Island? Adjacent to Smith Island is Old River. What is this type of feature called?
6. Explain the stippled brown areas in the meanders south of Campti. Are these areas on the inside or the outside of the meander bend?
7. How would you describe the features labeled as "bayou"?
8. Is this map area more typical of the upper, middle, or lower course of a river?
9. Although this is not a tectonically active region, how would the river change if it experienced tectonic uplift?

U.S. Department of Agriculture

Healthy green vegetation appears red on color infrared images, here showing the Red River and Smith Island, Louisiana.

Opposite:
Campti, Louisiana
Scale 1:62,500
Contour interval = 20 ft
U.S. Geological Survey

Epoch. While glaciers were advancing at high latitudes and in high-elevation mountain regions during the Pleistocene, precipitation was also greater than it is today in the now-arid basins, valleys, and plains of the middle and subtropical latitudes. At the same time, cooler temperatures for these regions meant that they also experienced lower evaporation rates. In many deserts, evidence of past wet periods includes deposits and wave-cut shorelines of now-extinct lakes (■ Fig. 15.2) and immense canyons occupied by streams that are now too small to have eroded such large valleys.

Running water is a highly effective geomorphic agent in deserts even though it operates only occasionally. In most desert areas, running water flows just during and shortly after rainstorms. Desert streams, therefore, are typically *ephemeral channels*, containing water only for brief intervals and remaining dry the rest of the time. In contrast to *perennial channels*, which flow all year and are typical of humid environments, ephemeral streams do not receive seepage from groundwater to sustain them between episodes of surface runoff. Ephemeral streams instead generally lose water to the groundwater system through infiltration into the channel bed. Because of the low weathering rates in desert environments, most arid region streams receive an abundance of coarse sediment that they must transport as bed load. As a result, *braided channels*, in which multiple threads of flow split and rejoin around temporary deposits of coarse-grained sediments, are common in deserts (■ Fig. 15.3).

Unlike the typical situation for humid region streams, many desert streams undergo a downstream decrease, rather than increase, in discharge. A discharge decrease occurs for two major reasons: (1) water losses by infiltration into the gravelly stream channel accumulate downstream, and (2) evaporation losses increase downstream due to warmer temperatures at lower elevations. As a result of the diminishing discharge, many desert streams terminate before reaching the ocean. The same mountains that contribute to aridity through the rain-shadow effect can effectively block desert streams from flowing to the sea. Without sufficient discharge to reach ultimate base level, desert streams terminate in depressions in the continental interior where they commonly form shallow, ephemeral lakes. Ephemeral lakes evaporate and disappear only to reappear when rain provides another episode of adequate inflow.

J. Petersen

■ **FIGURE 15.1** In almost all deserts the effects of running water are prominent in the landscape.
Why do you think the drainage density is so high on the terrain in this photo?

■ **FIGURE 15.2** Many desert basins have remnant shorelines that were created by lakes they contained during the Pleistocene. The linear feature extending across this mountain front was the highest shoreline of one of these ancient lakes.
Why does the terrain look so much smoother below than above the highest shoreline?

J. Petersen

During the cooler and wetter times of the Pleistocene Epoch, many closed basins in now-arid regions were filled with considerable amounts of water that in some cases formed large perennial freshwater lakes instead of the shallow ephemeral lakes that they contain today.

D. Sack

■ **FIGURE 15.3** This braided stream in Canyon de Chelly National Monument near Chinle, Arizona, splits and rejoins multiple times as it works to carry its extensive bed load of coarse sand.
Why do the number and position of the multiple channels sometimes change rapidly?

Where surface runoff drains into closed desert basins, sea level does not govern erosional base level as it does for streams that flow into the ocean and thereby attain *exterior drainage*. Desert drainage basins characterized by streams that terminate in interior depressions are known as basins of *interior drainage*; such streams are controlled by a **regional base level** instead of ultimate base level (■ Fig. 15.4). When sedimentation raises the elevation of the desert basin floor located at the stream's terminus, the stream's base level rises, causing a decrease in the stream's slope, velocity, and energy. If tectonic activity lowers the basin floor, the regional base level falls, which may lead to rejuvenation of the desert stream. Tectonism has even created some desert basins of interior drainage with floors below sea level, as in Death Valley, California, and the Dead Sea Basin in the Middle East.

Many streams found in deserts originate in nearby humid regions or in cooler, wetter mountain areas adjacent to the desert. Even these, however, rarely have sufficient discharge to sustain flow across a large desert.

■ **FIGURE 15.4** Owens Lake bed (white area) occupies a basin in the rain shadow of the Sierra Nevada to the west (top of photo) that is part of a large region of interior drainage which terminates in Death Valley. Any outflow from Owens Lake through the stream channel (at left) is conducted to even lower basins.

NASA/Earth Observations Lab/Johnson Space Center

National Park Service, Canyonlands

■ **FIGURE 15.8** On the Colorado Plateau in the southwestern United States, differential weathering and erosion of horizontal rock strata of varying resistance have created a stair-stepped topography that is easily seen in the sparsely vegetated, arid landscape.

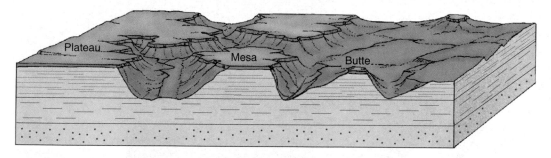

■ **FIGURE 15.9** Plateaus, mesas, and buttes are developed through weathering and erosion in areas, like the Colorado Plateau, of horizontal rock layers with a resistant caprock.

extending down the sides of buttes, mesas, and plateaus are related to the height of the cliff at the top, which is controlled by the thickness of the caprock in comparison to the overall height of the landform (■ Fig. 15.10).

Sheet wash and gully development generally accomplish extensive erosion of mountainsides and hillslopes fringing a desert basin or plain. Particularly in arid regions with exterior drainage or a sizable trunk stream on the basin or plain, this fluvial action, aided by weathering, can lead to the gradual erosional retreat of bedrock slopes. This retreat of the steep mountain front can leave behind a more gently sloping sur-

face of eroded bedrock, called a **pediment** (■ Fig. 15.11). Characteristically in desert areas, there tends to be a sharp break in slope between the base of steep hills or mountains, which rise at angles of 20° to 30° or steeper, and the gentle pediment, with a slope that is usually only 2° to 7°. Resistant knobs of the bedrock comprising the pediment may project up above the surface on some pediments. These resistant knobs are referred to as **inselbergs** (from German: *insel*, island; *berg*, mountain).

Geomorphologists do not agree on exactly how pediments form, and different processes may be responsible for

■ FIGURE 15.10   The caprock in Monument Valley, Arizona, is particularly thick and represents a rock layer that once covered the entire region. The prominent buttes are erosional remnants of that layer.

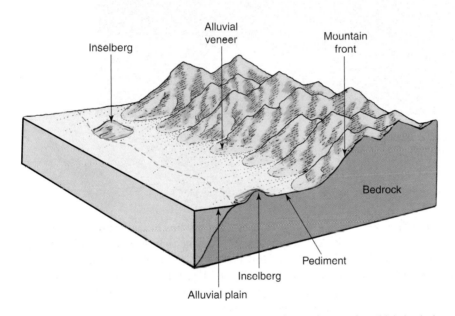

■ FIGURE 15.11   Pediments are gently sloping erosion surfaces, often thinly buried by alluvium, cut into bedrock beyond the present mountain front. They are most common in desert areas with exterior drainage that can remove some of the erosion products. Any remaining resistant knobs of bedrock sticking up above the pediment surface are inselbergs.

desert areas have wide expanses of alluvium deposited either in closed basins or at the base of mountains by streams as they lose water in the arid environment. As the flow of a stream diminishes, so does its *capacity*—the amount of load it can transport. Most landforms made by fluvial deposition in arid lands are not exclusive to desert regions but are particularly common and visible in drylands due to the sparse vegetation.

Alluvial Fans   Where streams, particularly ephemeral (sporadic) or *intermittent* (seasonal) ones, flow out of narrow upland canyons onto open lowland plains, the channels become wider and shallower. Because stream discharge (Q) equals channel cross-sectional width (w) and depth (d) multiplied by flow velocity (v), $Q = wdv$, the increase in width causes a decrease in flow velocity, reducing stream competence (maximum size of load) and stream capacity (maximum amount of load). Discharge also decreases as water seeps from the channel into coarse alluvium below. As a result, most of the sediment load carried by such streams is deposited along the base of the highlands.

Upstream of the mouth of the canyon, the channel is constrained by bedrock valley walls, but as it flows out of the canyon, the channel is free to not only widen but also shift its position laterally. Sediments are initially deposited just beyond the canyon mouth, building up the lowland area near the mountain front. Eventually this aggradation causes the channel to shift laterally where it begins to deposit and build up another zone close to the mountain front and adjacent to the first aggraded area. The canyon mouth serves as a pivot point anchoring the channel as it swings back and forth over the lowlands near the mountain front, leaving alluvium behind. This creates a fan-shaped depositional landform, called an **alluvial fan,** in which deposition takes place radially away from that pivot point, or **fan apex** (■ Fig. 15.12).

An important characteristic of an alluvial fan is the sorting of sediment that typically occurs on its surface. Coarse sediments, like boulders and cobbles, are deposited near the fan apex where the stream first undergoes a decrease in competence and capacity as it emerges from the confinement of the canyon. In part because of the large size of the clasts deposited there, the slope of an alluvial fan is steepest at its apex and gradually diminishes, along with grain size, with

their formation in different regions. However, there is general agreement that most pediments are erosional surfaces created or partially created by the action of running water. In some areas, weathering, perhaps when the climate was wetter in the past, may also have played a strong role in the development of pediments.

Arid Region Landforms of Fluvial Deposition

Deposition is as important as erosion in creating landform features in arid regions, and in many areas sedimentation by water does as much to level the land as does erosion. Many

J. Petersen

■ **FIGURE 15.15**   The floors of most bolsons contain large playas, which only occasionally hold water.

What evidence suggests that the playa in this photo is partially wet?

■ **FIGURE 15.16**   Salt deposits accumulate on playas that evaporate considerable groundwater. In the foreground, a mixture of salt and mud has created the playa microtopography known as puffy ground.

J. Petersen

The playa lake may persist for a day or for a few months (■ Fig. 15.15). Wind blowing over the playa lake moves the shallow water, along with its suspended and dissolved load, around on the playa surface. This helps to fill in any low spots on the playa, and contributes to making playas one of the flattest of all landforms on Earth. Playa lakes lose most of their water by evaporation to the desert air.

Although playas are very flat, considerable variation exists in the nature of playa surfaces. Playas that receive most of their water from surface runoff typically have a smooth clay surface, sometimes called a **clay pan,** baked hard by the desert sun when it is dry but extremely gooey and slippery when wet. In contrast, **salt-crust playas,** also known as **salt flats,** receive much of their water from groundwater, are damp most of the time, and are encrusted with salt mineral deposits crystallizing out of the evaporating groundwater (■ Fig. 15.16). Some playas, such as Utah's famous Bonneville Salt Flats, are the floors of desiccated, ancient lakes that occupied now-desert basins during the Pleistocene Epoch.

Wind as a Geomorphic Agent

Overall, wind is less effective than running water, waves, groundwater, moving ice, or mass movement in accomplishing geomorphic work. Under certain circumstances, however, wind can be a significant agent in the modification of topography. Landforms—whether in the desert or elsewhere—that are created by wind are called **eolian** (or **aeolian**) landforms, after Aeolus, the god of winds in classical Greek mythology. The three principal conditions necessary for wind to be effective as a geomorphic agent are a sparse vegetative cover, the presence of dry, loose materials at the surface, and a wind velocity that is high enough to pick up and move those surface materials. These three conditions occur most commonly in arid regions and on beaches.

A dense cover of vegetation reduces wind velocity near the surface by providing frictional resistance. It also prevents wind from being directed against the land surface and holds materials in place with its root network. Without such a protective cover, fine-grained and sufficiently dry surface materials are subject to removal by strong gusts of wind. If surface particles are damp, they tend to adhere together in wind-resistant aggregates due to increased cohesion provided by the water. The arid conditions of deserts, therefore, make these regions very susceptible to wind erosion.

Eolian processes have many things in common with fluvial processes because air and water are both fluids. Some important contrasts also exist, however, due to fundamental differences between gases and liquids. For example, rock-forming materials cannot dissolve in air, as some can in water; thus air does not erode by corrosion or move load in solution. Otherwise, the wind detaches and transports rock fragments in ways comparable to flowing water, but it does so with less overall effectiveness because air has a much lower density than water. Another difference is that, compared to streams, the wind has fewer lateral or vertical limitations on movement. As a result, the dissemination of material by the wind can be more widespread and unpredictable than that by streams.

A principal similarity between the geomorphic properties of wind and running water is that flow velocity controls their competence—that is, the size of particles each can pick up and carry. However, because of its low density, the competence of moving air is generally limited to rock fragments that are sand sized or smaller. Wind erosion selectively entrains small particles, leaving behind the coarser and heavier particles that it is not able to lift. Like water-laid sediments, wind deposits are stratified according to changes in the fluid's velocity although within a much narrower range of grain sizes than occurs with most alluvium.

Wind Transportation and Erosion

Strong winds blow frequently in arid regions, whipping up loose surface materials and transporting them within turbulent air currents. The finest particles transported by winds, clays and silts, are moved in *suspension,* buoyed by vertical currents (■ Fig. 15.17). Such particles essentially comprise a fine dust that will remain in suspension as long as the strength of upward air currents exceeds the tendency of the particles to settle out to the ground due to gravity. The sediments carried in suspension by the wind make up its suspended load. If the

wind velocity surpasses 16 kilometers (10 mi) per hour, surface sand grains can be put into motion. As with fluvial transportation, particles that are too large to be carried in eolian suspension are bounced along the ground as part of the bed load in the transportation process of *saltation.* When particles moving in eolian saltation, which are typically sand sized, bounce on the ground, they generally dislodge other particles that are then added to the wind's suspended or saltating load. Even larger sand grains too heavy to be lifted into the air are pushed forward along the ground surface by the impact of the saltating grains in a process called **surface creep.** Grains moving along the ground surface often become organized into small wave forms termed **ripples.**

Wind erodes surface materials by two main processes. **Deflation,** which is similar to the hydraulic force of running water, occurs when wind blowing fast enough or with enough turbulence over an area of loose sediment is able to pick up and remove on its own small fragments of rock. Like all of the geomorphic agents, once the wind obtains some load, it also erodes by *abrasion.* In eolian abrasion particles already being transported by the wind strike rocks or sediment and break off or dislodge additional rock fragments. Wind-driven solid particles are more effective than the wind alone in dislodging and entraining other grains and in breaking off new fragments of rock. Most eolian abrasion is quite literally sandblasting, and quartz sand, which is common in many desert areas, can be a very effective abrasive agent in eolian processes. Yet sand grains are typically the largest size of clast that the wind can move, and they rarely are lifted higher than 1 meter (3 ft) above the surface. Consequently, the effect of this natural sandblast is limited to a zone close to ground level.

Where loose fine-grained particles exist on the land surface, during strong winds they will be picked up primarily

■ **FIGURE 15.17**   Wind moves sediment in the transportation processes of suspension, saltation, and surface creep. Some of the bed load forms ripples, which can be seen moving forward when the wind is strong.

Why are grains larger than sand not generally moved by the wind?

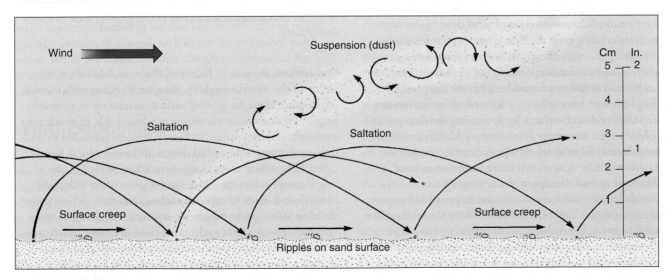

Copyright and photograph by Dr. Parvinder S. Sethi

■ **FIGURE 15.23**   A spectacular view of wind-deposited sand dunes in Death Valley, California.

■ **FIGURE 15.24**   Active and stabilized dunes. (a) Sand moves freely in active dunes, which typically have little or no vegetation. (b) Vegetation reduces wind speed over the dune and helps increase material cohesion, contributing to dune stabilization.
What effect might the trails that are visible on the stabilized dunes have on the dune system?

D. Sack

(a)

(b)

D. Sack

Sand dunes may be classified as either *active* or *stabilized* (■ Fig. 15.24). Active dunes change their shape or advance downwind as a result of wind action. Dunes may change their shape with variations in wind direction and/or wind strength. Dunes travel forward as the wind erodes sand from their upwind (windward) slope, depositing it on their downwind (leeward) side. Sand entrained on the upwind slope moves by saltation and surface creep up to and over the dune crest onto the steep leeward slope, which is the **slip face.** The slip face of a dune lies at the angle of repose for dry, loose sand. The angle of repose (about 34° for sand) is the steepest slope that a pile of dry, loose material can maintain without experiencing slipping or sliding down the slope. When wind direction and velocity are relatively constant, a dune can move forward while maintaining its general form by this downwind transfer of sediment from the windward slope to the slip face (■ Fig. 15.25). The speed at which active dunes move downwind varies greatly, but as with many processes, the movement is episodic; dunes advance only when the wind is strong enough to move sand from the upwind to the downwind side. Because of their greater height and especially their greater amount of sand, large dunes travel more slowly than smaller dunes, which may migrate up to 40 meters (130 ft) a year. During sandstorms, a dune may migrate more than 1 meter (3 ft) in a single day. Some dunes are affected by seasonal wind reversals so that they do not advance, but the crest at the top moves back and forth annually under the influence of seasonally opposing winds.

Stabilized dunes maintain their shape and position over time. Vegetation normally stabilizes dunes. If the vegetation cover becomes breached on a stabilized dune, perhaps due to the effects of range animals or off-road vehicles, the wind can then remove some of the sand, creating a **blowout.** In places where plants, including trees, lie in the path of an advancing dune, the sand cover may move over and smother the vegetation. Where invading dunes and blowing sands are a problem, attempts are frequently made to plant grasses or other vegetation to stabilize the dunes, halting their advance. Vegetation can stabilize a sand dune if plants can gain a foothold and send roots down to moisture deep within the dune. This task is difficult for most plants because sand offers limited nutrients and has high permeability.

One extensive area of stabilized dunes in North America is the Sand Hills of Nebraska. This region features large dunes that formed during a drier period between glacial advances in the Pleistocene Epoch. Subsequent climate change affected sand supply, wind patterns, and moisture availability so that these impressive dunes are now stabilized by grasses (■ Fig. 15.26).

Types of Sand Dunes   Sand dunes are classified according to their shape and their relationship to the wind direction.

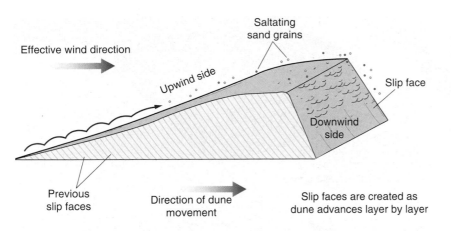

Effective wind direction

Saltating sand grains

Upwind side

Slip face

Downwind side

Slip face

Previous slip faces

Direction of dune movement

Slip faces are created as dune advances layer by layer

■ **FIGURE 15.25**   Active dunes migrate downwind as sediment from the upwind side is transported and deposited on the slip face.

Why does the inside of the migrating dune consist of former slip faces?

The different types are also related to the amount of available sand, which affects not only the size but also the shape of sand dunes.

Barchans are one kind of crescent-shaped dune (■ Fig. 15.27). The two arms of the crescent, called the dune's *horns*, point downwind (■ Fig. 15.28a). The main body of the crescent lies on the upwind side of the dune. From the desert floor at its upwind edge, the dune rises as a gentle slope up which the sand moves until it reaches the highest point, or crest, of the dune and, just beyond that, the slip face at the angle of repose. The slip face is oriented perpendicular to the barchan's horns. The horns extend downwind beyond the location of the slip face. Barchans form in areas of minimal sand supply where winds are strong enough to move sand downwind in a single prevailing direction. They may be most common in smaller desert basins surrounded by highlands where they tend to form near the downslope, sandy edge (toe) of alluvial fans, and adjacent to small playas. Although they sometimes form as isolated dunes, barchans often appear in small groups, called barchan fields.

Parabolic dunes are similar to barchans in that they are also crescent-shaped dunes, but their orientation is reversed from that of barchans (Fig. 15.28b). Here, the arms of the crescent are stabilized by vegetation, long, and point upwind, trailing behind the unvegetated main body and crest of the dune, rather than extending downwind from it. The main body of a parabolic dune points downwind, and the slip face along its downwind edge has a convex shape when viewed from above. Parabolic dunes commonly occur just inland of beaches and along the margin of active dune areas in deserts.

Transverse dunes are created where sand-transporting winds blow from a constant direction and the supply of sand is abundant (Fig. 15.28c). As with barchans, the upwind slope of a transverse dune ridge is gentle, while the steeper downwind slip face is at the angle of repose. In the downwind direction, transverse dunes form ridge after ridge separated from each other by low swales in a repeating wavelike fashion. Each dune ridge is laterally extensive perpendicular to the sand-transporting wind direction. The dune ridges, slip faces, and interdune swales trend perpendicular to the direction of prevailing winds, hence the name *transverse*. Abundant sand supply derives from such

© David B. Loope/ University of Nebraska

■ **FIGURE 15.26**   The rolling, grazing lands of the Sand Hills in central Nebraska were once a major region of active sand dunes.

Why are these dunes no longer active today?

■ **FIGURE 15.27**   A small barchan in southern Utah.

Why do smaller barchans migrate faster than larger barchans?

D. Sack

GEOGRAPHY'S ENVIRONMENTAL PERSPECTIVE

:: OFF-ROAD VEHICLE IMPACTS ON DESERT LANDSCAPES

Whether you prefer to call them off-road vehicles (ORVs) or all-terrain vehicles (ATVs), driving motorized vehicles of any kind over the desert landscape off established roadways has tremendous negative impacts on desert flora and fauna, the habitat of those organisms, the stability of landforms and landforming processes, and the aesthetic beauty of the natural desert environment. Deserts are particularly fragile environments primarily because of the low amount and high variability of precipitation. The climatic regime leads to slow rates of weathering, slow rates of soil formation, low density of vegetation, unusual species of plants and animals specially adapted to the hostile conditions of severe moisture stress, and rapid generation of surface runoff when precipitation does occur.

ORVs damage desert biota in several ways. ORVs kill and injure plants, animals (for example, birds, badgers, foxes, snakes, lizards, and tortoises, to name just a few), and insects when they hit or ride over top of them. Even careful ORV riders who drive slowly cannot avoid crushing small animals and insects that lie hidden under a loose cover of sand, or shallow roots that extend out far from a plant. Driving ORVs at night is particularly deadly for desert animals, which tend to be nocturnal. ORVs do not have to hit or ride over organisms to cause populations to decrease. Hearing loss experienced by animals in areas frequented by ORVs puts them at a major disadvantage for feeding, defense, and mating. ORVs crush and destroy burrows and other animal and insect homes and nesting sites, including plants. Oil and gas leaking from poorly maintained vehicles are another danger for desert dwellers, as are grass and range fires inadvertently started from such vehicles. No desert subenvironment, not even sand dunes, is immune from these negative impacts.

In addition to the direct impacts on organisms, physical properties of the landscape are also negatively affected by ORVs. One major problem is that ORVs compact the desert surface sediments and soil. Compaction causes a decrease in permeability, which greatly restricts the ability of water to infiltrate into the subsurface. More surface runoff translates to greater erosion of the already thin desert soils. In addition, even long after ORV use ceases in a disturbed area, soil compaction makes it very difficult for desert plants, animals, and insects to become reestablished there. Denuded, eroded swaths made by ORVs often remain visible as ugly scars in the landscape for decades, and act as sources of sediment, adding to the severity of dust storms. Where ORV trails traverse steep slopes, they have contributed to the occurrence of debris flows, mudflows, and other forms of mass wasting.

The attributes of desert environments that give them such stark beauty also make them very vulnerable to disturbance. With limited, but often torrential, precipitation, sparse and slow-growing vegetation, and slow rates of weathering and soil formation, deserts require long periods of time to recover from environmental disturbance. Recovery, moreover, will likely not return the landscape to the condition it would have been in if the disturbance had not occurred.

Motorcycles ridden on a desert hillslope in Utah have stripped the vegetation, compacted the soil, instigated accelerated erosion, and seriously marred the aesthetic beauty of the landscape.

Riding ORVs on active sand dunes harms the sensitive organisms that live there and interferes with the natural dune processes by compacting the sand.

Landscape Development in Deserts

Geomorphic landscape development in arid climates is comparable in many ways to that in humid climates, but not in all ways. Weathering and mass movement processes operate, and fluvial processes predominate, in both environments but at different rates and with the tendency to produce some different landform and landscape features in the two contrasting environments. Arid environments experience the added regionally or locally important effect of eolian processes, which are not very common in humid settings beyond the coastal zone. The major differences in the results of geomorphic work in arid climates, as compared to humid climates, are caused by the great expanses of exposed bedrock, a lack of continuous water flow, and a more active role of the wind in arid regions.

Some desert landscapes, such as most of those in the American Southwest, are found in regions of considerable topographic relief, whereas others, like much of the Sahara and interior Australia, occupy huge expanses of open terrain with few mountains. An excellent example of a typical desert landscape in a region of considerable structural relief is found in the Basin and Range region of western North America. The region extends from west Texas and northwestern Mexico to eastern Oregon. It includes all of Nevada and large portions of New Mexico, Arizona, Utah, and eastern California. Here, more than 200 mountain ranges, with basins between them, dominate the topography. The Great Basin—a large subregion centered on Nevada and characterized by interior drainage, numerous alternating mountain ranges and basins, and active tectonism—occupies much of the central and northern part of the Basin and Range.

Fault-block mountains in the Basin and Range region rise thousands of meters above the desert basins, and many form continuous ranges (■ Fig. 15.31). These high ranges encourage orographic rainfall. Fluvial erosion dissects the mountain blocks to carve canyons between peaks and cut washes between interfluves. Where active tectonism continues so that the uplift of mountain ranges matches or exceeds the rate of their erosion, as in the Great Basin, fluvial deposition constructs alluvial fans extending from canyon mouths outward toward the basin floor. In many basins, the alluvial fans have coalesced to create a bajada. In these tectonically active basins with interior drainage, playas often occupy the lowest part of the basin, beyond the toe of the alluvial fans (■ Fig. 15.32a). Although the mountain ranges create considerable roughness, the wind can remove sandy alluvium from the toe areas of alluvial fans, sand-sized aggregates of playa sediments, and, in some cases, sandy sediment from beaches left behind by ancient perennial lakes. These sediments may be transported relatively short distances before being deposited in local dune fields.

Pediments lie along the base of some mountains, particularly in the tectonically less active areas of exterior drainage beyond the southern boundary of the Great Basin. In these areas, mountains are being lowered and the pediments extended. Resistant inselbergs remain on some of the pediments. Sandy alluvium along streams and sandy deposits left

■ **FIGURE 15.31**   Satellite image of fault-block mountains and basins in northeastern Nevada.

NASA

MAP INTERPRETATION

EOLIAN LANDFORMS

The Map

Eolian processes formed the Sand Hills region, the largest expanse of sand in North America. The region covers over 52,000 square kilometers (20,000 sq mi) of central and western Nebraska.

The Sand Hills region was part of an extensive North American desert some 5000 years ago. The sand dunes here reached more than 120 meters (400 ft) high and inundated postglacial peat bogs and rivers. As the climate became wetter, vegetation growth invaded the dunes, greatly reducing eolian erosion and transportation. The vegetation anchored the sand, and the stabilized dunes developed a more rounded form. Underlying the Sand Hills is the Ogallala aquifer. The high water table of the aquifer supports the many lakes nestled between the dunes.

The Sand Hills area has a middle-latitude steppe climate (BSk) and receives about 50 centimeters (20 in.) of precipitation annually. Temperatures have a great annual range, from freezing winters to very hot summers. During summer

months, the region is often pelted by thunderstorms and hail; during the winter months, the area is subject to blinding blizzards.

Vegetation is mainly bunch grasses that can survive on the dry, sandy, and hilly slopes. Some species of bunch grasses have extensive root systems that may extend more than 1 meter (3 ft) into the sandy soil. The lakes and marshes in the interdunal valleys support a marsh plant community that in turn supports thousands of migratory and local birds. Currently, the main land use in the region is cattle grazing. Some scientists are predicting that the Sand Hills area will lose its protective grass cover if global warming continues and will revert to an active area of migrating desert sand dunes.

You can also compare this topographic map to the Google Earth presentation of the area. Find the map area by zooming in on these latitude and longitude coordinates: 42.368889°N, 101.626944°W.

Interpreting the Map

1. What is the approximate relief between the dune crests and the interdunal valleys?
2. What is the general linear direction in which the dunes and valleys trend?
3. To which direction do the steepest sides of the dunes generally face?
4. Knowing that the slip face is the steepest slope of a dune, determine what the prevailing wind direction was when the dunes were active.
5. Based on your answers to the previous three questions, determine what type of sand dune formed the Sand Hills.

6. Use the elevation of the lakes to determine the water table elevation at several points. Use that information to identify the general direction of groundwater flow in the aquifer beneath the Sand Hills.
7. Sketch a north–south profile across the middle of the map from School No. 94 to the eastern end of School Section Lake (in section 16). Label the following landform features: dune crests, dune slip faces, interdunal valleys, and lakes.
8. What cultural features on the map indicate the dominant land use for the region?

Location map of the Nebraska Sand Hills, which cover almost one third of the state.

Opposite:
Steverson Lake, Nebraska
Scale 1:62,500
Contour interval = 20 ft
U.S. Geological Survey

Glacial Systems and Landforms

:: Outline

The Byron Glacier near Portage Lake, Alaska.

Copyright and photograph by
Dr. Parvinder S. Sethi

:: Objectives

When you complete this chapter you should be able to:

- Appreciate the role of glaciers in Earth's hydrologic budget.
- Explain what a glacier is and how glacial ice forms.
- Differentiate the different types of alpine and continental glaciers.
- Discuss the ways in which glaciers flow.
- Draw a cross section of an alpine glacier identifying its major surface and subsurface zones.

- Understand the processes of glacial erosion.
- Define each of the major glacial erosional landforms.
- Describe the principal characteristics of glacial sediments.
- Distinguish among the various glacial depositional landforms.
- Discuss what Earth's glacial environment was like during the Pleistocene Epoch, and how that environment influences people today.

Glaciers, large masses of flowing ice, play several important roles in the Earth system. They are excellent climate indicators because certain environmental conditions are required for glaciers to exist, and they respond visibly to climate variation. Glaciers become established, expand, contract, and disappear in response to changes in climate. Their long-term storage of freshwater as ice has a tremendous impact on the hydrologic cycle and the oceans, and the accumulation of ice by glaciers provides a record of past climates that can be studied in ice cores. Where glaciers once existed or where they were once much larger than they are today, much evidence can be found concerning past climatic conditions.

The processes of erosion, transportation, and deposition by glaciers, whether ongoing or in the past, leave a distinctive stamp on a landscape. Some of the most beautiful and rugged terrain in the world exists in mountainous and other highland regions that have been sculpted by glaciers. Virtually every high-mountain region in the world displays glacial landscapes, including the Alps, the Rocky Mountains, the Himalayas, and the Andes. Glaciers have also carved impressive steep-sided coastal valleys in Norway, Chile, New Zealand, and Alaska. Rugged mountain peaks rising high above lake-filled valleys or narrow and deep ocean embayments create the ultimate in scenic appeal for many people.

Masses of moving ice have transformed the appearance of high mountains, as well as large portions of continental plains, into distinctive glacial landscapes. The flowing ice of glaciers is an effective and spectacular geomorphic agent on major portions of Earth's surface.

Glacier Formation and the Hydrologic Cycle

Glaciers are masses of flowing ice that have accumulated on land in areas where the annual input of frozen precipitation, especially snowfall, has exceeded its yearly loss by melting and other processes. Snow falls as hexagonal ice crystals that form flakes of intricate beauty and variety. Snowflakes have

a low density (mass per unit volume) of about 0.1 grams per cubic centimeter (0.06 oz/in.³). Once snow accumulates on the land, it becomes transformed by compaction along with melting and refreezing at pressure points into a mass of smaller, rounded grains (■ Fig. 16.1). Density increases as the air space around this more granular snow continues to decrease by compaction, melting, and refreezing. Through melting, refreezing, and pressure caused by the increasing weight from burial under newer snowfalls, the granular snow compacts further into a denser, crystalline granular stage, known as **firn**. Over time, the small firn granules grow together into larger interlocked ice crystals through pressure, partial melting, and refreezing. When the ice is deep enough and has a density up to about 0.9 grams per cubic centimeter (0.52 oz/in.³), it becomes glacial ice. Pressure from burial under many layers of snow, firn, and ice causes the glacial ice below to become plastic and flow outward or downward away from the area of greatest snow and ice accumulation.

■ **FIGURE 16.1** The transformation of frozen water from snow to glacial ice.

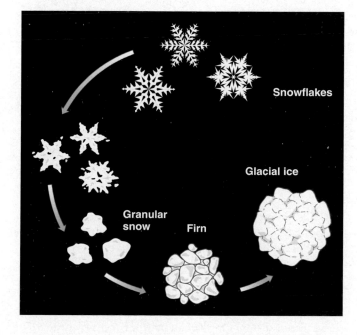

GEOGRAPHY'S PHYSICAL SCIENCE PERSPECTIVE

:: GLACIAL ICE IS BLUE

When we make ice in our freezers, clear colorless water turns to relatively clear ice cubes. The ice cubes may contain some white crystalline forms and air bubbles, but in general the ice is clear. In nature, the process of making ice is very different from that of an ice maker in a refrigerator-freezer. As snow falls at colder, higher latitudes and elevations, it forms a layer of snow on the surface. Each successive snowfall makes another layer as it piles onto the previous snowpack. The weight of the successive layers of snow creates pressure that compresses the older layers beneath. Through time, the layers of low-density snow become layers of much denser solid ice. Some of this change is due to compaction, but

pressure also causes some melting and refreezing of the ice. The temperature at which ice melts is 0°C (32°F) at atmospheric pressure, but ice can melt at lower temperatures if it is under enough pressure. Compaction, as well as pressure melting and refreezing, work to reduce the amount of air in the frozen mass and thereby increase the material's density.

Objects that appear white to the human eye reflect all wavelengths of light with equal intensity, and this is what the hexagonal crystalline structure of a snowflake does. As the snow strata under great pressure in a glacier become compacted over the years (sometimes hundreds or thousands of years) the ice becomes denser. Basically, under this pressure,

more ice crystals are squeezed into the same volume. As the density of ice increases, it reflects increasing amounts of shorter wavelengths of light, which is the blue part of the spectrum. The denser the ice, the bluer it appears. Ice density can be influenced by factors other than time, though, so we must be careful not to assume that deeper blue layers in a glacier are necessarily the older layers. For example, the packing of higher density wetter snow as opposed to lower density drier snow can affect the density of specific layers. Nevertheless, what is certain when looking at massive glacial ice accumulations in nature, such as in ice caps and ice sheets, is that the ice will appear as shades of blue.

An iceberg revealing very old layers of glacial ice.

Glaciers are open systems with input, storage, and output of material. Any addition of frozen water to a glacier is termed **accumulation.** Most accumulation consists of winter snowfall onto a glacier, but there are many ways that frozen water can accumulate, including other forms of precipitation onto the ice, atmospheric water vapor freezing directly onto the ice (gas to solid), and transportation of snow to a glacier surface from surrounding terrain by wind and avalanches.

Ablation is the opposite of accumulation, representing any removal of frozen water mass from a glacier. Although most ablation is accomplished in summer through melting, a glacier may lose mass by direct change from ice to water vapor (solid to gas) and by other processes. **Calving,** for example, refers to the loss of large chunks of ice from a glacier to an adjacent lake or the ocean, where the chunks float away as icebergs (■ Fig. 16.2).

■ **FIGURE 16.2** Chunks of ice calving from a glacier into the sea become icebergs off the shore of Ellesmere Island, Canada.

A glacial system is controlled by two basic climatic conditions: frozen precipitation and freezing temperatures. First, to establish a glacier, there must be sufficient mass input (accumulation) to exceed the annual loss through ablation. Some very cold polar regions in Alaska and Siberia and a few valleys in Antarctica have no glaciers because the climate is too dry. Yet mountains along mid-latitude coastlines and high mountains near the equator can support glaciers because of heavy orographic snowfalls, despite intense sunshine and warm climates in the surrounding lowlands. Summer temperatures must not be high for too long, or all of the accumulation from the previous winter will be lost through ablation (primarily melting). Surplus snowfall is essential for glacier formation because it allows the pressure from years of accumulated snow layers to transform older buried snow first into firn, and then into glacial ice. When the ice reaches a depth of about 30 meters (100 ft), a pressure threshold is reached that enables the solid, glacial ice to flow.

Glaciers are an important part of Earth's hydrologic cycle and are second only to the ocean in the amount of water they contain. Approximately 2.25% of Earth's total water is currently frozen in glaciers. This frozen water, moreover, makes up about 70% of the world's *fresh* water, with the vast majority stored on Greenland and Antarctica. The amount is so impressive that if all of the world's glaciers were to melt, sea level would rise about 74 meters (243 ft), changing the planet's

geography considerably. In contrast, during the last major glacial advance, sea level fell about 120 meters (400 ft).

Unlike the water in a stream system, much of which returns rapidly to the sea or atmosphere, glaciers may store water as ice for hundreds or even hundreds of thousands of years before it is released as meltwater. Yet, glacial ice is not stagnant. It moves slowly but with tremendous energy across the land. Glaciers reshape the landscape by engulfing, eroding, pushing, dragging, carrying, and finally depositing rock debris, often in places far from its original location. Long after glaciers recede from a landscape, glacial landforms remain as a reminder of the energy of the glacial system and as evidence of past climates.

Glaciers have not existed on the planet during most of Earth history. A period during which significant areas of the middle latitudes are covered by glaciers is an *ice age*. Glaciers today cover about 10% of Earth's land surface, at high latitudes and high elevations on all continents except Australia. During recent Earth history, from about 2.6 million years ago to about 10,000 years before the present, during the Pleistocene Epoch, glaciers periodically covered nearly a third of Earth's land area. Other ice ages occurred in the much more distant geologic past.

Types of Glaciers

The two major categories of glaciers are alpine and continental. **Alpine glaciers** exist where the precipitation and temperature conditions required for glacier formation result from high elevation. Alpine glaciers are fed by ice and snow in mountain areas, and usually occupy valleys that were previously created by stream erosion. The ice masses flow downslope because of their own weight, that is, due to the force of gravity. Alpine glaciers that occupy former stream valleys are **valley glaciers** (■ Fig. 16.3). They are known as **piedmont glaciers** when the ice extends to lower elevations beyond the mouth of a canyon, where it spreads out over more open terrain. The smallest type of glacier are the alpine **cirque glaciers** restricted to distinctive steep-sided amphitheater-like depressions, called **cirques,** which are eroded by the ice near high peaks at valley heads (■ Fig. 16.4). Alpine glaciers begin as cirque glaciers at the start of an ice age, become valley glaciers when they expand into the valley below the cirque, and may eventually become piedmont glaciers as the ice age intensifies. Most cirque glaciers today represent small remnants of previously larger alpine glaciers.

Alpine glaciers created the characteristic rugged scenery of much of the world's high-mountain regions, including the Himalayas, Sierra Nevada, Rockies, Andes, and Alps. The largest alpine glaciers currently in existence are located in Alaska

and temperatures warmer. As a general rule, temperate alpine glaciers flow much faster than the cold polar continental glaciers.

The flow of an individual glacier varies from time to time with changes in its mass balance, and from place to place because of variations in the gradient over which it flows or differences in the friction encountered with adjacent rock. Within an alpine glacier, the rate of movement is greatest on the glacier surface toward the middle of the ice because this location experiences accumulated movement from layers of plastic flow below and is farthest from frictional resistance with the valley sides.

Sometimes a glacier's velocity will increase by many times its normal rate, causing the glacier to advance hundreds of meters per year. The reasons for such enormous *glacial surges* are not completely clear, although meltwater reducing friction at the base of the ice is probably involved.

Glaciers as Geomorphic Agents

Because a deep, and therefore heavy, accumulation of ice is required for glaciers to flow, even the smaller alpine glaciers are particularly powerful geomorphic agents able to perform great amounts of geomorphic work. Whether it is an alpine glacier carving out a trough-shaped valley or a continental glacier gouging out the basins of the North American Great Lakes, the work done by glaciers is impressive.

Glaciers remove and entrain rock particles by two erosional processes. Glacial **plucking** is the process by which moving ice freezes onto loosened rocks and sediments, incorporating them into the flow. Weathering, particularly the freezing of water in bedrock joints and fractures, breaks rock fragments loose, encouraging plucking. Once load is entrained at the base and sides of the ice, moving glaciers are armed with clastic particles that are very effective tools for scraping and gouging out more rock material by the erosional process of *abrasion.* Bedrock obstructions subjected to intense glacial abrasion are typically smoother and more rounded than those produced by plucking.

Unlike the situation with liquid water in streams, volume and velocity of flow do not directly determine the particle sizes that plastically flowing, solid ice can erode and transport. Plucking and abrasion provide the bottom and sides of glaciers with a chaotic load of rock fragments of all sizes, from clay-sized crushed rock, called *rock flour,* to giant boulders. Mass wasting along steep mountain slopes, especially above alpine glaciers, contributes sediment, also of a variety of grain sizes, to the ice surface and sides. Also in contrast to streams, little sorting of sediment by size is accomplished by glaciers during transportation and deposition. This lack of sorting makes glacial deposits look very different from accumulations of stream deposits. Because of this contrast in the two types of sediment, it is logical that they are

referred to by two different terms. Whereas stream-deposited sediment is called *alluvium,* sediment deposited directly by moving ice is **till.**

Alpine Glaciers

From a mass balance perspective, alpine glaciers consist of two functional parts, or zones (■ Fig. 16.8). The colder, snowier upslope portion of a glacier, where annual accumulation (input) exceeds annual ablation (output), is the **zone of accumulation.** In contrast, annual ablation exceeds annual accumulation in the warmer downslope portion of an alpine glacier, the **zone of ablation.** Winter is the dominant accumulation season and summer is the dominant ablation season. Alpine glaciers change size, sometimes quite dramatically, over the course of a year. The toe of a glacier lies farthest downvalley near the end of winter and farthest upvalley at the end of the ablation season, near the end of summer. The **equilibrium line** marks the boundary between the zones of accumulation and ablation on an alpine glacier, thus it indicates the cross-sectional position at which annual accumulation equals annual ablation for the glacier.

Several factors influence the location of the equilibrium line. The interaction between latitude and elevation, both of which affect temperature, is an important factor. On mountains near the equator, the equilibrium line lies at very high elevations. Elevation of the equilibrium line decreases with increasing latitude until it coincides with sea level in the polar regions. Equally important to temperature in determining the position of the equilibrium line is the amount of snowfall received during winter. With colder temperatures and greater snowfall, the equilibrium line will drop in elevation; it retreats to higher elevations if the climate warms. Other attributes causing variations in the equilibrium-line elevation include the amount of insolation. A shady mountain slope will have a lower equilibrium line than one that receives more insolation. Wind is another factor because it produces snowdrifts on the leeward side of mountain ranges. In the middle latitudes of the Northern Hemisphere, the equilibrium line is lower on the north (shaded) and east (leeward) slopes of mountains. Consequently, the most significant glacier development in this region is on north-facing and east-facing slopes.

At the farthest upslope edge of an alpine glacier, the upslope end of the zone of accumulation, lies the *glacier's head.* The head of the glacier abuts the steep bedrock cliff that comprises the *cirque headwall.* Ice within an alpine glacier flows downslope from the zone of accumulation to the zone of ablation. This downslope movement is sometimes evidenced by a large crevasse, known as a *bergschrund,* which may develop between the head of the glacier and the cirque headwall (see again Fig. 16.8). Presence of a bergschrund shows that the ice mass is pulling away from the confining rock walls of the cirque. The downslope end of a glacier is called its *terminus,* or the *glacier's toe.*

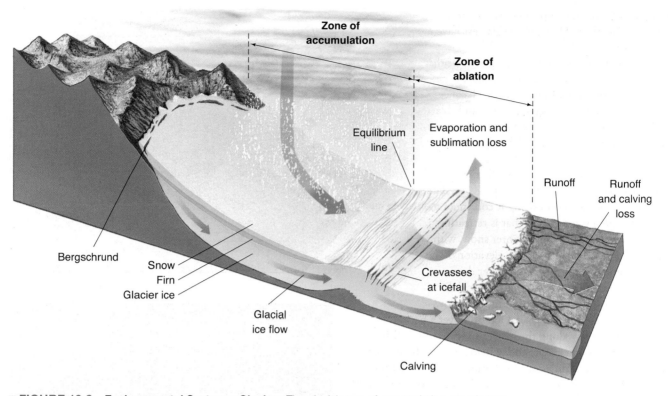

Zone of accumulation

Zone of ablation

Equilibrium line

Evaporation and sublimation loss

Runoff

Runoff and calving loss

Bergschrund

Snow

Firn

Glacier ice

Glacial ice flow

Crevasses at icefall

Calving

■ FIGURE 16.8 Environmental Systems: Glaciers The glacial zone of accumulation experiences greater annual input of frozen water (accumulation) than it loses (ablation) during the year. At lower elevations, in the zone of ablation, annual loss exceeds accumulation. The equilibrium line marks the elevation on the glacier where annual accumulation equals annual ablation. If the glacier as a whole experiences more accumulation than ablation in a year, ice thickness increases and the toe of the glacier advances farther downvalley. If ablation exceeds accumulation for the year, the glacier loses mass and retreats. Equilibrium exists if annual accumulation equals annual ablation for the glacier as a whole. Internal plastic flow transports ice from the zone of accumulation toward the toe of the glacier in the zone of ablation, whether or not the glacier as a whole is advancing or retreating.

Equilibrium and the Glacial Budget

Because the location of the toe of a glacier changes throughout the year with accumulation and ablation, to determine if an alpine glacier is growing or shrinking over a period of years requires annually noting the location of its terminus at the same time of year (■ Fig. 16.9). Typically this is done at the end of the ablation season when the glacier is at its minimum size for the year. If a glacier received more input of frozen water (accumulation) during a year than was removed from it (ablation) that year, it experienced net accumulation. The result of net accumulation is a larger glacier, and as alpine glaciers grow they *advance*, that is, their toes extend farther downvalley. A glacier that undergoes net ablation, more removal than addition of frozen water for the year, shrinks in size

■ FIGURE 16.9 The Jacobshavn Glacier in Greenland has been retreating since scientists began monitoring it in 1850. The colored lines mark the position of the glacier's terminus at various years.

Why is the rapid recession of this glacier of concern to scientists?

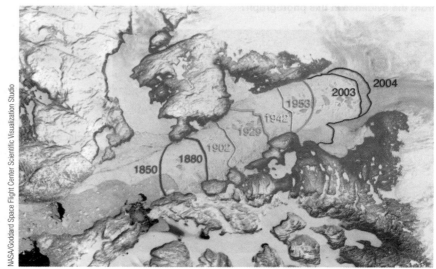

2004

2003

1953

1942

1929

1902

1880

1850

NASA/Goddard Space Flight Center Scientific Visualization Studio

(a)

(b)

■ **FIGURE 16.27** (a) A kettle lake fills a kettle in the Northwest Territories of Canada. (b) Because of Pleistocene continental glaciation, numerous large kettle lakes dot this region in Siberia.

■ **FIGURE 16.28** Drumlins, such as this one in Montana, are streamlined hills elongated in the direction of ice flow.

or more clustered together. Opinions among scientists differ regarding the origin of drumlins, particularly with respect to the relative importance of ice versus meltwater processes in their formation. Drumlins are well developed in Ireland, Canada, and the states of New York and Wisconsin. Boston's Bunker Hill, one of America's best-known historical sites, is a drumlin.

Eskers
An **esker** is a narrow and typically winding ridge composed of glaciofluvial sands and gravels (■ Fig. 16.29). Some eskers are as long as 200 kilometers (130 mi), although several kilometers is more typical of esker length. Most eskers probably formed by meltwater streams flowing in ice tunnels at the base of ice sheets. Many eskers are mined for their sand and gravel.

Kames
Roughly conical hills composed of sorted glaciofluvial deposits are known as **kames.** Kames may develop from sediments that accumulated in glacial ice pits, in crevasses, and among jumbles of detached ice blocks. Like eskers, kames are excellent sources for mining sand and gravel and are especially common in New England. *Kame terraces* are landforms resulting from accumulations of glaciofluvial sand and gravel along the margins of ice lobes that melted away in valleys of hilly regions.

Erratics
Large boulders scattered in and on the surface of glacial deposits or on glacially scoured bedrock are called **erratics** if the rock they consist of differs from the local bedrock (■ Fig. 16.30). Moving ice is capable of transporting large rocks very far from their source. The source region of an erratic can be identified by rock type, and this provides evidence of the direction of ice flow. Erratics are known to be of glacial origin because they are marked by glacial striations and are found only in glaciated terrain. Erratics can occur in association with alpine glaciers, but they are best known and more impressive when deposited by ice sheets, which have moved boulders weighing hundreds of tons over hundreds of

■ **FIGURE 16.29** An esker near Albert Lea, Minnesota.

© Robert B. Jorstad

■ **FIGURE 16.30** This huge boulder is an erratic, transported far from its point of origin by a continental glacier.
What does this erratic illustrate about the ability of flowing ice to modify the terrain?

NASA

■ **FIGURE 16.31** The New York Finger Lakes occupy ice-scoured basins excavated during the Pleistocene.
What characteristics of the bedrock contributed to the formation of these narrow lake basins?

kilometers. In Illinois, for example, glacially deposited erratics have come from source regions as far away as Canada.

Glacial Lakes

Many, many thousands of lakes exist today in the surface depressions found on the deposits of the continental glaciers that once covered much of North America and Eurasia. The Pleistocene ice sheets created numerous other lake basins by erosion, scooping out deep elongated basins along zones of weak rock or along former stream valleys. New York's Finger Lakes are excellent examples of lakes in elongated, ice-deepened basins (■ Fig. 16.31). Lakes are also common in areas that were impacted by erosion and deposition from alpine glaciers. Lakes in ice-free cirques and in glacial troughs are commonly contained on one side by end moraines. Evidence for many lakes that no longer exist is found in widespread *glaciolacustrine* (from *glacial*, ice; *lacustrine*, lake) deposits that prove the former existence and size of those lakes.

Some lakes formed while the Pleistocene ice was present where glacial deposition disrupted the surface drainage, or where a glacier prevented depressions from being drained of meltwater. These lakes usually accumulated where water became trapped between a large end moraine and the ice front, or where the land sloped toward, instead of away from, the ice front. In both situations, *ice-marginal lakes* filled with meltwater (see again 16.24). They drained and ceased to exist when the retreat of the ice front uncovered an outlet route for the water body.

During their existence, fine-grained sediment accumulated on the floors of these ice-marginal lakes, filling in topographic irregularities. As a result of this sedimentation, extremely flat surfaces characterize glacial plains where they consist of glaciolacustrine deposits. An outstanding example

of such a plain is the valley of the Red River in North Dakota, Minnesota, and Manitoba. The plain was created by deposition within a vast Pleistocene lake held between the front of the receding continental ice sheet on the north and moraine dams and higher topography to the south. This ancient body of water is named Lake Agassiz for the Swiss scientist who early on championed the theory of an ice age. The Red River flows northward eventually into the last remnant of Lake Agassiz, Lake Winnipeg, which occupies the deepest part of an ice-scoured and sediment-filled lowland.

Another ice-marginal lake in North America produced much more spectacular landscape features, but not in the area of the lake itself. In northern Idaho, a glacial lobe moving southward from Canada blocked the valley of a major tributary of the Columbia River, creating an enormous ice-dammed lake known as Lake Missoula. This lake covered almost 7800 square kilometers (3000 sq mi) and was 610 meters (2000 ft) deep at the ice dam. On occasions when the ice dam failed, Lake Missoula emptied in tremendous floods that engulfed much of eastern Washington. The racing floodwaters scoured the basaltic terrain, producing Washington's channeled scablands consisting of intertwining steep-sided troughs (*coulees*), dry waterfalls, scoured-out basins, and other features quite unlike those associated with normal stream erosion, particularly because of their gigantic size.

The Great Lakes of the eastern United States and Canada make up the world's largest lake system. Lakes Superior, Michigan, Huron, Erie, and Ontario occupy former river valleys that were vastly enlarged and deepened by glacial erosion. All of the lake basins, except that of Lake Erie, have been gouged out to depths below sea level and have irregular bedrock floors lying beneath thick blankets of glacial till. The history of the Great Lakes is exceedingly complex because the fluctuating position of the ice front led to many changes in lake level and in the location of outlets.

the grinding erosive process of *abrasion*. Abrasion is the most effective form of erosion by each of the geomorphic agents, including waves.

Weathering is an important factor in the breakdown of rocks in the coastal zone, as in other environments, preparing pieces for removal by wave erosion. Water is a key element in most weathering processes, and in addition to normal precipitation, rocks near the shoreline are subjected to spray from breaking waves as well as high relative humidities and condensation. Salt weathering is particularly significant in preparing rocks for removal through chemical and physical weathering along the marine coast and coasts of salt lakes.

Coastal Erosional Landforms

Coasts of high relief are dominated by erosion (■ Fig. 17.12). *Sea cliffs* (or *lake cliffs*) are carved where waves pound directly against steep land. If a steep coastal slope continues deep beneath the water, the slope may reflect much of the incoming wave energy until corrosion and hydraulic action eventually take their toll on the rock. Tides present along marine coasts allow these processes to attack a range of shoreline elevations. Once a recess, or **notch,** has been carved out along the base of a cliff (■ Fig. 17.13a), weathering and rockfall within the shaded overhang supply clasts that can collect on the notch floor and be used by the water as tools for more efficient erosion by abrasion. Abrasion extends the notch landward, leaving the cliff above subject to rockfall and other forms of mass wasting. Stones used as tools in abrasion quickly become rounded and may accumulate at the base of the cliff as a **cobble beach.** Where the cliffs are well jointed but cohesive, wave erosion can create *sea caves* along the lines of weakness (Fig. 17.13b). *Sea arches* result where two caves meet from each side of a headland (Fig. 17.13c). When the top of an arch collapses or a sea cliff retreats and a resistant pillar is left standing, the remnant is called a **sea stack.**

Landward recession of a sea cliff leaves behind a wave-cut bench of rock, an **abrasion platform,** which is sometimes visible at lower water levels, such as at low tide (Fig. 17.13d). Abrasion platforms record the amount of cliff recession. In some cases, deposits accumulate as wave-built terraces just seaward of an abrasion platform. If tectonic activity uplifts these wave-cut benches and wave-built terraces above sea level out of the reach of wave action, they become **marine terraces** (Fig. 17.13e). Successive periods of uplift can create a coastal topography of marine terraces that resembles a series of steps. Each step represents a period of time that a terrace was at sea level. The Palos Verdes peninsula just south of Los Angeles has perhaps as many as ten marine terraces, each representing a period of platform formation separated by episodes of uplift.

Rates of coastal erosion are controlled by the interaction between wave energy and rock type. Coastal erosion is greatly accelerated during high-energy events, such as severe storms and tsunamis. Human actions can also accelerate coastal erosion rates. We commonly do so by interfering with coastal sediment and vegetation systems that would naturally protect some coastal segments from excessive erosion rates.

■ **FIGURE 17.12**   Some of the major landforms found along erosion-dominated coastlines.

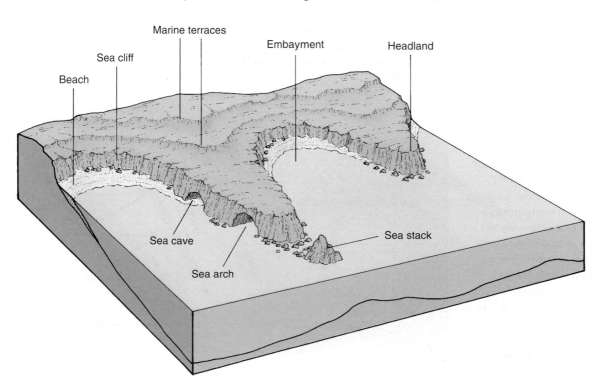

D. Sack

(a)

J. Petersen

(b)

NOAA Image Library

(c)

D. Sack

(d)

© Robert Cameron/Getty Images

(e)

■ **FIGURE 17.13** Examples of landforms created by wave erosion. (a) A notch exposed near the base of basaltic sea cliffs in Hawaii. (b) A sea cave carved in limestone along Italy's Mediterranean Sea coast. (c) A sea arch in Alaska. (d) An abrasion platform along the California coast exposed at low tide in steeply dipping sedimentary rocks. (e) The wide, beveled plain atop this eroding coastline is a marine terrace.

What other coastal erosional landforms do you see in photo (e)?

Coastal Deposition

Significant amounts of sediment accumulate along coasts where wave energy is low relative to the amount or size of sediment supplied. Embayments and settings where waves break at a distance from the shoreline, such as areas with gently shelving underwater topography, tend to sap wave energy and encourage deposition. Amount and size of sediment supplied to the coastal zone vary with rock type, weathering rates, and other elements of the climatic, biological, and geomorphic environment.

Sediment within coastal deposits comes from three principal sources. Most of it is delivered to the standing body of water by streams. At its mouth, the load of a stream may be deposited for the long term in a delta or within an *estuary*, a biologically very productive embayment that forms at some river mouths where salt and freshwater meet. Elsewhere, stream load may instead be delivered to the ocean or lake for continued transportation. Once in the standing body of water, fine-grained sediments that stay in suspension for long periods may be carried out to deep water where they eventually settle out onto the basin floor. Other clasts are transported by waves and currents in the coastal zone, being deposited when energy decreases and, if accessible, reentrained when wave energy increases. The same is true of the second major source of coastal sediment, coastal cliff erosion. Of less importance is sediment brought to the coast from offshore sources. Although we may tend to think of sand-sized sediment when we think of coastal deposits, coastal depositional landforms may be composed of silt, sand, or any size classes of gravel, from granules and pebbles through cobbles and boulders.

Coastal Depositional Landforms

The most common landform of coastal deposition is the **beach,** a wave-deposited feature that is contiguous with the mainland throughout its length (■ Fig. 17.14). Many beaches are sandy, but beaches of other grain sizes are also common, as for example, the cobble beach discussed earlier in the section on coastal erosion. In areas with high wave energy, particles tend to be larger and beaches steeper than where wave energy is low. Beach sediments come in a variety of colors depending on the rock and mineral types represented. Tan quartz, black basalt, white coral, and even green olivine beaches exist on Earth.

Any given stretch of beach may be a permanent feature, but much of the visible sediment deposited in it is not. Individual grains come and go with swash and backwash, wear away through abrasion, are washed offshore in storms, or move into, along, and out of the stretch of beach by littoral drifting. Because waves generated by closer storms tend to be higher than waves generated in distant storms, some beaches undergo seasonal changes in the amount and size of sediment present.

In the middle latitudes, beaches are generally narrower, steeper, and composed of coarser material in winter than they are in summer. The larger winter storm waves are more erosive

J. Petersen

(a)

NOAA / Captain Albert E. Theberge

(b)

U.S. Coast Guard

(c)

■ **FIGURE 17.14**   Beaches are the most common evidence of wave deposition and may be made of any material deposited by waves. (a) A sandy beach in Alameda, California, on San Francisco Bay. (b) A boulder beach in Acadia National Park, Maine (note the person for scale). (c) White sand beaches made of broken coral pieces are common on tropical islands with coral reefs.

and destructive, while the smaller summer waves, which often travel from the other hemisphere, are depositional and constructive. On the Pacific coast of the United States, summer beaches are generally temporary accumulations of sand deposited over coarser winter beach materials (■ Fig. 17.15).

© John Shelton

(a)   **(b)**

■ **FIGURE 17.15**   Because of seasonal variations in wave energy, the differences in a beach from summer to winter can be striking, particularly in the middle latitudes. (a) Waves in summer are generally mild and deposit sand on the beach. (b) Winter waves from storms nearer to and at the beach remove the sand, leaving boulders and bare bedrock in the beach area.

What attribute of waves represents the amount of energy they have?

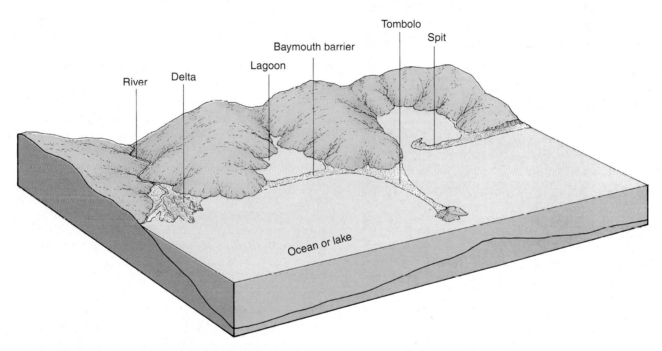

■ **FIGURE 17.16**   Some of the major landforms found along deposition-dominated coastlines.

Sand-sized sediment eroded from the beach in winter forms a deposit called a *longshore bar* that lies submerged parallel to shore, with the sediment returning to the beach in summer. On the Atlantic and Gulf coasts of the United States, the late summer to early fall hurricane season is also a time when beach erosion can be severe.

Whereas beaches are attached to the mainland along their entire length, **spits** are coastal depositional landforms connected to the mainland at just one end (■ Fig. 17.16). Spits project out into the water like peninsulas of sediment. They form where the mainland curves significantly inland while the trend of the longshore current remains at the original orientation. Sediments accumulate into a spit in the direction of the longshore current (■ Fig. 17.17a). Where similar processes deposit a strip of sediment connecting the mainland to an island, the landform is a **tombolo** (Fig. 17.17b).

Another category of coastal landforms are **barrier beaches,** elongate depositional features constructed parallel to the mainland. Barrier beaches act to protect the mainland from direct wave attack. All barrier beaches have restricted waterways, called **lagoons,** which lie between them and the mainland. Salinity in the lagoon varies from that of the open water body, depending on freshwater inflow and evaporation, and affects organisms living in the lagoon. Like beaches and

(a)

(b)

(c)

■ **FIGURE 17.17**   (a) A spit is attached to the mainland at one end. (b) A tombolo connects a nearby island with the mainland at Point Sur, California. (c) A baymouth barrier crosses the mouth of an embayment connecting to land at both ends.

spits, barrier beaches have a submerged part and a portion that is always above water, except in extreme storm conditions or extremely high tide. This contrasts with *bars,* like the longshore bars discussed above, which are submerged except in extreme conditions.

There are three kinds of barrier beaches. A **barrier spit** originated as a spit and thus is attached to the mainland at one

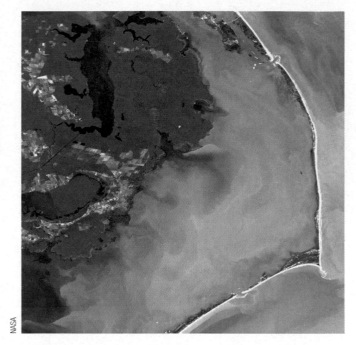

■ **FIGURE 17.18**   Barrier islands lie parallel to the mainland, but are not attached to it. This barrier island is located near Pamlico Sound on the North Carolina coast.
What feature separates a barrier island from the mainland?

end, but has extended almost completely across the mouth of an embayment to restrict the circulation of water between it and the ocean or lake. If the barrier spit crosses the mouth of the embayment to connect with the mainland at both ends, it becomes a **baymouth barrier** (Fig. 17.17c). Limited connection is maintained between the lagoon and the main water body through a breach or inlet cut across the barrier somewhere along its length. The position of inlets can change during storms. **Barrier islands** are likewise elongated parallel to the mainland and separate lagoons and the mainland from the open water body, but they are not attached to the mainland at all (■ Fig. 17.18).

Barrier islands are common features of low-relief coastlines. They dominate the Atlantic and Gulf coasts of the United States from New York to Texas. Some excellent examples of long barrier islands are Fire Island (New York), Cape Hatteras (North Carolina), Cape Canaveral and Miami Beach (Florida), and Padre Island (Texas).

Rising sea level since the Pleistocene appears to have played a major role in the formation of barrier islands. They migrate landward over long periods of time and may change drastically during severe storms, especially hurricanes (■ Fig. 17.19).

Beach systems are in equilibrium when input and output of sediment are in balance. People build artificial obstructions to the longshore current to increase the size of some beaches. A *groin* is an obstruction, usually a concrete or rock wall, built perpendicular to a beach to inhibit sediment removal while sediment input remains the same. This

USGS Coastal & Marine Geology Program

USGS Coastal & Marine Geology Program

(a)　　　　　　　　　　　　　　　　　　　　(b)

■ **FIGURE 17.19** Homes along the shore of a barrier island (a) before, and (b) after a hurricane.
How can this type of damage be prevented in the future?

USGS

■ **FIGURE 17.20** A series of groins on the Atlantic shoreline at Norfolk, Virginia, captures sand to maintain the beach along one stretch of the coastline.
How do you think the stretch of coast beyond the last groin would be impacted by these structures?

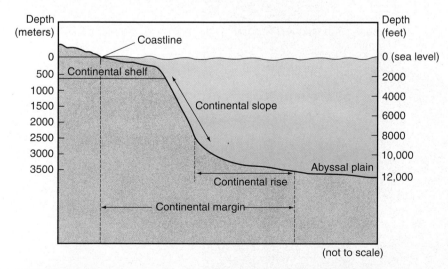

■ **FIGURE 17.21** The general nature of the boundary between the continents (continental crust) and ocean basins (oceanic crust). The gently sloping continental shelf varies in width along different coasts depending on plate tectonic history and proximity to plate margins.

obstruction, however, starves the adjacent downcurrent beach area of material input from upcurrent while it still has the usual rate of sediment removal (■ Fig. 17.20). Beach deposition is also often engineered to keep harbors free of sediment or to encourage growth of recreational beaches. When human actions deplete the natural sediment supply by damming rivers, beaches become narrow and lose some of their ability to protect the coastal region against storms. In Florida, New Jersey, and California, hundreds of millions of dollars have been spent to replenish sandy beaches. Beaches not only serve recreational needs but also help protect coastal settlements from erosion and flooding by storm waves.

Types of Coasts

Coasts are spectacular, dynamic, and complex systems that are influenced by tectonics, global sea-level change, storms, and marine and continental geomorphic processes. Because of this complexity, there is no single classification system for coasts. Each major way of classifying coasts focuses on a different characteristic of coastal systems; all aid our understanding of these natural, complex systems.

On a global scale, coastal classification is based on plate tectonic relationships. This system has two major coastal types: passive-margin coasts and active-margin coasts. **Passive-margin coasts** are tectonically quiescent, with little mountain-building or volcanic activity. Like the eastern seaboard of the United States, which is an excellent example, these coasts generally have low relief with broad coastal plains and wide continental shelves (■ Fig. 17.21), but passive-margin coasts that are relatively young, such as those of the Red Sea and Gulf of California, may have somewhat greater relief. Passive-margin coasts are typically dominated by depositional landforms, and are well represented by the coastal regions of continents along the Atlantic Ocean (■ Fig. 17.22). Most major tectonic activity within the Atlantic occurs in the center of the ocean along the Mid-Atlantic Ridge, rather than along its coastlines.

Active-margin coasts are best represented by coastal regions along the Pacific Ocean (■ Fig. 17.23). There, most tectonic activity occurs around the ocean margins because of active subduction and transform plate boundaries along the "Pacific Ring of Fire." Active-margin coasts are usually characterized by high relief with narrow coastal plains, narrow continental shelves, earthquake activity, and volcanism. These coasts tend to be erosional, having less time within Earth history for the development of marine or continental depositional features. The west coast of the United States is an excellent example of an active-margin coast.

On a regional scale, coasts can be classified as coastlines of emergence or coastlines of submergence. **Coastlines of emergence** occur where the water level has fallen or the land has risen in the coastal zone. In either case, land that was once below sea level has emerged above the water. Evidence for emergence includes marine terraces and relict sea cliffs, sea stacks, and beaches found above the reach of present wave action. Many coastlines of emergence would have existed during the glacial phases of the Pleistocene Epoch when sea level was about 120 meters (400 ft) lower

Stockbyte/Getty Images

■ **FIGURE 17.22** A sandy passive-margin coast on the Atlantic Ocean at Marconi Beach, Cape Cod National Seashore, Massachusetts.

■ **FIGURE 17.23** The rugged coast of Point Lobos, California, exemplifies an active-margin coastline, having experienced much tectonic activity.

Where else in the world might you expect active-margin coastlines?

J. Petersen

USGS

■ FIGURE 17.24   Cape Blanco, Oregon, represents an emergent coastline. The flat surface on which the lighthouse is built is a marine terrace.

than it is today. Features of emergence are common along active-margin coasts like those of California, Oregon, and Washington, where tectonic uplift has raised coastal landforms as much as 370 meters (1200 ft) above sea level (■ Fig. 17.24). Other emergent coastlines, such as around the Baltic Sea and Hudson Bay, are located where isostatic rebound has elevated the land following the retreat of the continental ice sheets.

Along **coastlines of submergence** many features of the former shore lie underwater and the present shoreline crosses land areas that are not fully adjusted to coastal processes. Coastlines of submergence were created as global sea level rose in response to the retreat of the Pleistocene ice sheets. Coastlines of submergence also occur where tectonic forces have lowered the level of the land, as in San Francisco Bay. Great thicknesses of river deposits and compaction of alluvial sediments, as along the Louisiana coast, can also cause coastal submergence. The specific features along a coastline of submergence are related to the character of coastal lands prior to submergence. Plains, for instance, will produce a far more regular shoreline upon submergence than will a mountainous region or an area of hills and valleys. When areas of low relief with soft sedimentary rocks are submerged, barrier islands form with shallow bays and lagoons behind them. The classic examples of this type of submerging coastline are the Atlantic and Gulf coasts of the United States.

Two special types of submerged coastlines are ria and fjord coasts. **Rias** are created where river valleys are drowned by a rise in sea level or a sinking of the coastal area (■ Fig. 17.25). These irregular coastlines result when valleys become narrow bays and the ridges form peninsulas. The Aegean coast of Greece and Turkey is an outstanding example of a ria coastline. *Fjords*, which are drowned glacial troughs, form spectacular scenic shorelines (■ Fig. 17.26). A coastline with fjords is highly irregular, with deep, steep-sided embayments penetrating far inland in the U-shaped

NASA

■ FIGURE 17.25   Submergent coasts like the Chesapeake Bay region are characterized by drowned river valleys (rias), which developed as sea level rose at the end of the Pleistocene.

■ FIGURE 17.26   Fjords, like this one in Greenland, are glaciated valleys that were drowned by the sea following the Pleistocene Epoch as the glaciers receded and sea level rose.

© Matt Ebiner

valleys originally deepened by glaciers. Tributary streams cascade down fjord sidewalls, which may be a few thousand meters high. In many fjords, the glaciers have retreated far inland, but glaciers reach the sea in other fjords, especially in Greenland and Alaska, where they calve icebergs into the cold fjord waters.

Some coastlines, such as those composed of river deltas, cannot be classified as either submerging or emerging. Actually, features of both submerged and emerged shorelines characterize many coastlines because the land elevation and the level of the ocean have changed many times during geologic history.

Islands and Coral Reefs

Within the ocean there exist three basic types of islands: continental, oceanic, and atolls. **Continental islands** are geologically part of the continent. Continental islands are usually found on the part of a continent submerged by the ocean, the continental shelf. Most large continental islands became separated from the continent because of global sea-level change or regional tectonic activity. A few large continental islands, such as New Zealand and Madagascar, are isolated "continental fragments" that separated from continents millions of years ago. The world's largest islands—Greenland, New Guinea, Borneo, and Great Britain—are continental. Smaller

continental islands include the barrier islands off the Atlantic and Gulf Coasts of the United States, New York's Long Island, California's Channel Islands, and Vancouver Island off the west coast of Canada.

Oceanic islands are volcanoes that rise from the deep-ocean floor and are geologically related to oceanic crust, not the continents. Most oceanic islands, such as the Aleutians, Tonga, and the Marianas, occur in island arcs along the edges of the associated trenches. Others, like Iceland and the Azores, are peaks of oceanic ridges rising above sea level. Many oceanic islands, like the Hawaiian Islands, occur in chains. The oceanic crust sliding over a stationary "hot spot" in the mantle creates these island chains. The Hawaiian Islands are moving northwestward with the Pacific lithospheric plate, and will eventually submerge. A new volcanic island, named Loihi, is forming to the southeast (■ Fig. 17.27). Evidence of the plate motion is indicated by the fact that the youngest islands of the Hawaiian chain, Hawaii and Maui, are to the southeast, while the older islands, such as Kauai and Midway, are located to the northwest.

An **atoll** is an island consisting of a ring of coral reefs that have grown up from a subsiding volcanic island and that encircle a central lagoon (■ Fig. 17.28a). In order to understand atolls, we must first consider coral reefs.

Coral reefs are shallow, wave-resistant structures made by the accumulation of remains of tiny sea animals that secrete a skeleton of calcium carbonate. Many other organisms, including algae, sponges, and mollusks, add material to the reef structure. Reef corals need special conditions

■ **FIGURE 17.27** The oceanic island of Hawaii was formed over a "hot spot" like the other Hawaiian Islands before it. Two large shield volcanoes, Mauna Loa and Mauna Kea, dominate the island. To the south, Loihi, an active submarine volcano, may reach sea level to become the newest Hawaiian island in about 50,000 years.

University of Hawaii, School of Ocean and Earth Science

© Tahiti Tourism Board

(a)

© David Hiser/Getty Images

(b)

© Paul Chelsey/Getty Images

(c)

■ **FIGURE 17.28** The major types of coral reefs are evident in the Society Island chain of French Polynesia. (a) Atolls are islands consisting of a ring of coral with no surface evidence remaining of its former volcanic core. (b) Fringing reefs are attached to mainland or island coasts. (c) Like coastal barriers in general, barrier reefs are separated from dry land by a lagoon.

MAP INTERPRETATION
PASSIVE-MARGIN COASTLINES

The Map

Eastport, New York, is located on the south shore of Long Island, 115 kilometers (70 mi) east of New York City. Long Island is part of the Atlantic Coastal Plain, which extends from Cape Cod, Massachusetts, to Florida. Water bodies, such as Delaware Bay, Chesapeake Bay, and Long Island Sound, embay much of the Atlantic Coastal Plain in the eastern United States. The south shore of Long Island has low relief, and its coastal location moderates its humid continental climate.

Although this is a coastal region, its recent glacial history influences the landforms that exist today. Two east–west trending glacial terminal moraines deposited during the Pleistocene glacial advances form Long Island. Between the coast and the south moraine is a sandy glacial outwash plain that forms the higher elevations at the northern part of the map area. As the glaciers melted, sea level rose, submerging the lowland now occupied by Long Island Sound. This water body separates Long Island from the mainland of the Atlantic Coastal Plain.

The Atlantic and Gulf coasts of the United States have nearly 300 barrier islands with a combined length of over 2500 kilometers (1600 mi). New York, especially the coastal zone of Long Island's south shore, has 15 barrier islands with a total length of over 240 kilometers (150 mi). Coastal barrier islands protect the mainland from storm waves, and they contribute to the formation of coastal wetlands on their landward side, which are a critical habitat for fish, shellfish, and birds.

You can compare this topographic map to the Google Earth presentation of the area. Find the map area by zooming in on these latitude and longitude coordinates: 40.796111°N, 72.666111°W.

Interpreting the Map

1. What type of landform is the entire long, narrow feature on which lies the straight shoreline labeled Westhampton Beach?

2. What is the highest elevation on the linear coastal feature? What do you think comprises the highest portions of this feature?

3. How much evidence of human use exists on the landform in Question 1? What problems might these cultural features be subjected to?

4. Behind the linear feature is a water body. Is it deep or shallow? Is it a high-energy or low-energy environment? What is a water body such as this called?

5. Is the coastline shown on the Eastport map predominantly erosional or depositional? What features support your answer? Is this a coastline of submergence or emergence?

6. Note Beaverdam Creek in the upper middle part of the map area. Does it have a steep or gentle gradient?

7. Based on its gradient, does Beaverdam Creek always flow seaward? If not, what might influence its flow?

8. What kind of topographic feature exists as indicated by the map symbols in the area surrounding Oneck?

9. How do you think geomorphic processes will likely modify the Eastport map region in the future? Which area will probably be modified the most? Explain.

10. What type of natural hazard is this coastal area of Long Island most susceptible to? Why?

NASA/GSFC/SVS

Satellite image of Long Island, New York.

Opposite:
Eastport, New York
Scale: 1:24,000
Contour interval = 10 ft
U.S. Geological Survey

Appendix A

International System of Units (SI), Abbreviations, and Conversions

Symbol	Multiply	By	To Find	Symbol
Area				
in.^{2}	square inches	645.2	square millimeters	mm^2
ft^2	square feet	0.093	square meters	m^2
yd^2	square yards	0.836	square meters	m^2
ac	acres	0.405	hectares	ha
mi^2	square miles	2.59	square kilometers	km^2
ac	acres	43,560	square feet	ft^2
mm^2	square millimeters	0.0016	square inches	in.^{2}
m^2	square meters	10.764	square feet	ft^2
m^2	square meters	1.195	square yards	yd^2
ha	hectares	2.47	acres	ac
km^2	square kilometers	0.386	square miles	mi^2
Mass				
oz	ounces	28.35	grams	g
lb	pounds	0.454	kilograms	kg
g	grams	0.035	ounces	oz
kg	kilograms	2.202	pounds	lb
Length				
in.	inches	25.4	millimeters	mm
ft	feet	0.305	meters	m
yd	yards	0.914	meters	m
mi	miles	1.61	kilometers	km
ft	feet	5280	mile	mi
mm	millimeters	0.039	inches	in.
m	meters	3.28	feet	ft
m	meters	1.09	yards	yd
km	kilometers	0.62	miles	mi
Volume				
gal	U.S. gallons	3.785	liters	l
ft^3	cubic feet	0.028	cubic meters	m^3
yd^3	cubic yards	0.765	cubic meters	m^3
l	liters	0.264	U.S. gallons	gal
m^3	cubic meters	35.30	cubic feet	ft^3
m^3	cubic meters	1.307	cubic yards	yd^3
Velocity				
mph	miles/hour	1.61	kilometers/hour	km/h
knot	nautical miles/hour	1.85	kilometers/hour	km/h
km/h	kilometers/hour	0.62	miles/hour	mph
km/h	kilometers/hour	0.54	nautical miles/hour	knots

International System of Units (SI), Abbreviations, and Conversions (continued)

Symbol	Multiply	By	To Find	Symbol
Pressure or Stress				
mb	millibars	0.75	millimeters of mercury	mm Hg
mb	millibars	0.02953	inches of mercury	in. Hg
mb	millibars	0.01450	pounds per square inch	(lb/in.^{2} or psi)
lbs/in.^{2}	pounds per square inch	6.89	kilopascals	kPa
in. Hg	inches of mercury	33.865	millibars	mb
kPa	kilopascals	0.145	pounds per square inch	(lb/in.^{2} or psi)

Standard Sea-Level Pressure

29.92 in. Hg
14.7 lb/in.^{2}
1013.2 mb
760 mm Hg

Temperature

°F	Fahrenheit	(°F − 32)/1.8	Celsius	°C
°C	Celsius	1.8°C + 32	Fahrenheit	°F
K	deg. Kelvin	K = °C + 273	Celsius	°C

Powers of Ten

nano	one billionth	$= 10^{-9}$	= 0.000000001
micro	one millionth	$= 10^{-6}$	= 0.000001
milli	one thousandth	$= 10^{-3}$	= 0.001
centi	one hundredth	$= 10^{-2}$	= 0.01
deci	one tenth	$= 10^{-1}$	= 0.1
hecto	one hundred	$= 10^{2}$	= 100
kilo	one thousand	$= 10^{3}$	= 1000
mega	one million	$= 10^{6}$	= 1,000,000
giga	one billion	$= 10^{9}$	= 1,000,000,000

Topographic Maps

Mapping has changed considerably in recent years, and with the ever-increasing capabilities of computers to store, retrieve, and display graphics, this trend will continue well into the future. In the United States the U.S. Geological Survey (USGS) produces the vast majority of topographic maps available. These maps have long been the tried and true tools of geographers and scientists in many other disciplines who study various aspects of the environment. Today virtually all USGS topographic maps are accessible in digital format, for downloading and printing, on computer disks, or for examining on a computer screen. Computers and the Internet have made maps much more available and accessible than they were just a few years ago. This availability makes it easy to print maps or map segments at home, school, or work. A logical question then would be, will paper maps become obsolete, given computer displays?

There are several reasons why paper maps will still be popular and useful, whether they are purchased from the source or downloaded and printed. Topographic maps are particularly important in fieldwork. Maps are highly portable, require no batteries or electrical power that could fail, and do not suffer from technology glitches. They are reliable and easy to use and they also provide a good base for making field notes, and marking routes.

Today the USGS is working to make and maintain a seamless database of the United States, with maps, imagery, and spatial data in digital form, so areas that once were split on adjacent maps can be printed on a single sheet. This is a great change from dividing the country into quadrangles (the roughly rectangular area a map displays). Topographic quadrangles (quads), however, will continue to be in use for a long time. There are several standard quadrangles, each with a specific scale, and many other special-purpose topographic maps.

7.5-minute quads—these are printed at a scale of 1:24,000 and cover 7.5 minutes of longitude and 7.5 minutes of latitude.

15-minute quads—these are printed at a scale of 1:62,500 and cover 15 minutes of longitude and 15 minutes of latitude.

1° × 2° quads—these are printed at a scale of 1:250,000 and cover 1 degree of latitude and 2 degrees of longitude.

1:100,000 metric quads—these are printed at a scale of 1:100,000, cover 30 minutes of longitude and 60 minutes of latitude, and use metric measurements for distance and elevation.

Brown topographic contours are used to show elevation differences and the terrain. Some rules for interpreting contours are given in Chapter 2.

Determining Distances on a Map, Distances from a Map, or an RF Scale
It is important to understand representative fractions, like 1:24,000, which means that any measurement on the map will represent 24,000 of the same measurements on the ground. This knowledge is particularly significant because reproduced maps may not be printed at the original size. On a map that is reduced or enlarged the bar scale will still be accurate, but the printed RF scale will not. Enlarging or reducing a map changes the RF scale from the original.

How to Find the RF of a Map (or Air Photo, or Satellite Image) of Unknown Scale
Here is the formula:

$$1/RFD = MD/GD$$

The numerator in an RF scale is always the number 1. The RFD is the denominator of the RF (such as 24,000). MD is map distance measured in any particular units on the map (cm, in.). GD is the true ground distance that the map distance represents (expressed in the same units used to measure the map distance). Never mix units in this calculation, but convert values into desired units afterward.

Example: How long in inches is a mile on a 1:24,000 scale map?

Important information: There are 63,360 inches in a mile. To find MD, for a known distance (mile) on a map of known scale, use this formula:

$$1/24,000 = MD/63,360 \text{ in.}$$

$$1 \text{ mile} = 2.64 \text{ inches at } 1:24,000$$

This statement has now been converted into a **stated scale** so units may be mixed.

1:24,000 Bar Scale
Here is a bar scale that can be used to make distance measurements directly from the 1:24,000 maps printed in the Map Interpretation sections. Note: Check the listed RF, because not all are at a 1:24,000 scale.

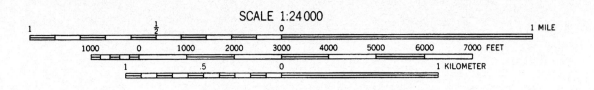

SCALE 1:24 000

BOUNDARIES

National ....

State or territorial ....

County or equivalent ....

Civil township or equivalent ....

Incorporated-city or equivalent ....

Park, reservation, or monument ....

Small park ....

LAND SURVEY SYSTEMS

U.S. Public Land Survey System:

Township or range line ....

Location doubtful ....

Section line ....

Location doubtful ....

Found section corner; found closing corner ...

Witness corner; meander corner ....

Other land surveys:

Township or range line ....

Section line ....

Land grant or mining claim; monument ......

Fence line ....

ROADS AND RELATED FEATURES

Primary highway ....

Secondary highway ....

Light duty road ....

Unimproved road ....

Trail ....

Dual highway ....

Dual highway with median strip ....

Road under construction ....

Underpass; overpass ....

Bridge ....

Drawbridge ....

Tunnel ....

BUILDINGS AND RELATED FEATURES

Dwelling or place of employment: small; large ...

School; church ....

Barn, warehouse, etc.: small; large ....

House omission tint ....

Racetrack ....

Airport ....

Landing strip ....

Well (other than water); windmill ....

Water tank: small; large ....

Other tank: small; large ....

Covered reservoir ....

Gaging station ....

Landmark object ....

Campground; picnic area ....

Cemetery: small; large ....

RAILROADS AND RELATED FEATURES

Standard gauge single track; station ....

Standard gauge multiple track ....

Abandoned ....

Under construction ....

Narrow gauge single track ....

Narrow gauge multiple track ....

Railroad in street ....

Juxtaposition ....

Roundhouse and turntable ....

TRANSMISSION LINES AND PIPELINES

Power transmission line; pole; tower ....

Telephone or telegraph line ....

Aboveground oil or gas pipeline ....

Underground oil or gas pipeline ....

CONTOURS

Topographic:

Intermediate ....

Index ....

Supplementary ....

Depression ....

Cut; fill ....

Bathymetric:

Intermediate ....

Index ....

Primary ....

Index Primary ....

Supplementary ....

MINES AND CAVES

Quarry or open pit mine ....

Gravel, sand, clay, or borrow pit ....

Mine tunnel or cave entrance ....

Prospect; mine shaft ....

Mine dump ....

Tailings ....

SURFACE FEATURES

Levee ....

Sand or mud area, dunes, or shifting sand ....

Intricate surface area ....

Gravel beach or glacial moraine ....

Tailings pond ....

VEGETATION

Woods ....

Scrub ....

Orchard ....

Vineyard ....

Mangrove ....

COASTAL FEATURES

Foreshore flat ....

Rock or coral reef ....

Rock bare or awash ....

Group of rocks bare or awash ....

Exposed wreck ....

Depth curve; sounding ....

Breakwater, pier, jetty, or wharf ....

Seawall ....

BATHYMETRIC FEATURES

Area exposed at mean low tide; sounding datum ...

Channel ....

Offshore oil or gas: well; platform ....

Sunken rock ....

RIVERS, LAKES, AND CANALS

Intermittent stream ....

Intermittent river ....

Disappearing stream ....

Perennial stream ....

Perennial river ....

Small falls; small rapids ....

Large falls; large rapids ....

Masonry dam ....

Dam with lock ....

Dam carrying road ....

Intermittent lake or pond ....

Dry lake ....

Narrow wash ....

Wide wash ....

Canal, flume, or aqueduct with lock ....

Elevated aqueduct, flume, or conduit ....

Aqueduct tunnel ....

Water well; spring or seep ....

GLACIERS AND PERMANENT SNOWFIELDS

Contours and limits ....

Form lines ....

SUBMERGED AREAS AND BOGS

Marsh or swamp ....

Submerged marsh or swamp ....

Wooded marsh or swamp ....

Submerged wooded marsh or swamp ....

Rice field ....

Land subject to inundation ....

The Köppen Climate Classification System

The best way to understand a classification system is to actually apply it to determine how it works. This is one reason why the Practical Application sections of both Chapters 7 and 8 are based on classifying climate data from selected locations around the world. Correctly classifying these sites in the modified Köppen system may seem complicated at first, but you will find that, after applying the system to a few of the sample locations, you will be able to easily determine the climate type and its correct letter symbol.

Before you begin, familiarize yourself with Table 1. You will note that there are precise definitions of temperatures or precipitation amounts that identify a site as one

of the five major climate categories in the Köppen system (*A*, tropical; *B*, arid; *C*, mesothermal; *D*, microthermal; *E*, polar). Also, note that the additional letters used to identify a correct climate type have precise definitions and can be determined by using the graphs in this Appendix. In other words, Table 1 is all you need to classify a site if monthly and annual means of precipitation and temperature are available. Table 1 should be used in a systematic fashion to determine first the major climate category and, once that is determined, the second and third letter symbols (if needed) that complete the classification. As you begin to classify, you should use the following procedure.

TABLE 1
Simplified Köppen Classification of Climates

First Letter	Second Letter	Third Letter
E Warmest month less than 10°C (50°F) POLAR CLIMATES *ET*–Tundra *EF*–Ice Sheet	*T* Warmest month between 10°C (50°F) and 0°C (32°F) *F* *Warmest* month below 0°C (32°F)	NO THIRD LETTER (with polar climates) SUMMERLESS
B Arid or semiarid climates ARID CLIMATES BS—Steppe BW—Desert	*S* Semiarid climate (see Graph 1) *W* Arid climate (see Graph 1)	*h* Mean annual temperature greater than 18°C (64.4°F) *k* Mean annual temperature

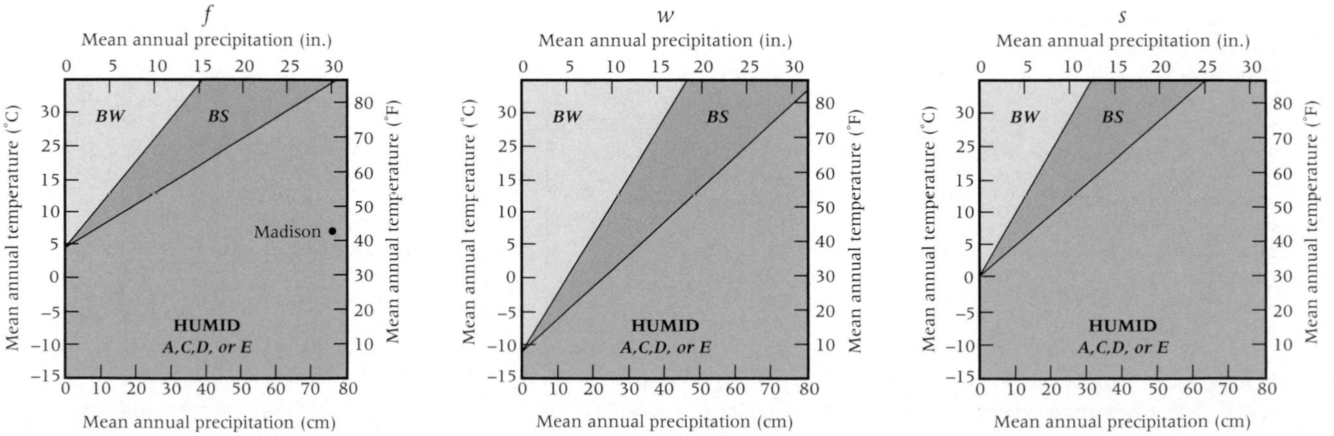

Graph 1 Humid/Dry Climate Boundaries

TABLE 1
Simplified Köppen Classification of Climates (*Continued*)

First Letter	Second Letter	Third Letter
A Coolest month greater than 18°C (64.4°F) TROPICAL CLIMATES *Am*—Tropical monsoon *Aw*—Tropical savanna *Af*—Tropical rainforest	*f* Driest month has at least 6 cm (2.4 in.) of precipitation *m* Seasonally, excessively moist (see Graph 2) *w* Dry winter, wet summer (see Graph 2)	NO THIRD LETTER (with tropical climates) WINTERLESS
C Coldest month between 18°C (64.4°F) and 0°C (32°F); at least one month over 10°C (50°F) MESOTHERMAL CLIMATES *Csa, Csb*—Mediterranean *Cfa, Cwa*—Humid subtropical *Cfb, Cfc*—Marine west coast	*s* (DRY SUMMER) Driest month in the summer half of the year, with less than 3 cm (1.2 in.) of precipitation and less than one third of the wettest winter month	*a* Warmest month above 22°C (71.6°F) *b* Warmest month below 22°C (71.6°F), with at least four months above 10°C (50°F)
D Coldest month less than 0°C (32°F); at least one month over 10°C (50°F) MICROTHERMAL CLIMATES *Dfa, Dwa*—Humid continental, hot summer *Dfb, Dwb*—Humid continental, mild summer *Dfc, Dwc, Dfd, Dwd*—Subarctic	*w* (DRY WINTER) Driest month in the winter half of the year, with less than one tenth the precipitation of the wettest summer month *f* (ALWAYS MOIST) Does not meet conditions for *s* or *w* above	*c* Warmest month below 22°C (71.6°F), with one to three months above 10°C (50°F) *d* Same as *c*, but coldest month is below −38°C (−36.4°F)

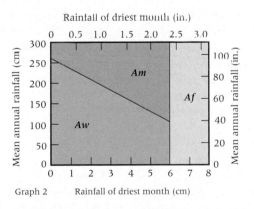

Graph 2   Rainfall of driest month (cm)

Graph 2 Rainfall of driest month (cm)

After a few examples you may find that you can omit some steps just by examining the climate statistics.

Step 1. Ask: Is this a polar climate (*E*)? Is the warmest month less than 10°C (50°F)? If so, is the warmest month between 10°C (50°F) and 0°C (32°F) (*ET*) or below 0°C (32°F) (*EF*)? If not, move on to:

Step 2. Ask: Is there a seasonal concentration of precipitation? Examine the monthly precipitation data for the driest and wettest summer and winter months for the site. Take careful note of temperature data as well because you must determine whether the site is located in the Northern or Southern Hemisphere. (April to September are summer months in the Northern Hemisphere but winter months in the Southern Hemisphere. Similarly, the Northern Hemisphere winter months of October to March are summer south of the equator.) As the table indicates, a site has a dry summer (*s*) if the driest month in summer has less than 3 cm (1.2 in.) of precipitation and less than one third of the precipitation of the wettest winter month. It has a dry winter (*w*) if the driest month in winter has less than one tenth the precipitation of the wettest summer month. If the site has neither a dry summer nor a dry winter, it is classified as having an even distribution of precipitation (*f*). Move on to:

Step 3. Ask: Is this an arid climate (*B*)? Use one of the small graphs (included in Graph 1) to decide. Based on your answer in Step 2, select one of the small graphs and compare mean annual temperature with mean annual precipitation. The graph will indicate whether the site is an arid (*B*) climate or not. If it is, the graph will indicate which one (*BW* or *BS*). You should further classify the site by adding *h* if the mean annual temperature is above 18°C (64.4°F) and *k* if it is below. If the site is neither *BW* nor *BS*, it is a humid climate (*A, C,* or *D*). Move on to:

Step 4a. Ask: Is this a tropical climate (*A*)? The site has a tropical climate if the temperature of the coolest month is higher than 18°C (64.4°F). If so, use Graph 2 in Table1 to determine which tropical climate the site represents. (Note that there are no additional lowercase letters required.) If not, move on to:

Step 4b. Ask: Which major middle-latitude climate group does that site represent, mesothermal (*C*) or

microthermal (*D*)? If the temperature of the coldest month is between 18°C (64.4°F) and 0°C (32°F), the site has a mesothermal climate. If is below 0°C (32°F), it has a microthermal climate. Once you have answered the question, move on to:

Step 5. Ask: What was the distribution of precipitation? This was determined back in Step 2. Add *s, w,* or *f* for a *C* climate or *w* or *f* for a *D* climate to the letter symbol for the climate. Then, move on to:

Step 6. Ask: What is needed to express the details of seasonal temperature for the site? Refer again to Table 1 and the definitions for the letter symbols. Add *a, b,* or *c* for the mesothermal (*C*) climates or *a, b, c,* or *d* for the microthermal (*D*) climates, and you have completed the classification of your climate. However, note that you may not have come this far because you might have completed your classification at Steps 1, 3, or 4a.

You should now be ready to try out the use of Table 1, following the recommended steps. The climate data for Madison, Wisconsin, will be used for our example.

	J	F	M	A	M	J	J	A	S	O	N	D	Year
T (°C)	−8	−7	−1	7	13	19	21	21	16	10	2	−6	7
P (cm)	3.3	2.5	4.8	6.9	8.6	11.0	9.6	7.9	8.6	5.6	4.8	3.8	77.0

The correct answer is derived below:

Step 1. Determine whether or not our site has an *E* climate. Because Madison has several months averaging above 10°C, it does not have an *E* climate.

Step 2. Determine if there is a seasonal concentration of precipitation. Because Madison is driest in winter, you must compare the 2.5 centimeters of February precipitation with the precipitation of June (¹⁄₁₀ of 11.0 cm, or 1.1 cm) and conclude that Madison has neither a dry summer nor a dry winter but instead has an even distribution of precipitation (*f*). (*Note:* The 2.5 cm of February precipitation is not less than ¹⁄₁₀ [1.1 cm] of June precipitation.)

Step 3. Next assess, through the use of Graphs 1(*f*), 1(*w*), or 1(*s*), whether this site is an arid climate (*BW, BS*) or a humid climate (*A, C,* or *D*). Because it has previously been determined that Madison has a relatively even distribution of precipitation, you will use Graph 1(*f*). Based on Madison's mean annual precipitation (77.0 cm) and mean annual temperature (7°C), you can conclude that Madison is a humid climate (*A, C,* or *D*).

Step 4. Now you must assess which humid climate type Madison falls under. Because the coldest month (−8°C) is below 18°C, Madison does *not* have an *A* climate. Although the warmest month (21°C) is above 10°C, the coldest month (−8°C) is not between 0°C and 18°C, so Madison does *not* have a *C* climate. Because the warmest month (21°C) is above 10°C and the coldest month is below 0°C, Madison *does* have a *D* climate.

Step 5. Because Madison has a *D* climate, the second letter will be *w* or *f*. Because precipitation in the driest month of winter (2.5 cm) is not less that one tenth of the amount of the wettest summer month (¹⁄₁₀ × 11.0 cm = 1.1 cm), Madison does *not* have a *Dw* climate. Madison therefore has a *Df* climate.

Step 6. Because Madison is a *Df* climate, the third letter will be *a, b, c,* or *d*. Because the average temperature of the warmest month (21°C) is not above 22°C, Madison does *not* have a *Dfa* climate. Because the average temperature of the warmest month is below 22°C, with at least 4 months above 10°C, Madison is a *Dfb* climate.

Appendix D

The 12 Soil Orders of the National Resource Conservation Service

The NCRS soil orders are generally based on and named by characteristics and processes that can be recognized by examining a soil and its profile.

Alfisols (*Al/Fe*--Aluminum/Iron): fertile soils formed mainly in hardwood forest regions.

Andisols (*Ande*s Mountains): soils developed on igneous deposits, mainly volcanic ash.

Aridisols (*arid*): soils of deserts and some semiarid regions.

Entisols (re*cent*): little or no soil development.

Gelisols (*gel*ere, Latin for freeze): soils formed over a permafrost layer.

Histosols (*histos*, Greek for organic tissues): bog or swamp soils mainly made of organic material.

Inceptisols (*incep*tion or beginning): young soils with mild development.

Mollisols (*mollis*, Latin for soft): fertile organic rich soils formed mainly in grasslands.

Oxisols (*oxide*): tropical soils, red or yellow, leached under warm wet conditions with high Al/ Fc oxides.

Spodosols (*podzol*): soils typically formed in sandy parent material and coniferous forest regions.

Ultisols (*ultimate*): reddish, heavily weathered clayey soils mainly in the forested humid subtropics.

Vertisols (*vertical*): clayey soils vertically churned by alternating wetting and drying.

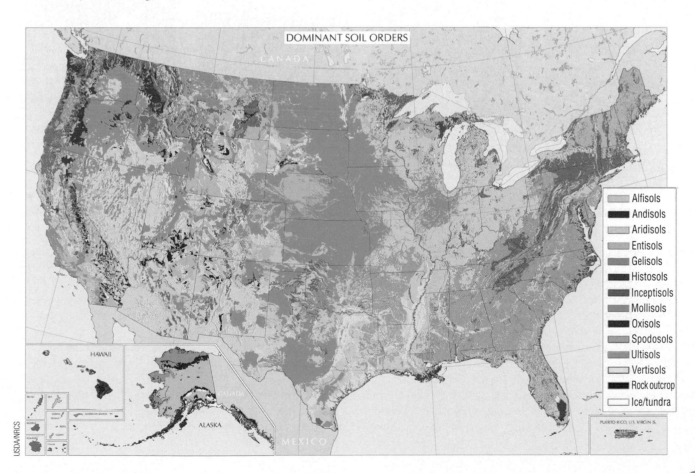

DOMINANT SOIL ORDERS

- Alfisols
- Andisols
- Aridisols
- Entisols
- Gelisols
- Histosols
- Inceptisols
- Mollisols
- Oxisols
- Spodosols
- Ultisols
- Vertisols
- Rock outcrop
- Ice/tundra

USDA/NRCS

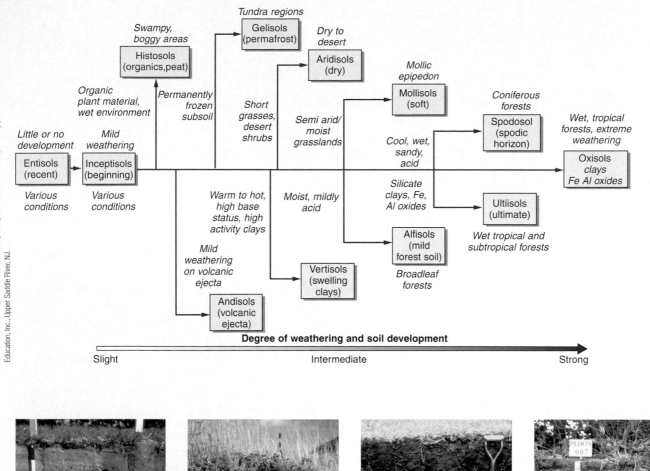

Degree of weathering and soil development

Slight   Intermediate   Strong

(a) Entisols   **(b)** Inceptisols   **(c)** Histosols   **(d)** Andisols

Entisols are recent soils that have undergone little soil development and lack horizons. They are often associated with continuing erosion on slopes, resistant rocks as parent material, or with recent deposits of alluvium by flooding, colluvium by mass wasting, or wind-blown sand and loess.

Inceptisols are young soils with weak horizon development that is just beginning, usually because of a very cold climate, repeated flood deposition, or soil erosion. In the United States, inceptisols are most common in Alaska, the Mississippi River floodplain, and the western Appalachians.

Histosols develop in swamps, meadows, and bogs. They mainly form from partly decomposed plant material (peat) under cold waterlogged conditions. Histosols are found worldwide in areas with poor drainage, but are common in the Arctic and alpine tundras, and in glaciated areas such as Alaska, Canada, and Northern Europe.

Andisols develop on volcanic materials, typically volcanic ash. They contain high proportions of glassy fragments and the weathering products of volcanic rocks. Repeated eruptions replenish these soils, so they are often fertile. In the U.S., andisols are common on volcanoes in the Pacific Northwest, Hawaii, and Alaska, and also downwind from the Cascade Range.

(a) Gelisols **(b)** Aridisols **(c)** Vertisols **(d)** Mollisols

(a) Alfisols **(b)** Spodosols **(c)** Ultisols **(d)** Oxisols

Gelisols experience frequent freezing and thawing, above permanently frozen subsoil. When the soil thaws, the waterlogged soil is pulled downward by gravity. Repeated freeze-thaw cycles mix the soil. Permafrost keeps water from infiltrating, so Gelisols are water-saturated when they are not frozen. Gelisols exist in tundra and subarctic regions in Alaska, Canada, and Siberia, and in highland tundra areas.

Aridisols form regions where potential evapotranspiration greatly exceeds precipitation. Aridisols have weak horizon development because of limited soil water movement, and often develop subsurface accumulations of calcium carbonate, or salt. Soil humus is minimal, so aridisols are usually light in color, and alkaline. Aridisols are the most common soils worldwide, because deserts cover large portions of the land areas.

Vertisols are clayey soils that develop in areas with strong wet-dry precipitation seasonality, and where parent materials produce clay-rich soils. Vertisols shrink during dry seasons and expand during wet seasons. This expansion-contraction causes vertical soil movements that damage roads, sidewalks, and foundations. Vertisols are dark, alkaline, and contain much organic material from their associated grassland or savanna vegetation.

Mollisols are grassland soils that are excellent for agriculture. Mollisols have thick, dark surface layers, rich in organic matter from the decay of abundant roots. Found in semiarid climates, mollisols are not heavily leached, and typically have high calcium contents. In the U.S., these soils are found in the Great Plains and the West, where natural grasslands were grazing lands for herds of antelope and bison.

Alfisols occur in a wide variety of settings, including humid or semiarid, and microthermal or mesothermal climate regions. For this reason they are often further divided into suborders related to their regional conditions of climate and vegetation. Alfisols share the characteristics of a clayey B horizon, medium to high alkalinity, and a thin, light-colored soil cover, both at and near the surface,

Spodosols are associated with podzolization in forested regions. They have well developed horizons, often a white or light-gray E horizon, covered with a thin, dark humus layer, underlain by a B horizon with iron and aluminum compounds. Many spodosols are acidic and form in porous parent materials like glacial deposits or sand. In the U.S. these soils are found in New England, the upper Midwest, the western mountains, and Florida.

Ultisols, like spodosols, develop in moist regions where the natural vegetation is forest. These soils represent high levels of weathering and soil development. Ultisols typically have a coarse surface texture, a subsurface clay horizon, and are often yellow or red because of iron and aluminum oxides. In North America, ultisols are mainly found in the southeastern United States.

Oxisols form in rainforest and savanna regions under high temperatures and high rainfall. Heavily leached, under the laterization process, they contain large amounts of iron and aluminum oxides. In the U.S., oxisols are only found in Hawaii. Oxisols lose their fertility if their vegetative cover is destroyed. Deforested regions with these soils typically become eroded and then will only support weeds, shrubs, and grasses.

Understanding and Recognizing Some Common Rocks

Rocks are aggregates of minerals, and although there are thousands of kinds of rocks on our planet, they can be classified into three fundamental kinds based on their origin: igneous, sedimentary, and metamorphic. The formation of rocks was outlined in Chapter 10. Having a solid knowledge of how the Rock Cycle operates (Fig. 10.18) as well as its components and its processes is essential to understanding the solid Earth. Specific rock types are mentioned in several chapters of this book. Although making a positive identification of a rock type requires examining several physical properties, having a mental image of what different rocks look like will be an aid to understanding Earth processes and landforms. The following is an illustrated guide to a few common rocks to help in their identification. Intrusive igneous rocks form from a molten state by cooling and crystallizing underground; they generally cool very slowly, which allows coarse crystals to form that are easy to see with the naked eye.

Igneous Rocks

Igneous rocks are subdivided into intrusive and extrusive, depending on whether they cooled within Earth or on its exterior.

Intrusive Igneous

Copyright and photograph by Dr. Parvinder S. Sethi

Granite

Granite forms deep in the crust, and has easily visible intergrown crystals of light and dark minerals, but is dominated by light-colored silicate minerals. Granitic rocks are typically gray or pink, and their mineral composition is similar to that of the continental crust.

Copyright and photograph by Dr. Parvinder S. Sethi

Diorite

Diorite is an intermediate intrusive rock, meaning that it has a roughly equal mix of easily visible light and dark minerals, which gives it a spotted appearance. Generally diorite is dark gray in color.

Copyright and photograph by Dr. Parvinder S. Sethi

Gabbro

Gabbro is a dark intrusive rock dominated by heavy, iron-rich silicate minerals. The crystals are coarse enough to be easily visible, but because of the overall dark tone, they tend to blend together. Gabbro is black and may contain some very dark green minerals.

Extrusive Igneous

Extrusive igneous rocks cool at or near the surface, and include lavas, as well as rocks made of tephra (pyroclastics), fragments blown out of a volcano. Extrusive rocks sometimes preserve gas bubble holes, and may contain visible crystals, but typically the grains are small. Cooling relatively rapidly at the surface produces fine-grained lavas. Clastic means made up of cemented rock fragments, such as clay, silt, sand, pebbles, cobbles, or even boulders. The sizes and shapes of clasts within a sedimentary rock provide clues about the environments under which the fragments were deposited (fluvial, eolian, glacial, coastal).

Copyright and photograph by Dr. Parvinder S. Sethi

Rhyolite

Rhyolite is a very thick lava when molten (much like melted glass), light in color, and high in silica. Colors of rhyolite vary widely, but light gray, very light brown, and pink or reddish are common. Rhyolite is the extrusive equivalent of granite in terms of mineral composition.

Copyright and photograph by Dr. Parvinder S. Sethi

Andesite

Andesite, named for the Andes, is an intermediate lava in terms of both mineral content and color. Associated with composite cone volcanoes, it is relatively thick when molten. Often mineral crystals are visible in a matrix of finer grains, and the color is gray to brown. Andesite is the extrusive equivalent of diorite.

Copyright and photograph by Dr. Parvinder S. Sethi

Basalt

Basalt is dark, typically black, and heavier than other lavas. Associated with fissure flows and shield volcanoes, basalt is relatively low in silica, so it has a lower viscosity than other lavas. Basalt tends to be hotter than other lavas, and is relatively thin when molten, thus it can flow for many kilometers before cooling enough to stop. Basalt is the rock of the oceanic crust and is the extrusive equivalent of gabbro.

Copyright and photograph by Dr. Parvinder S. Sethi

Tuff

Tuff is a rock made of fine tephra—volcanic ash—that was blown into the atmosphere by a volcanic eruption, and settled out in layers that blanketed the surface. Tephra (loose fragments) was converted into tuff by burial and compaction, or by being welded together from intense heat. Tuff is gray to tan.

Sedimentary Rocks

Sedimentary rocks can also be divided into two major categories, clastic and nonclastic.

Clastic Sedimentary

Clastic means made up of cemented rock fragments, such as clay, silt, sand, granules, pebbles, cobbles, or even boulders. The sizes and shapes of clasts within a sedimentary rock provide clues about the environments under which the fragments were deposited (fluvial, eolian, glacial, coastal).

Copyright and photograph by Dr. Parvinder S. Sethi

Shale
Shale is a fine-grained clastic rock that contains lithified clays, generally deposited in very thin layers. Shales represent a calm water environment, such as a sea or lake bottom. Shales vary widely in color, but most are gray or black, and they break up into smooth, flat surfaces.

Copyright and photograph by Dr. Parvinder S. Sethi

Sandstone
Sandstone is made of cemented fragments of sand size, typically made of quartz or other relatively hard mineral. Sandstones can be virtually any color, feel gritty, and may be banded or layered. Sandstone may represent ancient beaches, dunes, or fluvial deposits.

Copyright and photograph by Dr. Parvinder S. Sethi

Conglomerate
Conglomerate contains rounded pebbles cemented together by finer sediments. Conglomerate may represent deposits from a river's bed load, or a pebble beach.

Copyright and photograph by Dr. Parvinder S. Sethi

Breccia
Breccia is similar to conglomerate, but the cemented fragments are angular. Breccia is associated with mudflows, pyroclastic flows, and fragments deposited by mass wasting.

Nonclastic Sedimentary

Nonclastic sedimentary rocks consist of materials that are not rock fragments. Examples include chemical precipitates, such as limestone, evaporites such as rock salt, and deposits of organic materials, for example coal, or limestones made of shells and coral fragments. Nonclastic rocks also represent the ancient environment under which they were deposited.

Rock Salt

Rock salt consists of sodium chloride, table salt, often with a mix of other salts. Rock salt represents the deposits left behind by the evaporation of a saline inland lake, or an arm of the sea that was cut off from the ocean by sea-level change or tectonic activity. Common color is white.

Limestone

Limestone is made of calcium carbonate deposits (lime, $CaCO_3$). Limestone can represent a variety of environments and varies widely in color and appearance. Typical colors are white or gray, and the most common depositional environment was in shallow tropical seas, which were rich in lime. Many cave and spring deposits are also varieties of limestone.

Coal

Coal is a rock made of the carbonized remains of ancient plants. Coal deposits typically represent a swampy lowland environment that was invaded by sea-level rise, which killed and buried dense vegetation.

Metamorphic Rocks

Metamorphic rocks can also be divided into two general categories, foliated rocks and nonfoliated rocks.

Foliated

Foliated metamorphic rocks have either wavy, roughly parallel plates, or bands of light and dark minerals that formed under intense heat and pressure. The nature of these foliation indicates the degree of metamorphism or change from the rock's original state.

Copyright and photograph by Dr. Parvinder S. Sethi

Slate

Slate is metamorphosed shale, and looks much like shale, except it is harder and has very thin platy foliation. The most common color is black.

Copyright and photograph by Dr. Parvinder S. Sethi

Schist

Schist has very prominent, wavy, and platy foliation generally covered with mineral crystals that formed during metamorphism. Schist represents a high degree of metamorphism and could originally have been any of a wide variety of rocks, so colors and appearance vary greatly.

Copyright and photograph by Dr. Parvinder S. Sethi

Gneiss

Gneiss (pronounced "nice") is a banded metamorphic rock, with alternating bands of light and dark minerals. Gneiss represents extreme heat and pressure during metamorphism and also may originally have been any of a variety of rocks. Metamorphism of granites commonly produces gneiss.

Nonfoliated

Nonfoliated metamorphic rocks do not display regular patterns of banding or platy foliation. In general, non-foliated metamorphics represent a rock that has been changed by the fusing and recrystallization of minerals in the original, often identifiable, rock.

Quartzite

Quartzite is metamorphosed quartz sandstone in which the former sand grains have fused together to produce an extremely hard, resistant rock.

Marble

Marble is metamorphosed limestone that has been recrystallized. Many colors and patterns of marble exist, and its relative softness compared to other rocks makes it easy to cut and polish.

Glossary

aa a blocky, angular surface of a lava flow.

abiotic natural, nonliving component of an ecosystem.

ablation any removal of frozen water from the mass of a glacier.

abrasion (corrasion) erosion process in which particles already being carried by a geomorphic agent are used as tools to aid in eroding more Earth material.

abrasion platform wave-cut bench of rock just below the water level; indicates the landward extent of coastal cliff erosion.

absolute humidity mass of water vapor present per unit volume of air, expressed as grams per cubic meter, or grains per cubic foot.

absolute location location of an object on the basis of mathematical coordinates on an Earth grid.

accretion growth of a continent by adding large pieces of crust along its border by plate tectonic collision.

accumulation any addition of frozen water to the mass of a glacier.

acid mine drainage seepage to the surface of subsurface water that has become highly acidic by flowing through underground coal mines.

acid rain rain with a pH value of less than 5.6, the pH of natural rain; often linked to the pollution associated with the burning of fossil fuels.

active layer the upper soil zone that thaws in summer in regions underlain by permafrost.

active-margin coast coastal region characterized by active volcanic and tectonism.

actual evapotranspiration the actual amount of moisture loss through evapotranspiration measured from a surface.

adiabatic heating and cooling changes of temperature within a gas because of compression (resulting in heating) or expansion (resulting in cooling); no heat is added or subtracted from outside.

advection horizontal heat transfer within the atmosphere; air masses moved horizontally, usually by wind.

advection fog fog produced by the movement of warm, moist air across a cold sea or land surface.

aggradation building up of an area or landscape that results from more deposition than erosion over time.

air mass large portion of the atmosphere, sometimes subcontinental in size, that may move over Earth's surface as a distinct, relatively homogeneous entity.

albedo proportion of solar radiation reflected back from a surface, expressed as a percentage of radiation received on that surface.

Aleutian low center of low atmospheric pressure in the area of the Aleutian Islands, especially persistent in winter.

alluvial fan fan-shaped depositional landform, particularly common in arid regions, occurring where a stream emerges from a mountain canyon and deposits sediment on a plain.

alluvial plain extensive floodplain of very low relief.

alluvium general term for clastic particles deposited by a stream.

alpine glacier a mass of flowing ice that exists in a mountainous region due to climatic conditions resulting from high elevation.

altithermal an interval of time about 7000 years ago when the climate was hotter than it is today.

altitude heights of points above Earth's surface.

alto signifies a middle-level cloud (i.e., from 2000 to 6000 m in elevation).

analemma a diagram that shows the declination of the sun throughout the year.

angle of inclination tilt of Earth's polar axis at an angle of 23½° from the vertical to the plane of the ecliptic.

angle of repose maximum angle at which a slope of loose sediment can stand without particles tumbling or sliding downslope.

annual lag of temperature time lag between the maximum (minimum) of insolation during the solstices and the warmest (coldest) temperatures of the year.

annual march of temperature changes in monthly temperatures throughout the months of the year.

annual temperature range difference between the mean monthly temperatures for the warmest and coolest months of the year.

Antarctic circle latitude at 66½°S; the northern limit of the zone in the Southern Hemisphere that experiences a 24-hour period of sunlight and a 24-hour period of darkness at least once a year.

anticline the upfolded element of folded rock structure.

anticyclone an area of high atmospheric pressure, also known as a high.

aphelion position of Earth's orbit at farthest distance from the sun during each Earth revolution.

aquiclude rock layer that restricts flow and storage of groundwater; it is impermeable and nonporous.

aquifer rock layer that is a container and transmitter of groundwater; it is both porous and permeable.

Arctic circle latitude at 66½°N; the southern limit of the zone in the Northern Hemisphere that experiences a 24-hour period of sunlight and a 24-hour period of darkness at least once a year.

arête jagged, sawtooth spine or wall of rock separating two glacial valleys or cirques.

arid climates climate regions or conditions where annual potential evapotranspiration greatly exceeds annual precipitation.

artesian spring natural flow of groundwater to the surface from below due to pressure.

artesian well groundwater that flows to the surface under its own pressure through an artificial opening.

artificial recharge diverting surface water to permeable terrain for the purpose of replenishing groundwater supplies.

asteroid sometimes called a minor planet; any solar system body composed of rock and/or metal not exceeding 800 kilometers (500 mi) in diameter.

asthenosphere thick, plastic layer within Earth's upper mantle that flows in response to convection, instigating plate tectonic motion.

atmosphere blanket of air, composed of various gases, that envelops Earth.

atmospheric air pressure (barometric pressure) force per unit area that the atmosphere exerts on any surface at a particular elevation.

atmospheric disturbance refers to variation in the secondary circulation of the atmosphere that cannot correctly be classified as a storm; for example, front, air mass.

atoll ring of coral reefs and islands encircling a lagoon, with no inner island.

attrition the reduction of size in sediment as it is transported downstream.

auroras colorful interaction of solar radiation with ions in Earth's upper atmosphere; more commonly seen in higher latitudes. Called aurora borealis in the Northern Hemisphere (also known as the northern lights), and the aurora australis (southern lights) in the Southern Hemisphere.

autotroph organism that, because it is capable of photosynthesis, is at the foundation of a food web and is considered a basic producer.

avalanche density current of pulverized (powdered) Earth material traveling rapidly downslope by the pull of gravity.

axis an imaginary line between the geographic North Pole and South Pole, around which the planet rotates.

azimuth an angular direction of a line, point, or route measured clockwise from north 0–360°.

azimuthal map a projection that preserves the true direction from the map center to any other point on the map.

Azores high see Bermuda high.

backwash thin sheet of broken wave water that rushes back down the beach face in the swash zone.

badlands barren region of soft rock material intensely eroded into ridges and ravines by numerous gullies and washes.

bajada an extensive intermediate slope of adjacent, coalescing alluvial fans connecting a steep mountain front with a basin or plain.

bar a shallow, submerged, mounded accumulation of sediment in nearshore locations or in streams.

barchan crescent-shaped sand dune with arms pointing downwind.

barometer instrument for measuring atmospheric pressure.

barrier beach a category of elongate depositional coastal landforms that lie parallel to the mainland but separated from it by a lagoon.

barrier island a barrier beach constructed parallel to the mainland, but not attached to it; separated from the mainland by a lagoon.

barrier reef coral reef parallel to the coast and separated from it by a lagoon.

barrier spit a barrier beach attached to the mainland at one end like a spit, but enclosing a lagoon.

basalt a dark-colored, fine-grained extrusive igneous rock generally associated with the oceanic crust and oceanic volcanoes.

basaltic plateau elevated area of low surface relief consisting of horizontal layers of basaltic lava.

base flow groundwater that seeps into stream channels below the water surface; sustains perennial streams between storms.

base line east–west lines of division in the U.S. Public Lands Survey System.

base level elevation below which a stream cannot flow; most humid region streams flow to sea level (ultimate base level).

batholith a large, irregular mass of intrusive igneous rock (pluton).

baymouth barrier a barrier beach that extends across the mouth of an embayment, attached to the mainland at both ends, to form a lagoon.

beach coastal landform of wave-deposited sediment attached to dry land along its entire length.

beach drifting littoral drifting (coastal transportation) in which waves breaking at an angle to the shoreline move sediment along the beach in a zigzag fashion in the swash zone.

bearing an angular direction of a line, point, or route measured from north or from a current location to a desired location (often in 90° compass quadrants).

bed load solid particles moved by wind or water by bouncing, rolling, or sliding along the ground or streambed.

bedding plane boundary between different sedimentary layers.

Bergeron (ice crystal) process rain-forming process where cloud droplets begin as ice crystals and melt into rain as they fall toward the surface.

bergschrund the large crevasse at the head of an alpine glacier, beneath the cirque headwall.

Bermuda High persistent, high atmospheric pressure center located in the subtropics of the north Atlantic Ocean, also called the Azores High.

biomass amount of living material or standing crop in an ecosystem or at a particular trophic level within an ecosystem.

biome one of Earth's major terrestrial ecosystems, classified by the vegetation types that dominate the plant communities within the ecosystem.

biosphere the life forms, human, animal, or plant, of Earth that form one of the major Earth subsystems.

blizzard heavy snowstorm accompanied by strong winds (55 kph or greater) and that reduces visibility to less than half a kilometer.

blowout local, wind-eroded surface depression in an area dominated by wind-deposited sand.

bolson desert basin, surrounded by mountains, with no drainage outlet.

boreal forest (taiga) coniferous forest dominated by spruce, fir, and pine found growing in subarctic conditions around the world north of the 50th parallel of north latitude.

boulder a rock fragment greater than 256 millimeters (10 in.) in diameter.

braided channel stream channel composed of multiple subchannels of simultaneous flow that split and rejoin and frequently shift position.

breaker zone the part of the nearshore area in which waves break.

butte isolated erosional remnant of a tableland with a flat summit, bordered by steep-sided escarpments. Buttes are usually found in arid regions of flat-lying sediments and are somewhat taller than they are wide.

calcification soil-forming process of subhumid and semiarid climates. Soil types in the mollisol order, the typical end products of the process, are characterized by little leaching or eluviation and by the accumulation of both humus and mineral bases (especially calcium carbonate, $CaCO_3$).

caldera a large depression formed by a volcanic eruption.

caliche hardened layers of lime ($CaCO_3$) deposited at the surface of a soil by evaporating capillary water.

calorie amount of heat necessary to raise the temperature of 1 gram of water 1°C.

calving breaking off a mass of ice from the toe of a glacier at its junction with the ocean or a lake.

Canadian High high atmospheric pressure area that tends to develop over the central North American continent in winter.

capacity the maximum amount of water vapor that can be contained in a given quantity of air at a given temperature.

capillary action the upward movement of water through tiny cracks and pore spaces.

capillary water soil water that clings to soil peds and individual soil particles as a result of surface tension. Capillary water moves in all directions through the soil from areas of surplus water to areas of deficit.

caprock a resistant horizontal layer of rock that forms the flat top of a landform, such as a butte or a mesa.

carbonate a mineral group characterized by specific combinations of carbon and oxygen atoms.

carbonation carbon dioxide in water chemically combining with other substances to create new compounds.

carnivore animal that eats only other animals.

cartography the art and science of mapmaking.

catastrophism once-popular theory that Earth's landscapes developed in a relatively short time by cataclysmic events.

cavern (cave) natural void in rock created by solution that is large enough for people to enter.

Celsius (or centigrade) scale temperature scale in which 0° is the freezing point of water and 100° its boiling point at standard sea-level pressure.

centrifugal force force that pulls a rotating object away from the center of rotation.

channel roughness an expression of the frictional resistance to stream flow due to irregularities in a stream channel bed and sides.

chaparral sclerophyllous woodland vegetation found growing in the Mediterranean climate of the western United States; these seasonal, drought-resistant plants are low-growing, with small, hard-surfaced leaves and deep, water-probing roots.

chemical sedimentary rock rock created from dissolved minerals that have precipitated out of water.

chemical weathering breakdown of rock material by chemical reactions that change the rock's mineral composition (decomposition).

Chinook dry warm wind on the eastern slopes of the Rocky Mountains (see foehn wind).

cinder cone hill composed of fragments of volcanic rock (pyroclastics) erupted from a central vent.

circle of illumination line dividing the sunlit (day) hemisphere from the shaded (night) hemisphere; experienced by individuals on Earth's surface as sunrise and/or sunset.

cirque deep, sometimes steep-sided amphitheater formed at the head of an alpine valley by glacial ice erosion.

cirque glacier a generally small alpine glacier restricted to a high-elevation basin (cirque).

cirro signifies a high-level cloud (i.e., above 6000 m in elevation).

cirrus high, detached clouds consisting of ice particles. Cirrus clouds are white and feathery or fibrous in appearance.

Cl, O, R, P, T Hans Jenny's list of soil formation factors: Climate, Organics, Relief, Parent material, and Time.

clastic sedimentary rock rock formed by the compaction and/or cementation of preexisting rock, bone, or shell pieces.

clast solid broken piece of rock, bone, or shell.

clay (clayey) a very fine grained mineral particle with a size less than 0.004 millimeter, often the product of weathering.

clay pan an ephemeral lake bed composed mostly of fine-grained clastic particles.

climate accumulated and averaged weather patterns of a locality or region; the full description is based on long-term statistics and includes extremes or deviations from the norm.

climatology scientific study of climates of Earth and their distribution.

climax community the final step in the succession of plant communities that occupy a specific location.

climograph graph portraying information on mean monthly temperature and rainfall for a select location or station.

closed system system in which no substantial amount of matter and/or energy can cross its boundaries.

coastal zone general region of interaction between the land and a lake or the ocean.

coastline of emergence coast with formerly submerged land that is now above water, due either to uplift of the land or a drop in sea level.

coastline of submergence a coastal area that has undergone sinking or subsidence relative to sea level.

cloud forest rainforest that is produced by nearly constant light rain on the windward slopes of mountains.

cobble rock fragments ranging in diameter from 64 to 256 millimeters (2.5–10 in.).

cobble beach cobble-sized sediment deposited by waves along the shoreline, often found along the base of sea cliffs or lake cliffs.

col a glacially eroded pass between two mountain valleys.

cold current a flow of sea water that moves like a river through the ocean and is relatively colder than water in the ocean area that it flows through.

cold front leading edge of a relatively cooler, denser air mass that advances upon a warmer, less dense air mass.

collapse sinkhole topographic depression formed mainly by the cave-in of the land above a cavern.

collision–coalescence process rain-forming process where raindrops form by collision between cloud droplets.

column pillar-shaped speleothem resulting from the joining of a stalactite and a stalagmite.

columnar joints vertical, polygonal fractures caused by lava shrinking as it cools.

comet a small body of icy and dusty matter that revolves about the sun. When a comet comes near the sun, some of its material vaporizes, forming a large head and often a tail.

composite cone (stratovolcano) volcano formed from alternating layers of lava and pyroclastic materials; generally known for violent eruptions.

compressional force pushing together from opposite sides (convergence).

compromise projection maps that compromise true shape and true area in order to display both fairly well.

conceptual model image in the mind of an Earth feature or landscape as derived from personal experiences.

condensation process by which a vapor is converted to a liquid during which energy is released in the form of latent heat.

condensation nuclei minute particles in the atmosphere (e.g., dust, smoke, pollen, sea salt) on which condensation can take place.

conduction transfer of heat within a body or between adjacent matter by means of internal molecular movement.

conformal map projection a map projection that maintains the true shape of small areas on Earth's surface.

consumer an organism that consumes organic material from other life forms, including all animals, and parasitic plants (also called a heterotroph).

Continental Arctic (cA) very cold, very dry air mass originating from the arctic region.

continental collision the fusing together of landmasses as tectonic plates converge.

continental crust the less dense (avg. 2.7 g/cm³), thicker portion of Earth's crust; underlies the continents.

Continental Divide line of separation dividing runoff between the Pacific and Atlantic Oceans. In North America it generally follows the crest of the Rocky Mountains.

continental drift theory proposed by Alfred Wegener stating that the continents joined, broke apart, and moved on Earth's surface; it was later replaced by the theory of plate tectonics.

continental glacier a category of very large and thick masses of flowing ice that exist due to climatic conditions resulting from high latitude.

continental islands islands that are geologically part of a continent and are usually located on the continental shelf.

Continental Polar (cP) cold, dry air mass originating from landmasses approximately 40° to 60° N or S latitude.

continental shelf the gently sloping margin of a continent overlain by ocean water.

continental shield the ancient part of a continent that consists of crystalline rock.

Continental Tropical (cT) warm, dry air mass originating from subtropical landmasses.

continentality the distance a particular place is located in respect to a large body of water; the greater the distance, the greater the continentality.

continuous data numerical or locational representations of phenomena that are present everywhere—such as air pressure, temperature, elevation.

contour interval vertical distance represented by two adjacent contour lines on a topographic map.

contour map (topographic map) map that uses contour lines to show differences in elevation (topography).

control (temperature) factors that control temperatures around the world, such as latitude, proximity to water bodies, ocean currents, altitude, landform barriers, and human activities.

convection process by which circulation is produced within an air mass or fluid body (heated material rises, cooled material sinks); also, in tectonic plate theory, the method whereby heat is transferred towards Earth's surface from deep within the mantle.

convectional precipitation precipitation resulting from condensation of water vapor in an air mass that is rising convectionally as it is heated from below.

convective thunderstorm a thunderstorm produced by the convective uplift mechanism.

convergent wind circulation pressure-and-wind system where the airflow is inward toward the center, where pressure is lowest.

coordinate system a precise system of grid lines used to describe locations.

coral reef shallow, wave-resistant structure made by the accumulation of skeletal remains of tiny sea animals.

core extremely hot and dense, innermost portion of Earth's interior; the molten outer core is 2400 kilometers (1500 mi) thick; the solid inner core is 1120 kilometers (700 mi) thick.

Coriolis effect apparent effect of Earth's rotation on horizontally moving bodies, such as wind and ocean currents; such bodies tend to be deflected to the right in the Northern Hemisphere and to the left in the Southern Hemisphere.

corrosion chemical erosion of rock matter by water; the removal of ions from rock-forming minerals in water.

creep slow downslope movement of Earth material involving the lifting and falling action of sediment particles.

crevasse stress crack commonly found along the margins and at the toe of a glacier.

cross bedding thin layers within sedimentary rocks that were deposited at an angle to the dominant rock layering.

crust relatively thin, approximately 8–64 kilometers (5–40 mi) deep, low-density surface layer of Earth.

crustal warping gentle bending and folding of crustal rocks.

cumulus globular clouds, usually with a horizontal base and strong vertical development.

cumulonimbus towering rain-producing cloud with very strong convective uplift and a flat anvil-shaped top.

cut bank the steep slope found on the outside of a bend in a meandering stream channel.

cyclone center of low atmospheric pressure, also known as a low.

cyclonic (convergence) precipitation precipitation formed by cyclonic uplift.

daily march of temperature changes in daily temperatures as we go from an overnight low to a daytime high and back to the overnight low.

daily (diurnal) temperature range difference between the highest and lowest temperatures of the day (usually recorded hourly).

debris unconsolidated slope material with a wide range of grain sizes including at least 20% gravel (>2 mm).

debris flow rapid, gravity-induced downslope movement of wet, poorly sorted Earth material.

debris flow fan fan-shaped depositional landform, particularly common in arid regions, created where debris flows emerge onto a plain from a mountain canyon.

deciduous a plant, usually a tree or shrub, that drops its leaves (usually seasonally).

declination the latitude on Earth at which the noon sun is directly overhead.

decomposer (detritivore) organism that promotes decay by feeding on dead plant and animal material and returns mineral nutrients to the soil or water in a form that plants can utilize.

decomposition a term that refers to the processes of chemical weathering.

deep-water wave wave traveling in depth of water greater than or equal to half the wavelength.

deflation entrainment and removal of loose surface sediment by the wind.

deflation hollow a wind-eroded depression in an area not dominated by wind-deposited sand.

degradation landscape lowering that results from more erosion than deposition over time.

delta depositional landform constructed where a stream flows into a standing body of water (a lake or the ocean).

dendritic term used to describe a drainage pattern that is treelike with tributaries joining the main stream at acute angles.

dendrochronology method of determining past climatic conditions using tree rings.

deposition accumulation of Earth materials at a new site after being moved by gravity, water, wind, or glacial ice.

desert climate climate where the amount of precipitation received is less than one half of the potential evapotranspiration.

desert pavement (reg) desert surface mosaic of close-fitting stones that overlies a deposit of mostly fine-grained sediment.

dew   tiny droplets of water on ground surfaces, grass blades, or solid objects. Dew is formed by condensation when air at the surface reaches the dew point.

dew point   the temperature at which an air mass becomes saturated; any further cooling will cause condensation of water vapor in the air.

differential weathering and erosion   rock types vary in resistance to weathering and erosion, causing the processes to occur at different rates, often producing distinctive landform features.

digital elevation model (DEM)   three-dimensional views of topography.

digital image   an image made from computer data displayed like a mosaic of tiny squares, called pixels.

digital terrain model   a computer-generated graphic representation of topography.

dike   igneous intrusion with a vertical, wall-like shape.

dip   inclination of a rock layer from the horizontal; always measured at right angles to the strike.

dip-slip fault   a vertical fault where the movement is up and down the dip of the fault surface.

disappearing stream   stream that has its flow diverted entirely to the subsurface.

discharge (stream discharge)   rate of stream flow; measured as the volume of water flowing past a cross section of a stream per unit time (cubic meters or cubic feet per second).

discrete data   numerical or locational representations of phenomena that are present only at certain locations—such as earthquake epicenters, sinkholes, tornado paths.

dissolved load   soluble minerals or other chemical constituents carried in water as a solution.

distributary   a smaller stream that conducts flow away from the larger main channel, especially on deltas; the opposite of a tributary.

divergent wind circulation   pressure-and-wind system where the airflow is outward away from the center, where pressure is highest.

doldrums   zone of low pressure and calms along the equator.

doline   see sinkhole.

Doppler radar   advanced type of radar that can detect motion in storms, specifically motion toward and away from the radar signal.

drainage basin (watershed, catchment)   the region that provides runoff to a stream.

drainage density   the summed length of all stream channels per unit area in a drainage basin.

drainage divide   the outer boundary of a drainage basin.

drainage (stream) pattern   arrangement of the network of channels in a stream system in map view.

drift   sediment deposited by any means in association with glacial ice or its meltwater.

drizzle   fine mist or haze of very small water droplets with a barely perceptible falling motion.

drumlin   streamlined, elongated hill composed of glacial drift with a tapered end indicating direction of continental ice flow.

dry adiabatic lapse rate   rate at which a rising mass of air is cooled by expansion when no condensation is occurring (10°C/1000 m or 5.6°F/1000 ft).

dust storm   a moving cloud of wind-blown dust (typically silt).

dynamic equilibrium   constantly changing relationship among the variables of a system, which produces a balance between the amounts of energy and/or materials that enter a system and the amounts that leave.

earth (as a mass wasting material)   thick unit of unconsolidated, predominantly fine-grained slope material.

Earth system   set of interrelated components, variables, processes, and subsystems (e.g., atmosphere, lithosphere, biosphere, hydrosphere), which interact and function together to make up Earth.

earthquake   series of shock waves set in motion by sudden movement along a fault.

earthquake intensity   a measure of the impact of an earthquake on humans and their built environment.

earthquake magnitude   measurement representing an earthquake's size in terms of energy released.

easterly wave   trough-shaped, weak, low pressure cell that progresses slowly from east to west in the trade wind belt of the tropics; this type of disturbance sometimes develops into a hurricane.

eccentricity cycle   the change in Earth's orbit from slightly elliptical to more circular, and back to its earlier shape every 100,000 years.

ecological niche   combination of role and habitat as represented by a particular species in an ecosystem.

ecology   science that studies the interactions between organisms and their environment.

ecosystem   community of organisms functioning together in an interdependent relationship with the environment that they occupy.

ecotone   transition zone of varied natural vegetation occupying the boundary between two adjacent and differing plant communities.

effective precipitation   actual precipitation available to supply plants and soil with usable moisture; does not take into consideration storm runoff or evaporation.

El Niño   warm countercurrent that influences the central and eastern Pacific.

elastic solid   a solid that withstands stress with little deformation until a maximum value is reached, whereupon it breaks.

electromagnetic energy   all forms of energy that share the property of moving through space (or any medium) in a wavelike pattern of electric and magnetic fields; also called radiation.

elements (weather and climate)   the major elements include solar energy, temperature, pressure, winds, and precipitation.

elevation   vertical distance from mean sea level to a point or object on Earth's surface.

eluviation   downward removal of soil components by water.

end moraine   a ridge of till deposited at the toe or terminus of a glacier.

endogenic process   landforming process originating within Earth.

Enhanced Fujita Scale   A ranking (1–5) of the intensity of a tornado, based on wind speeds and specifically on the types of damage that occur.

entrenched stream   a stream that has eroded downward so that it flows in a relatively deep and steep-sided (trench-like) valley or canyon.

environment   surroundings, whether of humans or of any other living organism; includes physical, social, and cultural conditions that affect the development of that organism.

environmental lapse rate (normal lapse rate)   decrease in temperature with altitude under normal atmospheric conditions; approximately 6.5°C/1000 meters (3.6°F/1000 ft).

eolian (aeolian)   pertaining to the landforming work of the wind.

ephemeral flow   describes streams that conduct flow only occasionally, during or shortly after precipitation events, or due to ice or snowmelt.

ephemeral stream   a stream that flows only at certain times, when adequate discharge is supplied by precipitation events, ice or snowmelt, or irregular spring flow.

epicenter point on Earth's surface directly above the focus of an earthquake.

epiphyte a plant that grows on another plant, but does not take its nutrients from the host plant.

equal-area map projection a map projection on which any given areas of Earth's surface are shown in correct proportional sizes on the map.

equator great circle of Earth midway between the poles; the zero degree parallel of latitude that divides Earth into the Northern and Southern Hemispheres.

equatorial low (equatorial trough) zone of low atmospheric pressure centered more or less over the equator where heated air is rising; see also doldrums.

equidistance a property of some maps that depicts distances equally without scale variation.

equilibrium state of balance between the interconnected components of a system, an organized whole.

equilibrium line balance position on a glacier that separates the zone of accumulation from the zone of ablation.

equinox one of two times each year (approximately March 21 and September 22) when the position of the noon sun is overhead (and its vertical rays strike) at the equator; all over Earth, day and night are of equal length.

erosion removal of Earth materials from a site by gravity, water, wind, or glacial ice.

erratic large glacially transported boulder deposited on top of bedrock of different composition.

esker narrow, winding ridge of coarse sediment probably deposited in association with a meltwater tunnel at the base of a continental glacier.

estuary coastal embayment where salt and fresh water mix.

evaporation process by which a liquid is converted to the gaseous (vapor) state by the addition of latent heat.

evaporite mineral salts that are soluble in water and accumulate when water evaporates.

evapotranspiration combined water loss to the atmosphere from ground and water surfaces by evaporation and, from plants, by transpiration.

evergreen a plant, usually a tree or shrub, that does not seasonally drop its leaves.

exfoliation successive breaking away of outer rock sheets or slabs from a rock mass by weathering.

exfoliation dome large, smooth, convex (dome-shaped) mass of exposed rock undergoing exfoliation due to weathering by unloading.

exfoliation sheet relatively thin, outer layer of rock broken from the main rock mass by weathering.

exogenic process landforming process originating at or very near Earth's surface.

exotic stream (or river) stream that originates in a humid region and has sufficient water volume to flow across a desert region.

exposure direction of mountain slopes with respect to prevailing wind direction.

exterior drainage streams and stream systems that flow to the ocean.

extratropical disturbance see middle-latitude disturbance.

extrusive igneous rock rock solidified at Earth's surface from lava; also called volcanic rock.

Fahrenheit scale temperature scale in which 32° is the freezing point of water, and 212° its boiling point, at standard sea-level pressure.

fall type of fast mass wasting characterized by Earth material plummeting downward freely through air.

fan apex the most upflow point on an alluvial fan; where the fan-forming stream emerges from the mountain canyon.

fast mass wasting gravity-induced downslope movement of Earth material that people can witness directly.

fault breakage zone along which rock masses have slid past each other.

fault block discrete block-like region of crustal rocks bordered on two opposite sides by faults.

fault scarp (escarpment) the steep cliff or exposed face of a fault where one crustal block has been displaced vertically relative to another.

faulting the movement of rock masses past each other along either side of a fault.

feedback sequence of changes in the elements of a system, which ultimately affects the element that was initially altered to begin the sequence.

feedback loop path of change as its effects move through the variables of a system until the effects impact the variable originally experiencing change.

fertilization adding additional nutrients to the soil.

fetch distance over open water that winds blow without interruption.

firn compact granular snow formed by partial melting and refreezing due to overlying layers of snow.

fissure flow lava flow that emanates from a crack (fissure) in the surface rather than from the vent or crater of a volcano.

fjord deep, glacial trough along the coast invaded by the sea after the removal of the glacier.

flood stream water exceeding the amount that can be contained within its channel.

flood basalt massive outpouring of basaltic lava.

floodplain the low-gradient area adjacent to many stream channels that is subject to flooding and primarily composed of alluvium.

flow (mass wasting) rapid downslope movement of wet unconsolidated Earth material that experiences considerable mixing.

fluvial term used to describe landforms and processes associated with the work of streams.

fluvial geomorphology the study of streams as landforming agents.

focus point within Earth's crust or upper mantle where an earthquake originated.

foehn wind warm, dry, downslope wind on lee of mountain range, caused by adiabatic heating of descending air.

fog mass of suspended water droplets within the atmosphere that is in contact with the ground.

folding the bending or wrinkling of Earth's crust due to compressional tectonic forces.

foliation a banded, wavy, or platy structure in metamorphic rocks.

food chain sequence of levels in the feeding pattern of an ecosystem.

food web feeding mosaic formed by the interrelated and overlapping food chains of an ecosystem.

freeze–thaw weathering (ice wedging) breaking apart of rock by the expansive force of water freezing in cracks.

freezing rain rainfall that freezes into ice upon coming in contact with a surface or object that is colder than 0°C (32°F).

friction force that acts opposite to the direction of movement or flow; for example, turbulent resistance of Earth's surface on the flow of the atmosphere.

fringing reef coral reef attached to the coast.

front sloping boundary or contact surface between air masses that have different properties of temperature, moisture content, density, and atmospheric pressure.

frontal precipitation precipitation resulting from condensation of water vapor by the lifting or rising of warmer, lighter air above cooler, denser air along a frontal boundary.

frontal thunderstorm a thunderstorm produced by the frontal up-lift mechanism.

frost frozen condensation that occurs when air at ground level is cooled to a dew point of 0°C (32°F) or below; also any temperature near or below freezing that threatens sensitive plants.

galaxy a large assemblage of stars; a typical galaxy contains millions to hundreds of billions of stars.

galleria forest junglelike vegetation extending along and over streams in tropical forest regions.

gap an area within the territory occupied by a plant community when the climax vegetation has been destroyed or damaged by some natural process, such as a hurricane, forest fire, or landslide.

generalist species that can survive on a wide range of food supplies.

geographic grid lines of latitude and longitude form the geographic grid.

geographic information system (GIS) versatile computer software that combines the features of cartography and database management to produce new maps of data for solving spatial problems.

geography study of Earth phenomena; includes an analysis of distributional patterns and interrelationships among these phenomena.

geomorphic agent a medium that erodes, transports, and deposits Earth materials; includes water, wind, and glacial ice.

geomorphology the study of the origin and development of landforms.

geostrophic winds upper-level winds in which the Coriolis effect and pressure gradient are balanced, resulting in a wind flowing parallel to the isobars.

geothermal water water heated by contact with hot rocks in the subsurface.

geyser natural eruptive outflow of water that alternates between hot water and steam.

glacial outwash the fluvial deposits derived from glacial meltwater streams.

glacial trough a U-shaped valley carved by glacial erosion.

glaciation an interval of glacial activity.

glacier a large mass of ice that flows as a plastic solid.

glacier advance expansion of the toe or terminus of a glacier to a lower elevation or lower latitude due to an increase in size.

glacier head farthest upslope part of an alpine glacier.

glacier retreat withdrawal of the toe or terminus of a glacier to a higher elevation or higher latitude due to a decrease in size.

glacier terminus (toe) farthest downslope or lowest latitude part of a glacier.

glaciofluvial deposit sorted glacial drift deposited by meltwater.

glaciolacustrine deposit sorted glacial drift deposited by meltwater in lakes associated with the margins of glaciers.

gleization soil-forming process of poorly drained areas in cold, wet climates. The resulting soils have a heavy surface layer of humus with a water-saturated clay horizon directly beneath.

Global Positioning System (GPS) GPS uses satellites and computers to compute positions and travel routes anywhere on Earth.

global warming climate change that would cause Earth's temperatures to rise.

graben block of crustal rocks between two parallel normal faults that has slid downward relative to adjacent blocks.

gradient a term for slope often used to describe the angle of a streambed.

granite a coarse-grained intrusive igneous rock generally associated with continental crust.

granular disintegration weathering feature of coarse crystalline rocks in which visible individual mineral grains fall away from the main rock mass.

graphic (bar) scale a ruler-like device placed on maps for making direct measurements in ground distances.

gravel a general term for sediment sizes larger than sand (larger than 2.0 mm).

gravitational water meteoric water that passes through the soil under the influence of gravitation.

great circle any circle formed by a full circumference of the globe; the plane of a great circle passes through the center of the globe.

greenhouse effect warming of the atmosphere that occurs because shortwave solar radiation heats the planet's surface, but the loss of longwave heat radiation is hindered by CO_2 and other gases.

greenhouse gases atmospheric gases that hinder the escape of Earth's heat energy.

Greenwich mean time (GMT) time at zero degrees longitude used as the base time for Earth's 24 time zones; also called Universal Time or Zulu Time.

groin artificial structure extending out into the water at a right angle from a beach, built to inhibit loss of beach sediment.

ground moraine irregular, hummocky landscape of till deposited on Earth's surface by a wasting glacier.

ground-inversion fog see radiation fog.

groundwater subsurface water in the saturated zone below the water table.

gully steep-sided stream channel somewhat larger than a rill that even in humid climates flows only in direct response to precipitation events (ephemeral flow).

gyre broad circular patterns of major surface ocean currents produced by large subtropical high pressure systems.

habitat location within an ecosystem occupied by a particular organism.

hail form of precipitation consisting of pellets or balls of ice with a concentric layered structure usually associated with the strong convection of cumulonimbus clouds.

hanging valley tributary glacial trough that enters a main glaciated valley at a level high above the valley floor.

hardpan dense, compacted, clay-rich layer occasionally found in the subsoil (B horizon) that is an end product of excessive illuviation.

haystack hill (conical hill or hum) remnant hills of soluble rock remaining after adjacent rock has been dissolved away in karst areas.

headward erosion gullying and valley cutting that extends a stream channel in an upstream direction.

heat the total kinetic energy of all the atoms that make up a substance.

heat energy budget relationship between solar energy input, storage, and output within the Earth system.

heat island mass of warmer air overlying urban areas.

heaving various means by which particles are lifted perpendicular to a sloping surface, then fall straight down by gravity.

hemisphere half of a sphere; for example, the northern or southern half of Earth divided by the equator or the eastern and western half divided by two meridians, the 0° and 180° meridians.

herbivore an animal that eats only living plant material.

heterotroph organism that is incapable of producing its own food and that must survive by consuming other organisms.

high see anticyclone.

highland climates a general climate classification for regions of high, yet varying, elevations.

holistic approach considering and examining all phenomena relevant to a problem.

Holocene the most recent time interval of warm, relatively stable climate that began with the retreat of major glaciers about 10,000 years ago.

horn pyramid-like peak created where three or more expanding cirques meet at a mountain summit.

horst crustal block between two parallel normal faults that has slid upward relative to adjacent blocks.

hot spot a mass of hot molten rock material at a fixed location beneath a lithospheric plate.

hot spring natural outflow of geothermal groundwater to the surface.

human geography specialization in the systematic study of geography that focuses on the location, distribution, and spatial interaction of human (cultural) phenomena.

humid continental, hot-summer climate climate type characterized by hot, humid summers and mild, moist winters.

humid continental, mild-summer climate climate type characterized by mild, humid summers and cold, moist winters.

humidity amount of water vapor in an air mass at a given time.

humus organic matter found in the surface soil layers that is in various stages of decomposition as a result of bacterial action.

hurricane severe tropical cyclone of great size with nearly concentric isobars. Its torrential rains and high-velocity winds create unusually high seas and extensive coastal flooding; also called willy willies, tropical cyclones, baguios, and typhoons.

hydration rock weathering due to substances in cracks swelling and shrinking with the addition and removal of water molecules.

hydraulic action erosion resulting from the force of moving water.

hydrologic cycle circulation of water within the Earth system, from evaporation to condensation, precipitation, runoff, storage, and reevaporation back into the atmosphere.

hydrolysis water molecules chemically recombining with other substances to form new compounds.

hygroscopic water water in the soil that adheres to mineral particles.

hydrosphere major Earth subsystem consisting of the waters of Earth, including oceans, ice, freshwater bodies, groundwater, and water within the atmosphere and biomass.

ice age period of Earth history when much of Earth's surface was covered with massive continental glaciers. The most recent ice age occurred during the Pleistocene Epoch.

ice cap a continental glacier of regional size, less than 50,000 square kilometers.

ice fall portion of a glacier moving over and down a steep slope, creating a rigid white cascade, criss-crossed with deep crevasses.

ice sheet thick ice mass larger than 50,000 square kilometers that covers a major portion of a continent and buries all but the highest mountain peaks.

ice shelf large flat-topped plate of ice overlying the ocean but attached to land; a source of icebergs.

iceberg free-floating mass of ice broken off from a glacier where it flows into the ocean or a lake.

Icelandic Low center of low atmospheric pressure located in the north Atlantic, especially persistent in winter.

ice-marginal lake temporary lake formed by the disruption of meltwater drainage by deposition along a glacial margin, usually in the area of an end moraine.

ice-scoured plain a broad area of low relief and bedrock exposures eroded by a continental glacier.

ice-sheet climate climate type where the average temperature of every month of the year is below freezing.

igneous intrusion (pluton) a mass of igneous rock that cooled and solidified beneath Earth's surface.

igneous processes processes related to the solidification and eruption of molten rock matter.

igneous rock one of the three major categories of rock; formed from the cooling and solidification of molten rock matter.

illuviation deposition of fine soil components in the subsoil (B horizon) by gravitational water.

imaging radar radar systems designed to sense the ground and convert reflections into a map-like image.

infiltration water seeping downward into the soil or other surface materials.

infiltration capacity the greatest amount of infiltrated water that a surface material can hold.

inner core the solid, innermost portion of Earth's core, probably of iron and nickel, that forms the center of Earth.

inputs energy and material entering an Earth system.

inselberg remnant bedrock hill rising above a stream-eroded plain or pediment in an arid or semiarid region.

insolation incoming solar radiation, that is, energy received from the sun.

instability condition of air when it is warmer than the surrounding atmosphere and is buoyant with a tendency to rise; the lapse rate of the surrounding atmosphere is greater than that of unstable air.

interception the delay in arriving at the ground surface experienced by precipitation that strikes vegetation.

interfluve the land between two stream channels.

interglacial warmer period between glacial advances during which continental ice sheets and many valley glaciers retreat and disappear or are greatly reduced in size.

interior drainage streams and stream systems that flow within a closed basin and thus do not reach the ocean.

intermediate zone subsurface water layer between the zone of aeration above and the zone of saturation below; saturated only during times of ample precipitation.

intermittent flow describes streams that only flow seasonally.

International Date Line line roughly along the 180° meridian, where each day begins and ends; it is always a day later west of the line than east of the line.

Intertropical Convergence Zone (ITCZ) zone of low pressure and calms along the equator, where air carried by the trade winds from both sides of the equator converges and is forced to rise.

intrusive igneous rock rock that solidified within Earth from magma; also called plutonic rock.

inversion see temperature inversion.

ionosphere layer of ionized gasses concentrated between 80 kilometers (50 mi) and outer limits of the atmosphere.

isarithm line on a map that connects all points of the same numerical value, such as isotherms, isobars, and isobaths.

island arc a chain of volcanic islands along a deep oceanic trench; found near tectonic plate boundaries where subduction is occurring.

isobar line drawn on a map to connect all points with the same atmospheric pressure.

isoline a line on a map that represents equal values of some numerical measurement such as lines of equal temperature or elevation contours.

isostasy theory that holds that Earth's crust floats in hydrostatic equilibrium on the denser plastic layer of the mantle.

isotherm line drawn on a map connecting points of equal temperature.

jet stream high-velocity upper-air current with speeds of 120–640 kilometers per hour (75–250 mph).

jetty artificial structure, built in pairs, extending into a body of water; built to protect a harbor, inlet, or river mouth from excessive coastal sedimentation.

joint fracture or crack in rock.

joint set system of multiple parallel cracks (joints) in rock.

jungle dense tangle of trees and vines in areas where sunlight reaches the ground surface (not a true rainforest).

kame conical hill composed of sorted glaciofluvial deposits; presumed to have formed in contact with glacial ice when sediment accumulated in ice pits, crevasses, and among jumbles of detached ice blocks.

kame terraces landform resulting from accumulation of glaciofluvial sand and gravel along the margin of a glacier occupying a valley in an area of hilly relief.

karst unique landforms and landscapes derived by the solution of soluble rocks, particularly limestone.

katabatic wind downslope flow of cold, dense air that has accumulated in a high mountain valley or over an elevated plateau or ice cap.

Kelvin scale temperature scale developed by Lord Kelvin, equal to Celsius scale plus 273; no temperature can drop below absolute zero, or 0 degrees Kelvin.

kettle depression formed by the melting of an ice block buried in glacial deposits left by a retreating glacier.

kettle lake a small lake or pond occupying a kettle.

kinetic energy energy of motion; one half the mass (m) times velocity (v) squared, $E_k = \frac{1}{2}mv^2$.

Köppen system climate classification based on monthly and annual averages of temperature and precipitation; boundaries between climate classes are designed so that climate types coincide with vegetation regions.

La Niña cold sea-surface temperature anomaly in the Equatorial Pacific (opposite of El Niño).

laccolith massive igneous rock intrusion that bows overlying rock layers upwards in a domal fashion.

lagoon area of coastal water blocked from free circulation with the ocean or a lake by a barrier beach.

lahar rapid, gravity-driven downslope movement of wet, fine-grained volcanic sediment.

land breeze air flow at night from the land toward the sea, caused by the movement of air from a zone of higher pressure associated with cooler nighttime temperatures over the land.

landform a terrain feature, such as a mountain, valley, plateau, and so on.

Landsat a family of U.S. satellites that have been returning digital images since the 1970s.

landslide layperson's term for any fast mass wasting; used by some Earth scientists for massive slides that involve a variety of Earth materials.

lapse rate see normal lapse rate.

latent heat of condensation energy release in the form of heat, as water is converted from the gaseous (vapor) to the liquid state.

latent heat of evaporation amount of heat absorbed by water to evaporate from a surface (i.e., 590 calories/gram of water).

latent heat of fusion amount of heat transferred when liquid turns to ice and vice versa; this amounts to 80 calories/gram.

latent heat of sublimation amount of heat that is leased when ice turns to vapor without first going through the liquid phase; this amounts to 670 calories/gram.

lateral migration the sideways shift in the position of a stream channel over time.

lateral moraine a ridge of till deposited along the side margin of a glacier.

laterite iron, aluminum, and manganese rich layer in the subsoil (B horizon) that can be an end product of laterization in the wet-dry tropics (tropical savanna climate).

laterization soil-forming process of hot, wet climates. Oxisols, the typical end product of the process, are characterized by the presence of little or no humus, the removal of soluble and most fine soil components, and the accumulation of iron and aluminum compounds.

latitude angular distance (distance measured in degrees) north or south of the equator.

lava molten (melted) rock matter erupted onto Earth's surface; solidifies into extrusive igneous (volcanic) rocks.

lava flow erupted molten rock matter that oozed over the landscape and solidified.

leaching removal by gravitational water of soluble inorganic soil components from the surface layers of the soil.

leeward located on the side facing away from the wind.

legend key to symbols used on a map.

levee raised bank along margins of a stream or debris flow channel.

liana woody vine found in tropical forests that roots in the forest floor but uses trees for support as it grows upward toward available sunshine.

life-support system interacting and interdependent units (e.g., oxygen cycle, nitrogen cycle) that together provide an environment within which life can exist.

light year the distance light travels in 1 year—9.5 trillion kilometers (6 trillion mi).

lightning visible electrical discharge produced within a thunderstorm.

lithification processes of compaction and/or cementation that transform sediments into sedimentary rocks.

lithosphere (planetary structure) rigid and brittle outer layer of Earth consisting of the crust and uppermost mantle.

lithospheric plate Earth's exterior is broken into several of these large regions (plates) of rigid and brittle crust and upper mantle (lithosphere).

Little Ice Age an especially cold time interval from the early 14th century until the 19th century that had major impacts on civilizations in the Northern Hemisphere.

littoral drifting general term for sediment transport parallel to shore in the nearshore zone due to incomplete wave refraction.

loam soil a soil with a texture in which none of the three soil grades (sand, silt, or clay) predominate over the others.

loess wind-deposited silt; usually transported in dust storms and derived from arid or glaciated regions.

longitude angular distance (distance measured in degrees) east or west of the prime meridian.

longitudinal dune a linear ridge-like sand dune that is oriented parallel to the prevailing wind direction.

longitudinal profile the change in stream channel elevation with distance downstream from source to mouth.

longshore bar submerged feature of wave- and current-deposited sediment lying close to and parallel with the shore.

longshore current flow of water parallel to the shoreline just inside the breaker zone; caused by incomplete wave refraction.

longshore drifting transport of sediment parallel to shore by the longshore current.

longwave radiation electromagnetic radiation emitted by Earth in wavelengths longer than visible light, which includes heat reradiated by Earth's surface.

low see cyclone.

magma molten (melted) rock matter located beneath Earth's surface from which intrusive igneous rocks are formed.

magnetic declination horizontal angle between geographic north and magnetic north.

mantle moderately dense, relatively thick (2885 km/1800 mi) middle layer of Earth's interior that separates the crust from the outer core.

map projection any presentation of the spherical Earth on a flat surface.

maquis sclerophyllous woodland and plant community, similar to North American chaparral; can be found growing throughout the Mediterranean region.

marine terrace abrasion platform that has been elevated above sea level and thus abandoned from wave action.

marine west coast climate climate type characterized by cool wet conditions much of the year.

maritime relating to weather, climate, or atmospheric conditions in coastal or oceanic areas.

Maritime Equatorial (mE) hot and humid air mass that originates from the ocean region straddling the equator.

Maritime Polar (mP) cold, moist air mass originating from the oceans around 40° to 60° N or S latitude.

Maritime Tropical (mT) warm, moist air mass that originates from the tropical ocean regions.

mass a measure of the total amount of matter in a body.

mass wasting gravity-induced downslope movement of Earth material.

mathematical/statistical model computer-generated representation of an area or Earth system using statistical data.

matrix the dominant area of a mosaic (ecosystem supporting a particular plant community) where the major plant in the community is concentrated.

meander a broad, sweeping bend in a river or stream.

meander cut-off bend of a meandering stream that has become isolated from the active channel.

meandering channel stream channel with broadly sinuous banks that curve back and forth in sweeping bends.

medial moraine a central moraine in a large valley glacier formed where the interior lateral moraines of two tributary glaciers merge.

Mediterranean climate climate type characterized by warm, dry summers and cool, moist winters.

mental map conceptual model of special significance in geography because it consists of spatial information.

modified Mercalli Scale an earthquake intensity scale with Roman numerals from I to XII used to assess spatial variations in the degree of impact that a tremor generates.

Mercator projection mathematically produced, conformal map projection showing true compass bearings as straight lines.

meridian a line of longitude that forms half of a great circle on the globe and connects all points of equal longitude; all meridians connect the North and South Poles.

mesa flat-topped, steep-sided erosional remnant of a tableland, roughly as broad as tall, characteristic of arid regions with flat-lying sedimentary rocks.

mesothermal climates climate regions or conditions with hot, warm, or mild summers that do not have any months that average below freezing.

metamorphic rock one of the three major categories of rock; formed by heat and pressure changing a preexisting rock.

meteorite any fragment of a meteor that reaches Earth's surface.

meteorology study of the patterns and causes associated with short-term changes in the elements of the atmosphere.

microclimate climate associated with a small area at or near Earth's surface; the area may range from a few inches to 1 mile in size.

microplate terrane segment of crust of distinct geology added to a continent during tectonic plate collision.

microthermal climates climate regions or conditions with warm or mild summers that have winter months with temperatures averaging below freezing.

middle-latitude disturbance convergence of cold polar and warm subtropical air masses over the middle latitudes.

midoceanic ridge linear seismic mountain range that interconnects through all the major oceans; it is where new molten crustal material rises through the oceanic crust.

millibar unit of measurement for atmospheric pressure; 1 millibar equals a force of 1000 dynes per square centimeter; 1013.2 millibars is standard sea-level pressure.

mineral naturally occurring inorganic substance with a specific chemical composition and crystalline structure.

model a useful simplification of a more complex reality that permits prediction.

Mohorovičić discontinuity (Moho) zone marking the transition between Earth's crust and the denser mantle.

monsoon seasonal wind that reverses direction during the year in response to a reversal of pressure over a large landmass. The classic monsoons of Southeast Asia blow onshore in response to low pressure over Eurasia in summer and offshore in response to high pressure in winter.

moraine a category of glacial landform, with multiple subtypes, deposited beneath or along the margins of an ice mass.

mosaic a plant community and the ecosystem on which it is based, viewed as a landscape of interlocking parts by ecologists.

mountain breeze air flow downslope from mountains toward valleys during the night.

mouth downflow terminus of a stream.

mud wet, fine-grained sediment, particularly clay and silt sizes.

mudflow rapid mass wasting of wet, fine-grained sediment; may deposit levees and lobate (tongue-shaped) masses.

multispectral remote sensing using a number of energy wavelength bands to create images.

muskeg poorly drained vegetation-rich marshes or swamps usually overlying permafrost areas of polar climatic regions.

natural levees banks of a stream channel (or margins of a mass wasting flow channel) raised by deposition from flood (or flow) deposits; artificial levees are sometimes built along stream banks for flood control.

natural resource any element, material, or organism existing in nature that may be useful to humans.

natural vegetation vegetation that has been allowed to develop naturally without obvious interference from or modification by humans.

navigation the science of location and finding one's way, position, or direction.

neap tide the smaller than average tidal range that occurs during the first and third quarter moon.

near-infrared (NIR) film photographic film that makes pictures using near-infrared light that is not visible to the human eye.

nearshore zone area from the seaward or lakeward edge of breaking waves to the landward limit of broken wave water.

needleleaf tree a tree with very thin pointed leaves that are shaped like needles.

negative feedback reaction to initial change in a system that counteracts the initial change and leads to dynamic equilibrium in the system.

nekton classification of marine organisms that swim in the oceans.

nimbo a prefix for cloud types that means rain-producing.

nimbus term used in cloud description to indicate precipitation; thus cumulonimbus is a cumulus cloud from which rain is falling.

normal fault breakage zone with rock on one side sliding down relative to rock on the other side because of tensional forces; footwall up, hanging wall down.

North Atlantic Oscillation oscillating pressure tendencies between the Azores High and the Icelandic Low.

North Pole maximum north latitude (90°N), at the point marking the axis of rotation.

northeast trades see trade winds.

notch a recess, relatively small in height, eroded by wave action along the base of a coastal cliff.

oblate spheroid Earth's shape—a slightly flattened sphere.

obliquity cycle the change in the tilt of the Earth's axis relative to the plane of the ecliptic over a 41,000-year period.

occluded front boundary between a rapidly advancing cold air mass and an uplifted warm air mass cut off from Earth's surface; denotes the last stage of a middle-latitude cyclone.

ocean current horizontal movement of ocean water, usually in response to major patterns of atmospheric circulation.

oceanic crust the denser (avg. 3.0 g/cm^3), thinner, basaltic portion of Earth's crust; underlies the ocean basins.

oceanic islands volcanic islands that rise from the deep ocean floor.

oceanic trench (trench) long, narrow depression on the seafloor usually associated with an island arc. Trenches mark the deepest portions of the oceans and are associated with subduction of oceanic crust.

offshore zone the expanse of open water lying seaward or lakeward of the breaker zone.

omnivore animal that can feed on both plants and other animals.

open system system in which energy and/or materials can freely cross its boundaries.

organic sedimentary rock rock created from deposits of organic material, such as carbon from plants (coal).

orographic precipitation precipitation resulting from condensation of water vapor in an air mass that is forced to rise over a mountain range or other raised landform.

orographic thunderstorm a thunderstorm produced by the orographic uplift mechanism.

outcrop a mass of bedrock exposed at Earth's surface that is not concealed by regolith or soil.

outer core the upper portion of Earth's core; considered to be composed of molten iron liquefied by Earth's internal heat.

outlet glacier a valley glacier that flows outward from the main mass of a continental glacier.

outputs energy and material leaving an Earth system.

outwash glacial drift deposited beyond the terminus of a glacier by glacial meltwater.

outwash plain extensive, relatively smooth plain covered with sorted deposits carried forward by the meltwater from a continental glacier.

overthrust low-angle fault with rocks on one side pushed a considerable distance over those of the opposite side by compressional forces; the wedge of rocks that have overridden others in this way.

oxbow lake a lake or pond found in a meander cut-off on a floodplain.

oxidation union of oxygen with other elements to form new chemical compounds.

oxide a mineral group composed of oxygen combining with other Earth elements, especially metals.

oxygen-isotope analysis a dating method used to reconstruct climate history; it is based on the varying evaporation rates of different oxygen isotopes and the changing ratio between the isotopes revealed in foraminifera fossils.

ozone gas molecule consisting of three atoms of oxygen (O_3); forms a layer in the upper atmosphere that serves to screen out ultraviolet radiation harmful at Earth's surface.

ozonosphere also known as the ozone layer; this is a concentration of ozone gas in a layer between 20 and 50 kilometers (13–50 mi) above Earth's surface.

Pacific High persistent cell of high atmospheric pressure located in the subtropics of the North Pacific Ocean, also called the Hawaiian High.

pahoehoe a smooth, ropy surface on a lava flow.

paleogeography study of the past geographical distribution of environments.

paleomagnetism the historic record of changes in Earth's magnetic field.

palynology method of determining past climatic conditions using pollen analysis.

Pangaea ancient continent that consisted of all of today's continental landmasses.

parabolic dune crescent-shaped sand dune with arms pointing upwind.

parallel circle on the globe connecting all points of equal latitude.

parallelism tendency of Earth's polar axis to remain parallel to itself at all positions in its orbit around the sun.

parent material residual (derived from bedrock directly beneath) or transported (by water, wind, or ice) mineral matter from which soil is formed.

passive-margin coast coastal region that is far removed from the volcanism and tectonism associated with lithospheric plate boundaries.

patch a gap or area within a matrix (territory occupied by a dominant plant community) where the dominant vegetation is not supported due to natural causes.

paternoster lakes chain of lakes connected by a postglacial stream occupying the trough of a glaciated mountain valley.

patterned ground natural, repeating, often-polygonal designs of sorted sediment seen on the surface in periglacial environments.

ped naturally forming soil aggregate or clump with a distinctive shape that characterizes a soil's structure.

pediment gently sloping surface of eroded bedrock, thinly covered with fluvial sediments, found at the base of an arid-region mountain.

perched water table a minor zone of saturation overlying an aquiclude that exists above the regional water table.

percolation subsurface water moving downward to lower zones by the pull of gravity.

perennial flow describes streams that conduct flow continuously all year.

periglacial pertaining to cold-region landscapes that are impacted by intense frost action but not covered by year-round snow or ice.

perihelion position of Earth at closest distance to sun during each Earth revolution.

permafrost permanently frozen subsoil and underlying rock found in climates where summer thaw penetrates only the surface soil layer.

permeability characteristic of soil or bedrock that determines the ease with which water moves through Earth material.

pH scale scale from 0 to 14 that describes the acidity or alkalinity of a substance and that is based on a measurement of hydrogen ions; pH values below 7 indicate acidic conditions; pH values above 7 indicate alkaline conditions.

photosynthesis the process by which carbohydrates (sugars and starches) are manufactured in plant cells; requires carbon dioxide, water, light, and chlorophyll (the green substance in plants).

physical geography specialization in the systematic study of geography that focuses on the location, distribution, and spatial interaction of physical (environmental) phenomena.

physical model three-dimensional representation of all or a portion of Earth's surface.

physical (mechanical) weathering breakdown of rocks into smaller fragments (disintegration) by physical forces without chemical change.

phytoplankton tiny plants, algae and bacteria, that float and drift with currents in water bodies.

pictorial/graphic model representation of a portion of Earth's surface by means of maps, photographs, graphs, or diagrams.

piedmont alluvial plain a plain created by stream deposits at the base of an upland, such as a mountain, a hilly region, or a plateau.

piedmont fault scarp steep cliff due to movement along a fault that has offset unconsolidated sediment in the transition zone between mountains and basins.

piedmont glacier an alpine glacier that extends beyond a mountain valley spreading out onto lower flatter terrain.

pixel the smallest area that can be resolved in a digital image. Pixels, short for "picture element," are much like pieces in a mosaic, fitted together in a grid to make an image.

plane of the ecliptic plane of Earth's orbit about the sun and the apparent annual path of the sun along the stars.

plankton passively drifting or weakly swimming marine organisms, including both phytoplankton (plants) and zooplankton (animals).

plant community variety of individual plants living in harmony with each other and the surrounding physical environment.

plastic solid any solid material that changes its shape under stress, and retains that deformed shape after the stress is relieved.

plate convergence movement of lithospheric plates toward each other.

plate divergence movement of lithospheric plates away from each other.

plate tectonics theory that superseded continental drift and is based on the idea that the lithosphere is composed of a number of segments or plates that move independently of one another, at varying speeds, over Earth's surface.

plateau an extensive, flat-topped landform or region characterized by relatively high elevation, but low relief.

playa dry lake bed in a desert basin; typically fine-grained clastic (clay pan) or saline (salt crust).

playa lake a temporary lake that forms on a playa from runoff after a rainstorm or during a wet season.

Pleistocene the name given to the most recent "ice age" or period of Earth history experiencing cycles of continental glaciation; it commenced approximately 2.6 million years ago.

plucking erosion process by which a glacier pulls rocks and sediment from the ground along its bed and into the flowing ice.

plug dome a steep-sided, explosive type of volcano with its central vent or vents plugged by the rapid congealing of its highly acidic lava.

plunge pool a depression at the base of a waterfall formed by the impact of cascading water.

pluton see igneous intrusion.

plutonic rock see intrusive igneous rock.

plutonism the processes associated with the formation of rocks from magma cooling deep beneath Earth's surface.

pluvial rainy time period, usually pertaining to glacial periods when deserts were wetter than at present.

podzolization soil-forming process of humid climates with long cold winter seasons. Spodosols, the typical end product of the process, are characterized by the surface accumulation of raw humus, strong acidity, and the leaching or eluviation of soluble bases and iron and aluminum compounds.

point bar deposit of alluvium found on the inside of a bend in a meandering stream channel.

polar referring to the North or South Polar regions.

polar climates climate regions that do not have a warm season and are frozen much or all of the year.

polar easterlies easterly surface winds that move out from the polar highs toward the subpolar lows.

polar highs high pressure systems located near the poles where air is settling and diverging.

polarity reversals times in geologic history when the south magnetic pole became the north magnetic pole and vice versa.

pollution alteration of the physical, chemical, or biological balance of the environment that has adverse effects on the normal functioning of all life forms, including humans.

porosity characteristic of soil or bedrock that relates to the amount of pore space between individual peds or soil and rock particles and that determines the water storage capacity of Earth material.

positive feedback reaction to initial change in a system that reinforces the initial change and leads to imbalance in the system.

potential evapotranspiration hypothetical rate of evapotranspiration if at all times there is a more than adequate amount of soil water for growing plants.

pothole bedrock depression in a streambed drilled by the spinning of abrasive rocks in swirling flow.

prairie grassland regions of the middle latitudes. Tall-grass prairie varied from 0.5 to 3 meters (2 to 10 ft) in height and was native to areas of moderate rainfall; short-grass prairie of lesser height remains common in subhumid and semiarid (steppe) environments.

precession cycle changes in the time (date) of the year that perihelion occurs; the date is determined on the basis of a major period 21,000 years in length and a secondary period 19,000 years in length.

precipitation water in liquid or solid form that falls from the atmosphere and reaches Earth's surface.

pressure belts zones of high or low pressure that tend to circle Earth parallel to the equator in a theoretical model of world atmospheric pressure.

pressure gradient rate of change of atmospheric pressure horizontally with distance, measured along a line perpendicular to the isobars on a map of pressure distribution.

prevailing wind direction from which the wind for a particular location blows during the greatest proportion of the time.

primary coastline coast that has developed its present form primarily from land-based processes, especially fluvial and glacial processes.

primary productivity see autotrophs, and productivity.

prime meridian (Greenwich meridian) half of a great circle that connects the North and South Poles and marks zero degrees longitude. By international agreement the meridian passes through the Royal Observatory at Greenwich, England.

producer an organism that, because it is capable of photosynthesis, is at the foundation of a food web (also called an autotroph).

productivity rate at which new organic material is created at a particular trophic level. Primary productivity through photosynthesis by autotrophs is at the first trophic level; secondary productivity is by heterotrophs at subsequent trophic levels.

profile a graph of changes in height over a linear distance, such as a topographic profile.

punctuated equilibrium concept that periods of relative stability in many Earth systems are interrupted by short bursts of intense action causing major change.

pyroclastic flow airborne density current of hot gases and rock fragments unleashed by an explosive volcanic eruption.

pyroclastic material (tephra) pieces of volcanic rock, including cinders and ash, solidified from molten material erupted into the air.

RADAR RAdio Detection And Ranging.

radiation emission of electromagnetic wave energy through space; see also shortwave radiation and longwave radiation.

radiation fog fog produced by cooling of air in contact with a cold ground surface.

rain falling droplets of liquid water.

rain shadow dry, leeward side of a mountain range, resulting from the adiabatic warming of descending air.

recessional moraine end moraine deposited behind the terminal moraine marking a pause in the overall retreat of a glacier.

recharge replenishing the amount of stored water, particularly in the subsurface.

recumbent fold a fold in rock pushed over onto one side by asymmetric compressional forces; the axial plane of the fold is horizontal rather than vertical.

recurrence interval average length of time between events, such as floods, equal to or exceeding a given magnitude.

regional base level the lowest level to which a stream system in a basin of interior drainage can flow.

regional geography specialization in the systematic study of geography that focuses on the location, distribution, and spatial interaction

of phenomena organized within arbitrary areas of Earth space designated as regions.

regions areas identified by certain characteristics they contain that make them distinctive and separates them from surrounding areas.

regolith weathered rock material; usually covers bedrock.

rejuvenated stream a stream that has deepened its channel by erosion because of tectonic uplift in the drainage basin or lowering of its base level.

relative humidity ratio between the amount of water vapor in air of a given temperature and the maximum amount of water vapor that the air could hold at that temperature, if saturated; usually expressed as a percentage.

relative location location of an object in respect to its position relative to some other object or feature.

relief a measurement or expression of the difference in elevation between the highest and lowest location in a specified area.

remote sensing collection of information about the environment from a distance, usually from aircraft or spacecraft, for example, photography, radar, infrared.

representative fraction (RF) scale a map scale presented as a fraction or ratio between the size of a unit on the map to the size of the same unit on the ground, as in 1/24,000 or 1:24,000.

reservoir an artificial lake impounded behind a dam.

residual parent material rock fragments that form a soil and have accumulated in place through weathering.

resolution (spatial resolution) size of an area on Earth that is represented by a single pixel.

reverse fault high-angle break with rocks on one side pushed up relative to those on the other side by compressional forces; hanging wall up, footwall down.

revolution (Earth) motion of Earth along a path, or orbit, around the sun. One complete revolution requires approximately 365¼ days and determines an Earth year.

rhumb line line of true compass bearing (heading).

ria a narrow coastal embayment that occupies a submerged river valley.

rift valley major lowland consisting of one or more crustal blocks downfaulted as a result of tensional tectonic forces.

rill a tiny stream channel that even in a humid climate conducts flow only during precipitation events.

rime ice crystals formed along the windward side of tree branches, airplane wings, and the like, under conditions of supercooling.

rip current strong, narrow surface current flowing away from shore. It is produced by the return flow of water piled up near shore by incoming waves.

ripples small (centimeter-scale) wave forms in water or sediment.

roche moutonnée bedrock hill subjected to intense glacial abrasion on its up-ice side, with some plucking evident on the down-ice side.

rock a solid, natural aggregate of one or more minerals or particles of other rocks.

rock cycle a representation of the processes and pathways by which Earth material becomes different types of rocks.

rock flour rock fragments finely ground between the base of a glacier and the underlying bedrock surface.

rock structure the orientation, inclination, and arrangement of rock layers in Earth's crust.

rockfall nearly vertical drop through the air of individual rocks or a rock mass pulled downward by the force of gravity.

rockslide rock unit moving rapidly downslope by gravity in continuous contact with the surface below.

Rossby waves horizontal undulations in the flow of the upper air winds of the middle and upper latitudes.

rotation (Earth) turning of Earth on its polar axis; one complete rotation requires 24 hours and determines one Earth day.

runoff flow of water over the land surface, generally in the form of streams and rivers.

Saffir-Simpson Hurricane Scale A ranking (1–5) of the intensity of a hurricane, based on wind speeds and on the types of damage that occur, or may occur.

salinas see salt-crust playa.

salinization soil-forming process of low-lying areas in desert regions; the resulting soils are characterized by a high concentration of soluble salts as a result of the evaporation of surface water.

salt crystal growth weathering by the expansive force of salts growing in cracks in rocks; common in arid and coastal regions.

salt-crust playa (salt flat) an ephemeral lake bed composed mostly of salt minerals.

saltation the transportation by running water or wind of particles too large to be carried in suspension; the particles are bounced along on the surface or streambed by repeated lifting and deposition.

sand (sandy) sediment particles ranging in size from 0.05 millimeter to 2.0 millimeters.

sand dune mound or hill of sand-sized sediment deposited and shaped by the wind.

sand sea (erg) an extensive area covered by sand dunes.

sandstorm strong winds blowing a large amount of sand along the ground surface.

Santa Ana very dry downslope wind occurring in Southern California; see also foehn wind.

saturation (saturated air) point at which sufficient cooling has occurred so that an air mass contains the maximum amount of water vapor it can hold. Further cooling produces condensation of excess water vapor.

savanna tropical vegetation consisting primarily of coarse grasses associated with scattered low-growing trees.

scale ratio of the distance as measured on a map, globe, or other representation of Earth, to the actual distance on Earth.

sclerophyllous vegetation type commonly associated with the Mediterranean climate; characterized by tough surfaces, deep roots, and thick, shiny leaves that resist moisture loss.

sea steep, choppy, chaotic waves still forming under the influence of a storm.

sea arch span of rock extending from a coastal cliff under which the ocean or lake water freely moves.

sea breeze air flow by day from the sea toward the land; caused by the movement of air toward a zone of lower pressure associated with higher daytime temperatures over the land.

sea cave large wave-eroded opening formed near the water level in coastal cliffs.

sea cliff (lake cliff) steep slope of land eroded at its base by wave action.

sea level average position of the ocean shoreline.

sea stack resistant pillar of rock projecting above water close to shore along an erosion-dominated coast.

seafloor spreading movement of oceanic crust in opposite directions away from the midoceanic ridges, associated with the formation of new crust at the ridges and subduction of old crust at ocean margins.

secondary coastline coast that has developed its present form primarily through the action of coastal processes (waves, currents, and/or coral reefs).

secondary productivity the formation of new organic matter by heterotrophs, consumers of other life forms; see productivity.

section a square parcel of land with an area of 1 square mile as defined by the U.S. Public Lands Survey System.

sedimentary rock one of three major rock categories; formed by compaction and cementation of rock fragments, organic remains, or chemical precipitates.

seif a large, long, somewhat sinuous sand dune elongated parallel to the prevailing wind direction.

seismic wave traveling wave of energy released during an earthquake or other shock.

seismograph instrument used to measure amplitude of passing seismic waves.

selva characteristic tropical rainforest comprising multistoried, broad-leaf evergreen trees with relatively little undergrowth.

sextant navigation instrument used to determine latitude by star and sun positions.

shearing force a force that works to move two objects past each other in opposite directions.

sheet wash thin sheet of unchannelized water flowing over land.

shield volcano dome-shaped accumulation of multiple successive lava flows extruded from one or more vents or fissures.

shoreline exact contact between the edge of a standing body of water and dry land.

shortwave radiation energy radiated in wavelengths of less than about 1 micrometer (one millionth of a meter); includes X-rays, gamma rays, ultraviolet rays, visible light and near infrared light.

Siberian high intensively developed center of high atmospheric pressure located in northern central Asia in winter.

side-looking airborne radar (SLAR) a radar system that is used for making maps of terrain features.

silicate the largest mineral group, composed of oxygen and silica and forming most of the Earth's crust.

sill a horizontal sheet of igneous rock intruded and solidified between other rock layers.

silt (silty) sediment particles with a grain size between 0.002 millimeter and 0.05 millimeter.

sinkhole (doline) roughly circular surface depression related to the solution of rock in karst areas.

slash-and-burn also called shifting cultivation; typical subsistence agriculture of indigenous societies in the tropical rainforest. Trees are cut, the branches are burned, and crops are planted between the larger trees or stumps before rapid deterioration of the soil forces a move to a new area.

sleet form of precipitation produced when raindrops freeze as they fall through a layer of cold air; may also, locally, refer to a mixture of rain and snow.

slide fast mass wasting in which Earth material moves downslope in continuous contact with a discrete surface below.

slip face the steep, downwind side of a sand dune.

slope aspect direction a mountain slope faces in respect to the sun's rays.

slow mass wasting gravity-induced downslope movement of Earth material occurring so slowly that people cannot observe it directly.

slump thick unit of unconsolidated fine-grained material sliding downslope on a concave, curved slip plane.

small circle any circle that is not a full circumference of the globe. The plane of a small circle does not pass through the center of the globe.

smog combination of chemical pollutants and particulate matter in the lower atmosphere, typically over urban industrial areas.

snow precipitation in the form of ice crystals.

snowstorm storm situation where precipitation falls in the form of snow.

specialist an animal that can only survive on a single or very limited food type for its nutrition.

soil a dynamic, natural layer on Earth's surface that is a complex mixture of inorganic minerals, organic materials, microorganisms, water, and air.

soil (as a mass wasting material) relatively thin unit of unconsolidated fine-grained slope material.

soil fertilization adding nutrients to the soil to meet the conditions that certain plants require.

soil grade classification of soil texture by particle size: clay (less than 0.002 mm), silt (0.002–0.05 mm), and sand (0.05–2.0 mm) are soil grades.

soil horizon distinct soil layer characteristic of vertical zonation in soils; horizons are distinguished by their general appearances and their specific chemical and physical properties.

soil order largest classification of soils based on development and composition of soil horizons.

soil ped see ped.

soil profile vertical cross section of a soil that displays the various horizons or soil layers that characterize it; used for classification.

soil survey a publication that includes maps showing the distribution of soils within a given area, in the U.S. usually a county.

soil taxonomy the classification and naming of soils.

soil texture the distribution of particle sizes in a soil that give it a distinctive "feel."

soil water water in the zone of aeration, the uppermost subsurface water layer.

soil-forming regime processes that create soils.

solar constant rate at which insolation is received just outside Earth's atmosphere on a surface at right angles to the incoming radiation.

solar energy see insolation.

solar noon the time of day when the sun angle is at a maximum above the horizon (zenith).

solar system the system of the sun and the planets, their satellites, comets, meteoroids, and other objects revolving around the sun.

solifluction slow movement of saturated soil downslope by the pull of gravity; especially common in permafrost areas.

solstice one of two times each year when the position of the noon sun is overhead at its farthest distance from the equator; this occurs when the sun is overhead at the Tropic of Cancer (about June 21) and the Tropic of Capricorn (about December 21).

solution dissolving material in a fluid, such as water, or the liquid containing dissolved material; water transports dissolved load in solution.

solution sinkhole topographic depression formed mainly by the solution and removal of soluble rock at the surface.

sonar a system that uses sound waves for location and mapping underwater.

source location high in the drainage basin, near the drainage divide, where a stream system's flow begins.

source region nearly homogeneous surface of land or ocean over which an air mass acquires its temperature and humidity characteristics.

South Pole maximum south latitude (90°S), at the point marking the axis of rotation.

southeast trades see trade winds.

Southern Oscillation the systematic variation in atmospheric pressure between the eastern and western Pacific Ocean.

spatial distribution location and extent of an area or areas where a feature exists.

spatial interaction process whereby different phenomena are linked or interconnected, and, as a result, impact one another through Earth space.

spatial pattern arrangement of a feature as it is distributed through Earth space.

spatial science term used when defining geography as the science that examines phenomena as they are located, distributed, and interact with other phenomena throughout Earth space.

specific humidity mass of water vapor present per unit mass of air, expressed as grams per kilogram of moist air.

speleology the scientific study of caverns.

speleothem general term for any cavern feature made by secondary (later) precipitation of minerals from subsurface water.

spheroidal weathering rounded shape of rocks often caused by preferential weathering along joints in cross-jointed rocks.

spit coastal landform of wave- and current-deposited sediment attached to dry land at one end.

spring natural outflow of groundwater at the surface.

spring tide the larger than average tidal range that occurs during new and full moon.

squall line narrow line of rapidly advancing storm clouds, strong winds, and heavy precipitation; usually develops in front of a fast-moving cold front.

stability condition of air when it is cooler than the surrounding atmosphere and resists the tendency to rise; the lapse rate of the surrounding atmosphere is less than that of stable air.

stalactite spire-shaped speleothem that hangs from the ceiling of a cavern.

stalagmite spire-shaped speleothem that rises up from a cavern floor.

star dune a large pyramid-shaped sand dune with multiple slip faces due to changes in wind direction.

stationary front boundary between air masses of nearly equal strength; produces stagnation over one location for an extended period of time.

steppe climate characterized by middle-latitude semiarid vegetation, treeless and dominated by short bunch grasses.

stock an irregular mass of intrusive igneous rock (pluton) smaller than a batholith.

storm local atmospheric disturbance often associated with rain, hail, snow, sleet, lightning, or strong winds.

storm surge rise in sea level due to wind and reduced air pressure during a hurricane or other severe storm.

storm track path frequently traveled by cyclonic storms as they move in a generally eastward direction from their point of origin.

strata (stratification) distinct layers within sedimentary rocks.

strato signifies a low-level cloud (i.e., from the surface to 2000 m in elevation).

stratopause upper limit of stratosphere, separating it from the mesosphere.

stratosphere layer of atmosphere lying above the troposphere and below the mesosphere, characterized by fairly constant temperatures and ozone concentration.

stratovolcano see composite cone.

stratus uniform layer of low sheet-like clouds, frequently grayish in appearance.

stream general term for any natural, channelized flow of water regardless of size.

stream capacity the maximum amount of load that a stream can carry; varies with the stream's velocity.

stream competence the largest particle size that a stream can carry; varies with a stream's velocity.

stream discharge volume of water flowing past a point in a stream channel in a unit of time.

stream gradient vertical drop in a streambed over a given horizontal distance, generally expressed in meters per kilometer or feet per mile.

stream hydrograph plot showing changes in the amount of stream flow over time.

stream load material being transported by a stream at a given instant; includes bed load, suspended load, and dissolved load.

stream order numerical index expressing the position of a stream channel within the hierarchy of a stream system.

stream terrace former floor of a stream valley now abandoned and perched above the present valley floor and stream channel.

stress (pressure) force per unit area.

striations gouges, grooves, and scratches carved in bedrock by abrading rock particles imbedded in a glacier.

strike compass direction of the line formed at the intersection of a tilted rock layer and a horizontal plane.

strike-slip fault a fault with horizontal motion, where movement takes place along the strike of the fault.

structure the descriptive physical characteristics and arrangement of bedrock, such as folded, faulted, layered, fractured, massive.

subarctic climate climate type that produces a tundra landscape.

subduction process associated with plate tectonic theory whereby an oceanic crustal plate is forced downward into the mantle beneath a lighter continental or oceanic plate when the two converge.

sublimation direct change of state of a material, such as water, from solid to gas or gas to solid.

subpolar lows east–west trending belts or cells of low atmospheric pressure located in the upper middle latitudes.

subsurface water general term for all water that lies beneath Earth's surface, including soil water and groundwater.

subsystem separate system operating within the boundaries of a larger Earth system.

subtropical highs cells of high atmospheric pressure generally centered over the eastern portions of the oceans in the vicinity of 30°N and 30°S latitude; source of the westerlies poleward and the trades equatorward.

subtropical jet stream high-velocity air current flowing above the sinking air of the subtropical high pressure cells; most prominent in the winter season.

succession progression of natural vegetation from one plant community to the next until a final stage of equilibrium has been reached with the natural environment.

sunspots visible dark (cooler) spots on the surface of the sun; their numbers seem to follow an approximate 11-year cycle.

supercooled water liquid water that exists below the freezing point of 0°C or 32°F.

surf zone the part of the nearshore area that consists of a turbulent bore of broken wave water.

surface creep wind-generated transportation consisting of pushing and rolling sediment downwind in continuous contact with the surface.

surface of discontinuity three-dimensional surface with length, width, and height separating two different air masses; also referred to as a front.

surface runoff liquid water flowing over Earth's land surface.

surge (glacial) sudden shift downslope of glacial ice, possibly caused by a reduction of basal friction with underlying bedrock.

suspended load solid particles that are small enough to be transported considerable distances while remaining buoyed up in a moving air or water column.

suspension transportation process that moves small solids, often considerable distances, while buoyed up by turbulence in the moving air or water.

swallow hole the site where a surface stream is diverted to the subsurface, such as into a cavern system.

swash thin sheet of broken wave water that rushes up the beach face in the swash zone.

swash zone the most landward part of the nearshore zone; where a thin sheet of water rushes up, then back down, the beach face.

swell orderly lake or ocean waves of rounded form that have traveled beyond the storm zone of wave generation.

symbiotic relationship relationship between two organisms that benefits both organisms.

syncline the downfolded element of folded rock structure.

system group of interacting and interdependent units that together form an organized whole.

systems analysis determining the parts of a system, the processes involved, and studying how changes in interactions among those parts and processes may affect the system and its operation.

taiga the northern coniferous forest of subarctic regions on the Eurasian landmass.

talus (talus slope, talus cone) slope (sometimes cone-shaped) of angular, broken rocks at the base of a cliff deposited by rockfall.

tarn mountain lake in a glacial cirque.

tectonic forces forces originating within Earth that break and deform Earth's crust.

tectonic processes processes that derive their energy from within Earth's interior and serve to create landforms by elevating, disrupting, and roughening Earth's surface.

temperature degree of heat or cold and its measurement.

temperature gradient rate of change of temperature with distance in any direction from a given point; refers to rate of change horizontally; a vertical temperature gradient is referred to as the lapse rate.

temperature inversion reverse of the normal pattern of vertical distribution of air temperature; in the case of inversion, temperature increases rather than decreases with increasing altitude.

tensional force a force that works to pull an object apart in opposite directions.

tephra (pyroclastic material) pieces of volcanic rock, including cinders and ash, solidified in the air during an explosive eruption.

terminal moraine end moraine that marks the farthest advance of an alpine or continental glacier.

terminus the lower end of a glacier.

terra rossa characteristic calcium-rich (developed over limestone bedrock) red-brown soils of the climate regions surrounding the Mediterranean Sea.

terrestrial planets the four closest planets to the sun—Mercury, Venus, Earth, and Mars.

thematic map a map designed to present information or data about a specific theme, as in a population distribution map, a map of climate or vegetation.

thermal expansion and contraction notion that rocks can weather due to expansion and contraction effects of alternating heating and cooling.

thermal infrared (TIR) scanning images made with scanning equipment that produces an image of heat differences.

thermosphere highest layer of atmosphere extending from the mesopause to outer space.

Thornthwaite system climate classification based on moisture availability and of greatest use at the local level; climate types are distinguished by examining and comparing potential and actual evapotranspiration.

threshold condition within a system that causes dramatic and often irreversible change for long periods of time to all variables in the system.

thrust fault low-angle break with rocks on one side pushed over those of the other side by compressional forces.

thunder sound produced by the rapidly expanding, heated air along the channel of a lightning discharge.

thunderstorm intense convectional storm characterized by thunder and lightning, short in duration and often accompanied by heavy rain, hail, and strong winds.

tidal interval the time between successive high tides, or between successive low tides.

tidal range elevation difference between water levels at high tide and low tide.

tide periodic rise and fall of sea level in response to the gravitational interaction of the moon, sun, and Earth.

till sediment deposited directly by glacial ice.

till plain a broad area of low relief covered by glacial deposits.

tilted fault block crustal block between two parallel normal faults that has been uplifted along one fault and relatively downdropped along the other.

time zone Earth is divided into 24 time zones (24 h) to coordinate time with Earth's rotation.

tolerance ability of a species to survive under specific environmental conditions.

tombolo strip of wave- and current-deposited sediment connecting the mainland to an island.

topographic contour line line on a map connecting points that are the same elevation above mean sea level.

topography the arrangement of high and low elevations on a land surface.

tornado small, intense, funnel-shaped cyclonic storm of very low pressure, violent updrafts, and converging winds of enormous velocity.

tornado outbreak when a thunderstorm(s) produce more than one tornado.

tower karst high, steep-sided hills formed by solution of limestone or other soluble rocks in karst areas.

trace less than a measurable amount of rain or snow (i.e., less than 1 mm or 0.01 in.).

traction transportation process in moving water that drags, rolls, or slides heavy particles along in continuous contact with the bed.

trade winds consistent surface winds blowing in low latitudes from the subtropical highs toward the intertropical convergence zone; labeled northeast trades in the Northern Hemisphere and southeast trades in the Southern Hemisphere.

transform movement horizontal sliding of tectonic plates alongside and past each other.

transpiration transfer of moisture from living plants to the atmosphere by the emission of water vapor, primarily from leaf pores.

transportation movement of Earth materials from one site to another by gravity, water, wind, or glacial ice.

transported parent material rock fragments that form a soil and originated elsewhere and then were transported and deposited in the new location.

transverse dune a linear ridge-like sand dune that is oriented at right angles to the prevailing wind direction.

transverse stream a stream that flows across the general orientation or "grain" of the topography, such as mountains or ridges.

travertine calcium carbonate (limestone) deposits resulting from the Secondary calcium carbonate deposts in caves or caverns.

tree line elevation in mountain regions above which cold temperatures and wind stress prohibit tree growth.

tributary stream channel that delivers its water to another, larger channel.

trophic level number of feeding steps that a given organism is removed from the autotrophs (e.g., green plant—first level, herbivore—second level, carnivore—third level, etc.).

trophic structure organization of an ecosystem based on the feeding patterns of the organisms that comprise the ecosystem.

Tropic of Cancer parallel of latitude at 23½°N; the northern limit to the migration of the sun's vertical rays throughout the year.

Tropic of Capricorn parallel of latitude at 23½°S; the southern limit to the migration of the sun's vertical rays throughout the year.

tropical region on Earth lying between the Tropic of Cancer (23½°N latitude), and the Tropic of Capricorn (23½°S latitude).

tropical climates climate regions that are warm all year.

tropical easterlies winds that blow from the east in tropical regions.

tropical monsoon climate climate characterized by hot temperatures and alternating rainy and dry seasons.

tropical rainforest climate hot wet climate that promotes the growth of rainforests.

tropical savanna climate warm, semidry climate that promotes tall grasslands.

tropopause boundary between the troposphere and stratosphere.

troposphere lowest layer of the atmosphere, exhibiting a steady decrease in temperature with increasing altitude and containing virtually all atmospheric dust and water vapor.

trough elongated area or "belt" of low atmospheric pressure; also glacial trough, a U-shaped valley carved by a glacier.

trunk stream the largest channel in a drainage system; receives inflow from tributaries.

tsunami wave caused when an earthquake, volcanic eruption, or other sudden event displaces ocean water; builds to dangerous heights in shallow coastal waters.

tundra high-latitude or high-altitude environments or climate regions that are not able to support tree growth because the growing season is too cold or too short.

tundra climate characterized by treeless vegetation of polar regions and very high mountains, consisting of mosses, lichens, and low-growing shrubs and flowering plants.

turbulence chaotic, mixing flow of fluids, often with an upward component.

typhoon a tropical cyclone found in the western Pacific, the same as a hurricane.

unconformity an interruption in the accumulation of different rock layers; often represents a period of erosion.

uniformitarianism widely accepted theory that Earth's geological processes operate today as they have in the past.

unloading physical weathering process whereby removal of overlying weight leads to rock expansion and breakage.

uplift mechanisms methods of lifting surface air aloft, they are orographic, frontal, convergence (cyclonic), and convectional.

upslope fog type of fog where upward flowing air cools to form fog that hugs the slope of mountains.

upwelling upward movement of colder, nutrient-rich, subsurface ocean water, replacing surface water that is pushed away from shore by winds.

urban heat island see heat island.

U.S. Public Lands Survey System a method for locating and dividing land, used in much of the Midwest and western United States. This system divides land into 6- by 6-mile-square townships consisting of 36 sections of land (each 1 sq mi). Sections can also be subdivided into halves, quarter sections, and quarter-quarter-sections.

uvala (valley sink) large surface depression resulting from coalescing of sinkholes in karst areas.

valley breeze air flow upslope from the valleys toward the mountains during the day.

valley glacier an alpine glacier that extends beyond the zone of high mountain peaks into a confining mountain valley below.

valley train outwash deposit from glacial meltwater, resembling an alluvial fan confined by valley walls.

variable one of a set of objects and/or characteristics of objects, which are interrelated in such a way that they function together as a system.

varve a pairing of organic-rich summer sediments and organic poor winter sediments found in exposed lake beds; because each pair represents 1 year of time, counting varves is useful as a dating technique for recent Earth history.

veering wind shift the change in wind direction clockwise around the compass; for example, east to southeast to south, to southwest, to west, and northwest.

vent pipe-like conduit through which volcanic rock material is erupted.

ventifact rock displaying distinctive wind-abraded faces, pits, grooves, and polish.

verbal scale stating the scale of a map using words such as "one centimeter represents one kilometer."

vertical exaggeration a technique that stretches the height representation of terrain in order to emphasize topographic detail.

vertical rays sun's rays that strike Earth's surface at a 90° angle.

visualization a wide array of computer techniques used to vividly illustrate a place or concept, or the illustration produced by one of these techniques.

volcanic ash erupted fragments of volcanic rock of sand size or smaller (<2.0 mm).

volcanic neck vertical igneous intrusion that solidified in the vent of a volcano.

volcanism the eruption of molten rock matter onto Earth's surface.

volcano mountain or hill created from the accumulation of erupted rock matter.

V-shaped valley the typical shape of a stream valley cross profile where the gradient is steep.

warm current a flow of sea water that moves like a river through the ocean and is relatively warmer than water in the ocean area that it flows through.

warm front leading edge of a relatively warmer, less dense air mass advancing upon a cooler, denser air mass.

warping broad and general uplift or settling of Earth's crust with little or no local distortion.

wash (arroyo, wadi) an ephemeral stream channel in an arid climate.

water budget relationship between evaporation, condensation, and storage of water within the Earth system.

water mining taking more groundwater out of an aquifer through pumping than is being replaced by natural processes in the same period of time.

water table upper limit of the zone of saturation below which all pore spaces are filled with water.

water vapor water in its gaseous form.

wave base water depth equal to half the length of a given wave; at smaller depths the wave interacts with the underwater substrate.

wave crest the highest part of a wave form.

wave height vertical distance between the trough and adjacent crest of a wave.

wave period time it takes for one wavelength to pass a given point.

wave refraction bending of waves as seen in map view as they approach the shore, resulting from changes in velocity as waves interact with the underwater topography.

wave steepness ratio of wave height to wavelength for a given wave.

wave trough the lowest part of a wave form.

wavelength horizontal distance between two successive crests of a given wave.

weather atmospheric conditions, at a given time, in a specific location.

weather radar radar that is used to track thunderstorms, tornados, and hurricanes.

weathering physical (mechanical) fragmentation and chemical decomposition of rocks and minerals at and near Earth's surface.

well artificial opening that reaches the zone of saturation for the purpose of extracting groundwater.

westerlies surface winds flowing from the polar portions of the subtropical highs, carrying fronts, storms, and variable weather conditions from west to east through the middle latitudes.

wet adiabatic lapse rate rate at which a rising mass of air is cooled by expansion when condensation is taking place. The rate varies but averages 5°C/1000 meters (3.2°F/1000 ft).

white frost a heavy coating of white crystalline frost.

wind air in motion from areas of higher pressure to areas of lower pressure; movement is generally horizontal, relative to the ground surface.

windward location on the side that faces toward the wind and is therefore exposed or unprotected; usually refers to mountain and island locations.

xerophytic vegetation type that has genetically evolved to withstand the extended periods of drought common to arid regions.

yardang aerodynamically shaped remnant ridge of wind-eroded bedrock or partly consolidated sediments.

yazoo stream a stream tributary that flows parallel to the main stream for a considerable distance before joining it.

zone of ablation the lower elevation part of a glacier; where more frozen water is removed than added during the year.

zone of accumulation (glacial) the higher elevation part of a glacier; where more frozen water is added than removed during the year.

zone of aeration uppermost layer of subsurface water where pore spaces typically contain both air and water.

zone of depletion top layer, or a horizon, of a soil, characterized by the removal of soluble and insoluble soil components through leaching and eluviation by gravitational water.

zone of saturation subsurface water zone in which all voids in rock and soil are always filled with water; the top of this zone is the water table.

zone of transition an area of gradual change from one region to another.

zooplankton tiny animals that float and drift with currents in water bodies.

Index

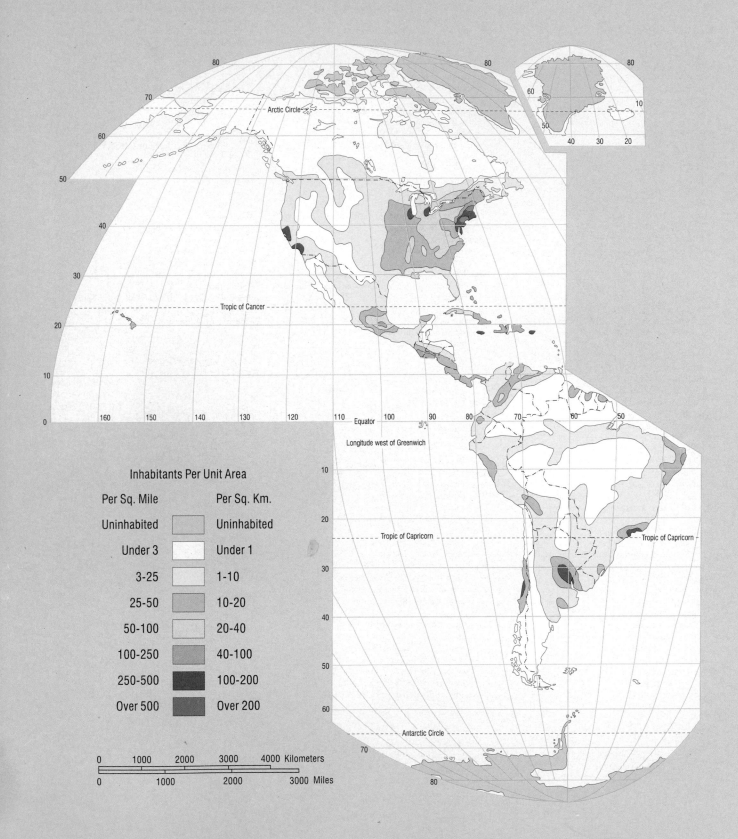

Inhabitants Per Unit Area

Per Sq. Mile		Per Sq. Km.
Uninhabited		Uninhabited
Under 3		Under 1
3-25		1-10
25-50		10-20
50-100		20-40
100-250		40-100
250-500		100-200
Over 500		Over 200

Arctic Circle

Tropic of Cancer

Equator

Longitude west of Greenwich

Tropic of Capricorn

Tropic of Capricorn

Antarctic Circle

World Map of Population Density